IAS/PARK CITY MATHEMATICS SERIES

Volume 1

Geometry and Quantum Field Theory

Daniel S. Freed
Karen K. Uhlenbeck
Editors

American Mathematical Society
Institute for Advanced Study

IAS/Park City Mathematics Institute runs mathematics education programs that bring together high school mathematics teachers, researchers, graduate students, and undergraduates to participate in distinct but overlapping programs of research and education. This volume contains the lecture notes from the Graduate Summer School program on the Geometry and Topology of Manifolds and Quantum Field Theory, held June 22–July 20, 1991, in Park City, Utah.

Supported by the National Science Foundation.

1991 *Mathematics Subject Classification*. Primary 58–XX, 81–XX, 70–XX, 53–XX; Secondary 16–XX, 18–XX, 22–XX, 34–XX, 35–XX, 49–XX.

Library of Congress Cataloging-in-Publication Data
Geometry and quantum field theory : June 22–July 20, 1991, Park City, Utah / Daniel S. Freed, Karen K. Uhlenbeck, editors.
    p. cm. — (IAS/Park City mathematics series, ISSN 1079-5634; v. 1)
Includes bibliographical references.
ISBN 0-8218-0400-6 (acid-free)
   1. Lie groups—Congresses. 2. Symplectic groups—Congresses. 3. Quantum field theory—Congresses. 4. Mathematical physics—Congresses. I. Freed, Daniel S. II. Uhlenbeck, Karen K. III. Series.
QC20.7.L54G46   1995
530.1$'$43$'$0151604–dc20
                                                                                                94-46731
                                                                                                CIP

Copying and reprinting. Material in this book may be reproduced by any means for educational and scientific purposes without fee or permission with the exception of reproduction by services that collect fees for delivery of documents and provided that the customary acknowledgment of the source is given. This consent does not extend to other kinds of copying for general distribution, for advertising or promotional purposes, or for resale. Requests for permission for commercial use of material should be addressed to the Assistant to the Publisher, American Mathematical Society, P. O. Box 6248, Providence, Rhode Island 02940-6248. Requests can also be made by e-mail to reprint-permission@ams.org.

Excluded from these provisions is material in articles for which the author holds copyright. In such cases, requests for permission to use or reprint should be addressed directly to the author(s). (Copyright ownership is indicated in the notice in the lower right-hand corner of the first page of each article.)

© Copyright 1995 by the American Mathematical Society. All rights reserved.
The American Mathematical Society retains all rights
except those granted to the United States Government.
Printed in the United States of America.
∞ The paper used in this book is acid-free and falls within the guidelines
established to ensure permanence and durability.
Portions of this volume were typeset by the authors using $\mathcal{A}_{\mathcal{M}}\mathcal{S}$-TEX and $\mathcal{A}_{\mathcal{M}}\mathcal{S}$-LATEX, the American Mathematical Society's TEX macro systems.

10 9 8 7 6 5 4 3   00 99 98 97

Preface

The IAS/Park City Mathematics Institute (PCMI) was founded in 1991 as part of the "Regional Geometry Institute" initiative of the National Science Foundation. In mid 1993 it found an institutional home at the Institute for Advanced Study (IAS) in Princeton. The PCMI will continue to hold summer programs in both Park City and in Princeton.

The IAS/Park City Mathematics Institute encourages research and education in mathematics and fosters interaction between the two. The month-long summer institute offers programs for researchers and postdoctoral scholars, graduate students, undergraduates, and high school teachers. One of our main goals is to make all of the participants aware of the total spectrum of activities that occur in mathematics education and research: we wish to involve professional mathematicians in education and to bring modern concepts in mathematics to the attention of educators. To that end the summer institute features general sessions designed to encourage interaction among the various groups. In-year activities at sites around the country form an integral part of the Program for High School Teachers.

Each summer a different topic is chosen as the focus of the Research Program and Graduate Summer School. (Activities in the Undergraduate Program deal with this topic as well.) Lecture notes from the Graduate Summer School will be published each year in this series. This first volume contains notes from the 1991 Summer School on the *Geometry and Topology of Manifolds and Quantum Field Theory*. Volumes from the 1992 Summer School *Nonlinear Partial Differential Equations in Differential Geometry*, the 1993 Summer School *Higher Dimensional Algebraic Geometry*, and the 1994 Summer School *Gauge Theory and the Topology of Four-Manifolds* are in preparation. The 1995 Research Program and Graduate Summer School topic is *Nonlinear Wave Phenomena*.

We plan to publish material from other parts of the IAS/Park City Mathematics Institute in the future. Unfortunately, the initial volumes in this series may not reflect the interaction between research and education which is a primary focus of the PCMI. At the summer institute late afternoons are devoted to seminars of common interest to all participants. Many deal with current issues in education; others treat mathematical topics at a level which encourages broad participation. Several popular evening programs are also well attended. These include lectures, panel discussions, computer demonstrations, and videos. The PCMI has also spawned interactions between universities and high schools at a local level. We hope to share these activities with a wider audience in future volumes.

Dan Freed, Series Editor
October, 1994

Contents

Introduction	1
Robert L. Bryant, An Introduction to Lie Groups and Symplectic Geometry	5
Introduction	7
Background Material and Basic Terminology	8
Lecture 1. Introduction: Symmetry and Differential Equations	13
Symmetry and conservation laws	13
Classical integration techniques	14
Lecture 2. Lie Groups and Lie Algebras	19
Lie groups	19
Lie algebras	30
Left-invariant forms and the structure equations	35
The 1-, 2-, and 3-dimensional Lie algebras	36
Exercises	41
Lecture 3. Group Actions on Manifolds	47
Lie group actions	47
Group actions and vector fields	53
Equations of Lie type	55
Appendix: Lie's transformation groups, I	61
Appendix: connections and curvature	63
Exercises	66
Lecture 4. Symmetries and Conservation Laws	71
Variational problems	71
Symmetries	75
Hamiltonian form	83
The cotangent bundle	85
Poincaré recurrence	86
Exercises	89
Lecture 5. Symplectic Manifolds, I	93
Symplectic Algebra	93
Symplectic Manifolds	97
Exercises	108
Lecture 6. Symplectic Manifolds, II	113

Obstructions	113
Deformations of symplectic structures	116
Submanifolds of symplectic manifolds	118
Appendix: Lie's transformation groups, II	122
Lie pseudo-groups	124
Exercises	125
Lecture 7. Classical Reduction	**129**
Symplectic group actions	129
The moment map	132
Reduction	134
Exercises	138
Lecture 8. Recent Applications of Reduction	**143**
Riemannian holonomy	143
Kähler manifolds and algebraic geometry	145
Kähler reduction	149
Hyperkähler manifolds	153
Exercises	160
Lecture 9. The Gromov School of Symplectic Geometry	**163**
Soft techniques in symplectic manifolds	163
Hard techniques in symplectic manifolds	167
Epilogue	174
Exercises	175
Bibliography	**179**

Jeffrey M. Rabin, Introduction to Quantum Field Theory for Mathematicians — 183

Introduction	185
Acknowledgements	186
Lecture 1. Classical Mechanics	**187**
The Geometry of Hamiltonian Mechanics	192
Lecture 2. Classical Field Theory	**195**
Lecture 3. The Lorentz Group and Spinors	**203**
Lecture 4. Quantum Mechanics	**209**
Symmetries in Quantum Mechanics	213
Heisenberg Picture	213
Lecture 5. The Simple Harmonic Oscillator	**215**
Lecture 6. The Path Integral Formulation of Quantum Mechanics	**219**
Lecture 7. The Harmonic Oscillator via Path Integrals	**227**
Lecture 8. Quantum Field Theory: Free Fields	**233**
Axioms	239
The Dirac Field	241

Lecture 9. Interacting Fields, Feynman Diagrams, and Renormalization	245
Exercises	257
Self-adjointness and time evolution in quantum mechanics	257
Gaussian measure problems	259
Miscellaneous physical problems	264
Bibliography	269

Orlando Alvarez, Lectures on Quantum Mechanics and the Index Theorem

	271
Introduction	273
References	274
Lecture 1. Canonical Quantization of Bosonic Systems	275
Lagrangian and Hamiltonian mechanics	275
Naive quantization	277
Noether's theorem	277
Review of the simple harmonic oscillator	278
Lecture 2. Canonical Quantization of Fermionic Systems	281
The basics	281
The fermionic oscillator	283
Quantization of fermionic systems	284
Clifford algebra fermions	287
Lecture 3. Supersymmetry	293
Getting spin 1/2 particles to move	293
The simplest supersymmetry	295
The Dirac equation on a manifold	298
Lecture 4. The Index of Operators	301
The basics	301
Supersymmetry and the index	301
Lecture 5. Path Integrals	303
Path integral for the simple harmonic oscillator	303
Theory of fermionic integration	309
Lecture 6. The Atiyah-Singer formula	311
Atiyah-Singer formula for the Dirac operator	311
Lecture 7. Global Topological Issues	319
Characteristic classes	319
Global properties of the path integral	321

Frank Quinn, Lectures on Axiomatic Topological Quantum Field Theory

	323
Outline	326
Lecture 1. A rough idea	329

Lecture 2. Topological opportunities for TQFT	331
Lecture 3. The Euler theory	335
Euler characteristics	335
Lecture 4. Finite total homotopy TQFT	339
The skew theory	342
Witten's integral for finite gauge group	342
Calculations for finite groups	343
Lecture 5. Twisted finite homotopy TQFT	345
The answer	347
Calculations for finite groups	348
The construction	355
Induced homomorphisms	358
Dependence on the cocycle	360
The Chern-Simons TQFT	361
Lecture 6. Axioms	365
Definition of domain category	365
Examples	366
Bordism categories	368
Definition of TQFT	369
Examples	370
Lecture 7. Elementary structure	373
Pairings	373
Ambialgebras	378
Classification on surfaces and graphs	380
Lecture 8. Modular examples	383
The idea	383
The Euler theory	386
The finite total homotopy TQFT	387
The twisted finite homotopy TQFT	390
The endomorphism TQFT	393
Lecture 9. Modular axioms and structure	395
Modular domain categories	395
Modular bordisms and TQFT	397
The corner ambialgebras	398
Classification on surfaces	401
Module structures on $Z(Y)$	403
Nondegenerate pairings	404
The rank TQFT	405
Morita equivalence and rational conformal field theories	406
The Verlinde basis and link invariants	407
The structure of $Z(T^2)$	408
Lecture 10. Categorical constructions	411
Tensor categories	412

Semisimple tensor categories	413
Examples of finite semisimple tensor categories	414
Example: $SL(2)$ mod 5	415
State modules	416
Glueing trees	417
The corner ambialgebra	419
State modules of graphs	420
Slices through 2-complexes	422
Beginning and ending a presentation	423
Beginning and ending a relation	423
The circulator	426
The Interchange	428
Doing the calculation	430
Extensions	431
Appendix A. Algebra	435
Tensor products	435
Pairings and traces	437
Ambialgebras	439
Examples of ambialgebras	442
Complex ambialgebras	447
Bibliography	451

Introduction

From earliest times geometry has been based on physical ideas. In Egypt geometry was created to measure the size of land. Later the Alexandrian school developed trigonometry to make astronomical calculations and to assist in more down-to-earth matters such as the telling of time, navigation, and geography. Certainly observations of classical mechanics (apples!), as well as Kepler's laws of planetary motion, motivated Newton in his creation of the calculus. Gauss' theory of surfaces and Riemann's fundamental ideas about higher dimensional space were in part motivated by physics, as was much of the subsequent development of Riemannian geometry. Symplectic geometry originates with Hamilton's formulation of optical laws. The most important influence on differential geometry in this century has been Einstein's general theory of relativity. Other parts of differential geometry developed independently but relate in a fundamental way to classical physics. Thus the geometry of connections and the nonlinear gauge theory of Yang and Mills are intimately related, despite their different origins. While general relativity and the classical theory of fields have permeated our geometric thinking, one great advance in 20th century physics—Quantum Theory—which stimulated 50 years of growth in functional analysis and representation theory, has not had the same influence on geometry.

The recent interaction of geometers and quantum field theorists began in the late 1970s with the application of index theory and algebraic geometry to the instanton problem in Yang-Mills theory. On the mathematical side this eventually led to, among other things, Donaldson's work on 4-manifolds.[1] Although the physical motivation to study instantons comes from *quantum* Yang-Mills theory, the mathematical problem studies the spaces of solutions to the *classical* field equations. Another striking use of modern geometry and topology in quantum field theory came in the early 1980s when anomalies were shown to have topological significance and to be computable via the Atiyah-Singer index theorem. This also led to new insight into the index theorem itself, which physicists derived from a supersymmetric version of quantum mechanics. (This is the subject of Orlando Alvarez's lectures in this volume.) Subsequently Ed Witten introduced the elliptic genus, a new index of manifolds coming from a 2 dimensional quantum field theory. More recently *topological* quantum field theories have shed light on the new invariants of low dimensional manifolds which geometers and topologists discovered in the 1980s (the Donaldson invariants of 4-manifolds, the Jones invariants of links in 3-space, the Casson invariant and Floer homology of 3-manifolds). Other field theories have shed light on the theory of infinite dimensional Lie algebras, loop

---

[1] Instantons and 4-manifolds are the subject of Volume 4 in this series.

groups, geometric quantization, quantum groups and Poisson-Lie structures, and much more! Another exciting recent development—this time coming from string theory—is the pioneering work on "mirror symmetry" by Philip Candelas and his collaborators. The mathematical implication is a count of rational curves of arbitrary degree in certain Calabi-Yau manifolds.[2] This has stimulated much recent activity among mathematicians.

In 1990 when we began to organize the Summer School it was certainly clear that quantum field theory has much to say about geometry and topology. However, in many instances the physical ideas involved just scratch the surface of quantum field theory. Often only formal properties of symmetry, quantization, and the path integral are used to draw mathematical conclusions. More recent developments indicate that deep analytic ideas in quantum field theory will also play a major role. For example, Ed Witten recently derived a formula for Donaldson invariants on Kähler manifolds using a twisted version of supersymmetric Yang-Mills theory in four dimensions. His argument depends on the existence of a mass gap, cluster decomposition, spontaneous symmetry breaking, asymptotic freedom, and gluino condensation. While none of this is rigorous by mathematical standards, the final formula is correct in all cases which can be checked against rigorous mathematical computations. This is only one example of recent work which indicates to mathematicians the ultimate necessity of incorporating more than the formal structure of quantum field theory into mathematics. We can safely assert that the physical intuition physicists bring to bear on these problems, even in models whose relevance to the real world seems tenuous at best, has not yet penetrated at all into mathematics. The challenge for mathematicians is not so much to rigorize and axiomitize what the physicists are doing, but rather to develop a framework for it which fits into mathematics and allows us to make further discoveries. Many high energy theoretical physicists state as an important problem: What is the geometry which underlies String Theory? Although Quantum Field Theory is less of a mystery, and certainly physicists have the intuition to deal with the infinities which plagued their predecessors, it may still be worth posing the same question for quantum field theory.

The goal of the Summer School is to introduce mathematicians to some of the ideas which enter into the work described above. The five lecture series which comprise the Summer School are:

Robert Bryant	*Lectures on Lie Groups and Symplectic Geometry*
Jeff Rabin	*Introduction to Quantum Field Theory for Mathematicians*
Orlando Alvarez	*Applied Functional Integrals*
Frank Quinn	*Topology and Quantum Field Theory*
Isadore Singer	*Path Integrals*

The lecture notes from the first four lecture series make up this volume. At first glance the different lecture series may seem quite different, even unrelated. However, there are broad themes which connect these five topics. Two such fundamental connections are symmetry and the relationship between the Lagrangian and Hamiltonian viewpoints.

Bryant's lectures provide a mathematical introduction to Lie groups, which enter as symmetries in many ways in field theories. Bryant shows how symmetries lead to conservation laws, an idea which occurs in Rabin's and Alvarez's lectures as

---

[2] This is the subject of lectures by David Morrison in Volume 3 of this series.

well. Bryant demonstrates how to use such conservation laws to solve differential equations. Rabin emphasizes the physical content of the conservation laws in both the classical and quantum context. Here the Lorentz group and the Poincaré group, the symmetries of Minkowski spacetime, lead to the basic conservation laws of momentum and energy. Alvarez adds a new twist, or rather a square root, by looking at fermionic systems with supersymmetry. In his one dimensional case the supersymmetry generator is the square root of ordinary translation, and the associated conserved current is the Dirac operator, which is a square root of the Laplace operator. Quinn's lectures abstract properties of theories with a much larger symmetry group than the Poincaré group. These theories are invariant under all diffeomorphisms, so are termed "topological quantum field theories." He also finds that in low dimensions these theories are related to "quantum groups," which he treats in terms of categorical constructions.

There are two main approaches to both the classical and quantum theories—the Lagrangian and the Hamiltonian—and the interplay between them underlies all of the lectures as well as much of the mathematics and physics in this area. In classical mechanics the Lagrangian viewpoint is expressed via the calculus of variations. Bryant, Rabin, and Alvarez all discuss how to pass from this formulation to the Hamiltonian formulation, whose mathematical expression is in terms of symplectic geometry. Bryant uses this as a springboard to a more general discussion of symplectic geometry, including recent developments by the Gromov school. Rabin looks at the quantum theory as well. Here the Hamiltonian side of the story is the theory of canonical quantization and the quantum Hilbert space. The Lagrangian version is Feynman's path integral. (This is the subject of Singer's lectures, which are unfortunately not available in this volume.) Rabin treats these two approaches in both quantum mechanics and quantum field theory. He also introduces the ideas of perturbation theory and the renormalization group in quantum field theory. Alvarez introduces canonical quantization and the path integral for fermionic systems. In his example of supersymmetric quantum mechanics the relationship between the two approaches combine with the supersymmetry to yield the index theorem for Dirac operators! Quinn does not discuss Lagrangian and Hamiltonian field theory in the usual terms. Rather, his axioms for topological quantum field theory abstract some basic common properties of the path integral and the quantum Hilbert space in a very general context. In this way he investigates aspects of the structures in the intersection of the Hamiltonian and Lagrangian formulations of quantum field theory.

As organizers, we can be satisfied that three years after they were given, these summer school lectures are still a superb background for one of the most active areas in mathematics: the interface between theoretical mathematics and geometry. None of the basic questions has been answered. The subject matter has shifted a bit, but the basic mathematics remains the same. A mathematical understanding of symmetry (Lie groups), classical field theory (calculus of variations and symplectic geometry), axiomatization (topological quantum field theory), the basics of perturbation theory and renormalization (ordinary quantum field theory), and supersymmetry are now and probably will remain for a long time prerequisites for a mathematician entering this arena.

An Introduction to Lie Groups and Symplectic Geometry

Robert L. Bryant

An Introduction to Lie Groups and Symplectic Geometry

Robert L. Bryant

Introduction

These are the lecture notes for a short course on Lie groups and symplectic geometry which I gave at the 1991 Regional Geometry Institute at Park City, Utah beginning on 24 June and ending on 11 July.

The course really was designed to be an introduction, aimed at an audience of students who were familiar with basic constructions in differential topology and rudimentary differential geometry and who wanted to get a feel for Lie groups and symplectic geometry. My purpose was not to provide an exhaustive treatment of either Lie groups, which would have impossible even if I had had an entire year, or of symplectic manifolds, which has lately undergone something of a revolution. Instead, I tried to provide an introduction to what I regard as the basic concepts of the two subjects, with an emphasis on examples which drove the development of the theory.

I included a few topics which are not often treated in introductions to these subjects, such as Lie's reduction of order for differential equations and its relation with the notion of a solvable group on the one hand and integration of ODE by quadrature on the other. I also tried, in the later lectures, to introduce the reader to some of the global methods which are now becoming so important in symplectic geometry. However, a full treatment of these topics in the space of nine lectures beginning at the elementary level was beyond my abilities.

After the lectures were over, I contemplated reworking these notes into a more comprehensive introduction to modern symplectic geometry but, after some soul-searching, finally decided against doing so. Thus, I have contented myself with making only minor modifications and corrections. It is my hope that an interested person could read these notes in a few weeks, get some sense of what the subject is about, and then pass on to articles and books written by the experts in the field.

The exercise sets are an essential part of the course. Each set begins with elementary material and works up to more involved and intricate problems. My

---

[1] Duke University, Durham, NC
E-mail address: bryant@math.duke.edu

object was to provide a path to understanding of the material which could be entered at several different levels and so the exercises vary greatly in difficulty. Many of these exercise sets are obviously too long for any person to do them during the three weeks of the course, so I provided extensive hints to aid the student in completing the exercises after the course was over.

I want to take this opportunity to thank the many people who made helpful suggestions for these notes both during and after the course. Particular thanks goes to Karen Uhlenbeck and Dan Freed, who invited me to give an introductory set of lectures at the RGI, and to my course assistant, Tom Ivey, who provided invaluable help and criticism in the early stages of the notes and who tirelessly helped the students with the exercises. The faults of the presentation are entirely my own, but without the help, encouragement, and proofreading contributed by these folks and others, neither these notes nor the course could ever have been completed.

Background Material and Basic Terminology

In these lectures, I assume that the reader is familiar with the basic notions of manifolds, vector fields, and differential forms. All manifolds will be assumed to be both second countable and Hausdorff. Also, unless I say otherwise, I generally assume that all maps and manifolds are smooth (i.e., C^∞).

Since it came up several times in the course of the lectures, it is probably worth emphasizing the following point: A *submanifold* of a smooth manifold X is, by definition, a pair (S, f) where S is a smooth manifold and $f: S \to X$ is a one-to-one immersion. In particular, f need not be an embedding.

Notation

The notation I use for smooth manifolds and mappings is fairly standard, but with a few slight variations:

If $f: X \to Y$ is a smooth mapping, then $f': TX \to TY$ denotes the induced mapping on tangent bundles, with $f'(x)$ denoting its restriction to $T_x X$. (However, I follow tradition when $X = \mathbb{R}$ and let $f'(t)$ stand for $f'(t)(\partial/\partial t)$ for all $t \in \mathbb{R}$. I trust that this abuse of notation will not cause confusion.)

For any vector space V, I generally use $A^p(V)$ (instead of, say, $\Lambda^p(V^*)$) to denote the space of alternating (or exterior) p-forms on V. For a smooth manifold M, I denote the space of smooth, alternating p-forms on M by $\mathcal{A}^p(M)$. The algebra of all (smooth) differential forms on M is denoted by $\mathcal{A}^*(M)$.

I generally reserve the letter d for the exterior derivative $d: \mathcal{A}^p(M) \to \mathcal{A}^{p+1}(M)$.

For any vector field X on M, I will denote *left-hook with X* (often called *interior product with X*) by the symbol $X \lrcorner$. This is the graded derivation of degree -1 of $\mathcal{A}^*(M)$ which satisfies $X \lrcorner (df) = Xf = df(X)$ for all smooth functions f on M. For example, the Cartan formula for the Lie derivative of differential forms is written in the form

$$\mathcal{L}_X \phi = X \lrcorner d\phi + d(X \lrcorner \phi).$$

Jets

Occasionally, it will be convenient to use the language of jets in describing certain constructions. Jets provide a coordinate free way to talk about the Taylor expansion of some mapping up to a specified order. No detailed knowledge about these objects will be needed in these lectures, so the following comments should suffice:

If f and g are two smooth maps from a manifold X^m to a manifold Y^n, we say that f and g *agree to order k at $x \in X$* if, first, $f(x) = g(x) = y \in Y$ and, second, when $u: U \to \mathbb{R}^m$ and $v: V \to \mathbb{R}^n$ are local coordinate systems centered on x and y respectively, the functions $F = v \circ f \circ u^{-1}$ and $G = v \circ g \circ u^{-1}$ have the same Taylor series at $0 \in \mathbb{R}^m$ up to and including order k. Using the Chain Rule, it is not hard to show that this condition is independent of the choice of local coordinates u and v centered at x and y respectively.

The notation $f \equiv_{x,k} g$ will mean that f and g agree to order k at x. This is easily seen to define an equivalence relation. Denote the $\equiv_{x,k}$-equivalence class of f by $j^k(f)(x)$, and call it the *k-jet of f at x*.

For example, knowing the 1-jet at x of a map $f: X \to Y$ is equivalent to knowing both $f(x)$ and the linear map $f'(x): T_x \to T_{f(x)}Y$.

The set of k-jets of maps from X to Y is usually denoted by $J^k(X, Y)$. It is not hard to show that $J^k(X, Y)$ can be given a unique smooth manifold structure in such a way that, for any smooth $f: X \to Y$, the obvious map $j^k(f): X \to J^k(X, Y)$ is also smooth.

These jet spaces have various functorial properties which we shall not need at all. The main reason for introducing this notion is to give meaning to concise statements like "The critical points of f are determined by its 1-jet", "The curvature at x of a Riemannian metric g is determined by its 2-jet at x", or, from Lecture 8, "The integrability of an almost complex structure $J: TX \to TX$ is determined by its 1-jet". Should the reader wish to learn more about jets, I recommend the first two chapters of [GG].

Basic and Semi-Basic

Finally, I use the following terminology: If $\pi: V \to X$ is a smooth submersion, a p-form $\phi \in \mathcal{A}^p(V)$ is said to be *π-basic* if it can be written in the form $\phi = \pi^*(\varphi)$ for some $\varphi \in \mathcal{A}^p(X)$ and *π-semi-basic* if, for any π-vertical* vector field X, we have $X \lrcorner \phi = 0$. When the map π is clear from context, the terms "basic" or "semi-basic" are used.

It is an elementary result that if the fibers of π are connected and ϕ is a p-form on V with the property that both ϕ and $d\phi$ are π-semi-basic, then ϕ is actually π-basic.

Basic Theorems

In the early lectures at least, we will need very little in the way of major theorems, but we will make extensive use of the following results:

---

*A vector field X is π-vertical with respect to a map $\pi: V \to X$ if and only if $\pi'(X(v)) = 0$ for all $v \in V$

The implicit function theorem. If $f: X \to Y$ is a smooth map of manifolds and $y \in Y$ is a regular value of f, then $f^{-1}(y) \subset X$ is a smooth embedded submanifold of X, with
$$T_x f^{-1}(y) = \ker(f'(x): T_x X \to T_y Y).$$

Existence and uniqueness of solutions of ODE. If X is a vector field on a smooth manifold M, then there exists an open neighborhood U of $\{0\} \times M$ in $\mathbb{R} \times M$ and a smooth mapping $F: U \to M$ with the following properties:
1. $F(0, m) = m$ for all $m \in M$.
2. For each $m \in M$, the slice $U_m = \{t \in \mathbb{R} \mid (t, m) \in U\}$ is an open interval in $\mathbb{R}$ (containing 0) and the smooth mapping $\phi_m: U_m \to M$ defined by $\phi_m(t) = F(t, m)$ is an integral curve of X.
3. (*Maximality*) If $\phi: I \to M$ is any integral curve of X where $I \subset \mathbb{R}$ is an interval containing 0, then $I \subset U_{\phi(0)}$ and $\phi(t) = \phi_{\phi(0)}(t)$ for all $t \in I$.

The mapping F is called the (local) *flow* of X and the open set U is called the domain of the flow of X. If $U = \mathbb{R} \times M$, then we say that X is *complete*.

Two useful properties of this flow are easy consequences of this existence and uniqueness theorem. First, the interval $U_{F(t,m)} \subset \mathbb{R}$ is simply the interval U_m translated by $-t$. Second, $F(s+t, m) = F(s, F(t, m))$ whenever t and $s+t$ lie in U_m.

The simultaneous flow-box theorem. If $X_1, X_2, \ldots, X_r$ are smooth vector fields on M which satisfy the Lie bracket identities $[X_i, X_j] = 0$ for all i and j, and if $p \in M$ is a point where the r vectors $X_1(p), X_2(p), \ldots, X_r(p)$ are linearly independent in $T_p M$, then there is a coordinate system $x^1, x^2, \ldots, x^n$ on an open neighborhood U of p so that, on U,

$$X_1 = \frac{\partial}{\partial x^1}, \qquad X_2 = \frac{\partial}{\partial x^2}, \qquad \ldots, \qquad X_r = \frac{\partial}{\partial x^r}.$$

The simultaneous flow-box theorem has two particularly useful consequences. Before describing them, we introduce an important concept.

Let M be a smooth manifold and let $E \subset TM$ be a smooth subbundle of rank p. We say that E is *integrable* if, for any two vector fields X and Y on M which are sections of E, the Lie bracket $[X, Y]$ is also a section of E.

The local Frobenius theorem. If M^n is a smooth manifold and $E \subset TM$ is a smooth, integrable sub-bundle of rank r, then every p in M has a neighborhood U on which there exist local coordinates $x^1, \ldots, x^r, y^1, \ldots, y^{n-r}$ so that the sections of E over U are spanned by the vector fields

$$\frac{\partial}{\partial x^1}, \qquad \frac{\partial}{\partial x^2}, \qquad \ldots, \qquad \frac{\partial}{\partial x^r}.$$

Associated to this local theorem is the following global version:

The global Frobenius theorem. Let M be a smooth manifold and let $E \subset TM$ be a smooth, integrable subbundle of rank r. Then for any $p \in M$, there exists a connected r-dimensional submanifold $L \subset M$ which contains p, which satisfies

$T_q L = E_q$ for all $q \in S$, and which is maximal in the sense that any connected r'-dimensional submanifold $L' \subset M$ which contains p and satisfies $T_q L' \subset E_q$ for all $q \in L'$ is a submanifold of L.

The submanifolds L provided by this theorem are called the *leaves* of the sub-bundle E. (Some books call a sub-bundle $E \subset TM$ a *distribution* on M, but I will avoid this since "distribution" already has a well-established meaning in analysis.)

LECTURE 1
Introduction: Symmetry and Differential Equations

Symmetry and conservation laws

Consider the classical equations of motion for a particle in a conservative force field

$$\ddot{x} = -\text{grad } V(x),$$

where $V: \mathbb{R}^n \to \mathbb{R}$ is some function on $\mathbb{R}^n$. If V is proper (i.e. the inverse image under V of a compact set is compact, as when $V(x) = |x|^2$), then, to a first approximation, V is the potential for the motion of a ball of unit mass rolling around in a cup of the form $y = V(x)$, moving only under the influence of gravity (directed downward along the y-axis). For a general function V we have only the grossest knowledge of how the solutions to this equation ought to behave.

Nevertheless, we can say a few things. The total energy (= kinetic plus potential) is given by the formula $E = \frac{1}{2}|\dot{x}|^2 + V(x)$ and is easily shown to be constant on any solution (just differentiate $E(x(t))$ and use the equation). When V is proper and non-negative, it follows that x must stay inside the compact set $V^{-1}([0, E(x(0))])$, and so the orbits are bounded. Without knowing any more about V, one can still show that the motion has a certain "recurrent" behavior: The trajectory resulting from "most" initial positions and velocities tends to return, infinitely often, to a small neighborhood of the initial position and velocity (see Lecture 4 for a precise statement). Beyond this, very little is known is known about the behavior of the trajectories for generic V.

Suppose now that the potential function V is rotationally symmetric, i.e. that V depends only on the distance from the origin. For the sake of simplicity, let us take $n = 3$ as well. This is classically called the case of a central force field in space. If we let $V(x) = \frac{1}{2}v(|x|^2)$, then the equations of motion become

$$\ddot{x} = -v'(|x|^2)\, x.$$

The energy E is still a conserved quantity, i.e., a function of the position and velocity which stay constant on any solution of the equation, but is it easy to see that the vector-valued function $x \times \dot{x}$ is also conserved, since

$$\frac{d}{dt}(x \times \dot{x}) = \dot{x} \times \dot{x} - x \times \left(v'(|x|^2)\, x\right) = 0.$$

Call this vector-valued function μ. We can think of E and μ as functions on the phase space $\mathbb{R}^6$. For generic values of E_0 and μ_0, the simultaneous level set

$$\Sigma_{E_0,\mu_0} = \{\, (x,\dot{x}) \mid E(x,\dot{x}) = E_0,\ \mu(x,\dot{x}) = \mu_0 \,\}$$

of these functions cut out a surface $\Sigma_{E_0,\mu_0} \subset \mathbb{R}^6$ and any integral of the equations of motion must lie in one of these surfaces. Since we know a great deal about integrals of ODEs on surfaces, this problem is very tractable. (See Lecture 4 and its exercises for more details on this.)

The function μ, known as the *angular momentum*, is called a *first integral* of the second-order ODE for $x(t)$, and somehow seems to correspond to the rotational symmetry of the original ODE. This vague relationship will be considerably sharpened and made precise in the upcoming lectures.

The relationship between symmetry and solvability in differential equations is profound and far reaching. The subjects which are now known as Lie groups and symplectic geometry got their beginnings from the study of symmetries of systems of ordinary differential equations and of integration techniques for them.

By the middle of the nineteenth century, Galois theory had clarified the relationship between the solvability of polynomial equations by radicals and the group of "symmetries" of the equations. Sophus Lie set out to do the same thing for differential equations and their symmetries.

Here is a "dictionary" showing the (rough) correspondence which Lie developed between these two achievements of nineteenth century mathematics.

Galois theory	infinitesimal symmetries
finite groups	continuous groups
polynomial equations	differential equations
solvable by radicals	solvable by quadrature

Although the full explanation of these correspondences must await the later lectures, we can at least begin the story in the simplest examples as motivation for developing the general theory. This is what I shall do for the rest of today's lecture.

Classical integration techniques

The very simplest ordinary differential equation that we ever encounter is the equation

(1) $$\dot{x}(t) = \alpha(t)$$

LECTURE 1. INTRODUCTION: SYMMETRY AND DIFFERENTIAL EQUATIONS

where α is a known function of t. The solution of this differential equation is simply

$$x(t) = x_0 + \int_0^x \alpha(\tau)\, d\tau.$$

The process of computing an integral was known as "quadrature" in the classical literature (a reference to the quadrangles appearing in what we now call Riemann sums), so it was said that (1) was "solvable by quadrature". Note that, once one finds a particular solution, all of the others are got by simply translating the particular solution by a constant, in this case, by x_0. Alternatively, one could say that the equation (1) itself was invariant under "translation in x".

The next most trivial case is the homogeneous linear equation

$$(2) \qquad \dot{x} = \beta(t)\, x.$$

This equation is invariant under scale transformations $x \mapsto rx$. Since the mapping $\log\colon \mathbb{R}^+ \to \mathbb{R}$ converts scaling to translation, it should not be surprising that the differential equation (2) is also solvable by a quadrature:

$$x(t) = x_0 e^{\int_0^t \beta(\tau)\, d\tau}.$$

Note that, again, the symmetries of the equation suffice to allow us to deduce the general solution from the particular.

Next, consider an equation where the right hand side is an affine function of x,

$$(3) \qquad \dot{x} = \alpha(t) + \beta(t)\, x.$$

This equation is still solvable in full generality, using two quadratures. For, if we set

$$x(t) = u(t) e^{\int_0^t \beta(\tau)\, d\tau},$$

then u satisfies $\dot{u} = \alpha(t) e^{-\int_0^t \beta(\tau)\, d\tau}$, which can be solved for u by another quadrature. It is not at all clear why one can somehow "combine" equations (1) and (2) and get an equation which is still solvable by quadrature, but this will be explained in Lecture 3.

Now consider an equation with a quadratic right-hand side, the so-called Riccati equation:

$$(4) \qquad \dot{x} = \alpha(t) + 2\beta(t)x + \gamma(t)x^2.$$

It can be shown that there is no method for solving this by quadratures and algebraic manipulations alone. However, there is a way of obtaining the general solution from a particular solution. If $s(t)$ is a particular solution of (4), try the *ansatz* $x(t) = s(t) + 1/u(t)$. The resulting differential equation for u has the form (3) and hence is solvable by quadratures.

The Riccati equation (4) has an extensive history and we will return to it often. Its remarkable property, that given one solution we can obtain the general solution, should be contrasted with the case of

$$\dot{x} = \alpha(t) + \beta(t)x + \gamma(t)x^2 + \delta(t)x^3. \tag{5}$$

For equation (5), knowing one solution does not usually help you find the rest of the solutions. There is in fact a world of difference between this and the Riccati equation, although this is far from evident looking at them.

Before leaving these simple ODE, we note the following curious progression: If x_1 and x_2 are solutions of an equation of type (1), then clearly the difference $x_1 - x_2$ is constant. Similarly, if x_1 and $x_2 \neq 0$ are solutions of an equation of type (2), then the ratio x_1/x_2 is constant. Furthermore, if x_1, x_2, and $x_3 \neq x_1$ are solutions of an equation of type (3), then the expression $(x_1 - x_2)/(x_1 - x_3)$ is constant. Finally, if x_1, x_2, $x_3 \neq x_1$, and $x_4 \neq x_2$ are solutions of an equation of type (4), then the *cross-ratio*

$$\frac{(x_1 - x_2)(x_4 - x_3)}{(x_1 - x_3)(x_4 - x_2)}$$

is constant. There is no such corresponding expression (for any number of particular solutions) for equations of type (5). The reason for this will be made clear in Lecture 3. For right now, we just want to remark on the fact that the linear fractional transformations of the real line, a group isomorphic to $SL(2, \mathbb{R})$, are exactly the transformations which leave fixed the cross-ratio of any four points. As we shall see, the group $SL(2, \mathbb{R})$ is closely connected with the Riccati equation and it is this connection which accounts for many of the special features of this equation.

I will conclude this lecture by discussing the group of rigid motions in Euclidean 3-space. These are transformations of the form

$$T(\mathbf{x}) = \mathbf{R}\,\mathbf{x} + \mathbf{t},$$

where $\mathbf{R}$ is a rotation in $\mathbb{E}^3$ and $\mathbf{t} \in \mathbb{E}^3$ is any vector. It is easy to check that the set of rigid motions form a group under composition which is, in fact, isomorphic to the group of 4-by-4 matrices

$$\left\{ \begin{pmatrix} \mathbf{R} & \mathbf{t} \\ 0 & 1 \end{pmatrix} \,\middle|\, {}^t\mathbf{R}\,\mathbf{R} = I_3, \ \mathbf{t} \in \mathbb{R}^3 \right\}.$$

(Topologically, the group of rigid motions is just the product $O(3) \times \mathbb{R}^3$.)

Now, suppose that one wants to solve for a unit speed curve $\mathbf{x} \colon \mathbb{R} \to \mathbb{R}^3$ with a prescribed curvature $\kappa(t)$ and torsion $\tau(t)$. If $\mathbf{x}$ were such a curve, then the curvature and torsion are calculated by defining an oriented orthonormal basis $(\mathbf{e}_1, \mathbf{e}_2, \mathbf{e}_3)$ along the curve, satisfying $\dot{x} = \mathbf{e}_1$, $\dot{\mathbf{e}}_1 = \kappa \mathbf{e}_2$, and $\dot{\mathbf{e}}_2 = -\kappa \mathbf{e}_1 + \tau \mathbf{e}_3$.

(Think of the torsion as measuring how e_2 falls away from the e_1e_2-plane.) Form the 4-by-4 matrix

$$X = \begin{pmatrix} e_1 & e_2 & e_3 & x \\ 0 & 0 & 0 & 1 \end{pmatrix},$$

(vectors in $\mathbb{R}^3$ are columns of height 3). Then one can express the ODE for prescribed curvature and torsion as

$$\dot{X} = X \begin{pmatrix} 0 & -\kappa & 0 & 1 \\ \kappa & 0 & -\tau & 0 \\ 0 & \tau & 0 & 0 \\ 0 & 0 & 0 & 0 \end{pmatrix}.$$

This is a linear system of equations for a curve $X(t)$ in the space of 4-by-4 matrices. It is not hard to show that if $X(0)$ lies in the group of rigid motions, then the entire curve $X(t)$ lies in the group of rigid motions.

It is going to turn out that, just as in the case of the Riccati equation, the prescribed curvature and torsion equations cannot be solved by algebraic manipulations and quadrature alone. However, once we know one solution, all other solutions for that particular $(\kappa(t), \tau(t))$ can be obtained by rigid motions. In fact, though, we are going to see that one does not have to know a solution to the full set of equations before finding the rest of the solutions by quadrature, but only a solution to an equation connected to SO(3) just in the same way that the Riccati equation is connected to $SL(2, \mathbb{R})$, the group of transformations of the line which fix the cross-ratio of four points.

LECTURE 2
Lie Groups and Lie Algebras

In this Lecture, I define and develop some of the basic properties of the central objects of interest in these lectures: Lie groups and Lie algebras.

Lie groups

Definition 1. A *Lie group* is a pair (G, μ) where G is a smooth manifold and $\mu \colon G \times G \to G$ is a smooth mapping which gives G the structure of a group.

When the multiplication μ is clear from context, we usually just say "G is a Lie group." Also, for the sake of notational sanity, I will follow the practice of writing $\mu(a, b)$ simply as ab whenever this will not cause confusion. I will usually denote the multiplicative identity by $e \in G$ and the multiplicative inverse of $a \in G$ by $a^{-1} \in G$.

Most of the algebraic constructions in the theory of abstract groups have straightforward analogues for Lie groups:

Definition 2. A *Lie subgroup* of a Lie group G is a subgroup $H \subset G$ which is also a submanifold of G. A *Lie group homomorphism* is a group homomorphism $\phi \colon H \to G$ which is also a smooth mapping of the underlying manifolds.

Here is the prototypical example of a Lie group:

Example: The general linear group. The (real) *general linear group* in dimension n, denoted $\mathrm{GL}(n, \mathbb{R})$, is the set of invertible n-by-n real matrices regarded as an open submanifold of the n^2-dimensional vector space of all n-by-n real matrices with multiplication map μ given by matrix multiplication: $\mu(a, b) = ab$. Since the matrix product ab is defined by a formula which is polynomial in the matrix entries of a and b, it is clear that $\mathrm{GL}(n, \mathbb{R})$ is a Lie group.

Actually, if V is any finite dimensional real vector space, then $\mathrm{GL}(V)$, the set of bijective linear maps $\phi \colon V \to V$, is an open subset of the vector space $\mathrm{End}(V) = V \otimes V^*$ and becomes a Lie group when endowed with the multiplication $\mu \colon \mathrm{GL}(V) \times \mathrm{GL}(V) \to \mathrm{GL}(V)$ given by composition of maps: $\mu(\phi_1, \phi_2) = \phi_1 \circ \phi_2$. If $\dim(V) = n$, then $\mathrm{GL}(V)$ is isomorphic (as a Lie group) to $\mathrm{GL}(n, \mathbb{R})$, though not canonically.

The advantage of considering abstract vector spaces V rather than just $\mathbb{R}^n$ is mainly conceptual, but, as we shall see, this conceptual advantage is great. In fact,

Lie groups of linear transformations are so fundamental that a special terminology is reserved for them:

Definition 3. A *(linear) representation* of a Lie group G is a Lie group homomorphism $\rho\colon G \to \mathrm{GL}(V)$ for some vector space V called the *representation space*. Such a representation is said to be *faithful* (resp., *almost faithful*) if ρ is one-to-one (resp., has 0-dimensional kernel).

It is a consequence of a theorem of Ado and Iwasawa (see Theorem 4 below) that every 1-connected Lie group has an almost faithful, finite-dimensional representation. (One of the later exercises constructs a 1-connected Lie group which has no faithful finite-dimensional representation, so *almost faithful* is the best that can be expected.)

Example: vector spaces. Any vector space over $\mathbb{R}$ becomes a Lie group when the group multiplication μ is taken to be vector addition.

Example: matrix Lie groups. The Lie subgroups of $\mathrm{GL}(n,\mathbb{R})$ are called *matrix Lie groups* and play an important role in the theory. Not only are they the most frequently encountered, but, because of the theorem of Ado and Iwasawa, practically anything which is true for matrix Lie groups has an analog for a general Lie group. In fact, for the first pass through, the reader can simply imagine that all of the Lie groups mentioned are matrix Lie groups. Here are a few simple examples:
1. Let A_n be the set of diagonal n-by-n matrices with positive entries on the diagonal.
2. Let N_n be the set of upper triangular n-by-n matrices with all diagonal entries all equal to 1.
3. ($n = 2$ only) Let $\mathbb{C}^\bullet = \left\{ \begin{pmatrix} a & -b \\ b & a \end{pmatrix} \Big| a^2 + b^2 > 0 \right\}$. Then $\mathbb{C}^\bullet$ is a matrix Lie group diffeomorphic to $S^1 \times \mathbb{R}$. (You should check that this is actually a subgroup of $\mathrm{GL}(2,\mathbb{R})$!)
4. Let $\mathrm{GL}_+(n,\mathbb{R}) = \{a \in \mathrm{GL}(n,\mathbb{R}) \mid \det(a) > 0\}$

There are more interesting examples, of course. A few of these are

$$\mathrm{SL}(n,\mathbb{R}) = \{a \in \mathrm{GL}(n,\mathbb{R}) \mid \det(a) = 1\}$$
$$\mathrm{O}(n) = \{a \in \mathrm{GL}(n,\mathbb{R}) \mid {}^t\!a\, a = I_n\}$$
$$\mathrm{SO}(n,\mathbb{R}) = \{a \in \mathrm{O}(n) \mid \det(a) = 1\}$$

which are known respectively as the *special linear group*, the *orthogonal group*, and the *special orthogonal group* in dimension n. In each case, one must check that the given subset is actually a subgroup and submanifold of $\mathrm{GL}(n,\mathbb{R})$. These are exercises for the reader. (See the problems at the end of this lecture for hints.)

A Lie group can have "wild" subgroups which cannot be given the structure of a Lie group. For example, $(\mathbb{R}, +)$ is a Lie group which contains totally disconnected, uncountable subgroups. Since all of our manifolds are second countable, such subgroups (by definition) cannot be given the structure of a (0-dimensional) Lie group.

LECTURE 2. LIE GROUPS AND LIE ALGEBRAS

It can be shown [Wa, pg. 110] that any *closed* subgroup of a Lie group G is an embedded submanifold of G and hence is a Lie subgroup. However, for reasons which will become apparent, it is disadvantageous to consider only closed subgroups.

Example: a non-closed subgroup. Even $GL(n, \mathbb{R})$ can have Lie subgroups which are not closed. Here is a simple example: Let λ be any irrational real number and define a homomorphism $\phi_\lambda \colon \mathbb{R} \to GL(4, \mathbb{R})$ by the formula

$$\phi_\lambda(t) = \begin{pmatrix} \cos t & -\sin t & 0 & 0 \\ \sin t & \cos t & 0 & 0 \\ 0 & 0 & \cos \lambda t & -\sin \lambda t \\ 0 & 0 & \sin \lambda t & \cos \lambda t \end{pmatrix}$$

Then ϕ_λ is easily seen to be a one-to-one immersion so its image is a submanifold $G_\lambda \subset GL(4, \mathbb{R})$ which is therefore a Lie subgroup. It is not hard to see that

$$\overline{G_\lambda} = \left\{ \begin{pmatrix} \cos t & -\sin t & 0 & 0 \\ \sin t & \cos t & 0 & 0 \\ 0 & 0 & \cos s & -\sin s \\ 0 & 0 & \sin s & \cos s \end{pmatrix} \middle| \, s, t \in \mathbb{R} \right\}.$$

Note that G_λ is diffeomorphic to $\mathbb{R}$ while its closure in $GL(4, \mathbb{R})$ is diffeomorphic to $S^1 \times S^1$!

It is also useful to consider matrix Lie groups with complex coefficients. However, complex matrix Lie groups are really no more general than real matrix Lie groups (though they may be more convenient to work with). To see why, note that we can write a complex n-by-n matrix $A + Bi$ (where A and B are real n-by-n matrices) as the $2n$-by-$2n$ matrix $\begin{pmatrix} A & -B \\ B & A \end{pmatrix}$. In this way, we can embed $GL(n, \mathbb{C})$, the space of n-by-n invertible complex matrices, as a closed submanifold of $GL(2n, \mathbb{R})$. The reader should check that this mapping is actually a group homomorphism.

Among the more commonly encountered complex matrix Lie groups are the *complex special linear group*, denoted by $SL(n, \mathbb{C})$, and the *unitary* and *special unitary groups*, denoted, respectively, as

$$U(n) = \{ a \in GL(n, \mathbb{C}) \mid {}^*a\, a = I_n \}$$
$$SU(n) = \{ a \in U(n) \mid \det{}_\mathbb{C}(a) = 1 \}$$

where ${}^*a = {}^t\bar{a}$ is the Hermitian adjoint of a. These groups will play an important role in what follows. The reader may want to familiarize himself with these groups by doing some of the exercises for this section.

General properties

Translations and inverse as smooth mappings. If G is a Lie group with $a \in G$, let $L_a, R_a \colon G \to G$ denote the smooth mappings defined by

$$L_a(b) = ab \quad \text{and} \quad R_a(b) = ba.$$

Proposition 1. *For any Lie group G, the maps L_a and R_a are diffeomorphisms, the map $\mu: G \times G \to G$ is a submersion, and the inverse mapping $\iota: G \to G$ defined by $\iota(a) = a^{-1}$ is smooth.*

Proof. By the axioms of group multiplication, $L_{a^{-1}}$ is both a left and right inverse to L_a. Since $(L_a)^{-1}$ exists and is smooth, L_a is a diffeomorphism. The argument for R_a is similar.

In particular, $L'_a: TG \to TG$ induces an isomorphism $T_b G \xrightarrow{\sim} T_{ab} G$ for all $b \in G$ and $R'_a: TG \to TG$ induces an isomorphism $T_b G \xrightarrow{\sim} T_{ba} G$ for all $b \in G$. Using the natural identification $T_{(a,b)} G \times G \simeq T_a G \oplus T_b G$, the formula for $\mu'(a,b): T_{(a,b)} G \times G \to T_{ab} G$ is readily seen to be

$$\mu'(a,b)(v,w) = L'_a(w) + R'_b(v)$$

for all $v \in T_a G$ and $w \in T_b G$. In particular $\mu'(a,b)$ is surjective for all $(a,b) \in G \times G$, so $\mu: G \times G \to G$ is a submersion.

Thus, by the implicit function theorem, $\mu^{-1}(e)$ is a closed, embedded submanifold of $G \times G$ whose tangent space at (a,b), by the above formula, is

$$T_{(a,b)} \mu^{-1}(e) = \{(v,w) \in T_a G \times T_b G \,|\, L'_a(w) + R'_b(v) = 0\}.$$

Meanwhile, the group axioms imply that

$$\mu^{-1}(e) = \{(a, a^{-1}) \,|\, a \in G\},$$

which is precisely the graph of $\iota: G \to G$. Since L'_a and R'_a are isomorphisms at every point, it easily follows that the projection on the first factor $\pi_1: G \times G \to G$ restricts to $\mu^{-1}(e)$ to be a diffeomorphism of $\mu^{-1}(e)$ with G. Its inverse is therefore also smooth and is simply the graph of ι. It follows that ι is smooth, as desired.

For any Lie group G, let $G^\circ \subset G$ denote the connected component of G which contains e. This is usually called the *identity component* of G.

Proposition 2. *For any Lie group G, the set G° is an open, normal subgroup of G. Moreover, if U is any open neighborhood of e in G°, then G° is the union of the 'powers' U^n defined inductively by $U^1 = U$ and $U^{k+1} = \mu(U^k, U)$ for $k > 0$.*

Proof. Since G is a manifold, its components are open and path-connected, so G° is open and path-connected. If $\alpha, \beta: [0,1] \to G$ are two continuous maps with $\alpha(0) = \beta(0) = e$, then $\gamma: [0,1] \to G$ defined by $\gamma(t) = \alpha(t)\beta(t)^{-1}$ is a continuous path from e to $\alpha(1)\beta(1)^{-1}$, so G° is closed under multiplication and inverse, and hence is a subgroup. It is a normal subgroup since, for any $a \in G$, the map

$$C_a = L_a \circ (R_a)^{-1}: G \to G$$

(conjugation by a) is a diffeomorphism which clearly fixes e and hence fixes its connected component G° also.

Finally, let $U \subset G^\circ$ be any open neighborhood of e. For any $a \in G^\circ$, let $\gamma\colon [0,1] \to G$ be a path with $\gamma(0) = e$ and $\gamma(1) = a$. The open sets $\{L_{\gamma(t)}(U)\,|\,t \in [0,1]\}$ cover $\gamma([0,1])$, so the compactness of $[0,1]$ implies (via the Lebesgue Covering Lemma) that there is a finite subdivision $0 = t_0 < t_1 \cdots < t_n = 1$ so that $\gamma([t_k, t_{k+1}]) \subset L_{\gamma(t_k)}(U)$ for all $0 \le k < n$. But then each of the elements

$$a_k = \left(\gamma(t_k)^{-1}\right)\gamma(t_{k+1})$$

lies in U and $a = \gamma(1) = a_0 a_1 \cdots a_{n-1} \in U^n$.

A useful consequence of Proposition 2 is that, for a connected Lie group H, any Lie group homomorphism $\phi\colon H \to G$ is determined by its behavior on any open neighborhood of $e \in H$. We are soon going to show an even more striking fact, namely that, for connected H, any homomorphism $\phi\colon H \to G$ is determined by $\phi'(e)\colon T_e H \to T_e G$.

The adjoint representation. It is conventional to denote the tangent space at the identity of a Lie group by an appropriate lower case gothic letter. Thus, the vector space $T_e G$ is denoted $\mathfrak{g}$, the vector space $T_e \mathrm{GL}(n, \mathbb{R})$ is denoted $\mathfrak{gl}(n, \mathbb{R})$, etc.

For example, one can easily compute the tangent spaces at e of the Lie groups defined so far. Here is a sample:

$$\mathfrak{sl}(n, \mathbb{R}) = \{a \in \mathfrak{gl}(n, \mathbb{R})\,|\,\mathrm{tr}(a) = 0\}$$
$$\mathfrak{so}(n, \mathbb{R}) = \{a \in \mathfrak{gl}(n, \mathbb{R})\,|\,a + {}^t a = 0\}$$
$$\mathfrak{u}(n, \mathbb{R}) = \{a \in \mathfrak{gl}(n, \mathbb{C})\,|\,a + {}^t \bar{a} = 0\}$$

Definition 4. For any Lie group G, the *adjoint mapping* is the mapping $\mathrm{Ad}\colon G \to \mathrm{End}(\mathfrak{g})$ defined by

$$\mathrm{Ad}(a) = \left(L_a \circ (R_a)^{-1}\right)'(e)\colon T_e G \to T_e G.$$

As an example, for $G = \mathrm{GL}(n, \mathbb{R})$ it is easy to see that

$$\mathrm{Ad}(a)(x) = axa^{-1}$$

for all $a \in \mathrm{GL}(n, \mathbb{R})$ and $x \in \mathfrak{gl}(n, \mathbb{R})$. Of course, this formula is valid for any matrix Lie group.

The following proposition explains why the adjoint mapping is also called the adjoint representation.

Proposition 3. *The adjoint mapping is a linear representation* $\mathrm{Ad}\colon G \to \mathrm{GL}(\mathfrak{g})$.

Proof. For any $a \in G$, let $C_a = L_a \circ R_{a^{-1}}$. Then $C_a\colon G \to G$ is a diffeomorphism which satisfies $C_a(e) = e$. In particular, $\mathrm{Ad}(a) = C_a'(e)\colon \mathfrak{g} \to \mathfrak{g}$ is an isomorphism and hence belongs to $\mathrm{GL}(\mathfrak{g})$.

The associative property of group multiplication implies $C_a \circ C_b = C_{ab}$, so the Chain Rule implies that $C'_a(e) \circ C'_b(e) = C'_{ab}(e)$. Hence, $\mathrm{Ad}(a)\mathrm{Ad}(b) = \mathrm{Ad}(ab)$, so Ad is a homomorphism.

It remains to show that Ad is smooth. However, if $C\colon G \times G \to G$ is defined by $C(a,b) = aba^{-1}$, then by Proposition 1, C is a composition of smooth maps and hence is smooth. It follows easily that the map $c\colon G \times \mathfrak{g} \to \mathfrak{g}$ given by $c(a,v) = C'_a(e)(v) = \mathrm{Ad}(a)(v)$ is a composition of smooth maps. The smoothness of the map c clearly implies the smoothness of $\mathrm{Ad}\colon G \to \mathfrak{g} \otimes \mathfrak{g}^*$.

Left-invariant vector fields. Because L'_a induces an isomorphism from $\mathfrak{g}$ to $T_a G$ for all $a \in G$, it is easy to show that the map $\Psi\colon G \times \mathfrak{g} \to TG$ given by

$$\Psi(a,v) = L'_a(v)$$

is actually an isomorphism of vector bundles which makes the following diagram commute.

$$\begin{array}{ccc} G \times \mathfrak{g} & \xrightarrow{\Psi} & TG \\ \pi_1 \downarrow & & \downarrow \pi \\ G & \xrightarrow{id} & G \end{array}$$

Note that, in particular, G is a parallelizable manifold. This implies, for example, that the only compact surface which can be given the structure of a Lie group is the torus $S^1 \times S^1$.

For each $v \in \mathfrak{g}$, we may use Ψ to define a vector field X_v on G by the rule $X_v(a) = L'_a(v)$. Note that, by the Chain Rule and the definition of X_v, we have

$$L'_a(X_v(b)) = L'_a(L'_b(v)) = L'_{ab}(v) = X_v(ab).$$

Thus, the vector field X_v is invariant under left translation by any element of G. Such vector fields turn out to be extremely useful in understanding the geometry of Lie groups, and are accorded a special name:

Definition 5. If G is a Lie group, a *left-invariant vector field* on G is a vector field X on G which satisfies $L'_a(X(b)) = X(ab)$.

For example, consider $\mathrm{GL}(n,\mathbb{R})$ as an open subset of the vector space of n-by-n matrices with real entries. Here, $\mathfrak{gl}(n,\mathbb{R})$ is just the vector space of n-by-n matrices with real entries itself and one easily sees that

$$X_v(a) = (a, av).$$

(Since $\mathrm{GL}(n,\mathbb{R})$ is an open subset of a vector space, namely, $\mathfrak{gl}(n,\mathbb{R})$, we are using the standard identification of the tangent bundle of $\mathrm{GL}(n,\mathbb{R})$ with $\mathrm{GL}(n,\mathbb{R}) \times \mathfrak{gl}(n,\mathbb{R})$.)

The following proposition determines all of the left-invariant vector fields on a Lie group.

Proposition 4. *Every left-invariant vector field X on G is of the form $X = X_v$ where $v = X(e)$ and hence is smooth. Moreover, such an X is complete, i.e., the flow Φ associated to X has domain $\mathbb{R} \times G$.*

Proof. That every left-invariant vector field on G has the stated form is an easy exercise for the reader. It remains to show that the flow of such an X is complete, i.e., that for each $a \in G$, there exists a smooth curve $\gamma_a \colon \mathbb{R} \to G$ so that $\gamma_a(0) = a$ and $\gamma_a'(t) = X(\gamma_a(t))$ for all $t \in \mathbb{R}$.

It suffices to show that such a curve exists for $a = e$, since we may then define

$$\gamma_a(t) = a\, \gamma_e(t)$$

and see that γ_a satisfies the necessary conditions: $\gamma_a(0) = a\,\gamma_e(0) = a$ and

$$\gamma_a'(t) = L_a'(\gamma_e'(t)) = L_a'(X(\gamma_e(t))) = X(a\gamma_e(t)) = X(\gamma_a(t)).$$

Now, by the ODE existence theorem, there is an $\varepsilon > 0$ so that such a γ_e can be defined on the interval $(-\varepsilon, \varepsilon) \subset \mathbb{R}$. If γ_e could not be extended to all of $\mathbb{R}$, then there would be a maximum such ε. I will now show that there is no such maximum ε.

For each $s \in (-\varepsilon, \varepsilon)$, the curve $\alpha_s \colon (-\varepsilon + |s|, \varepsilon - |s|) \to G$ defined by

$$\alpha_s(t) = \gamma_e(s+t)$$

clearly satisfies $\alpha_s(0) = \gamma_e(s)$ and

$$\alpha_s'(t) = \gamma_e'(s+t) = X(\gamma_e(s+t)) = X(\alpha_s(t)),$$

so, by the ODE uniqueness theorem, $\alpha_s(t) = \gamma_e(s)\gamma_e(t)$. In particular, we have

$$\gamma_e(s+t) = \gamma_e(s)\gamma_e(t)$$

for all s and t satisfying $|s| + |t| < \varepsilon$.

Thus, I can extend the domain of γ_e to $(-\tfrac{3}{2}\varepsilon, \tfrac{3}{2}\varepsilon)$ by the rule

$$\gamma_e(t) = \begin{cases} \gamma_e(-\tfrac{1}{2}\varepsilon)\gamma_e(t + \tfrac{1}{2}\varepsilon) & \text{if } t \in (-\tfrac{3}{2}\varepsilon, \tfrac{1}{2}\varepsilon); \\ \gamma_e(+\tfrac{1}{2}\varepsilon)\gamma_e(t - \tfrac{1}{2}\varepsilon) & \text{if } t \in (-\tfrac{1}{2}\varepsilon, \tfrac{3}{2}\varepsilon). \end{cases}$$

By the previous arguments, this extended γ_e is still an integral curve of X, contradicting the assumption that $(-\varepsilon, \varepsilon)$ was maximal.

As an example, consider the flow of the left-invariant vector fields on $\mathrm{GL}(n, \mathbb{R})$ (or any matrix Lie group, for that matter): For any $v \in \mathfrak{gl}(n, \mathbb{R})$, the differential equation which γ_e satisfies is simply

$$\gamma_e'(t) = \gamma_e(t)\, v.$$

This is a matrix differential equation and, in elementary ODE courses, we learn that the "fundamental solution" is

$$\gamma_e(t) = e^{tv} = I_n + \sum_{k=1}^{\infty} \frac{v^k}{k!} t^k$$

and that this series converges uniformly on compact sets in $\mathbb{R}$ to a smooth matrix-valued function of t.

Matrix Lie groups are by far the most commonly encountered and, for this reason, the notation $\exp(tv)$ or even e^{tv} is often used for the integral curve $\gamma_e(t)$ associated to X_v in a general Lie group G. (Actually, in order for this notation to be unambiguous, it has to be checked that if $tv = uw$ for $t, u \in \mathbb{R}$ and $v, w \in \mathfrak{g}$, then $\gamma_e(t) = \delta_e(u)$ where γ_e is the integral curve of X_v with initial condition e and δ_e is the integral curve of X_w initial condition e. However, this is an easy exercise in the use of the Chain Rule.)

It is worth remarking explicitly that for any $v \in \mathfrak{g}$ the formula for the flow of the left invariant vector field X_v on G is simply

$$\Phi(t, a) = a \exp(tv) = a\, e^{tv}.$$

(Warning: many beginners make the mistake of thinking that the formula for the flow of the left invariant vector field X_v should be $\Phi(t, a) = \exp(tv)\, a$, instead. It is worth pausing for a moment to think why this is not so.)

It is now possible to describe all of the homomorphisms from the Lie group $(\mathbb{R}, +)$ into any given Lie group:

Proposition 5. *Every Lie group homomorphism $\phi \colon \mathbb{R} \to G$ is of the form $\phi(t) = e^{tv}$ where $v = \phi'(0) \in \mathfrak{g}$.*

Proof. Let $v = \phi'(0) \in \mathfrak{g}$, and let X_v be the associated left-invariant vector field on G. Since $\phi(0) = e$, by ODE uniqueness, it suffices to show that ϕ is an integral curve of X_v. However, $\phi(s+t) = \phi(s)\phi(t)$ implies $\phi'(s) = L'_{\phi(s)}(\phi'(0)) = X_v(\phi(s))$, as desired.

The exponential map

We are now ready to introduce one of the principal tools in the study of Lie groups.

Definition 6. For any Lie group, the *exponential mapping* of G is the mapping $\exp \colon \mathfrak{g} \to G$ defined by $\exp(v) = \gamma_e(1)$ where γ_e is the integral curve of the vector field X_v with initial condition e.

It is an exercise for the reader to show that $\exp \colon \mathfrak{g} \to G$ is smooth and that

$$\exp'(0) \colon \mathfrak{g} \to T_e G = \mathfrak{g}$$

is just the identity mapping.

Example. As has already been remarked, for $GL(n, \mathbb{R})$ (or $GL(V)$ in general for that matter), the formula for the exponential mapping is just the usual power series:

$$e^x = I + x + \tfrac{1}{2}x^2 + \tfrac{1}{6}x^3 + \cdots.$$

This formula works for all matrix Lie groups as well, and can simplify considerably in certain special cases. For example, the group N_3 defined earlier (usually called the Heisenberg group) has its Lie algebra given by

$$\mathfrak{n}_3 = \left\{ \begin{pmatrix} 0 & x & z \\ 0 & 0 & y \\ 0 & 0 & 0 \end{pmatrix} \;\bigg|\; x, y, z \in \mathbb{R} \right\}.$$

Note that $v^3 = 0$ for all $v \in \mathfrak{n}_3$, thus

$$\exp\left(\begin{pmatrix} 0 & x & z \\ 0 & 0 & y \\ 0 & 0 & 0 \end{pmatrix} \right) = \begin{pmatrix} 1 & x & z + \tfrac{1}{2}xy \\ 0 & 1 & y \\ 0 & 0 & 1 \end{pmatrix}.$$

The Lie bracket. Now, the mapping exp is not generally a homomorphism from $\mathfrak{g}$ (with its additive group structure) to G, although, in a certain sense, it comes as close as possible, since, by construction, it *is* a homomorphism when restricted to any one-dimensional linear subspace $\mathbb{R}v \subset \mathfrak{g}$. We now want to spend a few moments considering what the multiplication map on G "looks like" when pulled back to $\mathfrak{g}$ via exp.

Since $\exp'(0) \colon \mathfrak{g} \to T_e G = \mathfrak{g}$ is the identity mapping, it follows from the Implicit Function Theorem that there is a neighborhood U of $0 \in \mathfrak{g}$ so that $\exp \colon U \to G$ is a diffeomorphism onto its image. Moreover, there must be a smaller open neighborhood $V \subset U$ of 0 so that $\mu\bigl(\exp(V) \times \exp(V)\bigr) \subset \exp(U)$. It follows that there is a unique smooth mapping $\nu \colon V \times V \to U$ such that

$$\mu\left(\exp(x), \exp(y)\right) = \exp\left(\nu(x, y)\right).$$

Since exp is a homomorphism restricted to each line through 0 in $\mathfrak{g}$, it follows that ν satisfies

$$\nu(\alpha x, \beta x) = (\alpha + \beta)x$$

for all $x \in V$ and $\alpha, \beta \in \mathbb{R}$ such that $\alpha x, \beta x \in V$.

Since $\nu(0,0) = 0$, the Taylor expansion to second order of ν about $(0,0)$ is of the form,

$$\nu(x, y) = \nu_1(x, y) + \tfrac{1}{2}\nu_2(x, y) + R_3(x, y)$$

where ν_i is a $\mathfrak{g}$-valued polynomial of degree i on the vector space $\mathfrak{g} \oplus \mathfrak{g}$ and R_3 is a $\mathfrak{g}$-valued function on V which vanishes to at least third order at $(0,0)$.

Since $\nu(x,0) = \nu(0,x) = x$, it easily follows that $\nu_1(x,y) = x + y$ and that $\nu_2(x,0) = \nu_2(0,y) = 0$. Thus, the quadratic polynomial ν_2 is linear in each $\mathfrak{g}$-variable separately.

Moreover, since $\nu(x,x) = 2x$ for all $x \in V$, substituting this into the above expansion and comparing terms of order 2 yields that $\nu_2(x,x) \equiv 0$. Of course, this implies that ν_2 is actually skew-symmetric since

$$0 = \nu_2(x+y, x+y) - \nu_2(x,x) - \nu_2(y,y) = \nu_2(x,y) + \nu_2(y,x).$$

Definition 7. The skew-symmetric, bilinear multiplication $[,]\colon \mathfrak{g} \times \mathfrak{g} \to \mathfrak{g}$ defined by
$$[x,y] = \nu_2(x,y)$$
is called the *Lie bracket* in $\mathfrak{g}$. The pair $(\mathfrak{g}, [,])$ is called the *Lie algebra* of G.

Using this notation, there is a formula

$$\exp(x)\exp(y) = \exp\bigl(x + y + \tfrac{1}{2}[x,y] + R_3(x,y)\bigr)$$

valid for all x and y in some fixed open neighborhood of 0 in $\mathfrak{g}$.

One might think of the term involving $[,]$ as the first deviation of the Lie group multiplication from being just vector addition. In fact, it is clear from this formula that, if G is abelian, then $[x,y] = 0$ for all $x,y \in \mathfrak{g}$. For this reason, a Lie algebra in which all brackets vanish is called an *abelian* Lie algebra. (In fact, (see the Exercises) $\mathfrak{g}$ being abelian implies that G°, the identity component of G, is abelian.)

Example. If $G = \mathrm{GL}(n, \mathbb{R})$, then it is easy to see that the induced bracket operation on $\mathfrak{gl}(n, \mathbb{R})$, the vector space of n-by-n matrices, is just the matrix "commutator"

$$[x,y] = xy - yx.$$

In fact, the reader can verify this by examining the following second order expansion:

$$\begin{aligned}
e^x e^y &= (I_n + x + \tfrac{1}{2}x^2 + \cdots)(I_n + y + \tfrac{1}{2}y^2 + \cdots) \\
&= (I_n + x + y + \tfrac{1}{2}(x^2 + 2xy + y^2) + \cdots) \\
&= (I_n + (x + y + \tfrac{1}{2}[x,y]) + \tfrac{1}{2}(x + y + \tfrac{1}{2}[x,y])^2 + \cdots)
\end{aligned}$$

Moreover, this same formula is easily seen to hold for any x and y in $\mathfrak{gl}(V)$ where V is any finite dimensional vector space.

Theorem 1. *If $\phi\colon H \to G$ is a Lie group homomorphism, then $\varphi = \phi'(e)\colon \mathfrak{h} \to \mathfrak{g}$ satisfies*
$$\exp_G(\varphi(x)) = \phi(\exp_H(x))$$
for all $x \in \mathfrak{h}$. In other words, the diagram

$$\begin{array}{ccc}
\mathfrak{h} & \xrightarrow{\varphi} & \mathfrak{g} \\
{\scriptstyle \exp_H}\downarrow & & \downarrow{\scriptstyle \exp_G} \\
H & \xrightarrow{\phi} & G
\end{array}$$

commutes. Moreover, for all x and y in $\mathfrak{h}$,

$$\varphi([x,y]_H) = [\varphi(x), \varphi(y)]_G.$$

Proof. The first statement is an immediate consequence of Proposition 5 and the Chain Rule since, for every $x \in \mathfrak{h}$, the map $\gamma \colon \mathbb{R} \to G$ given by $\gamma(t) = \phi(e^{tx})$ is clearly a Lie group homomorphism with initial velocity $\gamma'(0) = \varphi(x)$ and hence must also satisfy $\gamma(t) = e^{t\varphi(x)}$.

To get the second statement, let x and y be elements of $\mathfrak{h}$ which are sufficiently close to zero. Then, using self-explanatory notation,

$$\phi(\exp_H(x) \exp_H(y)) = \phi(\exp_H(x))\phi(\exp_H(y)),$$

so

$$\phi(\exp_H(x + y + \tfrac{1}{2}[x,y]_H + R_3^H(x,y))) = \exp_G(\varphi(x))\exp_G(\varphi(y)),$$

and thus

$$\exp_G(\varphi(x + y + \tfrac{1}{2}[x,y]_H + R_3^H(x,y))) = \exp_G(\varphi(x) + \varphi(y) + \tfrac{1}{2}[\varphi(x),\varphi(y)]_G \\ + R_3^G(\varphi(x),\varphi(y))),$$

finally giving

$$\varphi(x + y + \tfrac{1}{2}[x,y]_H + R_3^H(x,y)) = \varphi(x) + \varphi(y) + \tfrac{1}{2}[\varphi(x),\varphi(y)]_G \\ + R_3^G(\varphi(x),\varphi(y)).$$

Now using the fact that φ is linear and comparing second order terms gives the desired result.

On account of this theorem, it is usually not necessary to distinguish the map exp or the bracket [,] according to the group in which it is being applied, and I will follow this practice. Henceforth, these symbols will be used without group decorations whenever confusion seems unlikely.

Theorem 1 has many useful corollaries. Among them is

Proposition 6. *If H is a connected Lie group and $\phi_1, \phi_2 \colon H \to G$ are two Lie group homomorphisms which satisfy $\phi'_1(e) = \phi'_2(e)$, then $\phi_1 = \phi_2$.*

Proof. There is an open neighborhood U of 0 in $\mathfrak{h}$ so that $\exp_H$ is smoothly invertible on this neighborhood. In particular, $\exp_H(U)$ is an open neighborhood of e in H. By Theorem 1, for $a \in \exp_H(U)$,

$$\phi_i(a) = \exp_G(\varphi_i(\exp_H^{-1}(a))).$$

Since $\varphi_1 = \varphi_2$, it follows that $\phi_1 = \phi_2$ on $\exp_H(U)$. By Proposition 2, every element of H can be written as a finite product of elements of $\exp_H(U)$, so $\phi_1 = \phi_2$ everywhere.

We also have the following fundamental result:

Proposition 7. *Let $\mathrm{Ad} \colon G \to GL(\mathfrak{g})$ be the adjoint representation. Then the formula for $\mathrm{ad} = \mathrm{Ad}'(e) \colon \mathfrak{g} \to \mathfrak{gl}(\mathfrak{g})$ is $\mathrm{ad}(x)(y) = [x,y]$.*

Proof. This is simply a matter of unwinding the definitions. By definition, $\mathrm{Ad}(a) = C'_a(e)$ where $C_a\colon G \to G$ is defined by $C_a(b) = aba^{-1}$. In order to compute $C'_a(e)(y)$ for $y \in \mathfrak{g}$, we may just compute $\gamma'(0)$ where γ is the curve $\gamma(t) = a\exp(ty)a^{-1}$. Moreover, since $\exp'(0)\colon \mathfrak{g} \to \mathfrak{g}$ is the identity, we may as well compute $\beta'(0)$ where $\beta = \exp^{-1}\circ\gamma$. Now, assuming $a = \exp(x)$, we compute

$$\begin{aligned}\beta(t) &= \exp^{-1}(\exp(x)\exp(ty)\exp(-x))\\ &= \exp^{-1}(\exp(x+ty+\tfrac{1}{2}[x,ty]+\cdots)\exp(-x))\\ &= \exp^{-1}(\exp((x+ty+\tfrac{1}{2}[x,ty])+(-x)+\tfrac{1}{2}[x+ty,-x]+\cdots)\\ &= ty + t[x,y] + E_3(x,ty)\end{aligned}$$

where the omitted terms and the function E_3 vanish to order at least 3 at $(x,y) = (0,0)$. (Note that I used the identity $[y,x] = -[x,y]$.) It follows that

$$\mathrm{Ad}(\exp(x))(y) = \beta'(0) = y + [x,y] + E'_3(x,0)y$$

where $E'_3(x,0)$ denotes the derivative of E_3 with respect to y evaluated at $(x,0)$ and is hence a function of x which vanishes to order at least 2 at $x = 0$. On the other hand, since, by the first part of Theorem 1, we have

$$\mathrm{Ad}(\exp(x)) = \exp(\mathrm{ad}(x)) = I + \mathrm{ad}(x) + \tfrac{1}{2}(\mathrm{ad}(x))^2 + \cdots .$$

Comparing the x-linear terms in the last two equations clearly gives the desired result.

Lie algebras

For any Lie group G, the algebra structure induced on $\mathfrak{g} = T_e G$ by the multiplication $[,]$ turns out to be very important.

The Jacobi identity

In light of Proposition 7, applying the second part of Theorem 1 to Proposition 3 yields the identity
$$\mathrm{ad}([x,y]) = [\mathrm{ad}(x),\mathrm{ad}(y)],$$
known as the *Jacobi identity*.

The Jacobi identity is often presented differently. Using Proposition 7, the reader can verify that the equation $\mathrm{ad}([x,y]) = [\mathrm{ad}(x),\mathrm{ad}(y)]$ where $\mathrm{ad}(x)(y) = [x,y]$ is equivalent to the condition that

$$[[x,y],z] + [[y,z],x] + [[z,x],y] = 0 \quad \text{for all} \quad x,y,z \in \mathfrak{g}.$$

This is the form in which the Jacobi identity is often stated. Unfortunately, although this is a very symmetric form of the identity, it somewhat obscures its importance and meaning.

LECTURE 2. LIE GROUPS AND LIE ALGEBRAS

The Jacobi identity is so important that the class of algebras in which it holds is given a name:

Definition 8. A *Lie algebra* is a pair $(\mathfrak{g}, [\,,\,])$ where $\mathfrak{g}$ is a vector space and $[\,,\,] \colon \mathfrak{g} \times \mathfrak{g} \to \mathfrak{g}$ is a skew-symmetric bilinear pairing (called the *bracket*) which satisfies the Jacobi identity, i.e., $\mathrm{ad}([x, y]) = [\mathrm{ad}(x), \mathrm{ad}(y)]$, where $\mathrm{ad}\colon \mathfrak{g} \to \mathfrak{gl}(\mathfrak{g})$ is defined as $\mathrm{ad}(x)(y) = [x, y]$. A *Lie subalgebra* of $\mathfrak{g}$ is a linear subspace $\mathfrak{h} \subset \mathfrak{g}$ which is closed under bracket. A *homomorphism* of Lie algebras is a linear mapping of vector spaces $\varphi \colon \mathfrak{h} \to \mathfrak{g}$ which satisfies

$$\varphi([x, y]) = [\varphi(x), \varphi(y)].$$

The only examples of Lie algebras encountered so far in these notes are the ones provided by Proposition 7, namely, the Lie algebras of Lie groups. This is not accidental, for, as will be seen, every finite dimensional Lie algebra is the Lie algebra of some Lie group.

Lie brackets of vector fields

There is another notion of Lie bracket, namely the Lie bracket of smooth vector fields on a smooth manifold. This bracket is also skew-symmetric and satisfies the Jacobi identity, so it is reasonable to ask how it might be related to the notion of Lie bracket that has just been defined. Since the Lie bracket of vector fields commutes with diffeomorphisms, it easily follows that the Lie bracket of two left-invariant vector fields on a Lie group G is also a left-invariant vector field on G. The following result is, perhaps then, to be expected.

Proposition 8. *The identity* $[X_x, X_y] = X_{[x,y]}$ *holds for any* $x, y \in \mathfrak{g}$.

Proof. This is a direct calculation. For simplicity, I will use the following characterization of the Lie bracket for vector fields: If Φ_x and Φ_y are the flows associated to the vector fields X_x and X_y, then for any function f on G one has the formula:

$$([X_x, X_y]f)(a) = \lim_{t \to 0^+} \frac{f(\Phi_y(-\sqrt{t}, \Phi_x(-\sqrt{t}, \Phi_y(\sqrt{t}, \Phi_x(\sqrt{t}, a))))) - f(a)}{t}.$$

Now, as has been seen, the formulas for the flows of X_x and X_y are given by $\Phi_x(t, a) = a \exp(tx)$ and $\Phi_y(t, a) = a \exp(ty)$. This implies that the general formula above simplifies to

$$([X_x, X_y]f)(a) = \lim_{t \to 0^+} \frac{f\bigl(a \exp(\sqrt{t}x) \exp(\sqrt{t}y) \exp(-\sqrt{t}x) \exp(-\sqrt{t}y)\bigr) - f(a)}{t}.$$

Now

$$\exp(\pm\sqrt{t}x) \exp(\pm\sqrt{t}y) = \exp(\pm\sqrt{t}(x + y) + \tfrac{t}{2}[x, y] + \cdots)$$

so $\exp(\sqrt{t}x)\exp(\sqrt{t}y)\exp(-\sqrt{t}x)\exp(-\sqrt{t}y)$ simplifies to $\exp(t[x,y]+\cdots)$ where the omitted terms vanish to higher t-order than t itself. Thus,

$$([X_x, X_y]f)(a) = \lim_{t \to 0^+} \frac{f\big(a\,\exp(t[x,y]+\cdots)\big) - f(a)}{t}.$$

Since $[X_x, X_y]$ must be a left-invariant vector field and since

$$(X_{[x,y]}f)(a) = \lim_{t \to 0^+} \frac{f\big(a\,\exp(t[x,y])\big) - f(a)}{t},$$

the desired result follows.

Subgroups, subalgebras, and homomorphisms

I can now prove the following fundamental result.

Theorem 2. *For each Lie subgroup H of a Lie group G, the subspace $\mathfrak{h} = T_e H$ is a Lie subalgebra of $\mathfrak{g}$. Moreover, every Lie subalgebra $\mathfrak{h} \subset \mathfrak{g}$ is $T_e H$ for a unique connected Lie subgroup H of G.*

Proof. Suppose that $H \subset G$ is a Lie subgroup. Then the inclusion map is a Lie group homomorphism and Theorem 1 thus implies that the inclusion map $\mathfrak{h} \hookrightarrow \mathfrak{g}$ is a Lie algebra homomorphism. In particular, $\mathfrak{h}$, when considered as a subspace of $\mathfrak{g}$, is closed under the Lie bracket in G and hence is a subalgebra.

Suppose now that $\mathfrak{h} \subset \mathfrak{g}$ is a subalgebra.

First, I will show that there is at most one connected Lie subgroup of G with Lie algebra $\mathfrak{h}$. Suppose that there were two, say H_1 and H_2. Then by Theorem 1, $\exp_G(\mathfrak{h})$ is a subset of both H_1 and H_2 and contains an open neighborhood of the identity element in each of them. However, since, by Proposition 2, each of H_1 and H_2 are generated by finite products of the elements in any open neighborhood of the identity, it follows that $H_1 \subset H_2$ and $H_2 \subset H_1$, so $H_1 = H_2$, as desired.

Second, to prove the existence of a subgroup H with $T_e H = \mathfrak{h}$, I call on the Global Frobenius Theorem. Let $r = \dim(\mathfrak{h})$ and let $E \subset TG$ be the rank r sub-bundle spanned by the vector fields X_x where $x \in \mathfrak{h}$. Note that $E_a = L'_a(E_e) = L'_a(\mathfrak{h})$ for all $a \in G$, so E is left-invariant. Since $\mathfrak{h}$ is a subalgebra of $\mathfrak{g}$, Proposition 8 implies that E is an integrable sub-bundle on G. By the Global Frobenius Theorem, there is an r-dimensional leaf of E through e. Call this submanifold H.

It remains is to show that H is closed under multiplication and inverse. Inverse is easy: Let $a \in H$ be fixed. Then, since H is path-connected, there exists a smooth curve $\alpha: [0,1] \to H$ so that $\alpha(0) = e$ and $\alpha(1) = a$. Now consider the curve $\bar{\alpha}$ defined on $[0,1]$ by $\bar{\alpha}(t) = a^{-1}\alpha(1-t)$. Because E is left-invariant, $\bar{\alpha}$ is an integral curve of E and it joins e to a^{-1}. Thus a^{-1} must also lie in H. Multiplication is only slightly more difficult: Now suppose in addition that $b \in H$ and let $\beta: [0,1] \to H$ be a smooth curve so that $\beta(0) = e$ and $\beta(1) = b$. Then the piecewise smooth curve

$\gamma\colon [0,2] \to G$ given by

$$\gamma(t) = \begin{cases} \alpha(t) & \text{if } 0 \le t \le 1; \\ a\beta(t-1) & \text{if } 1 \le t \le 2, \end{cases}$$

is an integral curve of E joining e to ab. Hence ab belongs to H, as desired.

Theorem 3. *If H is a connected and simply connected Lie group, then, for any Lie group G, each Lie algebra homomorphism $\varphi\colon \mathfrak{h} \to \mathfrak{g}$ is of the form $\varphi = \phi'(e)$ for some unique Lie group homomorphism $\phi\colon H \to G$.*

Proof. In light of Theorem 1 and Proposition 6, all that remains to be proved is that for each Lie algebra homomorphism $\varphi\colon \mathfrak{h} \to \mathfrak{g}$ there exists a Lie group homomorphism ϕ satisfying $\phi'(e) = \varphi$.

I do this as follows: Suppose that $\varphi\colon \mathfrak{h} \to \mathfrak{g}$ is a Lie algebra homomorphism. Consider the product Lie group $H \times G$. Its Lie algebra is $\mathfrak{h} \oplus \mathfrak{g}$ with Lie bracket given by $[(h_1, g_1), (h_2, g_2)] = ([h_1, h_2], [g_1, g_2])$, as is easily verified. Now consider the subspace $\widehat{\mathfrak{h}} \subset \mathfrak{h} \oplus \mathfrak{g}$ spanned by elements of the form $(x, \varphi(x))$ where $x \in \mathfrak{h}$. Since φ is a Lie algebra homomorphism, $\widehat{\mathfrak{h}}$ is a Lie subalgebra of $\mathfrak{h} \oplus \mathfrak{g}$ (and happens to be isomorphic to $\mathfrak{h}$). In particular, by Theorem 2, it follows that there is a connected Lie subgroup $\widehat{H} \subset H \times G$, whose Lie algebra is $\widehat{\mathfrak{h}}$. I will now show that $\widehat{H}$ is the graph of the desired Lie group homomorphism $\phi\colon H \to G$.

Note that since $\widehat{H}$ is a Lie subgroup of $H \times G$, the projections $\pi_1\colon \widehat{H} \to H$ and $\pi_2\colon \widehat{H} \to G$ are Lie group homomorphisms. The associated Lie algebra homomorphisms $\varpi_1\colon \widehat{\mathfrak{h}} \to \mathfrak{h}$ and $\varpi_2\colon \widehat{\mathfrak{h}} \to \mathfrak{g}$ are clearly given by $\varpi_1(x, \varphi(x)) = x$ and $\varpi_2(x, \varphi(x)) = \varphi(x)$.

Now, I claim that π_1 is actually a surjective covering map: It is surjective since $\varpi_1\colon \widehat{\mathfrak{h}} \to \mathfrak{h}$ is an isomorphism so $\pi_1(\widehat{H})$ contains a neighborhood of the identity in H and hence, by Proposition 2 and the connectedness of H, must contain all of H. It remains to show that, under π_1, points of H have evenly covered neighborhoods.

Let $\widehat{Z} = \ker(\pi_1)$. Then $\widehat{Z}$ is a closed discrete subgroup of $\widehat{H}$. Let $\widehat{U} \subset \widehat{H}$ be a neighborhood of the identity to which π_1 restricts to be a smooth diffeomorphism onto a neighborhood U of e in H. Then the reader can easily verify that for each $a \in \widehat{H}$ the map $\sigma_a\colon \widehat{Z} \times \widehat{U} \to \widehat{H}$ given by $\sigma_a(z, u) = azu$ is a diffeomorphism onto $(\pi_1)^{-1}(L_{\pi_1(a)}(U))$ which commutes with the appropriate projections and hence establishes the even covering property.

Finally, since $\widehat{H}$ is connected and, by hypothesis, H is simply connected, it follows that π_1 must actually be a one-to-one and onto diffeomorphism. The map $\phi = \pi_2 \circ \pi_1^{-1}$ is then the desired homomorphism.

An existence theorem

As the last general theorem for this Lecture, I state, without proof, the following existence result.

Theorem 4. *For each finite dimensional Lie algebra $\mathfrak{g}$, there exists a Lie group G whose Lie algebra is isomorphic to $\mathfrak{g}$.*

Unfortunately, this theorem is surprisingly difficult to prove. It would suffice, by Theorem 2, to show that every Lie algebra $\mathfrak{g}$ is isomorphic to a subalgebra of the Lie algebra of a Lie group. In fact, an even stronger statement is true. A theorem of Ado asserts that every finite dimensional Lie algebra is isomorphic to a subalgebra of $\mathfrak{gl}(n, \mathbb{R})$ for some n. Thus, to prove Theorem 4, it would be enough to prove Ado's theorem. Unfortunately, this theorem also turns out to be rather delicate (see [Va] for a proof). However, there are many interesting examples of $\mathfrak{g}$ for which a proof can be given by elementary means (see the Exercises).

On the other hand, this abstract existence theorem is not used very often anyway. It is rare that a (finite dimensional) Lie algebra arises in practice which is not readily representable as the Lie algebra of some Lie group.

The reader may be wondering about uniqueness: How many Lie groups are there whose Lie algebras are isomorphic to a given $\mathfrak{g}$? Since the Lie algebra of a Lie group G only depends on the identity component G, it is reasonable to restrict to the case of connected Lie groups. Now, as you are asked to show in the Exercises, the universal cover $\tilde{G}$ of a connected Lie group G can be given a unique Lie group structure for which the covering map $\tilde{G} \to G$ is a homomorphism. Thus, there always exists a connected and simply connected Lie group, say $G(\mathfrak{g})$, whose Lie algebra is isomorphic to $\mathfrak{g}$. A simple application of Theorem 3 shows that if G' is any other Lie group with Lie algebra $\mathfrak{g}$, then there is a homomorphism $\phi \colon G(\mathfrak{g}) \to G$ which induces an isomorphism on the Lie algebras. It follows easily that, up to isomorphism, there is only one simply connected and connected Lie group with Lie algebra $\mathfrak{g}$. Moreover, every other connected Lie group with Lie algebra G is isomorphic to a quotient of $G(\mathfrak{g})$ by a discrete subgroup of G which lies in the center of $G(\mathfrak{g})$ (see the Exercises).

The Structure Constants

Our work so far has shown that the problem of classifying the connected Lie groups up to isomorphism is very nearly the same thing as classifying the (finite dimensional) Lie algebras. (See the Exercises for a clarification of this point.) This is a remarkable state of affairs, since, *a priori*, Lie groups involve the topology of smooth manifolds and it is rather surprising that their classification can be reduced to what is essentially an algebra problem. It is worth taking a closer look at this algebra problem itself.

Let $\mathfrak{g}$ be a Lie algebra of dimension n, and let $x_1, x_2, \ldots, x_n$ be a basis for $\mathfrak{g}$. Then there exist constants c_{ij}^k so that (using the summation convention)

$$[x_i, x_j] = c_{ij}^k \, x_k.$$

(These quantities c are called the structure constants of $\mathfrak{g}$ relative to the given basis.) The skew-symmetry of the Lie bracket is is equivalent to the skew-symmetry of c in its lower indices:

$$c_{ij}^k + c_{ji}^k = 0.$$

The Jacobi identity is equivalent to the quadratic equations:

$$c_{ij}^\ell c_{k\ell}^m + c_{jk}^\ell c_{i\ell}^m + c_{ki}^\ell c_{j\ell}^m = 0.$$

Conversely, any set of n^3 constants satisfying these relations defines an n-dimensional Lie algebra by the above bracket formula.

Left-invariant forms and the structure equations

Dual to the left-invariant vector fields on a Lie group G, there are the left-invariant 1-forms, which are indispensable as calculational tools.

Definition 9. For any Lie group G, the $\mathfrak{g}$-valued 1-form on G defined by

$$\omega_G(v) = L'_{a^{-1}}(v) \qquad \text{for } v \in T_a G$$

is called the *canonical left-invariant 1-form* on G.

It is easy to see that ω_G is smooth. Moreover, ω_G is the unique left-invariant $\mathfrak{g}$-valued 1-form on G which satisfies $\omega_G(v) = v$ for all $v \in \mathfrak{g} = T_e G$.

By a calculation which is left as an exercise for the reader,

$$\phi^*(\omega_G) = \varphi(\omega_H)$$

for any Lie group homomorphism $\phi \colon H \to G$ with $\varphi = \phi'(e)$. In particular, when H is a subgroup of G, the pull back of ω_G to H via the inclusion mapping is just ω_H. For this reason, it is common to simply write ω for ω_G when there is no danger of confusion.

Example. If $G \subset \mathrm{GL}(n, \mathbb{R})$ is a matrix Lie group, then we may regard the inclusion $g \colon G \to \mathrm{GL}(n, \mathbb{R})$ as a matrix-valued function on G and compute that ω is given by the simple formula

$$\omega = g^{-1}\, dg.$$

From this formula, the left-invariance of ω is obvious.

In the matrix Lie group case, it is also easy to compute the exterior derivative of ω: Since $g\, g^{-1} = I_n$, we get

$$dg\, g^{-1} + g\, d(g^{-1}) = 0,$$

so

$$d(g^{-1}) = -g^{-1}\, dg\, g^{-1}.$$

This implies the formula

$$d\omega = -\omega \wedge \omega.$$

(Warning: Matrix multiplication is implicit in this formula!)

For a general Lie group, the formula for $d\omega$ is only slightly more complicated. To state the result, let me first define some notation. I will use $[\omega,\omega]$ to denote the $\mathfrak{g}$-valued 2-form on G whose value on a pair of vectors $v,w \in T_aG$ is

$$[\omega,\omega](v,w) = [\omega(v),\omega(w)] - [\omega(w),\omega(v)] = 2[\omega(v),\omega(w)].$$

Proposition 9. *For any Lie group G,* $d\omega = -\tfrac{1}{2}[\omega,\omega]$.

Proof. First, let X_v and X_w be the left-invariant vector fields on G whose values at e are v and w respectively. Then, by the usual formula for the exterior derivative

$$d\omega(X_v, X_w) = X_v\bigl(\omega(X_w)\bigr) - X_w\bigl(\omega(X_v)\bigr) - \omega\bigl([X_v, X_w]\bigr).$$

However, the $\mathfrak{g}$-valued functions $\omega(X_v)$ and $\omega(X_w)$ are clearly left-invariant and hence are constants and equal to v and w respectively. Moreover, by Proposition 8, $[X_v, X_w] = X_{[v,w]}$, so the formula simplifies to

$$d\omega(X_v, X_w) = -\omega\bigl(X_{[v,w]}\bigr).$$

The right hand side is, again, a left-invariant function, so it must equal its value at the identity, which is clearly $-[v,w]$, which equals $-[\omega(X_v), \omega(X_w)]$ Thus,

$$d\omega(X_v, X_w) = -\tfrac{1}{2}[\omega,\omega](X_v, X_w)$$

for any pair of left-invariant vector fields on G. Since any pair of vectors in T_aG can be written as $X_v(a)$ and $X_w(a)$ for some $v, w \in \mathfrak{g}$, the result follows.

The formula proved in Proposition 9 is often called the *structure equation* of Maurer and Cartan. It is also usually expressed slightly differently. If $x_1, x_2, \ldots, x_n$ is a basis for $\mathfrak{g}$ with structure constants c^i_{jk}, then ω can be written in the form

$$\omega = x_1\,\omega^1 + \cdots + x_n\,\omega^n$$

where the ω^i are $\mathbb{R}$-valued left-invariant 1-forms. Proposition 9 can then be expanded to give

$$d\omega^i = -\tfrac{1}{2} c^i_{jk}\, \omega^j \wedge \omega^k,$$

which is the most common form in which the structure equations are given. Note that the identity $d(d(\omega^i)) = 0$ is equivalent to the Jacobi identity.

The 1-, 2-, and 3-dimensional Lie algebras

To conclude this lecture, I want to outline the classification of the low-dimensional Lie algebras.

1-dimensional Lie algebras

It is clear that up to isomorphism, there is only one (real) Lie algebra of dimension 1, namely $\mathfrak{g} = \mathbb{R}$ with the zero bracket. This is the Lie algebra of the connected Lie groups $\mathbb{R}$ and S^1. (You are asked to prove in an exercise that these are, in fact, the only connected one-dimensional Lie groups.)

2-dimensional Lie algebras

The first interesting case, therefore, is dimension 2. If $\mathfrak{g}$ is a 2-dimensional Lie algebra with basis x_1, x_2, then the entire Lie algebra structure is determined by the bracket $[x_1, x_2] = a^1 x_1 + a^2 x_2$. If $a^1 = a^2 = 0$, then all brackets are zero, and the algebra is abelian. If one of a^1 or a^2 is non-zero, then, by switching x_1 and x_2 if necessary, we may assume that $a^1 \neq 0$. Then, considering the new basis $y_1 = a^1 x_1 + a^2 x_2$ and $y_2 = (1/a^1) x_2$, we get $[y_1, y_2] = y_1$. Since the Jacobi identity is easily verified for this Lie bracket, this does define a Lie algebra. Thus, up to isomorphism, there are only two distinct 2-dimensional Lie algebras.

The abelian example is, of course, the Lie algebra of the vector space $\mathbb{R}^2$ (as well as the Lie algebra of $S^1 \times \mathbb{R}$, and the Lie algebra of $S^1 \times S^1$).

An example of a Lie group of dimension 2 with a non-abelian Lie algebra is the matrix Lie group

$$G = \left\{ \begin{pmatrix} a & b \\ 0 & 1 \end{pmatrix} \;\middle|\; a \in \mathbb{R}^+,\ b \in \mathbb{R} \right\}.$$

In fact, it is not hard to show that, up to isomorphism, this is the only connected non-abelian Lie group of dimension 2 (see the Exercises).

3-dimensional Lie algebras

Now, let me pass on to the classification of the three dimensional Lie algebras. Here, the story becomes much more interesting. Let $\mathfrak{g}$ be a 3-dimensional Lie algebra, and let x_1, x_2, x_3 be a basis of $\mathfrak{g}$. Then the bracket relations can be written in matrix form as

$$(\ [x_2, x_3]\ \ [x_3, x_1]\ \ [x_1, x_2]\ ) = (\ x_1\ \ x_2\ \ x_3\ )\, C$$

where C is the 3-by-3 matrix of structure constants. How is this matrix affected by a change of basis? Well, let

$$(\ y_1\ \ y_2\ \ y_3\ ) = (\ x_1\ \ x_2\ \ x_3\ )\, A$$

where $A \in \mathrm{GL}(3, \mathbb{R})$. Then it is easy to compute that

$$(\ [y_2, y_3]\ \ [y_3, y_1]\ \ [y_1, y_2]\ ) = (\ [x_2, x_3]\ \ [x_3, x_1]\ \ [x_1, x_2]\ )\, \mathrm{Adj}(A)$$

where $\mathrm{Adj}(A)$ is the classical adjoint matrix of A, i.e., the matrix of 2-by-2 minors. Thus,

$$A^{-1} = (\det(A))^{-1}\, {}^t\mathrm{Adj}(A).$$

(Do not confuse this with the adjoint mapping defined earlier!) It then follows that

$$(\; [y_2, y_3] \;\; [y_3, y_1] \;\; [y_1, y_2] \;) = (\; y_1 \;\; y_2 \;\; y_3 \;) C',$$

where

$$C' = A^{-1} \, C \, \mathrm{Adj}(A) = \det(A) \, A^{-1} \, C \, {}^t\!A^{-1}.$$

It follows without too much difficulty that, writing $C = S + \hat{a}$, where S is a symmetric 3-by-3 matrix and

$$\hat{a} = \begin{pmatrix} 0 & -a^3 & a^2 \\ a^3 & 0 & -a^1 \\ -a^2 & a^1 & 0 \end{pmatrix} \quad \text{whenever} \quad a = \begin{pmatrix} a^1 \\ a^2 \\ a^3 \end{pmatrix},$$

then $C' = S' + \widehat{a'}$, where

$$S' = \det(A) \, A^{-1} \, S \, {}^t\!A^{-1} \quad \text{and} \quad a' = {}^t\!A a.$$

Now, I claim that the condition that the Jacobi identity hold for the bracket defined by the matrix C is equivalent to the condition $Sa = 0$. To see this, note first that

$$\begin{aligned}
&[[x_2, x_3], x_1] + [[x_3, x_1], x_2] + [[x_1, x_2], x_3] \\
&= [C_1^1 x_1 + C_1^2 x_2 + C_1^3 x_3, x_1] + [C_2^1 x_1 + C_2^2 x_2 + C_2^3 x_3, x_2] + [C_3^1 x_1 + C_3^2 x_2 + C_3^3 x_3, x_3] \\
&= (C_3^2 - C_2^3)[x_2, x_3] + (C_1^3 - C_3^1)[x_3, x_1] + (C_2^1 - C_1^2)[x_1, x_2] \\
&= 2a^1 [x_2, x_3] + 2a^2 [x_3, x_1] + 2a^3 [x_1, x_2] \\
&= 2 (\; [x_2, x_3] \;\; [x_3, x_1] \;\; [x_1, x_2] \;) \, a \\
&= 2 (\; x_1 \;\; x_2 \;\; x_3 \;) Ca,
\end{aligned}$$

and $Ca = (S + \hat{a})a = Sa$ since $\hat{a}\,a = 0$. Thus, the Jacobi identity applied to the basis x_1, x_2, x_3 implies that $Sa = 0$. However, if y_1, y_2, y_3 is any other triple of elements of $\mathfrak{g}$, then for some 3-by-3 matrix B,

$$(\; y_1 \;\; y_2 \;\; y_3 \;) = (\; x_1 \;\; x_2 \;\; x_3 \;) B,$$

and I leave it to the reader to check that

$$\begin{aligned}
&[[y_2, y_3], y_1] + [[y_3, y_1], y_2] + [[y_1, y_2], y_3] \\
&= \det(B) \, ([[x_2, x_3], x_1] + [[x_3, x_1], x_2] + [[x_1, x_2], x_3])
\end{aligned}$$

in this case. Thus, $Sa = 0$ implies the full Jacobi identity.

There are now two essentially different cases to treat. In the first case, if $a = 0$, then the Jacobi identity is automatically satisfied, and S can be any symmetric matrix. However, two such choices S and S' will clearly give rise to isomorphic Lie algebras if and only if there is an $A \in \mathrm{GL}(3, \mathbb{R})$ for which $S' = \det(A) \, A^{-1} \, S \, {}^t\!A^{-1}$. I leave as an exercise for the reader to show that every choice of S yields an algebra (with $a = 0$) which is equivalent to exactly one of the algebras made by one of the following six choices:

$$\begin{pmatrix} 0 & 0 & 0 \\ 0 & 0 & 0 \\ 0 & 0 & 0 \end{pmatrix} \quad \begin{pmatrix} 0 & 1 & 0 \\ 1 & 0 & 0 \\ 0 & 0 & 0 \end{pmatrix} \quad \begin{pmatrix} 0 & 1 & 0 \\ 1 & 0 & 0 \\ 0 & 0 & 1 \end{pmatrix}$$

$$\begin{pmatrix} 1 & 0 & 0 \\ 0 & 0 & 0 \\ 0 & 0 & 0 \end{pmatrix} \quad \begin{pmatrix} 1 & 0 & 0 \\ 0 & 1 & 0 \\ 0 & 0 & 0 \end{pmatrix} \quad \begin{pmatrix} 1 & 0 & 0 \\ 0 & 1 & 0 \\ 0 & 0 & 1 \end{pmatrix}$$

On the other hand, if $a \neq 0$, then by a suitable change of basis A, I can arrange that $a^1 = a^2 = 0$ and that $a^3 = 1$. Any change of basis A which preserves this normalization is seen to be of the form

$$A = \begin{pmatrix} A_1^1 & A_2^1 & A_3^1 \\ A_1^2 & A_2^2 & A_3^2 \\ 0 & 0 & 1 \end{pmatrix}.$$

Now, since $Sa = 0$ and since S is symmetric, it follows that S must be of the form

$$S = \begin{pmatrix} s_{11} & s_{12} & 0 \\ s_{12} & s_{22} & 0 \\ 0 & 0 & 0 \end{pmatrix}.$$

Moreover, a simple calculation shows that the result of applying a change of basis of the above form is to change the matrix S into the matrix

$$S' = \begin{pmatrix} s'_{11} & s'_{12} & 0 \\ s'_{12} & s'_{22} & 0 \\ 0 & 0 & 0 \end{pmatrix}$$

where

$$\begin{pmatrix} s'_{11} & s'_{12} \\ s'_{12} & s'_{22} \end{pmatrix} = \frac{1}{A_1^1 A_2^2 - A_2^1 A_1^2} \begin{pmatrix} A_2^2 & -A_2^1 \\ -A_1^2 & A_1^1 \end{pmatrix} \begin{pmatrix} s_{11} & s_{12} \\ s_{12} & s_{22} \end{pmatrix} \begin{pmatrix} A_2^2 & -A_1^2 \\ -A_2^1 & A_1^1 \end{pmatrix}.$$

It follows that $s'_{11} s'_{22} - (s'_{12})^2 = s_{11} s_{22} - (s_{12})^2$, so there is an "invariant" to be dealt with. I leave it to the reader to show that the upper left-hand 2-by-2 block of S can be brought by a change of basis of the above form into exactly one of the four forms

$$\begin{pmatrix} 0 & 0 \\ 0 & 0 \end{pmatrix} \quad \begin{pmatrix} 1 & 0 \\ 0 & 0 \end{pmatrix} \quad \begin{pmatrix} \sigma & 0 \\ 0 & \sigma \end{pmatrix} \quad \begin{pmatrix} \sigma & 0 \\ 0 & -\sigma \end{pmatrix}$$

where $\sigma > 0$ is a real positive number.

To summarize, every 3-dimensional Lie algebra is isomorphic to exactly one of the following Lie algebras: Either

$$\mathfrak{so}(3): \begin{aligned}[x_2, x_3] &= x_1 \\ [x_3, x_1] &= x_2 \\ [x_1, x_2] &= x_3\end{aligned} \quad \text{or} \quad \mathfrak{sl}(2, \mathbb{R}): \begin{aligned}[x_2, x_3] &= x_2 \\ [x_3, x_1] &= x_1 \\ [x_1, x_2] &= x_3\end{aligned}$$

or an algebra of the form

$$[x_2, x_3] = b_{11}x_1 + b_{12}x_2$$
$$[x_3, x_1] = b_{21}x_1 + b_{22}x_2$$
$$[x_1, x_2] = 0$$

where the 2-by-2 matrix B is one of the following

$$\begin{pmatrix} 0 & 0 \\ 0 & 0 \end{pmatrix} \quad \begin{pmatrix} 1 & 0 \\ 0 & 0 \end{pmatrix} \quad \begin{pmatrix} 1 & 0 \\ 0 & 1 \end{pmatrix} \quad \begin{pmatrix} 0 & 1 \\ 1 & 0 \end{pmatrix}$$
$$\begin{pmatrix} 0 & 1 \\ -1 & 0 \end{pmatrix} \quad \begin{pmatrix} 1 & 1 \\ -1 & 0 \end{pmatrix} \quad \begin{pmatrix} \sigma & 1 \\ -1 & \sigma \end{pmatrix} \quad \begin{pmatrix} \sigma & 1 \\ -1 & -\sigma \end{pmatrix}$$

and, in the latter two cases, σ is a positive real number. Each of these eight latter types can be represented as a subalgebra of $\mathfrak{gl}(3, \mathbb{R})$ in the form

$$\mathfrak{g} = \left\{ \begin{pmatrix} (1+b_{21})z & -b_{11}z & x \\ b_{22}z & (1-b_{12})z & y \\ 0 & 0 & z \end{pmatrix} \;\middle|\; x, y, z \in \mathbb{R} \right\}$$

I leave as an exercise for the reader to show that the corresponding subgroup of $\mathrm{GL}(3, \mathbb{R})$ is a closed, embedded, simply connected matrix Lie group whose underlying manifold is diffeomorphic to $\mathbb{R}^3$.

Exercises

1. Show that for any real vector space of dimension n, the Lie group $GL(V)$ is isomorphic to $GL(n,\mathbb{R})$. (Hint: Choose a basis **b** of V, use **b** to construct a mapping $\phi_{\mathbf{b}}\colon GL(V) \to GL(n,\mathbb{R})$, and then show that $\phi_{\mathbf{b}}$ is a smooth isomorphism.)

2. Let G be a Lie group and let H be an abstract subgroup. Show that if there is an open neighborhood U of e in G so that $H \cap U$ is a smooth embedded submanifold of G, then H is a Lie subgroup of G.

3. Show that $SL(n,\mathbb{R})$ is an embedded Lie subgroup of $GL(n,\mathbb{R})$. (Hint: Note that $SL(n,\mathbb{R}) = \det^{-1}(1)$.)

4. Show that $O(n)$ is an compact Lie subgroup of $GL(n,\mathbb{R})$. (Hint: $O(n) = F^{-1}(I_n)$, where F is the map from $GL(n,\mathbb{R})$ to the vector space of n-by-n symmetric matrices given by $F(A) = {}^t\! A\, A$. Taking note of Exercise 2, show that the Implicit Function Theorem applies. To show compactness, apply the Heine-Borel theorem.) Show also that $SO(n)$ is an open-and-closed, index 2 subgroup of $O(n)$.

5. Carry out the analysis in Exercise 3 for the complex matrix Lie group $SL(n,\mathbb{C})$ and the analysis in Exercise 4 for the complex matrix Lie groups $U(n)$ and $SU(n)$. What are the (real) dimensions of all of these groups?

6. Show that the map $\mu\colon O(n) \times A_n \times N_n \to GL(n,\mathbb{R})$ defined by matrix multiplication is a diffeomorphism although it is *not* a group homomorphism. (Hint: The map is clearly smooth, you must only compute an inverse. To get the first factor $\nu_1\colon GL(n,\mathbb{R}) \to O(n)$ of the inverse map, think of an element $\mathbf{b} \in GL(n,\mathbb{R})$ as a row of column vectors in $\mathbb{R}^n$ and let $\nu_1(\mathbf{b})$ be the row of column vectors which results from **b** by apply the Gram-Schmidt orthogonalization process. Why does this work and why is the resulting map ν_1 smooth?) Show, similarly that the map

$$\mu\colon SO(n) \times \big(A_n \cap SL(n,\mathbb{R})\big) \times N_n \to SL(n,\mathbb{R})$$

is a diffeomorphism. Are there similar 'factorizations' for the groups $GL(n,\mathbb{C})$ and $SL(n,\mathbb{C})$? (Hint: Consider unitary bases rather than orthogonal ones.)

7. Show that

$$SU(2) = \left\{ \begin{pmatrix} a & -\bar{b} \\ b & \bar{a} \end{pmatrix} \,\Big|\, a\bar{a} + b\bar{b} = 1 \right\}.$$

Conclude that $SU(2)$ is diffeomorphic to the 3-sphere and, using the previous exercise, that, $SL(2,\mathbb{C})$ is simply connected, while $\pi_1\big(SL(2,\mathbb{R})\big) \simeq \mathbb{Z}$.

8. Show that, for any Lie group G, the mappings L_a satisfy

$$L'_a(b) = L'_{ab}(e) \circ (L'_b(e))^{-1}.$$

where $L'_a(b)\colon T_b G \to T_{ab} G$. (This shows that the effect of left translation is completely determined by what it does at e.) State and prove a similar formula for the mappings R_a.

9. Let (G, μ) be a Lie group. Using the canonical identification $T_{(a,b)}(G \times G) = T_a G \oplus T_b G$, prove the formula

$$\mu'(a,b)(v,w) = R'_b(a)(v) + L'_a(b)(w)$$

for all $v \in T_a G$ and $w \in T_b G$.

10. Complete the proof of Proposition 3 by explicitly exhibiting the map c as a composition of known smooth maps. (Hint: if $f: X \to Y$ is smooth, then $f': TX \to TY$ is also smooth.)

11. Show that, for any $v \in \mathfrak{g}$, the left-invariant vector field X_v is indeed smooth. Also prove the first statement in Proposition 4. (Hint: Use Ψ to write the mapping $X_v: G \to TG$ as a composition of smooth maps. Show that the assignment $v \mapsto X_v$ is linear. Finally, show that if a left-invariant vector field on G vanishes anywhere, then it vanishes identically.)

12. Show that $\exp: \mathfrak{g} \to G$ is indeed smooth and that $\exp'(0): \mathfrak{g} \to \mathfrak{g}$ is the identity mapping. (Hint: Write down a smooth vector field Y on $\mathfrak{g} \times G$ such that the integral curves of Y are of the form $\gamma(t) = (v_0, a_0 e^{tv_0})$. Now use the flow of Y,

$$\Psi: \mathbb{R} \times \mathfrak{g} \times G \to \mathfrak{g} \times G,$$

to write exp as the composition of smooth maps.)

13. For the homomorphism $\det: \mathrm{GL}(n, \mathbb{R}) \to \mathbb{R}^\bullet$, show that $\det'(I_n)(x) = \mathrm{tr}(x)$, where tr denotes the trace function. Conclude, using Theorem 1 that, for any matrix a,

$$\det(e^a) = e^{\mathrm{tr}(a)}.$$

14. Prove that, for any $g \in G$ and any $x \in \mathfrak{g}$,

$$g \exp(x) g^{-1} = \exp(\mathrm{Ad}(g)(x)).$$

(Hint: Replace x by tx in the above formula and consider Proposition 5.) Use this to show that $\mathrm{tr}(\exp(x)) \geq -2$ for all $x \in \mathfrak{sl}(2, \mathbb{R})$. Conclude that $\exp: \mathfrak{sl}(2, \mathbb{R}) \to \mathrm{SL}(2, \mathbb{R})$ is *not* surjective. (Hint: show that every $x \in \mathfrak{sl}(2, \mathbb{R})$ is of the form gyg^{-1} for some $g \in \mathrm{SL}(2, \mathbb{R})$ and some y which is one of the matrices

$$\begin{pmatrix} 0 & \pm 1 \\ 0 & 0 \end{pmatrix}, \quad \begin{pmatrix} \lambda & 0 \\ 0 & -\lambda \end{pmatrix}, \quad \text{or} \quad \begin{pmatrix} 0 & -\lambda \\ \lambda & 0 \end{pmatrix}, \quad (\lambda > 0).$$

Also, remember that $\mathrm{tr}(aba^{-1}) = \mathrm{tr}(b)$.)

15. Using Theorem 1, show that if H_1 and H_2 are Lie subgroups of G, then $H_1 \cap H_2$ is also a Lie subgroup of G. (Hint: What should the Lie algebra of this intersection be? Be careful: $H_1 \cap H_2$ might have countably many distinct components even if H_1 and H_2 are connected!)

16. For any skew-commutative algebra $(\mathfrak{g}, [,])$, define the map $\mathrm{ad}: \mathfrak{g} \to \mathrm{End}(\mathfrak{g})$ by $\mathrm{ad}(x)(y) = [x, y]$. Verify that the validity of the Jacobi identity $[\mathrm{ad}(x), \mathrm{ad}(y)] =$

ad($[x,y]$) (where the bracket on End($\mathfrak{g}$) is the commutator) is equivalent to the validity of the identity

$$[[x,y],z] + [[y,z],x] + [[z,x],y] = 0$$

for all $x, y, z \in \mathfrak{g}$.

17. Show that, as $\lambda \in \mathbb{R}$ varies, all of the groups

$$G_\lambda = \left\{ \begin{pmatrix} a & b \\ 0 & a^\lambda \end{pmatrix} \,\Big|\, a \in \mathbb{R}^+,\ b \in \mathbb{R} \right\}$$

with $\lambda \neq 1$ are isomorphic, but are not conjugate in GL(2, $\mathbb{R}$). What happens when $\lambda = 1$?

18. Show that a connected Lie group G is abelian if and only if its Lie algebra satisfies $[x, y] = 0$ for all $x, y \in \mathfrak{g}$. Conclude that a connected abelian Lie group of dimension n is isomorphic to $\mathbb{R}^n/\mathbb{Z}^d$ where $\mathbb{Z}^d$ is some discrete subgroup of rank $d \leq n$. (Hint: To show "G abelian" implies "$\mathfrak{g}$ abelian", look at how $[,]$ was defined. To prove the converse, use Theorem 3 to construct a surjective homomorphism $\phi \colon \mathbb{R}^n \to G$ with discrete kernel.)

19. (*Covering Spaces of Lie groups*) Let G be a connected Lie group and let $\pi \colon \tilde{G} \to G$ be the universal covering space of G. (Recall that the points of $\tilde{G}$ can be regarded as the space of fixed-endpoint homotopy classes of continuous maps $\gamma \colon [0,1] \to G$ with $\gamma(0) = e$.) Show that there is a unique Lie group structure $\tilde{\mu} \colon \tilde{G} \times \tilde{G} \to \tilde{G}$ for which the homotopy class of the constant map $\tilde{e} \in \tilde{G}$ is the identity and so that π is a homomorphism. (Hints: Give $\tilde{G}$ the (unique) smooth structure for which π is a local diffeomorphism. The multiplication $\tilde{\mu}$ can then be defined as follows: The map $\bar{\mu} = \mu \circ (\pi \times \pi) \colon \tilde{G} \times \tilde{G} \to G$ is a smooth map and satisfies $\bar{\mu}(\tilde{e}, \tilde{e}) = e$. Since $\tilde{G} \times \tilde{G}$ is simply connected, the universal lifting property of the covering map π implies that there is a unique map $\tilde{\mu} \colon \tilde{G} \times \tilde{G} \to \tilde{G}$ which satisfies $\pi \circ \tilde{\mu} = \bar{\mu}$ and $\tilde{\mu}(\tilde{e}, \tilde{e}) = \tilde{e}$. Show that $\tilde{\mu}$ is smooth, that it satisfies the axioms for a group multiplication (associativity, existence of an identity, and existence of inverses), and that π is a homomorphism. You will want to use the universal lifting property of covering spaces a few times.)

The kernel of π is a discrete normal subgroup of $\tilde{G}$. Show that this kernel lies in the center of G. (Hint: For any $z \in \ker(\pi)$, the connected set $\{aza^{-1} \mid a \in G\}$ must also lie in $\ker(\pi)$.)

Show that the center of the simply connected Lie group

$$G = \left\{ \begin{pmatrix} a & b \\ 0 & 1 \end{pmatrix} \,\Big|\, a \in \mathbb{R}^+,\ b \in \mathbb{R} \right\}$$

is trivial, so any connected Lie group with the same Lie algebra is actually isomorphic to G.

(In the next Lecture, we will show that whenever K is a closed normal subgroup of a Lie group G, the quotient group G/K can be given the structure of a Lie group. Thus, in many cases, one can effectively list all of the connected Lie groups with a given Lie algebra.)

20. Show that $G = \widetilde{SL(2,\mathbb{R})}$ is not a matrix group. In fact, show that any homomorphism $\phi\colon G \to GL(n,\mathbb{R})$ factors through the projections $\pi\colon G \to SL(2,\mathbb{R})$. (Hint: Recall, from earlier exercises, that the inclusion map $SL(2,\mathbb{R}) \to SL(2,\mathbb{C})$ induces the zero map on π_1 since $SL(2,\mathbb{C})$ is simply connected.

    Now, any homomorphism $\phi\colon G \to GL(n,\mathbb{R})$ induces a Lie algebra homomorphism $\phi'(e)\colon \mathfrak{sl}(2,\mathbb{R}) \to \mathfrak{gl}(n,\mathbb{R})$ and this may clearly be complexified to yield a Lie algebra homomorphism $\phi'(e)^{\mathbb{C}}\colon \mathfrak{sl}(2,\mathbb{C}) \to \mathfrak{gl}(n,\mathbb{C})$. Since $SL(2,\mathbb{C})$ is simply connected, there must be a corresponding Lie group homomorphism $\phi^{\mathbb{C}}\colon SL(2,\mathbb{C}) \to GL(n,\mathbb{C})$. Now suppose that ϕ does not factor through $SL(2,\mathbb{R})$, i.e., that ϕ is non-trivial on the kernel of $\pi\colon G \to SL(2,\mathbb{R})$, and show that this leads to a contradiction.)

21. An *ideal* in a Lie algebra $\mathfrak{g}$ is a linear subspace $\mathfrak{h}$ which satisfies $[\mathfrak{h},\mathfrak{g}] \subset \mathfrak{h}$. Show that the kernel $\mathfrak{k}$ of a Lie algebra homomorphism $\varphi\colon \mathfrak{h} \to \mathfrak{g}$ is an ideal in $\mathfrak{h}$ and that the image $\varphi(\mathfrak{h})$ is a subalgebra of $\mathfrak{g}$. Conversely, show that if $\mathfrak{k} \subset \mathfrak{h}$ is an ideal, then the quotient vector space $\mathfrak{h}/\mathfrak{k}$ carries a unique Lie algebra structure for which the quotient mapping $\mathfrak{h} \to \mathfrak{h}/\mathfrak{k}$ is a homomorphism.

    Show that the subspace $[\mathfrak{g},\mathfrak{g}]$ of $\mathfrak{g}$ which is generated by all brackets of the form $[x,y]$ is an ideal in $\mathfrak{g}$. What can you say about the quotient $\mathfrak{g}/[\mathfrak{g},\mathfrak{g}]$?

22. Show that, for a connected Lie group G, a connected Lie subgroup H is normal if and only if $\mathfrak{h}$ is an ideal of $\mathfrak{g}$. (Hint: Use Proposition 7 and the fact that $H \subset G$ is normal if and only if $e^x H e^{-x} = H$ for all $x \in \mathfrak{g}$.)

23. For any Lie algebra $\mathfrak{g}$, let $\mathfrak{z}(\mathfrak{g}) \subset \mathfrak{g}$ denote the kernel of the homomorphism $\mathrm{ad}\colon \mathfrak{g} \to \mathfrak{gl}(\mathfrak{g})$. Use Theorem 2 and Exercise 16 to prove Theorem 4 for any Lie algebra $\mathfrak{g}$ for which $\mathfrak{z}(\mathfrak{g}) = 0$. (Hint: Look at the discussion after the statement of Theorem 4.)

    Show also that if $\mathfrak{g}$ is the Lie algebra of the connected Lie group G, then the connected Lie subgroup $Z(\mathfrak{g}) \subset G$ which corresponds to $\mathfrak{z}(\mathfrak{g})$ lies in the center of G. (In the next lecture, we will be able to prove that the center of G is a closed Lie subgroup of G and that $Z(\mathfrak{g})$ is actually the identity component of the center of G.)

24. For any Lie algebra $\mathfrak{g}$, there is a canonical bilinear pairing $\kappa\colon \mathfrak{g} \times \mathfrak{g} \to \mathbb{R}$, called the *Killing form*, defined by the rule:

$$\kappa(x,y) = \mathrm{tr}\bigl(\mathrm{ad}(x)\mathrm{ad}(y)\bigr).$$

(i) Show that κ is symmetric and, if $\mathfrak{g}$ is the Lie algebra of a Lie group G, then κ is Ad-invariant:

$$\kappa\bigl(\mathrm{Ad}(g)x, \mathrm{Ad}(g)y\bigr) = \kappa(x,y) = \kappa(y,x).$$

Show also that

$$\kappa\bigl([z,x],y\bigr) = -\kappa\bigl(x,[z,y]\bigr).$$

A Lie algebra $\mathfrak{g}$ is said to be *semi-simple* if κ is a non-degenerate bilinear form on $\mathfrak{g}$.

(ii) Show that, of all the 2- and 3-dimensional Lie algebras, only $\mathfrak{so}(3)$ and $\mathfrak{sl}(2, \mathbb{R})$ are semi-simple.
(iii) Show that if $\mathfrak{h} \subset \mathfrak{g}$ is an ideal in a semi-simple Lie algebra $\mathfrak{g}$, then the Killing form of $\mathfrak{h}$ as an algebra is equal to the restriction of the Killing form of $\mathfrak{g}$ to $\mathfrak{h}$. Show also that the subspace $\mathfrak{h}^\perp = \{x \in \mathfrak{g} \mid \kappa(x,y) = 0 \text{ for all } y \in \mathfrak{h}\}$ is also an ideal in $\mathfrak{g}$ and that $\mathfrak{g} = \mathfrak{h} \oplus \mathfrak{h}^\perp$ as Lie algebras. (Hint: For the first part, examine the effect of $\mathrm{ad}(x)$ on a basis of $\mathfrak{g}$ chosen so that the first $\dim \mathfrak{h}$ basis elements are a basis of $\mathfrak{h}$.)
(iv) Finally, show that a semi-simple Lie algebra can be written as a direct sum of ideals $\mathfrak{h}_i$, each of which has no proper ideals. (Hint: Apply (iii) as many times as you can find proper ideals of the summands found so far.)

A more general class of Lie algebras are the reductive ones. A Lie algebra is said to be *reductive* if there is a non-degenerate symmetric bilinear form $(\,,\,): \mathfrak{g} \times \mathfrak{g} \to \mathbb{R}$ which satisfies the identity $([z,x], y) + (x, [z,y]) = 0$. Using the above arguments, it is easy to see that a reductive algebra can be written as the direct sum of an abelian algebra and some number of simple algebras in a unique way.

25. Show that, if ω is the canonical left-invariant 1-form on G and Y_v is the right-invariant vector field on G satisfying $Y_v(e) = v$, then

$$\omega(Y_v(a)) = \mathrm{Ad}(a^{-1})(v).$$

LECTURE 3
Group Actions on Manifolds

In this Lecture, I turn from the abstract study of Lie groups to their realizations as "transformation groups."

Lie group actions

Definition 1. If (G, μ) is a Lie group and M is a smooth manifold, then a *left action* of G on M is a smooth mapping $\lambda\colon G \times M \to M$ which satisfies $\lambda(e, m) = m$ for all $m \in M$ and
$$\lambda(\mu(a,b), m) = \lambda(a, \lambda(b, m)).$$

Similarly, a *right action* of G on M is a smooth mapping $\rho\colon M \times G \to M$, which satisfies $\rho(m, e) = m$ for all $m \in M$ and
$$\rho(m, \mu(a,b)) = \rho(\rho(m, a), b).$$

For notational sanity, whenever the action (left or right) can be easily inferred from context, we will usually write $a \cdot m$ instead of $\lambda(a, m)$ or $m \cdot a$ instead of $\rho(m, a)$. Thus, for example, the axioms for a left action in this abbreviated notation are simply $e \cdot m = m$ and $a \cdot (b \cdot m) = ab \cdot m$.

For a given a left action $\lambda\colon G \times M \to M$, it is easy to see that for each fixed $a \in G$ the map $\lambda_a\colon M \to M$ defined by $\lambda_a(m) = \lambda(a, m)$ is a smooth diffeomorphism of M onto itself. Thus, G gets represented as a group of diffeomorphisms, or "transformations" of a manifold M. This notion of "transformation group" was what motivated Lie to develop his theory in the first place. See the Appendix to this Lecture for a more complete discussion of this point.

Equivalence of left and right actions. Note that every right action $\rho\colon M \times G \to M$ can be rewritten as a left action and vice versa. One merely defines
$$\tilde{\rho}(a, m) = \rho(m, a^{-1}).$$

(The reader should check that this $\tilde{\rho}$ is, in fact, a left action.) Thus, all theorems about left actions have analogues for right actions. The distinction between the two is mainly for notational and conceptual convenience. I will concentrate on left actions and only occasionally point out the places where right actions behave slightly differently (mainly changes of sign, etc.).

Stabilizers and orbits. A left action is said to be *effective* if $g \cdot m = m$ for all $m \in M$ implies that $g = e$. (Sometimes, the word *faithful* is used instead.) A left action is said to be *free* if $g \neq e$ implies that $g \cdot m \neq m$ for all $m \in M$.

A left action is said to be *transitive* if, for any $x, y \in M$, there exists a $g \in G$ so that $g \cdot x = y$. In this case, M is usually said to be *homogeneous* under the given action.

For any $m \in M$, the *G-orbit* of m is defined to be the set

$$G \cdot m = \{g \cdot m \mid g \in G\}$$

and the *stabilizer* (or *isotropy group*) of m is defined to be the subset

$$G_m = \{g \in G \mid g \cdot m = m\}.$$

Note that
$$G_{g \cdot m} = g\, G_m\, g^{-1}.$$

Thus, whenever $H \subset G$ is the stabilizer of a point of M, then all of the conjugate subgroups of H are also stabilizers. These results imply that

$$G_M = \bigcap_{m \in M} G_m$$

is a closed normal subgroup of G and consists of those $g \in G$ for which $g \cdot m = m$ for all $m \in M$. Often in practice, G_M is a discrete (in fact, usually finite) subgroup of G. When this is so, we say that the action is *almost effective*.

The following theorem says that orbits and stabilizers are particularly nice objects. Though the proof is relatively straightforward, it is a little long, so we will consider a few examples before attempting it.

Theorem 1. *Let $\lambda: G \times M \to M$ be a left action of G on M. Then, for all $m \in M$, the stabilizer G_m is a closed Lie subgroup of G. Moreover, the orbit $G \cdot m$ can be given the structure of a smooth submanifold of M in such a way that the map $\phi: G \to G \cdot m$ defined by $\phi(g) = \lambda(g, m)$ is a smooth submersion.*

Example 1. Any Lie group left-acts on itself by left multiplication. I.e., we set $M = G$ and define $\lambda: G \times M \to M$ to simply be μ. This action is both free and transitive.

Example 2. Given a homomorphism of Lie groups $\phi: H \to G$, define a smooth left action $\lambda: H \times G \to G$ by the rule $\lambda(h, g) = \phi(h)g$. Then $H_e = \ker(\phi)$ and $H \cdot e = \phi(H) \subset G$.

In particular, note that Theorem 1 implies that the kernel of a Lie group homomorphism is a (closed, normal) Lie subgroup of the domain group and the image of a Lie group homomorphism is a Lie subgroup of the range group.

Example 3. Any Lie group acts on itself by conjugation: $g \cdot g_0 = g g_0 g^{-1}$. This action is neither free nor transitive (unless $G = \{e\}$). Note that $G_e = G$ and, in general, G_g is the centralizer of $g \in G$. This action is effective (respectively, almost effective) if and only if the center of G is trivial (respectively, discrete). The orbits are the conjugacy classes of G.

Example 4. $\mathrm{GL}(n, \mathbb{R})$ acts on $\mathbb{R}^n$ as usual by $A \cdot v = Av$. This action is effective but is neither free nor transitive since $\mathrm{GL}(n, \mathbb{R})$ fixes $0 \in \mathbb{R}^n$ and acts transitively on $\mathbb{R}^n \backslash \{0\}$. Thus, there are exactly two orbits of this action, one closed and the other not.

Example 5. $\mathrm{SO}(n+1)$ acts on $S^n = \{x \in \mathbb{R}^{n+1} \,|\, x \cdot x = 1\}$ by the usual action $A \cdot x = Ax$. This action is transitive and effective, but not free (unless $n = 1$) since, for example, the stabilizer of e_{n+1} is clearly isomorphic to $\mathrm{SO}(n)$.

Example 6. Let $\mathcal{S}_n$ be the $n(n+1)/2$-dimensional vector space of n-by-n real symmetric matrices. Then $\mathrm{GL}(n, \mathbb{R})$ acts on $\mathcal{S}_n$ by $A \cdot S = A S\, {}^tA$. The orbit of the identity matrix I_n is $\mathcal{S}_+(n)$, the set of all positive-definite n-by-n real symmetric matrices (Why?). In fact, it is known that, if we define $I_{p,q} \in \mathcal{S}_n$ to be the matrix

$$I_{p,q} = \begin{pmatrix} I_p & 0 & 0 \\ 0 & -I_q & 0 \\ 0 & 0 & 0 \end{pmatrix},$$

(where the "0" entries have the appropriate dimensions) then $\mathcal{S}_n$ is the (disjoint) union of the orbits of the matrices $I_{p,q}$ where $0 \leq p, q$ and $p + q \leq n$ (see the Exercises).

The orbit of $I_{p,q}$ is open in $\mathcal{S}_n$ and only if $p + q = n$. The stabilizer of $I_{p,q}$ in this case is defined to be $\mathrm{O}(p, q) \subset \mathrm{GL}(n, \mathbb{R})$.

Note that the action is merely almost effective since $\{\pm I_n\} \subset \mathrm{GL}(n, \mathbb{R})$ fixes every $S \in \mathcal{S}_n$.

Example 7. Let $\mathcal{J} = \{J \in \mathrm{GL}(2n, \mathbb{R}) \,|\, J^2 = -I_{2n}\}$. Then $\mathrm{GL}(2n, \mathbb{R})$ acts on $\mathcal{J}$ on the left by the formula $A \cdot J = A J A^{-1}$. I leave as exercises for the reader to prove that $\mathcal{J}$ is a smooth manifold and that this action of $\mathrm{GL}(2n, \mathbb{R})$ is transitive and almost effective. The stabilizer of $J_0 = $ multiplication by i in $\mathbb{C}^n\ (= \mathbb{R}^{2n})$ is simply $\mathrm{GL}(n, \mathbb{C}) \subset \mathrm{GL}(2n, \mathbb{R})$.

Example 8. Let $M = \mathbb{RP}^1$, denote the projective line, whose elements are the lines through the origin in $\mathbb{R}^2$. We will use the notation $\begin{bmatrix} x \\ y \end{bmatrix}$ to denote the line in $\mathbb{R}^2$ spanned by the non-zero vector $\begin{pmatrix} x \\ y \end{pmatrix}$.

Let $G = \mathrm{SL}(2, \mathbb{R})$ act on $\mathbb{RP}^1$ on the left by the formula

$$\begin{pmatrix} a & b \\ c & d \end{pmatrix} \cdot \begin{bmatrix} x \\ y \end{bmatrix} = \begin{bmatrix} ax + by \\ cx + dy \end{bmatrix}.$$

This action is easily seen to be almost effective, with only $\pm I_2 \in \mathrm{SL}(2,\mathbb{R})$ acting trivially.

Actually, it is more common to write this action more informally by using the identification $\mathbb{RP}^1 = \mathbb{R} \cup \{\infty\}$ which identifies $\begin{bmatrix} x \\ y \end{bmatrix}$ when $y \neq 0$ with $x/y \in \mathbb{R}$ and $\begin{bmatrix} 1 \\ 0 \end{bmatrix}$ with ∞. With this convention, the action takes on the more familiar "linear fractional" form

$$\begin{pmatrix} a & b \\ c & d \end{pmatrix} \cdot x = \frac{ax+b}{cx+d}.$$

Note that this form of the action makes it clear that the so-called "linear fractional" action or "Möbius" action on the real line is just the projectivization of the usual linear representation of $\mathrm{SL}(2,\mathbb{R})$ on $\mathbb{R}^2$.

I now turn to the proof of Theorem 1.

Proof of Theorem 1. Fix $m \in M$ and define $\phi: G \to M$ by $\phi(g) = \lambda(g, m)$ as in the theorem. Since $G_m = \phi^{-1}(m)$, it follows that G_m is a closed subset of G. The axioms for a left action clearly imply that G_m is closed under multiplication and inverse, so it is a subgroup.

I claim that G_m is a submanifold of G. To see this, let $\mathfrak{g}_m \subset \mathfrak{g} = T_e G$ be the kernel of the mapping $\phi'(e): T_e G \to T_m M$. Since $\phi \circ L_g = \lambda_g \circ \phi$ for all $g \in G$, the Chain Rule yields a commutative diagram:

$$\begin{array}{ccc} \mathfrak{g} & \xrightarrow{L'_g(e)} & T_g G \\ \phi'(e) \downarrow & & \downarrow \phi'(g) \\ T_m M & \xrightarrow{\lambda'_g(m)} & T_{g \cdot m} M \end{array}$$

Since both $L'_g(e)$ and $\lambda'_g(m)$ are isomorphisms, it follows that $\ker(\phi'(g)) = L'_g(e)(\mathfrak{g}_m)$ for all $g \in G$. In particular, the rank of $\phi'(g)$ is independent of $g \in G$. By the Implicit Function Theorem (see Exercise 2), it follows that $\phi^{-1}(m) = G_m$ is a smooth submanifold of G.

It remains to show that the orbit $G \cdot m$ can be given the structure of a smooth submanifold of M with the stated properties. That is, that $G \cdot m$ can be given a second countable, Hausdorff, locally Euclidean topology and a smooth structure for which the inclusion map $G \cdot m \hookrightarrow M$ is a smooth immersion and for which the map $\phi: G \to G \cdot m$ is a submersion.

Before embarking on this task, it is useful to remark on the nature of the fibers of the map ϕ. By the axioms for left actions, $\phi(h) = h \cdot m = g \cdot m = \phi(g)$ if and only if $g^{-1} h \cdot m = m$, i.e., if and only if $g^{-1} h$ lies in G_m. This is equivalent to the condition that h lie in the left G_m-coset $g G_m$. Thus, the fibers of the map ϕ are the left G_m-cosets in G. In particular, the map ϕ establishes a bijection $\bar{\phi}: G/G_m \to G \cdot m$.

First, I specify the topology on $G \cdot m$ to be quotient topology induced by the surjective map $\phi: G \to G \cdot m$. Thus, a set U in $G \cdot m$ is open if and only if $\phi^{-1}(U)$ is open in G. Since $\phi: G \to M$ is continuous, the quotient topology on the image $G \cdot m$ is at least as fine as the subspace topology $G \cdot m$ inherits via inclusion into M. Since the subspace topology is Hausdorff, the quotient topology must be also. Moreover, the quotient topology on $G \cdot m$ is also second countable since the topology of G is. For the rest of the proof, "the topology on $G \cdot m$" means the quotient topology.

I will both establish the locally Euclidean nature of this topology and construct a smooth structure on $G \cdot m$ at the same time by finding the required neighborhood charts and proving that they are smooth on overlaps. First, however, I need a lemma establishing the existence of a "tubular neighborhood" of the submanifold $G_m \subset G$. Let $d = \dim(G) - \dim(G_m)$. Then there exists a smooth mapping $\psi: B^d \to G$ (where B^d is an open ball about 0 in $\mathbb{R}^d$) so that $\psi(0) = e$ and so that $\mathfrak{g}$ is the direct sum of the subspaces $\mathfrak{g}_m$ and $V = \psi'(0)(\mathbb{R}^d)$. By the Chain Rule and the definition of $\mathfrak{g}_m$, it follows that $(\phi \circ \psi)'(0): \mathbb{R}^d \to T_m M$ is injective. Thus, by restricting to a smaller ball in $\mathbb{R}^d$ if necessary, I may assume henceforth that $\phi \circ \psi: B^d \to M$ is a smooth embedding.

Consider the mapping $\Psi: B^d \times G_m \to G$ defined by $\Psi(x,g) = \psi(x)g$. I claim that Ψ is a diffeomorphism onto its image (which is an open set), say $U = \Psi(B^d \times G_m) \subset G$. (Thus, U forms a sort of "tubular neighborhood" of the submanifold G_m in G.)

To see this, first I show that Ψ is one-to-one: If $\Psi(x_1, g_1) = \Psi(x_2, g_2)$, then

$$(\phi \circ \psi)(x_1) = \psi(x_1) \cdot m = (\psi(x_1)g_1) \cdot m = (\psi(x_2)g_2) \cdot m = \psi(x_2) \cdot m = (\phi \circ \psi)(x_2),$$

so the injectivity of $\phi \circ \psi$ implies $x_1 = x_2$. Since $\psi(x_1)g_1 = \psi(x_2)g_2$, this in turn implies that $g_1 = g_2$.

Second, I must show that the derivative

$$\Psi'(x,g): T_x \mathbb{R}^d \oplus T_g G_m \to T_{\psi(x)g} G$$

is an isomorphism for all $(x,g) \in B^d \times G_m$. However, from the beginning of the proof, $\ker(\phi'(\psi(x)g)) = L'_{\psi(x)g}(e)(\mathfrak{g}_m)$ and this latter space is clearly $\Psi'(x,g)(0 \oplus T_g G_m)$. On the other hand, since $\phi(\Psi(x,g)) = \phi \circ \psi(x)$, it follows that

$$\phi'(\Psi(x,g))\big(\Psi'(x,g)(T_x \mathbb{R}^d \oplus 0)\big) = (\phi \circ \psi)'(x)(T_x \mathbb{R}^d)$$

and this latter space has dimension d by construction. Hence, $\Psi'(x,g)(T_x \mathbb{R}^d \oplus 0)$ is a d-dimensional subspace of $T_{\psi(x)g} G$ which is transverse to $\Psi'(x,g)(0 \oplus T_g G_m)$. Thus, $\Psi'(x,g): T_x \mathbb{R}^d \oplus T_g G_m \to T_{\psi(x)g} G$ is surjective and hence an isomorphism, as desired.

This completes the proof that Ψ is a diffeomorphism onto U. It follows that the inverse of Ψ is smooth and can be written in the form $\Psi^{-1} = \pi_1 \times \pi_2$ where $\pi_1: U \to B^d$ and $\pi_2: U \to G_m$ are smooth submersions.

Now, for each $g \in G$, define $\rho_g: B^d \to M$ by the formula $\rho_g(x) = \phi(g\psi(x))$. Then $\rho_g = \lambda_g \circ \phi \circ \psi$, so ρ_g is a smooth embedding of B^d into M. By construction, $U = \phi^{-1}(\phi \circ \psi(B^d)) = \phi^{-1}(\rho_e(B^d))$ is an open set in G, so it follows that $\rho_e(B^d)$ is an open neighborhood of $e \cdot m = m$ in $G \cdot m$. By the axioms for left actions, it follows that $\phi^{-1}(\rho_g(B^d)) = L_g(U)$ (which is open in G) for all $g \in G$. Thus, $\rho_g(B^d)$

is an open neighborhood of $g \cdot m$ in $G \cdot m$ (in the quotient topology). Moreover, contemplating the commutative square

$$\begin{array}{ccc} U & \xrightarrow{L_g} & L_g(U) \\ \pi_1 \downarrow & & \downarrow \phi \\ B^d & \xrightarrow{\rho_g} & \rho_g(B^d) \end{array}$$

whose upper horizontal arrow is a diffeomorphism which identifies the fibers of the vertical arrows (each of which is a topological identification map) implies that ρ_g is, in fact, a homeomorphism onto its image. Thus, the quotient topology is locally Euclidean.

Finally, I show that the "patches" ρ_g overlap smoothly. Suppose that

$$\rho_g(B^d) \cap \rho_h(B^d) \neq \emptyset.$$

Then, because the maps ρ_g and ρ_h are homeomorphisms,

$$\rho_g(B^d) \cap \rho_h(B^d) = \rho_g(W_1) = \rho_h(W_2)$$

where $W_i \neq \emptyset$ are open subsets of B^d. It follows that

$$L_g\big(\Psi(W_1 \times G_m)\big) = L_h\big(\Psi(W_2 \times G_m)\big).$$

Thus, if $\tau: W_1 \to W_2$ is defined by the rule $\tau = \pi_1 \circ L_{h^{-1}} \circ L_g \circ \psi$, then τ is a smooth map with smooth inverse $\tau^{-1} = \pi_1 \circ L_{g^{-1}} \circ L_h \circ \psi$ and hence is a diffeomorphism. Moreover, we have $\rho_g = \rho_h \circ \tau$, thus establishing that the patches ρ_g overlap smoothly and hence that the patches define the structure of a smooth manifold on $G \cdot m$.

That the map $\phi: G \to G \cdot m$ is a smooth submersion and that the inclusion $G \cdot m \hookrightarrow M$ is a smooth one-to-one immersion are now clear.

It is worth remarking that the proof of Theorem 1 shows that the Lie algebra of G_m is the subspace $\mathfrak{g}_m$. In particular, if $G_m = \{e\}$, then the map $\phi: G \to M$ is a one-to-one immersion.

The proof also brings out the fact that the orbit $G \cdot m$ can be identified with the left coset space G/G_m, which thereby inherits the structure of a smooth manifold. It is natural to wonder which subgroups H of G have the property that the coset space G/H can be given the structure of a smooth manifold for which the coset projection $\pi: G \to G/H$ is a smooth map. This question is answered by the following result. The proof is quite similar to that of Theorem 1, so I will only provide an outline, leaving the details as exercises for the reader.

Theorem 2. *If H is a closed subgroup of a Lie group G, then the left coset space G/H can be given the structure of a smooth manifold in a unique way so that the coset mapping $\pi: G \to G/H$ is a smooth submersion. Moreover, with this smooth structure, the left action $\lambda: G \times G/H \to G/H$ defined by $\lambda(g, hH) = ghH$ is a transitive smooth left action.*

Proof. (Outline.) If the coset mapping $\pi\colon G \to G/H$ is to be a smooth submersion, elementary linear algebra tells us that the dimension of G/H will have to be $d = \dim(G) - \dim(H)$. Moreover, for every $g \in G$, there will have to exist a smooth mapping $\psi_g\colon B^d \to G$ with $\psi_g(0) = g$ which is transverse to the submanifold gH at g and so that the composition $\pi \circ \psi\colon B^d \to G/H$ is a diffeomorphism onto a neighborhood of $gH \in G/H$. It is not difficult to see that this is only possible if G/H is endowed with the quotient topology. The hypothesis that H be closed implies that the quotient topology is Hausdorff. It is automatic that the quotient topology is second countable. The proof that the quotient topology is locally Euclidean depends on being able to construct the "tubular neighborhood" U of H as constructed for the case of a stabilizer subgroup in the proof of Theorem 1. Once this is done, the rest of the construction of charts with smooth overlaps follows the end of the proof of Theorem 1 almost verbatim.

Group actions and vector fields

A left action $\lambda\colon \mathbb{R} \times M \to M$ (where $\mathbb{R}$ has its usual additive Lie group structure) is, of course, the same thing as a flow. Associated to each flow on M is a vector field which generates this flow. The generalization of this association to more general Lie group actions is the subject of this section.

Let $\lambda\colon G \times M \to M$ be a left action. Then, for each $v \in \mathfrak{g}$, there is a flow Ψ_v^λ on M defined by the formula

$$\Psi_v^\lambda(t, m) = e^{tv} \cdot m.$$

This flow is associated to a vector field on M which we shall denote by Y_v^λ, or simply Y_v if the action λ is clear from context. This defines a mapping $\lambda_*\colon \mathfrak{g} \to \mathfrak{X}(M)$, where $\lambda_*(v) = Y_v^\lambda$.

Proposition 1. *For each left action $\lambda\colon G \times M \to M$, the mapping λ_* is a linear anti-homomorphism from $\mathfrak{g}$ to $\mathfrak{X}(M)$. In other words, λ_* is linear and*

$$\lambda_*([x, y]) = -[\lambda_*(x), \lambda_*(y)].$$

Proof. For each $v \in \mathfrak{g}$, let Y_v denote the *right* invariant vector field on G whose value at e is v. Then, according to Lecture 2, the flow of Y_v on G is given by the formula $\Psi_v(t, g) = \exp(tv)g$. As usual, let Φ_v denote the flow of the left invariant vector field X_v. Then the formula

$$\Psi_v(t, g) = \left(\Phi_{-v}(t, g^{-1})\right)^{-1}$$

is immediate. If $\iota_*\colon \mathfrak{X}(G) \to \mathfrak{X}(G)$ is the map induced by the diffeomorphism $\iota(g) = g^{-1}$, then the above formula implies

$$\iota_*(X_{-v}) = Y_v.$$

In particular, since ι_* commutes with Lie bracket, it follows that

$$[Y_x, Y_y] = -Y_{[x,y]}$$

for all $x, y \in \mathfrak{g}$.

Now, regard Y_v and Ψ_v as being defined on $G \times M$ in the obvious way, i.e., $\Psi_v(g, m) = (e^{tv}g, m)$. Then λ intertwines this flow with that of Ψ_v^λ:

$$\lambda \circ \Psi_v = \Psi_v^\lambda \circ \lambda.$$

It follows that the vector fields Y_v and Y_v^λ are λ-related. Thus, $[Y_x^\lambda, Y_y^\lambda]$ is λ-related to $[Y_x, Y_y] = -Y_{[x,y]}$ and hence must be equal to $-Y_{[x,y]}^\lambda$. Finally, since the map $v \mapsto Y_v$ is clearly linear, it follows that λ_* is also linear.

Warning: The appearance of the minus sign in the above formula is something of an annoyance and has led some authors (cf. [A]) to introduce a non-classical minus sign into either the definition of the Lie bracket of vector fields or the definition of the Lie bracket on $\mathfrak{g}$ in order to get rid of the minus sign in this theorem. As logical as this revisionism is, it has not been particularly popular. However, let the reader of other sources beware when comparing formulas.

Even *with* a minus sign, however, Proposition 1 implies that the subspace $\lambda_*(\mathfrak{g}) \subset \mathfrak{X}(M)$ is a (finite dimensional) Lie subalgebra of the Lie algebra of all vector fields on M.

Example: Linear Fractional Transformations. Consider the Möbius action introduced earlier of $SL(2, \mathbb{R})$ on $\mathbb{RP}^1$:

$$\begin{pmatrix} a & b \\ c & d \end{pmatrix} \cdot s = \frac{as+b}{cs+d}.$$

A basis for the Lie algebra $\mathfrak{sl}(2, \mathbb{R})$ is

$$x = \begin{pmatrix} 0 & 1 \\ 0 & 0 \end{pmatrix}, \quad h = \begin{pmatrix} 1 & 0 \\ 0 & -1 \end{pmatrix}, \quad y = \begin{pmatrix} 0 & 0 \\ 1 & 0 \end{pmatrix}$$

Thus, for example, the flow Ψ_y^λ is given by

$$\Psi_y^\lambda(t, s) = \exp \begin{pmatrix} 0 & 0 \\ t & 0 \end{pmatrix} \cdot s = \begin{pmatrix} 1 & 0 \\ t & 1 \end{pmatrix} \cdot s = \frac{s}{ts+1} = s - s^2 t + \cdots,$$

so $Y_y^\lambda = -s^2 \partial/\partial s$. In fact, it is easy to see that, in general,

$$\lambda_*(a_0 x + a_1 h + a_2 y) = (a_0 + 2a_1 s - a_2 s^2)\frac{\partial}{\partial s}.$$

The basic ODE existence theorem can be thought of as saying that every vector field $X \in \mathfrak{X}(M)$ arises as the "flow" of a "local" $\mathbb{R}$-action on M. There is a generalization of this to finite dimensional subalgebras of $\mathfrak{X}(M)$. To state it, we

LECTURE 3. GROUP ACTIONS ON MANIFOLDS

first define a *local left action* of a Lie group G on a manifold M to be an open neighborhood $U \subset G \times M$ of $\{e\} \times M$ together with a smooth map $\lambda\colon U \to M$ so that $\lambda(e, m) = m$ for all $m \in M$ and so that

$$\lambda\big(a, \lambda(b, m)\big) = \lambda(ab, m)$$

whenever this makes sense, i.e., whenever (b, m), (ab, m), and $\big(a, \lambda(b, m)\big)$ all lie in U.

It is easy to see that even a mere local Lie group action induces a map $\lambda_*\colon \mathfrak{g} \to \mathfrak{X}(M)$ as before. We can now state the following result, whose proof is left to the Exercises:

Proposition 2. *Let G be a Lie group and let $\varphi\colon \mathfrak{g} \to \mathfrak{X}(M)$ be a Lie algebra homomorphism. Then there exists a local left action (U, λ) of G on M so that $\lambda_* = -\varphi$.* □

For example, the linear fractional transformations of the last example could just as easily been regarded as a local action of $\operatorname{SL}(2, \mathbb{R})$ on $\mathbb{R}$, where the open set $U \subset \operatorname{SL}(2, \mathbb{R}) \times \mathbb{R}$ is just the set of pairs where $cs + d \neq 0$.

Equations of Lie type

Early in the development of the theory of Lie groups, a special family of ordinary differential equations was singled out for study which generalized the theory of linear equations and the Riccati equation. These have come to be known as *equations of Lie type*. We are now going to describe this class.

Given a Lie algebra homomorphism $\lambda_*\colon \mathfrak{g} \to \mathfrak{X}(M)$ where $\mathfrak{g}$ is the Lie algebra of a Lie group G, and a curve $A\colon \mathbb{R} \to \mathfrak{g}$, the ordinary differential equation

$$\gamma'(t) = \lambda_*\big(A(t)\big)\big(\gamma(t)\big)$$

for a curve $\gamma\colon \mathbb{R} \to M$ is known as an *equation of Lie type*.

Example: The Riccati equation. By our previous example, the classical Riccati equation

$$s'(t) = a_0(t) + 2a_1(t)s(t) + a_2(t)\big(s(t)\big)^2$$

is an equation of Lie type for the (local) linear fractional action of $\operatorname{SL}(2, \mathbb{R})$ on $\mathbb{R}$. The curve A is

$$A(t) = \begin{pmatrix} a_1(t) & a_0(t) \\ -a_2(t) & -a_1(t) \end{pmatrix}$$

Example: Linear Equations. Every linear equation is an equation of Lie type. Let G be the matrix Lie subgroup of $\operatorname{GL}(n+1, \mathbb{R})$,

$$G = \left\{ \begin{pmatrix} A & B \\ 0 & 1 \end{pmatrix} \,\Big|\, A \in \operatorname{GL}(n, \mathbb{R}) \text{ and } B \in \mathbb{R}^n \right\}.$$

Then G acts on $\mathbb{R}^n$ by the standard affine action:

$$\begin{pmatrix} A & B \\ 0 & 1 \end{pmatrix} \cdot x = Ax + B.$$

It is easy to verify that the inhomogeneous linear differential equation

$$x'(t) = a(t)x(t) + b(t)$$

is then a Lie equation, with

$$A(t) = \begin{pmatrix} a(t) & b(t) \\ 0 & 0 \end{pmatrix}.$$

The following proposition follows from the fact that a left action $\lambda: G \times M \to M$ relates the right invariant vector field Y_v to the vector field $\lambda_*(v)$ on M. Despite its simplicity, it has important consequences.

Proposition 3. *If $A: \mathbb{R} \to \mathfrak{g}$ is a curve in the Lie algebra of a Lie group G and $S: \mathbb{R} \to G$ is the solution to the equation $S'(t) = Y_{A(t)}(S(t))$ with initial condition $S(0) = e$, then on any manifold M endowed with a left G-action λ, the equation of Lie type*

$$\gamma'(t) = \lambda_*\big(A(t)\big)\big(\gamma(t)\big),$$

with initial condition $\gamma(0) = m$ has, as its solution, $\gamma(t) = S(t) \cdot m$.

The solution S of Proposition 3 is often called the *fundamental solution* of the Lie equation associated to $A(t)$. The most classical example of this is the fundamental solution of a linear system of equations:

$$x'(t) = a(t)x(t)$$

where a is an n-by-n matrix of functions of t and x is to be a column of height n. In ODE classes, we learn that every solution of this equation is of the form $x(t) = X(t)x_0$ where X is the n-by-n matrix of functions of t which solves the equation $X'(t) = a(t)X(t)$ with initial condition $X(0) = I_n$. Of course, this is a special case of Proposition 3 where $\mathrm{GL}(n, \mathbb{R})$ acts on $\mathbb{R}^n$ via the standard left action described in Example 4.

Lie's reduction method

I now want to explain Lie's method of analysing equations of Lie type. Suppose that $\lambda: G \times M \to M$ is a left action and that $A: \mathbb{R} \to \mathfrak{g}$ is a smooth curve. Suppose that we have found (by some method) a particular solution $\gamma: \mathbb{R} \to M$ of the equation of Lie type associated to A with $\gamma(0) = m$. Select a curve $g: \mathbb{R} \to G$ so that $\gamma(t) = g(t) \cdot m$. Of course, this g will not, in general be unique, but any other choice $\tilde{g}$ will be of the form $\tilde{g}(t) = g(t)h(t)$ where $h: \mathbb{R} \to G_m$.

I would like to choose h so that $\tilde{g}$ is the fundamental solution of the Lie equation associated to A, i.e., so that

$$\tilde{g}'(t) = Y_{A(t)}(\tilde{g}(t)) = R'_{\tilde{g}(t)}(A(t))$$

Unwinding the definitions, it follows that h must satisfy

$$R'_{g(t)h(t)}(A(t)) = L'_{g(t)}(h'(t)) + R'_{h(t)}(g'(t))$$

so

$$R'_{h(t)}\left(R'_{g(t)}(A(t))\right) = L'_{g(t)}(h'(t)) + R'_{h(t)}(g'(t))$$

Solving for $h'(t)$, we find that h must satisfy the differential equation

$$h'(t) = R'_{h(t)}\left(L'_{g(t)^{-1}}\left(R'_{g(t)}(A(t)) - g'(t)\right)\right).$$

If we set

$$B(t) = L'_{g(t)^{-1}}\left(R'_{g(t)}(A(t)) - g'(t)\right),$$

then B is clearly computable from g and A and hence may be regarded as known. Since $B = \left(R'_{h(t)}\right)^{-1}(h'(t))$ and since h is a curve in G_m, it follows that B must actually be a curve in $\mathfrak{g}_m$.

It follows that the equation

$$h'(t) = R'_{h(t)}(B(t))$$

is a Lie equation for h. In other words in order to find the fundamental solution of a Lie equation for G when the particular solution with initial condition $g(0) = m \in M$ is known, it suffices to solve a Lie equation in G_m!

This observation is known as *Lie's method of reduction*. It shows how knowledge of a particular solution to a Lie equation simplifies the search for the general solution. (Note that this is definitely *not* true of general differential equations.) Of course, Lie's method can be generalized. If one knows k particular solutions with initial values $m_1, \ldots, m_k \in M$, then it is easy to see that one can reduce finding the fundamental solution to finding the fundamental solution of a Lie equation in

$$G_{m_1,\ldots,m_k} = G_{m_1} \cap G_{m_2} \cap \cdots \cap G_{m_k}.$$

If one can arrange that this intersection is discrete, then one can explicitly compute a fundamental solution which will then yield the general solution.

Example: The Riccati equation again. Consider the Riccati equation

$$s'(t) = a_0(t) + 2a_1(t)s(t) + a_2(t)(s(t))^2$$

and suppose that we know a particular solution $s_0(t)$. Then let

$$g(t) = \begin{pmatrix} 1 & s_0(t) \\ 0 & 1 \end{pmatrix},$$

so that $s_0(t) = g(t) \cdot 0$ (we are using the linear fractional action of $SL(2, \mathbb{R})$ on $\mathbb{R}$). The stabilizer of 0 is the subgroup G_0 of matrices of the form:

$$\begin{pmatrix} u & 0 \\ v & u^{-1} \end{pmatrix}.$$

Thus, if we set, as usual,

$$A(t) = \begin{pmatrix} a_1(t) & a_0(t) \\ -a_2(t) & -a_1(t) \end{pmatrix},$$

then the fundamental solution of $S'(t) = A(t)S(t)$ can be written in the form

$$S(t) = g(t)h(t) = \begin{pmatrix} 1 & s_0(t) \\ 0 & 1 \end{pmatrix} \begin{pmatrix} u(t) & 0 \\ v(t) & (u(t))^{-1} \end{pmatrix}$$

Solving for the matrix $B(t)$ (which we know will have values in the Lie algebra of G_0), we find

$$B(t) = \begin{pmatrix} b_1(t) & 0 \\ b_2(t) & -b_1(t) \end{pmatrix} = \begin{pmatrix} a_1(t) + a_2(t)s_0(t) & 0 \\ -a_2(t) & -a_1(t) - a_2(t)s_0(t) \end{pmatrix},$$

and the remaining equation to be solved is

$$h'(t) = B(t)h(t),$$

which is solvable by quadratures in the usual way:

$$u(t) = \exp\left(\int_0^t b_1(\tau) d\tau\right),$$

and, once $u(t)$ has been found,

$$v(t) = (u(t))^{-1} \int_0^t b_2(\tau)(u(\tau))^2 d\tau.$$

Example: Linear Equations Again. Consider the general inhomogeneous n-by-n system
$$x'(t) = a(t)x(t) + b(t).$$
Let G be the matrix Lie subgroup of $\operatorname{GL}(n+1, \mathbb{R})$,
$$G = \left\{ \begin{pmatrix} A & B \\ 0 & 1 \end{pmatrix} \,\middle|\, A \in \operatorname{GL}(n, \mathbb{R}) \text{ and } B \in \mathbb{R}^n \right\}.$$
acting on $\mathbb{R}^n$ by the standard affine action as before. If we embed $\mathbb{R}^n$ into $\mathbb{R}^{n+1}$ by the rule
$$x \mapsto \begin{pmatrix} x \\ 1 \end{pmatrix},$$
then the standard affine action of G on $\mathbb{R}^n$ extends to the standard linear action of G on $\mathbb{R}^{n+1}$. Note that G leaves invariant the subspace $x^{n+1} = 0$, and solutions of the Lie equation corresponding to
$$A(t) = \begin{pmatrix} a(t) & b(t) \\ 0 & 0 \end{pmatrix}$$
which lie in this subspace are simply solutions to the homogeneous equation $x'(t) = a(t)x(t)$. Suppose that we knew a basis for the homogeneous solutions, i.e., the fundamental solution to $X'(t) = a(x)X(t)$ with $X(0) = I_n$. This corresponds to knowing the n particular solutions to the Lie equation on $\mathbb{R}^{n+1}$ which have the initial conditions $e_1, \ldots, e_n$. The simultaneous stabilizer of all of these points in $\mathbb{R}^{n+1}$ is the subgroup $H \subset G$ of matrices of the form
$$\begin{pmatrix} I_n & y \\ 0 & 1 \end{pmatrix}$$
Thus, we choose
$$g(t) = \begin{pmatrix} X(t) & 0 \\ 0 & 1 \end{pmatrix}$$
as our initial guess and look for the fundamental solution in the form:
$$S(t) = g(t)h(t) = \begin{pmatrix} X(t) & 0 \\ 0 & 1 \end{pmatrix} \begin{pmatrix} I_n & y(t) \\ 0 & 1 \end{pmatrix}.$$
Expanding the condition $S'(t) = A(t)S(t)$ and using the equation $X'(t) = a(t)X(t)$ then reduces us to solving the equation
$$y'(t) = (X(t))^{-1} b(t),$$
which is easily solved by integration. The reader will probably recognize that this is precisely the classical method of "variation of parameters".

Solution by quadrature

This brings us to an interesting point: Just how hard is it to compute the fundamental solution to a Lie equation of the form

$$\gamma'(t) = R'_{\gamma(t)}(A(t))?$$

One case where it is easy is if the Lie group is abelian. We have already seen that if T is a connected abelian Lie group with Lie algebra $\mathfrak{t}$, then the exponential map $\exp\colon \mathfrak{t} \to T$ is a surjective homomorphism. It follows that the fundamental solution of the Lie equation associated to $A\colon \mathbb{R} \to \mathfrak{t}$ is given in the form

$$S(t) = \exp\left(\int_0^t A(\tau)\, d\tau\right)$$

(Exercise: Why is this true?) Thus, the Lie equation for an abelian group is "solvable by quadrature" in the classical sense.

Another instance where one can at least reduce the problem somewhat is when one has a homomorphism $\phi\colon G \to H$ and knows the fundamental solution S_H to the Lie equation for $\varphi \circ A\colon \mathbb{R} \to \mathfrak{h}$. In this case, S_H is the particular solution (with initial condition $S_H(0) = e$) of the Lie equation on H associated to A by regarding ϕ as defining a left action on H. By Lie's method of reduction, therefore, we are reduced to solving a Lie equation for the group $\ker(\phi) \subset G$.

Example. Suppose that G is connected and simply connected. Let $\mathfrak{g}$ be its Lie algebra and let $[\mathfrak{g}, \mathfrak{g}] \subset \mathfrak{g}$ be the linear subspace generated by all brackets of the form $[x, y]$ where x and y lie in $\mathfrak{g}$. Then, by the Exercises of Lecture 2, we know that $[\mathfrak{g}, \mathfrak{g}]$ is an ideal in $\mathfrak{g}$ (called the *commutator ideal* of $\mathfrak{g}$). Moreover, the quotient algebra $\mathfrak{t} = \mathfrak{g}/[\mathfrak{g}, \mathfrak{g}]$ is abelian.

Since G is connected and simply connected, Theorem 3 from Lecture 2 implies that there is a Lie group homomorphism $\phi_0\colon G \to T_0 = \mathfrak{t}$ whose induced Lie algebra homomorphism $\varphi_0\colon \mathfrak{g} \to \mathfrak{t} = \mathfrak{g}/[\mathfrak{g}, \mathfrak{g}]$ is just the canonical quotient mapping. From our previous remarks, it follows that any Lie equation for G can be reduced, by one quadrature, to a Lie equation for $G_1 = \ker \phi_0$. It is not difficult to check that the group G_1 constructed in this argument is also connected and simply connected.

The desire to iterate this process leads to the following construction: Define the sequence $\{\mathfrak{g}_k\}$ of commutator ideals of $\mathfrak{g}$ by the rules $\mathfrak{g}_0 = \mathfrak{g}$ and and $\mathfrak{g}_{k+1} = [\mathfrak{g}_k, \mathfrak{g}_k]$ for $k \geq 0$. Then we have the following result:

Proposition 4. *Let G be a connected and simply connected Lie group for which the sequence $\{\mathfrak{g}_k\}$ of commutator ideals satisfies $\mathfrak{g}_N = (0)$ for some $N > 0$. Then any Lie equation for G can be solved by a sequence of quadratures.*

A Lie algebra with the property described in Proposition 4 is called "solvable". For example, the subalgebra of upper triangular matrices in $\mathfrak{gl}(n, \mathbb{R})$ is solvable, as the reader is invited to check.

While it may seem that solvability is a lot to ask of a Lie algebra, it turns out that this property is surprisingly common. The reader can also check that, of all of the two and three dimensional Lie algebras found in Lecture 2, only $\mathfrak{sl}(2,\mathbb{R})$ and $\mathfrak{so}(3)$ fail to be solvable.

This (partly) explains why the Riccati equation holds such an important place in the theory of ODE. In some sense, it is the first Lie equation which cannot be solved by quadratures. (See the exercises for an interpretation and "proof" of this statement.)

In any case, the sequence of subalgebras $\{\mathfrak{g}_k\}$ eventually stabilizes at a subalgebra $\mathfrak{g}_N$ whose Lie algebra satisfies $[\mathfrak{g}_N, \mathfrak{g}_N] = \mathfrak{g}_N$. A Lie algebra $\mathfrak{g}$ for which $[\mathfrak{g}, \mathfrak{g}] = \mathfrak{g}$ is called "perfect". Our analysis of Lie equations shows that, by Lie's reduction method, we can, by quadrature alone, reduce the problem of solving Lie equations to the problem of solving Lie equations associated to Lie groups with perfect algebras. Further analysis of the relation between the structure of a Lie algebra and the solvability by quadratures of any associated Lie equation leads to the development of the so-called Jordan-Hölder decomposition theorems, see [Va].

Appendix: Lie's transformation groups, I

When Lie began his study of symmetry groups in the nineteenth century, the modern concepts of manifold theory were not available. Thus, the examples that he had to guide him were defined as "transformations in n variables" which were often, like the Möbius transformations on the line or like conformal transformations in space, only defined "almost everywhere". Thus, at first glance, it might appear that Lie's concept of a "continuous transformation group" should correspond to what we have defined as a local Lie group action.

However, it turns out that Lie had in mind a much more general concept. For Lie, a set Γ of local diffeomorphisms in $\mathbb{R}^n$ formed a "continuous transformation group" if it was closed under composition and inverse and moreover, the elements of Γ were characterized as the solutions of some system of differential equations.

For example, the Möbius group on the line could be characterized as the set Γ of (non-constant) solutions $f(x)$ of the differential equation

$$2f'''(x)f'(x) - 3\big(f''(x)\big)^2 = 0.$$

As another example, the "group" of area preserving transformations of the plane could be characterized as the set of solutions $\big(f(x,y), g(x,y)\big)$ to the equation

$$f_x\, g_y - g_x\, f_y \equiv 1,$$

while the "group" of holomorphic transformations of the plane

$$\mathbb{R}^2$$

(regarded as $\mathbb{C}$) was the set of solutions $\big(f(x,y), g(x,y)\big)$ to the equations

$$f_x - g_y = f_y + g_x = 0.$$

Notice a big difference between the first example and the other two. In the first example, there is only a 3-parameter family of local solutions and each of these solutions patches together on $\mathbb{RP}^1 = \mathbb{R} \cup \{\infty\}$ to become an element of the global Lie group action of $\mathrm{SL}(2, \mathbb{R})$ on $\mathbb{RP}^1$. In the other two examples, there are many local solutions that cannot be extended to the entire plane, much less any "completion". Moreover in the volume preserving example, it is clear that no finite dimensional Lie group could ever contain all of the globally defined volume preserving transformations of the plane.

Lie regarded these latter two examples as "infinite continuous groups". Nowadays, we would call them "infinite dimensional pseudo-groups". I will say more about this point of view in an appendix to Lecture 6.

Since Lie did not have a group manifold to work with, he did not regard his "infinite groups" as pathological. Instead of trying to find a global description of the groups, he worked with what he called the "infinitesimal transformations" of Γ. We would say that, for each of his groups Γ, he considered the space of vector fields $\gamma \subset \mathfrak{X}(\mathbb{R}^n)$ whose (local) flows were 1-parameter "subgroups" of Γ. For example, the infinitesimal transformations associated to the area preserving transformations are the vector fields

$$X = f(x,y)\frac{\partial}{\partial x} + g(x,y)\frac{\partial}{\partial y}$$

which are divergence free, i.e., satisfy $f_x + g_y = 0$.

Lie "showed" that for any "continuous transformation group" Γ, the associated set of vector fields γ was actually closed under addition, scalar multiplication (by constants), and, most significantly, the Lie bracket. (The reason for the quotes around "showed" is that Lie was not careful to specify the nature of the differential equations which he was using to define his groups. Without adding some sort of constant rank or non-degeneracy hypotheses, many of his proofs are incorrect.)

For Lie, every subalgebra L of the algebra $\mathfrak{X}(\mathbb{R}^n)$ which could be characterized by some system of PDE was to be regarded the Lie algebra of some Lie group. Thus, rather than classify actual groups (which might not really be groups because of domain problems), Lie classified subalgebras of the algebra of vector fields.

In the case that L was finite dimensional, Lie actually proved that there *was* a "germ" of a Lie group (in our sense) and a local Lie group action which generated this algebra of vector fields. This is Lie's so-called Third Fundamental Theorem.

The case where L was infinite dimensional remained rather intractable. I will have more to say about this in Lecture 6. For now, though, I want to stress that there *is* a sort of analogue of actions for these "infinite dimensional Lie groups".

For example, if M is a manifold and $\mathrm{Diff}(M)$ is the group of (global) diffeomorphisms, then we can regard the natural (evaluation) map $\lambda\colon \mathrm{Diff}(M) \times M \to M$ given by $\lambda(\phi, m) = \phi(m)$ as a faithful Lie group action. If M is compact, then every vector field is complete, so, at least formally, the induced map $\lambda_*\colon T_{id}\mathrm{Diff}(M) \to \mathfrak{X}(M)$ ought to be an isomorphism of vector spaces. If our analogy with the finite dimensional case is to hold up, λ_* must reverse the Lie bracket.

Of course, since we have not defined a smooth structure on $\mathrm{Diff}(M)$, it is not immediately clear how to make sense of $T_{id}\mathrm{Diff}(M)$. I will prefer to proceed formally and simply define the Lie algebra $\mathrm{diff}(M)$ of $\mathrm{Diff}(M)$ to be the vector space $\mathfrak{X}(M)$ with the Lie algebra bracket given by the negative of the vector field Lie bracket.

With this definition, it follows that a left action $\lambda\colon G\times M\to M$ where G is finite dimensional can simply be regarded as a homomorphism $\Lambda\colon G\to \mathrm{Diff}(M)$ inducing a homomorphism of Lie algebras.

A modern treatment of this subject can be found in [SS].

Appendix: connections and curvature

In this appendix, I want briefly to describe the notions of connections and curvature on principal bundles in the language that I will be using in the examples in this Lecture.

Let G be a Lie group with Lie algebra $\mathfrak{g}$ and let ω_G be the canonical $\mathfrak{g}$-valued, left-invariant 1-form on G.

Principal bundles. Let M be an n-manifold and let P be a principal right G-bundle over M. Thus, P comes equipped with a submersion $\pi\colon P\to M$ and a free right action $\rho\colon P\times G\to P$ so that the fibers of π are the G-orbits of ρ.

The gauge group. The group $\mathrm{Aut}(P)$ of automorphisms of P is, by definition, the set of diffeomorphisms $\phi\colon P\to P$ which are compatible with the two structure maps, i.e.,

$$\pi\circ\phi=\pi\quad\text{and}\quad \rho_g\circ\phi=\phi\circ\rho_g\quad\text{for all }g\in G.$$

For reasons having to do with Physics, this group is nowadays referred to as the *gauge group* of P. Of course, $\mathrm{Aut}(P)$ is not a finite dimensional Lie group, but it would have been considered by Lie himself as a perfectly reasonable 'continuous transformation group' (although not a very interesting one for his purposes).

For any $\phi\in\mathrm{Aut}(P)$, there is a unique smooth map $\varphi\colon P\to G$ which satisfies $\phi(p)=p\cdot\varphi(p)$. The identity $\rho_g\circ\phi=\phi\circ\rho_g$ implies that φ satisfies $\varphi(p\cdot g)=g^{-1}\varphi(p)g$ for all $g\in G$. Conversely, any smooth map $\varphi\colon P\to G$ satisfying this identity defines an element of $\mathrm{Aut}(P)$. It follows that $\mathrm{Aut}(P)$ is the space of sections of the bundle $C(P)=P\times_C G$ where $C\colon G\times G\to G$ is the conjugation action $C(a,b)=aba^{-1}$.

Moreover, it easily follows that the set of vector fields on P whose flows generate 1-parameter subgroups of $\mathrm{Aut}(P)$ is identifiable with the space of sections of the vector bundle $\mathrm{Ad}(P)=P\times_{\mathrm{Ad}}\mathfrak{g}$.

Connections. Let $\mathfrak{A}(P)$ denote the space of *connections* on P. Thus, an element $A\in\mathfrak{A}(P)$ is, by definition, a $\mathfrak{g}$-valued 1-form A on P with the following two properties:
 1. For any $p\in P$, we have $\iota_p^*(A)=\omega_G$ where $\iota_p\colon G\to P$ is given by $\iota_p(g)=p\cdot g$.
 2. For all g in G, we have $\rho_g^*(A)=\mathrm{Ad}(g^{-1})(A)$ where $\rho_g\colon P\to P$ is right action by g.

It follows from Property 1 that, for any connection A on P, we have $A\bigl(\rho_*(x)\bigr)=x$ for all $x\in\mathfrak{g}$. It follows from Property 2 that $\mathcal{L}_{\rho_*(x)}A=-[x,A]$ for all $x\in\mathfrak{g}$.

If A_0 and A_1 are connections on P, then it follows from Property 1 that the difference $\alpha=A_1-A_0$ is a $\mathfrak{g}$-valued 1-form which is semi-basic (i.e., $\alpha(v)=0$ for all $v\in\ker\pi'$). Moreover, Property 2 implies that α satisfies $\rho_g^*(\alpha)=\mathrm{Ad}(g^{-1})(\alpha)$. Conversely, if α is any $\mathfrak{g}$-valued 1-form on P satisfying these latter two properties and $A \in \mathfrak{A}(P)$ is a connection, then $A + \alpha$ is also a connection. It is easy to see that a 1-form α with these two properties can be regarded as a 1-form on M with values in $\mathrm{Ad}(P)$.

Thus, $\mathfrak{A}(P)$ is an affine space modeled on the vector space $\mathcal{A}^1(\mathrm{Ad}(P))$. In particular, if we regard $\mathfrak{A}(P)$ as an "infinite dimensional manifold", the tangent space $T_A \mathfrak{A}(P)$ at any point A is naturally isomorphic to $\mathcal{A}^1(\mathrm{Ad}(P))$.

Curvature. The *curvature* of a connection A is the 2-form $F_A = dA + \frac{1}{2}[A, A]$. From our formulas above, it follows that

$$\rho_*(x) \lrcorner F_A = \rho_*(x) \lrcorner dA + [x, A] = \mathcal{L}_{\rho_*(x)} A + [x, A] = 0.$$

Since the vector fields $\rho_*(x)$ span the vertical tangent spaces of P, it follows that F_A is a semi-basic 2-form with values in $\mathfrak{g}$. Moreover, the Ad-equivariance of A implies that $\rho_g^*(F_A) = \mathrm{Ad}(g^{-1})(F_A)$. Thus, F_A may be regarded as a section of the bundle of 2-forms on M with values in the bundle $\mathrm{Ad}(P)$.

The group $\mathrm{Aut}(P)$ acts naturally on the *right* on $\mathfrak{A}(P)$ via pullback: $A \cdot \phi = \phi^*(A)$. In terms of the corresponding map $\varphi \colon P \to G$, we have

$$A \cdot \phi = \varphi^*(\omega_G) + \mathrm{Ad}(\varphi^{-1})(A).$$

It follows by direct computation that $F_{A \cdot \phi} = \phi^*(F_A) = \mathrm{Ad}(\varphi^{-1})(F_A)$.

We say that A is *flat* if $F_A = 0$. It is an elementary ODE result that A is flat if and only if, for every $m \in M$, there exists an open neighborhood U of m and a smooth map $\tau \colon \pi^{-1}(U) \to G$ which satisfies $\tau(p \cdot g) = \tau(p)g$ and $\tau^*(\omega_G) = A_{|U}$. In other words A is flat if and only if the bundle-with-connection (P, A) is locally diffeomorphic to the trivial bundle-with-connection $(M \times G, \omega_G)$.

Covariant Differentiation. The space $\mathcal{A}^p(\mathrm{Ad}(P))$ of p-forms on M with values in $\mathrm{Ad}(P)$ can be identified with the space of $\mathfrak{g}$-valued, p-forms β on P which are both semi-basic and Ad-equivariant (i.e., $\rho_g^*(\beta) = \mathrm{Ad}(g^{-1})(\beta)$ for all $g \in G$). Given such a form β, the expression $d\beta + [A, \beta]$ is easily seen to be a $\mathfrak{g}$-valued $(p+1)$-form on P which is also semi-basic and Ad-equivariant. It follows that this defines a first-order differential operator

$$d_A \colon \mathcal{A}^p(\mathrm{Ad}(P)) \to \mathcal{A}^{p+1}(\mathrm{Ad}(P))$$

called *covariant differentiation with respect to* A. It is elementary to check that

$$d_A(d_A \beta) = [F_A, \beta] = \mathrm{ad}(F_A)(\beta).$$

Thus, for a flat connection, $(\mathcal{A}^*(\mathrm{Ad}(P)), d_A)$ forms a complex over M.

We also have the *Bianchi identity* $d_A F_A = 0$.

In some expositions, "covariant differentiation" means only $d_A \colon \mathcal{A}^0(\mathrm{Ad}(P)) \to \mathcal{A}^1(\mathrm{Ad}(P))$.

Horizontal Lifts and Holonomy. Let A be a connection on P. If $\gamma \colon [0, 1] \to M$ is a C^1 curve and $p \in \pi^{-1}(\gamma(0))$ is chosen, then there exists a unique C^1 curve

$\tilde\gamma\colon [0,1] \to P$ which both "lifts" γ in the sense that $\gamma = \pi \circ \tilde\gamma$ and also satisfies the differential equation $\tilde\gamma^*(A) = 0$.

(To see this, first choose any lift $\bar\gamma\colon [0,1] \to P$ which satisfies $\bar\gamma(0) = p$. Then the desired lifting will then be given by $\tilde\gamma(t) = \bar\gamma(t) \cdot g(t)$ where $g\colon [0,1] \to G$ is the solution of the Lie equation $g'(t) = -R_{g(t)}\bigl(A(\bar\gamma'(t))\bigr)$ satisfying the initial condition $g(0) = e$.)

The resulting curve $\tilde\gamma$ is called a *horizontal lift* of γ. If γ is merely piecewise C^1, the horizontal lift can still be defined by piecing together horizontal lifts of the C^1-segments in the obvious way. Also, if $p' = p \cdot g_0$, then the horizontal lift of γ with initial condition p' is easily seen to be $\rho_{g_0} \circ \tilde\gamma$.

Let $p \in P$ be chosen and set $m = \pi(p)$. For every piecewise C^1 loop $\gamma\colon [0,1] \to M$ based at m, the horizontal lift $\tilde\gamma$ has the property that $\tilde\gamma(1) = p \cdot h(\gamma)$ for some unique $h(\gamma) \in G$. By definition, the *holonomy* of A at p, denoted by $H_A(p)$, is the set of all such elements $h(\gamma)$ of G where γ ranges over all of the piecewise C^1 closed loops based at m.

I leave it to the reader to show that $H_A(p \cdot g) = g^{-1} H_A(p) g$ and that, if p and p' can be joined by a horizontal curve in P, then $H_A(p) = H_A(p')$. Thus, the conjugacy class of $H_A(p)$ in G is independent of p if M is connected.

A basic theorem due to Borel and Lichnerowitz (see [KN]) asserts that $H_A(p)$ is always a Lie subgroup of G.

Exercises

1. Verify the claim made in the lecture that every right (respectively, left) action of a Lie group on a manifold can be rewritten as a left (respectively, right) action. Is the assumption that a left action $\lambda: G \times M \to M$ satisfy $\lambda(e, m) = m$ for all $m \in M$ really necessary?

2. Show that if $f: X \to Y$ is a map of smooth manifolds for which the rank of $f'(x): T_x X \to T_{f(x)} Y$ is independent of x, then $f^{-1}(y)$ is a (possibly empty) closed, smooth submanifold of X for all $y \in Y$. Note that this properly generalizes the usual Implicit Function Theorem, which requires $f'(x)$ to be a surjection everywhere in order to conclude that $f^{-1}(y)$ is a smooth submanifold.

   (Hint: Suppose that the rank of $f'(x)$ is identically k. You want to show that $f^{-1}(y)$ (if non-empty) is a submanifold of X of codimension k. To do this, let $x \in f^{-1}(y)$ be given and construct a map $\psi: V \to \mathbb{R}^k$ on a neighborhood V of y so that $\psi \circ f$ is a submersion near x. Then show that $(\psi \circ f)^{-1}(\psi(y))$ (which, by the Implicit Function Theorem, is a closed codimension k submanifold of the open set $f^{-1}(V) \subset X$) is actually equal to $f^{-1}(y)$ on some neighborhood of x. Where do you need the constant rank hypothesis?)

3. This exercise concerns the automorphism groups of Lie algebras and Lie groups.

   (i) Show that, for any Lie algebra $\mathfrak{g}$, the group of automorphisms $\text{Aut}(\mathfrak{g})$ defined by

   $$\text{Aut}(\mathfrak{g}) = \{a \in \text{End}(\mathfrak{g}) \mid [a(x), a(y)] = a([x, y]) \text{ for all } x, y \in \mathfrak{g}\}$$

   is a closed Lie subgroup of $\text{GL}(\mathfrak{g})$. Show that its Lie algebra is

   $$\mathfrak{der}(\mathfrak{g}) = \{a \in \text{End}(\mathfrak{g}) \mid a([x, y]) = [a(x), y] + [x, a(y)] \text{ for all } x, y \in \mathfrak{g}\}.$$

   (Hint: Show that $\text{Aut}(\mathfrak{g})$ is the stabilizer of some point in some representation of the Lie group $\text{GL}(\mathfrak{g})$.)

   (ii) Show that if G is a connected and simply connected Lie group with Lie algebra $\mathfrak{g}$, then the group of (Lie) automorphisms of G is isomorphic to $\text{Aut}(\mathfrak{g})$.

   (iii) Show that $\text{ad}: \mathfrak{g} \to \text{End}(\mathfrak{g})$ actually has its image in $\mathfrak{der}(\mathfrak{g})$, and that this image is an ideal in $\mathfrak{der}(\mathfrak{g})$. What is the interpretation of this fact in terms of "inner" and "outer" automorphisms of G? (Hint: Use the Jacobi identity.)

   (iv) Show that if the Killing form of $\mathfrak{g}$ is non-degenerate, then $[\mathfrak{g}, \mathfrak{g}] = \mathfrak{g}$. (Hint: Suppose that $[\mathfrak{g}, \mathfrak{g}]$ lies in a proper subspace of $\mathfrak{g}$. Then there exists an element $y \in \mathfrak{g}$ so that $\kappa([x, z], y) = 0$ for all $x, z \in \mathfrak{g}$. Show that this implies that $[x, y] = 0$ for all $x \in \mathfrak{g}$, and hence that $\text{ad}(y) = 0$.)

   (v) Show that if the Killing form of $\mathfrak{g}$ is non-degenerate, then $\mathfrak{der}(\mathfrak{g}) = \text{ad}(\mathfrak{g})$. This shows that all of the automorphisms of a simple Lie algebra are "inner". (Hint: Show that the set $\mathfrak{p} = \{a \in \mathfrak{der}(\mathfrak{g}) \mid \text{tr}(a \, \text{ad}(x)) = 0 \text{ for all } x \in \mathfrak{g}\}$ is also an ideal in $\mathfrak{der}(\mathfrak{g})$ and hence that $\mathfrak{der}(\mathfrak{g}) = \mathfrak{p} \oplus \text{ad}(\mathfrak{g})$ as algebras.

Show that this forces $\mathfrak{p} = 0$ by considering what it means for elements of $\mathfrak{p}$ (which, after all, are derivations of $\mathfrak{g}$) to commute with elements in $\mathrm{ad}(\mathfrak{g})$.)

4. Consider the 1-parameter group which is generated by the flow of the vector field X in the plane

$$X = \cos y \frac{\partial}{\partial x} + \sin^2 y \frac{\partial}{\partial y}.$$

Show that this vector field is complete and hence yields a free $\mathbb{R}$-action on the plane. Let $\mathbb{Z}$ also act on the plane by the action

$$m \cdot (x, y) = ((-1)^m x, y + m\pi).$$

Show that these two actions commute, and hence together define a free action of $G = \mathbb{R} \times \mathbb{Z}$ on the plane. Sketch the orbits and show that, even though the G-orbits of this action are closed, and the quotient space is Hausdorff, the quotient space is not a manifold. (The point of this problem is to warn the student not to make the common mistake of thinking that the quotient of a manifold by a free Lie group action is a manifold if it is Hausdorff.)

5. Show that if $\rho: M \times G \to M$ is a right action, then the induced map $\rho_*: \mathfrak{g} \to \mathfrak{X}(M)$ satisfies $\rho_*([x, y]) = [\rho_*(x), \rho_*(y)]$.

6. Prove Proposition 2. (Hint: you are trying to find an open neighborhood U of $\{e\} \times M$ in $G \times M$ and a smooth map $\lambda: U \to M$ with the requisite properties. To do this, look for the graph of λ as a submanifold $\Gamma \subset G \times M \times M$ which contains all the points (e, m, m) and is tangent to a certain family of vector fields on $G \times M \times M$ constructed using the left invariant vector fields on G and the corresponding vector fields on M determined by the Lie algebra homomorphism $\phi: \mathfrak{g} \to \mathfrak{X}(M)$.)

7. Show that, if $A: \mathbb{R} \to \mathfrak{g}$ is a curve in the Lie algebra of a Lie group G, then there exists a unique solution to the ordinary differential equation $S'(t) = R_{S(t)}(A(t))$ with initial condition $S(0) = e$. (It is clear that a solution exists on some interval $(-\varepsilon, \varepsilon)$ in $\mathbb{R}$. The problem is to show that the solution exists on all of $\mathbb{R}$.)

8. Show that, under the action of $\mathrm{GL}(n, \mathbb{R})$ on the space of symmetric n-by-n matrices defined in the Lecture, every symmetric n-by-n matrix is in the orbit of an $I_{p,q}$.

9. This problem examines the geometry of the classical second order equation for one unknown.

(i) Rewrite the second-order ODE

$$\frac{d^2 x}{dt^2} = F(t) x$$

as a system of first-order ODEs of Lie type for an action of $\mathrm{SL}(2, \mathbb{R})$ on $\mathbb{R}^2$.

(ii) Suppose in particular that $F(t)$ is of the form $(f(t))^2 + f'(t)$, where $f(0) \neq 0$. Use the solution
$$x(t) = \exp\left(\int_0^t f(\tau)\,d\tau\right)$$
to write down the fundamental solution for this Lie equation up in $\mathrm{SL}(2,\mathbb{R})$.

(iii) Explain why the (more general) second order linear ODE
$$x'' = a(t)x' + b(t)x$$
is solvable by quadratures once we know a single solution with either $x(0) \neq 0$ or $x'(0) \neq 0$. (Hint: all two-dimensional Lie groups are solvable.)

10. * Show that the general equation of the form $y''(x) = f(x)\,y(x)$ is not integrable by quadratures. Specifically, show that there do not exist "universal" functions F_0 and F_1 of two and three variables respectively so that the function y defined by taking the most general solution of
$$u'(x) = F_0(x, f(x))$$
$$y'(x) = F_1(x, f(x), u(x))$$
is the general solution of $y''(x) = f(x)\,y(x)$. Note that this shows that the general solution cannot be got by two quadratures, which one might expect to need since the general solution must involve two constants of integration. However, it can be shown that no matter how many quadratures one uses, one cannot get even a particular solution of $y''(x) = f(x)\,y(x)$ (other than the trivial solution $y \equiv 0$) by quadrature. (If one could get a (non-trivial) particular solution this way, then, by two more quadratures, one could get the general solution.)

11. The point of this exercise is to prove Lie's theorem (stated below) on (local) group actions on $\mathbb{R}$. This theorem "explains" the importance of the Riccati equation, and why there are so few actions of Lie groups on $\mathbb{R}$. Let $\mathfrak{g} \subset \mathfrak{X}(\mathbb{R})$ be a *finite dimensional* Lie algebra of vector fields on $\mathbb{R}$ with the property that, at every $x \in \mathbb{R}$, there is at least one $X \in \mathfrak{g}$ so that $X(x) \neq 0$. (Thus, the (local) flows of the vector fields in $\mathfrak{g}$ do not have any common fixed point.)

(i) For each $x \in \mathbb{R}$, let $\mathfrak{g}_x^k \subset \mathfrak{g}$ denote the subspace of vector fields which vanish to order at least $k+1$ at x. (Thus, $\mathfrak{g}_x^{-1} = \mathfrak{g}$ for all x.) Let $\mathfrak{g}_x^\infty \subset \mathfrak{g}$ denote the intersection of all the $\mathfrak{g}_x^k$. Show that $\mathfrak{g}_x^\infty = 0$ for all x. (Hint: Fix $a \in \mathbb{R}$ and choose an $X \in \mathfrak{g}$ so that $X(a) \neq 0$. Make a local change of coordinates near a so that $X = \partial/\partial x$ on a neighborhood of a. Note that $[X, \mathfrak{g}_a^\infty] \subset \mathfrak{g}_a^\infty$. Now choose a basis $Y_1, \ldots, Y_N$ of $\mathfrak{g}_a^\infty$ and note that, near a, we have $Y_i = f_i\,\partial/\partial x$ for some functions f_i. Show that the f_i must satisfy some differential equations and then apply ODE uniqueness. Now go on from there.)

---

*This exercise is somewhat difficult, but you should enjoy seeing what is involved in trying to prove that an equation is not solvable by quadratures.

(ii) Show that the dimension of $\mathfrak{g}$ is at most 3. (Hint: First, show that $[\mathfrak{g}_x^j, \mathfrak{g}_x^k] \subseteq \mathfrak{g}_x^{j+k}$. Now, by part (i), you know that there is a smallest integer N (which may depend on x) so that $\mathfrak{g}_x^{N+1} = 0$. Show that if $X \in \mathfrak{g}$ does not vanish at x and $Y_N \in \mathfrak{g}$ vanishes to exactly order N at x, then $Y_{N-1} = [X, Y_N]$ vanishes to order exactly $N-1$. Conclude that the vectors $X, Y_0, \ldots, Y_N$ (where $Y_{i-1} = [X, Y_i]$ for $i > 0$) form a basis of $\mathfrak{g}$. Now, what do you know about $[Y_{N-1}, Y_N]$?).

(iii) (Lie's Theorem) Show that, if $\dim(\mathfrak{g}) = 2$, then $\mathfrak{g}$ is isomorphic to the (unique) non-abelian Lie algebra of that dimension and that there is a local change of coordinates so that

$$\mathfrak{g} = \{(a+bx)\partial/\partial x \,|\, a, b \in \mathbb{R}\}.$$

Show also that, if $\dim(\mathfrak{g}) = 3$, then $\mathfrak{g}$ is isomorphic to $\mathfrak{sl}(2, \mathbb{R})$ and that there exist local changes of coordinates so that

$$\mathfrak{g} = \{(a+bx+cx^2)\partial/\partial x \,|\, a, b, c \in \mathbb{R}\}.$$

(In the second case, after you have shown that the algebra is isomorphic to $\mathfrak{sl}(2, \mathbb{R})$, show that, at each point of $\mathbb{R}$, there exists a element $X \in \mathfrak{g}$ which does not vanish at the point and which satisfies $(\mathrm{ad}(X))^2 = 0$. Now put it in the form $X = \partial/\partial x$ for some local coordinate x and ask what happens to the other elements of $\mathfrak{g}$.)

(iv) (This is somewhat harder.) Show that if $\dim(\mathfrak{g}) = 3$, then there is a diffeomorphism of $\mathbb{R}$ with an open interval $I \subset \mathbb{R}$ so that $\mathfrak{g}$ gets mapped to the algebra

$$\mathfrak{g} = \left\{ (a + b\cos x + c\sin x)\frac{\partial}{\partial x} \,\Big|\, a, b, c \in \mathbb{R} \right\}.$$

In particular, this shows that every local action of $\mathrm{SL}(2, \mathbb{R})$ on $\mathbb{R}$ is the restriction of the Möbius action on $\mathbb{RP}^1$ after "lifting" to its universal cover. Show that two intervals $I_1 = (0, a)$ and $I_2 = (0, b)$ are diffeomorphic in such a way as to preserve the Lie algebra $\mathfrak{g}$ if and only if either $a = b = 2n\pi$ for some positive integer n or else $2n\pi < a, b < (2n+2)\pi$ for some positive integer n. (Hint: Show, by a local analysis, that any vector field $X \in \mathfrak{g}$ which vanishes at any point of $\mathbb{R}$ must have $\kappa(X, X) \geq 0$. Now choose an X so that $\kappa(X, X) = -2$ and choose a global coordinate $x \colon \mathbb{R} \to \mathbb{R}$ so that $X = \partial/\partial x$. You must still examine the effect of your choices on the image interval $x(\mathbb{R}) \subset \mathbb{R}$.)

Lie and his coworkers attempted to classify of the finite dimensional transitive Lie subalgebras of the vector fields on $\mathbb{R}^k$, for $k \leq 5$, since (they thought) this would give a classification of all of the equations of Lie type for at most 5 unknowns. However, this classification became extremely complex and lengthy by dimension 5 and it was abandoned. On the other hand, the project of classifying the *abstract* finite dimensional Lie algebras has enjoyed a great deal of success. In fact, one of the triumphs of nineteenth century mathematics was the classification, by Killing and Cartan, of all of the finite dimensional simple Lie algebras over $\mathbb{C}$ and $\mathbb{R}$.

LECTURE 4
Symmetries and Conservation Laws

In this Lecture, I will introduce a particular set of variational problems, the so-called 'first-order particle Lagrangian problems', which will serve as a link to the symplectic geometry to be developed in the next Lecture.

Variational problems

Lagrangians

Definition 1. A *Lagrangian* on a manifold M is a smooth function $L\colon TM \to \mathbb{R}$. For any smooth curve $\gamma\colon [a,b] \to M$, define

$$\mathcal{F}_L(\gamma) = \int_a^b L\bigl(\dot\gamma(t)\bigr)\, dt.$$

$\mathcal{F}_L$ is called the *functional* associated to L.

(The use of the word "functional" here is classical. The reader is supposed to think of the set of all smooth curves $\gamma\colon [a,b] \to M$ as a sort of infinite dimensional manifold and of $\mathcal{F}_L$ as a function on it.)

I have deliberately chosen to avoid the (mild) complications caused by allowing less smoothness for L and γ, though for some purposes, it is essential to do so. The geometric points that I want to make, however will be clearest if we do not have to worry about determining the optimum regularity assumptions.

Also, some sources only require L to be defined on some open set in TM. Others allow L to "depend on t", i.e., take L to be a function on $\mathbb{R} \times TM$. Though I will not go into any of these (slight) extensions, the reader should be aware that they exist. For example, see [A].

Example. Suppose that $L\colon TM \to \mathbb{R}$ restricts to each T_xM to be a positive definite quadratic form. Then L defines what is usually called a Riemannian metric on M. For a curve γ in M, the functional $\mathcal{F}_L(\gamma)$ is then twice what is usually called the "action" of γ. This example is, by far, the most commonly occurring Lagrangian in differential geometry. We will have more to say about this below.

For a Lagrangian L, one is usually interested in finding the curves $\gamma\colon [a,b] \to M$ with given "endpoint conditions" $\gamma(a) = p$ and $\gamma(b) = q$ for which the functional $\mathcal{F}_L(\gamma)$ is a minimum. For example, in the case where L defines a Riemannian metric on M, the curves with fixed endpoints of minimum "action" turn out also to be the *shortest* curves joining those endpoints. From calculus, we know that the way to find minima of a function on a manifold is to first find the critical points of the function and then look among those for the minima. As mentioned before, the set of curves in M can be thought of as a sort of "infinite dimensional" manifold, but I won't go into details on this point. What I will do instead is describe what ought to be the set of "curves" in this space (classically called "variations") if it were a manifold.

Given a curve $\gamma\colon [a,b] \to M$, a *(smooth) variation of γ with fixed endpoints* is, by definition, a smooth map

$$\Gamma : [a,b] \times (-\varepsilon, \varepsilon) \to M$$

for some $\varepsilon > 0$ with the property that $\Gamma(t,0) = \gamma(t)$ for all $t \in [a,b]$ and that $\Gamma(a,s) = \gamma(a)$ and $\Gamma(b,s) = \gamma(b)$ for all $s \in (-\varepsilon, \varepsilon)$.

In this Lecture, "variation" will always mean "smooth variation with fixed endpoints".

If L is a Lagrangian on M and Γ is a variation of $\gamma\colon [a,b] \to M$, then we can define a function $\mathcal{F}_{L,\Gamma}\colon (-\varepsilon, \varepsilon) \to \mathbb{R}$ by setting

$$\mathcal{F}_{L,\Gamma}(s) = \mathcal{F}_L(\gamma_s)$$

where $\gamma_s(t) = \Gamma(t,s)$.

Definition 2. A curve $\gamma\colon [a,b] \to M$ is *L-critical* if $\mathcal{F}'_{L,\Gamma}(0) = 0$ for all variations of γ.

It is clear from calculus that a curve which minimizes $\mathcal{F}_L$ among all curves with the same endpoints will have to be L-critical, so the search for minimizers usually begins with the search for the critical curves.

Canonical coordinates. I want to examine what the problem of finding L-critical curves "looks like" in local coordinates. If $U \subset M$ is an open set on which there exists a coordinate chart $x\colon U \to \mathbb{R}^n$, then there is a canonical extension of these coordinates to a coordinate chart $(x,p)\colon TU \to \mathbb{R}^n \times \mathbb{R}^n$ with the property that, for any curve $\gamma\colon [a,b] \to U$, with coordinates $y = x \circ \gamma$, the p-coordinates of the curve $\dot{\gamma}\colon [a,b] \to TU$ are given by $p \circ \dot{\gamma} = \dot{y}$. We shall call the coordinates (x,p) on TU, the *canonical coordinates* associated to the coordinate system x on U.

It is only fair to warn the reader that this notation is far from standard. I am deliberately avoiding the more standard notation $(q, \dot{q})$, which is common in texts on mechanics and physics. I find the standard notation confusing because, in these coordinates, an arbitrary curve in the tangent bundle has to be denoted $\bigl(q(t), \dot{q}(t)\bigr)$ in spite of the fact that the second n functions need not be the t-derivatives of the first n functions. (In fact, they will be so only if the curve in TU is the tangential lift of a curve in U.) Students of mechanics do learn not to be confused by this clash of notation, but I am not trying to produce mechanicians.

The Euler-Lagrange equations. In a canonical coordinate system (x,p) on TU where U is an open set in M, the function L can be expressed as a function $L(x,p)$ of x and p. For a curve $\gamma\colon [a,b] \to M$ which happens to lie in U, the functional $\mathcal{F}_L$ becomes simply

$$\mathcal{F}_L(\gamma) = \int_a^b L\bigl(y(t), \dot y(t)\bigr)\, dt.$$

I will now derive the classical conditions for such a γ to be L-critical: Let $h\colon [a,b] \to \mathbb{R}^n$ be any smooth map which satisfies $h(a) = h(b) = 0$. Then, for sufficiently small ε, there is a variation Γ of γ which is expressed in (x,p)-coordinates as

$$(x,p) \circ \Gamma = (y+sh, \dot y + s\dot h).$$

Then, by the classic integration-by-parts method,

$$\begin{aligned}
\mathcal{F}'_{L,\Gamma}(0) &= \frac{d}{ds}\bigg|_{s=0} \left(\int_a^b L\bigl(y(t)+sh(t), \dot y(t)+s\dot h(t)\bigr)\, dt \right) \\
&= \int_a^b \left(\frac{\partial L}{\partial x^k}\bigl(y(t),\dot y(t)\bigr) h^k(t) + \frac{\partial L}{\partial p^k}\bigl(y(t),\dot y(t)\bigr) \dot h^k(t) \right) dt \\
&= \int_a^b \left[\frac{\partial L}{\partial x^k}\bigl(y(t),\dot y(t)\bigr) - \frac{d}{dt}\left(\frac{\partial L}{\partial p^k}\bigl(y(t),\dot y(t)\bigr) \right) \right] h^k(t)\, dt.
\end{aligned}$$

This formula is valid for any $h\colon [a,b] \to \mathbb{R}^n$ which vanishes at the endpoints. It follows without difficulty that the curve γ is L-critical if and only if $y = x \circ \gamma$ satisfies the n differential equations

$$\frac{\partial L}{\partial x^k}\bigl(y(t),\dot y(t)\bigr) - \frac{d}{dt}\left(\frac{\partial L}{\partial p^k}\bigl(y(t),\dot y(t)\bigr) \right) = 0, \qquad \text{for } 1 \le k \le n.$$

These are the famous *Euler-Lagrange equations*.

The main drawback of the Euler-Lagrange equations in this form is that they only give necessary and sufficient conditions for a curve to be L-critical if it lies in a coordinate neighborhood U. Now, it is not hard to show that if $\gamma\colon [a,b] \to M$ is L-critical, then its restriction to any subinterval $[a',b'] \subset [a,b]$ is also L-critical. In particular, a necessary condition for γ to be L-critical is that it satisfy the Euler-Lagrange equations on any subcurve which lies in a coordinate system. However, it is *not* clear that these 'local' conditions are sufficient.

Another drawback is that, as derived, the equations depend on the choice of coordinates and it is not clear that one's success in solving them might not depend on a clever choice of coordinates.

In what follows, we want to remedy these defects. First, though, here are a couple of examples.

Example: Riemannian Metrics. Consider a Riemannian metric $L\colon TM \to \mathbb{R}$. Then, in local canonical coordinates,

$$L(x,p) = g_{ij}(x)p^i p^j.$$

where $g(x)$ is a positive definite symmetric matrix of functions. (Remember, the summation convention is in force.) In this case, the Euler-Lagrange equations are

$$\frac{\partial g_{ij}}{\partial x^k}\bigl(y(t)\bigr)\dot y^i(t)\dot y^j(t) = \frac{d}{dt}\bigl(2g_{kj}\bigl(y(t)\bigr)\dot y^j(t)\bigr) = 2\frac{\partial g_{kj}}{\partial x^i}\bigl(y(t)\bigr)\dot y^i(t)\dot y^j(t) + 2g_{kj}\bigl(y(t)\bigr)\ddot y^j(t).$$

Since the matrix $g(x)$ is invertible for all x, these equations can be put in more familiar form by solving for the second derivatives to get

$$\ddot y^i = -\Gamma^i_{jk}(y)\dot y^j \dot y^k$$

where the functions $\Gamma^i_{jk} = \Gamma^i_{kj}$ are given by the formula so familiar to geometers:

$$\Gamma^i_{jk} = \frac{g^{i\ell}}{2}\left(\frac{\partial g_{\ell j}}{\partial x^k} + \frac{\partial g_{\ell k}}{\partial x^j} - \frac{\partial g_{jk}}{\partial x^\ell}\right)$$

where the matrix (g^{ij}) is the inverse of the matrix (g_{ij}).

Example: 1-forms. Another interesting case is when L is linear on each tangent space, i.e., $L = \omega$ where ω is a smooth 1-form on M. In local canonical coordinates,

$$L = a_i(x)\,p^i$$

for some functions a_i and the Euler-Lagrange equations become:

$$\frac{\partial a_i}{\partial x^k}\bigl(y(t)\bigr)\dot y^i(t) = \frac{d}{dt}\bigl(a_k\bigl(y(t)\bigr)\bigr) = \frac{\partial a_k}{\partial x^i}\bigl(y(t)\bigr)\dot y^i(t)$$

or, simply,

$$\left(\frac{\partial a_i}{\partial x^k}(y) - \frac{\partial a_k}{\partial x^i}(y)\right)\dot y^i = 0.$$

This last equation should look familiar. Recall that the exterior derivative of ω has the coordinate expression

$$d\omega = \frac{1}{2}\left(\frac{\partial a_j}{\partial x^i} - \frac{\partial a_i}{\partial x^j}\right)dx^i \wedge dx^j.$$

If $\gamma\colon [a,b] \to U$ is $\mathcal{F}_\omega$-critical, then for every vector field v along γ the Euler-Lagrange equations imply that

$$d\omega\bigl(\dot\gamma(t), v(t)\bigr) = \frac{1}{2}\left(\frac{\partial a_j}{\partial x^i}\bigl(y(t)\bigr) - \frac{\partial a_i}{\partial x^j}\bigl(y(t)\bigr)\right)\dot y^i(t) v^j(t) = 0.$$

In other words, $\dot\gamma(t) \lrcorner\, d\omega = 0$. Conversely, if this identity holds, then γ is clearly ω-critical.

This leads to the following global result:

Proposition 1. *A curve $\gamma\colon [a,b] \to M$ is ω-critical for a 1-form ω on M if and only if it satisfies the first order differential equation*

$$\dot\gamma(t) \lrcorner\, d\omega = 0.$$

Proof. A straightforward integration-by-parts on M yields the coordinate-free formula

$$\mathcal{F}'_{\omega,\Gamma}(0) = \int_a^b d\omega\!\left(\dot\gamma(t), \tfrac{\partial \Gamma}{\partial s}(t,0)\right) dt$$

where Γ is any variation of γ and $\tfrac{\partial \Gamma}{\partial s}$ is the "variation vector field" along γ. Since this vector field is arbitrary except for being required to vanish at the endpoints, we see that "$d\omega(\dot\gamma, v) = 0$ for all vector fields v along γ" is the desired condition for ω-criticality.

Symmetries

The way is now paved for what will seem like a trivial observation, but, in fact, turns out to be of fundamental importance: It is the 'seed' of Noether's Theorem.

Proposition 2. *Suppose that ω is a 1-form on M and that X is a vector field on M whose (local) flow leaves ω invariant. Then the function $\omega(X)$ is constant on all ω-critical curves.*

Proof. The condition that the flow of X leave ω invariant is just that $\mathcal{L}_X(\omega) = 0$. However, by the Cartan formula,

$$0 = \mathcal{L}_X(\omega) = d(X \lrcorner\, \omega) + X \lrcorner\, d\omega,$$

so for any curve γ in M, we have

$$d\omega\bigl(\dot\gamma(t), X(\gamma(t))\bigr) = -d\omega\bigl(X(\gamma(t)), \dot\gamma(t)\bigr) = -(X \lrcorner\, d\omega)\bigl(\dot\gamma(t)\bigr) = d(X \lrcorner\, \omega)\bigl(\dot\gamma(t)\bigr)$$

and this last expression is clearly the derivative of the function $X \lrcorner\, \omega = \omega(X)$ along γ. Now apply Proposition 1.

It is worth pausing a moment to think about what Proposition 2 means. The condition that the flow of X leave ω invariant is essentially saying that the flow of X is a "symmetry" of ω and hence of the functional $\mathcal{F}_\omega$. What Proposition 2 says is that a certain kind of symmetry of the functional gives rise to a "first integral" (sometimes called "conservation law") of the equation for ω-critical curves. If the function $\omega(X)$ is not a constant function on M, then saying that the ω-critical curves lie in its level sets is useful information about these critical curves.

Now, this idea can be applied to the general Lagrangian with symmetries. The only trick is to find the appropriate 1-form on which to evaluate "symmetry" vector fields.

Proposition 3. *For any Lagrangian $L: TM \to \mathbb{R}$, there exist a unique function E_L on TM and a unique 1-form ω_L on TM which, relative to any local coordinate system $x: U \to \mathbb{R}$, have the expressions*

$$E_L = p^i \frac{\partial L}{\partial p^i} - L \quad \text{and} \quad \omega_L = \frac{\partial L}{\partial p^i} dx^i.$$

Moreover, if $\gamma: [a,b] \to M$ is any curve, then γ satisfies the Euler-Lagrange equations for L in every local coordinate system if and only if its canonical lift $\dot\gamma: [a,b] \to TM$ satisfies

$$\ddot\gamma(t) \lrcorner\, d\omega_L = -dE_L(\dot\gamma(t)).$$

Proof. This will mainly be a sequence of applications of the Chain Rule.

There is an invariantly defined vector field R on TM which is simply the radial vector field on each subspace $T_m M$. It is expressed in canonical coordinates as $R = p^i \partial/\partial p^i$. Now, using this vector field, the quantity E_L takes the form

$$E_L = -L + dL(R).$$

Thus, it is clear that E_L is well-defined on TM.

Now we check the well-definition of ω_L. If $z: U \to \mathbb{R}$ is any other local coordinate system, then $z = F(x)$ for some $F: \mathbb{R}^n \to \mathbb{R}^n$. The corresponding canonical coordinates on TU are (z, q) where $q = F'(x)p$. In particular,

$$\begin{pmatrix} dz \\ dq \end{pmatrix} = \begin{pmatrix} F'(x) & 0 \\ G(x,p) & F'(x) \end{pmatrix} \begin{pmatrix} dx \\ dp \end{pmatrix}.$$

where G is some matrix function whose exact form is not relevant. Then writing L_z for $\left(\frac{\partial L}{\partial z^1}, \ldots, \frac{\partial L}{\partial z^n}\right)$, etc., yields

$$\begin{aligned} dL &= L_z\, dz + L_q\, dq \\ &= \left(L_z F'(x) + L_q G(x,p)\right) dx + L_q F'(x)\, dp \\ &= L_x\, dx + L_p\, dp. \end{aligned}$$

Comparing dp-coefficients yields $L_p = L_q F'(x)$, so $L_p\, dx = L_q F'(x)\, dx = L_q\, dz$. In particular, as we wished to show, there exists a well-defined 1-form ω_L on TM whose coordinate expression in local canonical coordinates (x,p) is $L_p\, dx$.

The remainder of the proof is a coordinate calculation. The reader will want to note that I am using the expression $\ddot\gamma$ to denote the velocity of the curve $\dot\gamma$ in TM. The curve $\dot\gamma$ is described in U as $(x,p) = (y, \dot y)$ and its velocity vector $\ddot\gamma$ is simply $(\dot x, \dot p) = (\dot y, \ddot y)$.

Now, the Euler-Lagrange equations are just

$$\frac{\partial L}{\partial x^i}(y,\dot{y}) = \frac{d}{dt}\left(\frac{\partial L}{\partial p^i}(y,\dot{y})\right) = \frac{\partial^2 L}{\partial p^i \partial p^j}(y,\dot{y})\ddot{y}^j + \frac{\partial^2 L}{\partial p^i \partial x^j}(y,\dot{y})\dot{y}^j.$$

Meanwhile,

$$d\omega_L = \frac{\partial^2 L}{\partial p^i \partial p^j} dp^j \wedge dx^i + \frac{\partial^2 L}{\partial p^i \partial x^j} dx^j \wedge dx^i,$$

so

$$\ddot{\gamma} \lrcorner d\omega_L = \frac{\partial^2 L}{\partial p^i \partial p^j}(y,\dot{y})\left(\ddot{y}^j dx^i - \dot{y}^i dp^j\right) + \frac{\partial^2 L}{\partial p^i \partial x^j}(y,\dot{y})\left(\dot{y}^j dx^i - \dot{y}^i dx^j\right).$$

On the other hand, an easy computation yields

$$-dE_L(\dot{\gamma}) = \left(\frac{\partial L}{\partial x^i}(y,\dot{y}) - \frac{\partial^2 L}{\partial p^j \partial x^i}(y,\dot{y})\dot{y}^j\right) dx^i - \frac{\partial^2 L}{\partial p^i \partial p^j}(y,\dot{y})\dot{y}^i dp^j.$$

Comparing these last two equations, the condition $\ddot{\gamma} \lrcorner d\omega_L = -dE_L(\dot{\gamma})$ is seen to be the Euler-Lagrange equations, as desired.

Conservation of energy

One important consequence of Proposition 3 is that the function E_L is constant along the curve $\dot{\gamma}$ for any L-critical curve $\gamma\colon [a,b] \to M$. This follows since, for such a curve,

$$dE_L\left(\ddot{\gamma}(t)\right) = -d\omega_L\left(\ddot{\gamma}(t), \ddot{\gamma}(t)\right) = 0.$$

E_L is generally interpreted as the *energy** of the Lagrangian L, and this constancy of E_L on L-critical curves is often called the principle of Conservation of Energy.

Definition 3. If $L\colon TM \to \mathbb{R}$ is a Lagrangian on M, a diffeomorphism $f\colon M \to M$ is said to be a *symmetry* of L if L is invariant under the induced diffeomorphism $f'\colon TM \to TM$, i.e., if $L \circ f' = L$. A vector field X on M is said to be an *infinitesimal symmetry* of L if the (local) flow Φ_t of X is a symmetry of L for all t.

It is perhaps necessary to make a remark about the last part of this definition. For a vector field X which is not necessarily complete, and for any $t \in \mathbb{R}$, the "time t" local flow of X is well-defined on an open set $U_t \subset M$. The local flow of X then gives a well-defined diffeomorphism $\Phi_t\colon U_t \to U_{-t}$. The requirement for X is that, for each t for which $U_t \neq \emptyset$, the induced map $\Phi'_t\colon TU_t \to TU_{-t}$ should satisfy $L \circ \Phi'_t = L$. (Of course, if X is complete, then $U_t = M$ for all t, so symmetry has its usual meaning.)

---

*Some sources define E_L as $L - dL(R)$. My choice was to have E_L agree with the classical energy in the classical problems.

Let X be any vector field on M with local flow Φ. This induces a local flow on TM which is associated to a vector field X' on TM. If, in a local coordinate chart, $x: U \to \mathbb{R}^n$, the vector field X has the expression

$$X = a^i(x) \frac{\partial}{\partial x^i},$$

then the reader may check that, in the associated canonical coordinates on TU,

$$X' = a^i \frac{\partial}{\partial x^i} + p^j \frac{\partial a^i}{\partial x^j} \frac{\partial}{\partial p^i}.$$

The condition that X be an infinitesimal symmetry of L is then that L be invariant under the flow of X', i.e., that

$$dL(X') = a^i \frac{\partial L}{\partial x^i} + p^j \frac{\partial a^i}{\partial x^j} \frac{\partial L}{\partial p^i} = 0.$$

Noether's Theorem

The following theorem is now a simple calculation. Nevertheless, it is the foundation of a vast theory. It usually goes by the name "Noether's Theorem", though, in fact, Noether's Theorem is more general.

Theorem 1. *If X is an infinitesimal symmetry of the Lagrangian L, then the function $\omega_L(X')$ is constant on $\dot\gamma: [a,b] \to TM$ for every L-critical path $\gamma: [a,b] \to M$.*

Proof. Since the flow of X' fixes L it should not be too surprising that it also fixes E_L and ω_L. These facts are easily checked by the reader in local coordinates, so they are left as exercises. In particular,

$$\mathcal{L}_{X'} \omega_L = d(X' \lrcorner \omega_L) + X' \lrcorner d\omega_L = 0 \quad \text{and} \quad \mathcal{L}_{X'} E_L = dE_L(X') = 0.$$

Thus, for any L-critical curve γ in M,

$$\begin{aligned} d\big(\omega_L(X')\big)\big(\ddot\gamma(t)\big) = d(X' \lrcorner \omega_L)\big(\ddot\gamma(t)\big) &= -(X' \lrcorner d\omega_L)\big(\ddot\gamma(t)\big) \\ &= d\omega_L\big(\ddot\gamma(t), X'(\dot\gamma(t))\big) = \big(\ddot\gamma(t) \lrcorner d\omega_L\big)\big(X'(\dot\gamma(t))\big) \\ &= -dE_L\big(X'(\dot\gamma(t))\big) = 0. \end{aligned}$$

Hence, the function $\omega_L(X')$ is constant on $\dot\gamma$, as desired.

Of course, the formula for $\omega_L(X')$ in local canonical coordinates is simply

$$\omega_L(X') = a^i \frac{\partial L}{\partial p^i},$$

LECTURE 4. SYMMETRIES AND CONSERVATION LAWS

and the constancy of this function on the solution curves of the Euler-Lagrange equations is not difficult to check directly.

The principle

Symmetry $\implies$ **Conservation Law**

is so fundamental that whenever a new system of equations is encountered an enormous effort is expended to determine its symmetries. Moreover, the intuition is often expressed that "every conservation law ought to come from some symmetry", so whenever conserved quantities are observed in Nature (or, more accurately, our models of Nature) people nowadays look for a symmetry to explain it. Even when no symmetry is readily apparent, in many cases a sort of 'hidden' symmetry can be found.

Example: Motion in a Central Force Field. Consider the Lagrangian of "kinetic minus potential energy" for an particle (of mass $m \neq 0$) moving in a "central force field". Here, we take $\mathbb{R}^n$ with its usual inner product and a function $V(|x|^2)$ (called the potential energy) which depends only on distance from the origin. The Lagrangian is

$$L(x,p) = \tfrac{m}{2}|p|^2 - V(|x|^2).$$

The function E_L is given by

$$E_L(x,p) = \tfrac{m}{2}|p|^2 + V(|x|^2),$$

and $\omega_L = m\, p^i\, dx^i = m\, p \cdot dx$.

The Lagrangian L is clearly symmetric with respect to rotations about the origin. For example, the rotation in the ij-plane is generated by the vector field

$$X_{ij} = x^j \frac{\partial}{\partial x^i} - x^i \frac{\partial}{\partial x^j}.$$

According to Noether's Theorem, then, the functions

$$\mu_{ij} = \omega_L(X'_{ij}) = m\left(x^j p^i - x^i p^j\right)$$

are constant on all solutions. These are usually called the "angular momenta". It follows from their constancy that the bivector $\xi = y(t) \wedge \dot{y}(t)$ is constant on any solution $x = y(t)$ of the Euler-Lagrange equations and hence that $y(t)$ moves in a fixed 2-plane. Thus, we are essentially reduced to the case $n = 2$. In this case, for constants E_0 and μ_0, the equations

$$\tfrac{m}{2}|p|^2 + V(|x|^2) = E_0 \quad \text{and} \quad m(x^1 p^2 - x^2 p^1) = \mu_0$$

will generically define a surface in $T\mathbb{R}^2$. The solution curves to the Euler-Lagrange equations

$$\dot{x} = p \quad \text{and} \quad \dot{p} = -\frac{2}{m}V'(|x|^2)x$$

which lie on this surface can then be analysed by phase portrait methods. (In fact, they can be integrated by quadrature.)

Example: Riemannian metrics with Symmetries. As another example, consider the case of a Riemannian manifold with infinitesimal symmetries. If the flow of X on M preserves a Riemannian metric L, then, in local coordinates,

$$L = g_{ij}(x) p^i p^j$$

and

$$X = a^i(x) \frac{\partial}{\partial x^i}.$$

According to Conservation of Energy and Noether's Theorem, the functions

$$E_L = g_{ij}(x) p^i p^j \quad \text{and} \quad \omega_L(X') = 2 g_{ij}(x) a^i(x) p^j$$

are first integrals of the geodesic equations.

For example, if a surface $S \subset \mathbb{R}^3$ is a surface of revolution, then the induced metric can locally be written in the form

$$I = E(r)\, dr^2 + 2\, F(r) dr\, d\theta + G(r)\, d\theta^2$$

where the rotational symmetry is generated by the vector field $X = \partial/\partial\theta$. The following functions are then constant on solutions of the geodesic equations:

$$E(r)\, \dot r^2 + 2\, F(r) \dot r\, \dot\theta + G(r)\, \dot\theta^2 \quad \text{and} \quad F(r)\dot r + G(r)\, \dot\theta.$$

This makes it possible to integrate by quadratures the geodesic equations on a surface of revolution, a classical accomplishment. (See the Exercises for details.)

Subexample: Left Invariant Metrics on Lie Groups. Let G be a Lie group and let $\omega^1, \omega^2, \ldots, \omega^n$ be any basis for the left-invariant 1-forms on G. Consider the Lagrangian

$$L = (\omega^1)^2 + \cdots + (\omega^n)^2,$$

which defines a left-invariant metric on G. Since left translations are symmetries of this metric and since the flows of the *right*-invariant vector fields Y_i leave the left-invariant 1-forms fixed, we see that these generate symmetries of the Lagrangian L. In particular, the functions $E_L = L$ and

$$\mu_i = \omega^1(Y_i)\omega^1 + \cdots + \omega^n(Y_i)\omega^n$$

are functions on TG which are constant on all of the geodesics of G with the metric L. I will return to this example several times in future lectures.

LECTURE 4. SYMMETRIES AND CONSERVATION LAWS

Subsubexample: The Motion of Rigid Bodies. A special case of the Lie group example is particularly noteworthy, namely the theory of the rigid body.

A *rigid body* (in $\mathbb{R}^n$) is a (finite) set of points $\mathbf{x}_1, \ldots, \mathbf{x}_N$ with masses $m_1, \ldots, m_N$ such that the distances $d_{ij} = |\mathbf{x}_i - \mathbf{x}_j|$ are fixed (hence the name "rigid"). The free motion of such a body is governed by the "kinetic energy" Lagrangian

$$L = \frac{m_1}{2}|\mathbf{p}_1|^2 + \cdots + \frac{m_N}{2}|\mathbf{p}_N|^2.$$

where $\mathbf{p}_i$ represents the velocity of the i'th point mass. Here is how this can be converted into a left-invariant Lagrangian variational problem on a Lie group:

Let G be the matrix Lie group

$$G = \left\{ \begin{pmatrix} A & b \\ 0 & 1 \end{pmatrix} \;\middle|\; A \in O(n), b \in \mathbb{R}^n \right\}.$$

Then G acts as the space of isometries of $\mathbb{R}^n$ with its usual metric and thus also acts on the N-fold product

$$Y_N = \mathbb{R}^n \times \mathbb{R}^n \times \cdots \times \mathbb{R}^n$$

by the "diagonal" action. It is not difficult to show that G acts transitively on the simultaneous level sets of the functions $f_{ij}(\mathbf{x}) = |\mathbf{x}_i - \mathbf{x}_j|$. Thus, for each symmetric matrix $\Delta = (d_{ij})$, the set

$$M_\Delta = \{ \mathbf{x} \in Y_N \mid |\mathbf{x}_i - \mathbf{x}_j| = d_{ij} \}$$

is an orbit of G (and hence a smooth manifold) when it is not empty. The set M_Δ is said to be the "configuration space" of the rigid body. (Question: Can you determine a necessary and sufficient condition on the matrix Δ so that M_Δ is not empty? In other words, which rigid bodies are possible?)

Let us suppose that M_Δ is not empty and let $\bar{\mathbf{x}} \in M_\Delta$ be a "reference configuration" which, for convenience, we shall suppose has its center of mass at the origin:

$$m_k \bar{\mathbf{x}}_k = 0.$$

(This can always be arranged by a simultaneous translation of all of the point masses.) Now let $\gamma: [a, b] \to M_\Delta$ be a curve in the configuration space. (Such curves are often called "trajectories".) Since M_Δ is a G-orbit, there is a curve $g: [a, b] \to G$ so that $\gamma(t) = g(t) \cdot \bar{\mathbf{x}}$. Let us write

$$\gamma(t) = (\mathbf{x}_1(t), \ldots, \mathbf{x}_N(t))$$

and let

$$g(t) = \begin{pmatrix} A(t) & b(t) \\ 0 & 1 \end{pmatrix}.$$

The value of the canonical left invariant form on g is

$$g^{-1}\dot g = \begin{pmatrix} \alpha & \beta \\ 0 & 0 \end{pmatrix} = \begin{pmatrix} A^{-1}\dot A & A^{-1}\dot b \\ 0 & 0 \end{pmatrix}.$$

The kinetic energy along the trajectory γ is then

$$\tfrac{1}{2}\sum_k m_k |\dot{\mathbf{x}}_k|^2 = \tfrac{1}{2}\sum_k m_k (\dot{\mathbf{x}}_k \cdot \dot{\mathbf{x}}_k) = \tfrac{1}{2}\sum_k m_k \left(\dot A \bar{\mathbf{x}}_k + \dot b\right) \cdot \left(\dot A \bar{\mathbf{x}}_k + \dot b\right).$$

Since A is a curve in $O(n)$, this becomes

$$= \tfrac{1}{2}\sum_k m_k \left((A^{-1}(\dot A \bar{\mathbf{x}}_k + \dot b))\right) \cdot \left((A^{-1}(\dot A \bar{\mathbf{x}}_k + \dot b))\right)$$

$$= \tfrac{1}{2}\sum_k m_k (\alpha \bar{\mathbf{x}}_k + \beta) \cdot (\alpha \bar{\mathbf{x}}_k + \beta).$$

Using the center-of-mass normalization, this simplifies to

$$= \tfrac{1}{2}\sum_k m_k \left(-{}^t\bar{\mathbf{x}}_k \alpha^2 \bar{\mathbf{x}}_k + |\beta|^2\right).$$

With a slight rearrangement, this takes the simple form

$$L(\dot\gamma(t)) = -\operatorname{tr}\left((\alpha(t))^2 \mu\right) + \tfrac{1}{2}m |\beta(t)|^2$$

where $m = m_1 + \cdots + m_N$ is the total mass of the body and μ is the positive semi-definite symmetric n-by-n matrix

$$\mu = \tfrac{1}{2}\sum_k m_k \bar{\mathbf{x}}_k \,{}^t\bar{\mathbf{x}}_k.$$

It is clear that we can interpret L as a left-invariant Lagrangian on G. Actually, even the formula we have found so far can be simplified: If we write $\mu = R\delta\,{}^tR$ where δ is diagonal and R is an orthogonal matrix (which we can always do), then right acting on G by the element

$$\begin{pmatrix} R & 0 \\ 0 & 1 \end{pmatrix}$$

will reduce the Lagrangian to the form

$$L(\dot g(t)) = -\operatorname{tr}\left((\alpha(t))^2 \delta\right) + \tfrac{1}{2}m |\beta(t)|^2.$$

Thus, only the eigenvalues of the matrix μ really matter in trying to solve the equations of motion of a rigid body. This observation is usually given an interpretation like "the motion of any rigid body is equivalent to the motion of its 'ellipsoid of inertia' ".

Hamiltonian form

Let us return to the consideration of the Euler-Lagrange equations. As we have seen, in expanded form, the equations in local coordinates are

$$\frac{\partial^2 L}{\partial p^i \partial p^j}(y,\dot{y})\ddot{y}^j + \frac{\partial^2 L}{\partial p^i \partial x^j}(y,\dot{y})\dot{y}^j - \frac{\partial L}{\partial x^i}(y,\dot{y}) = 0.$$

Non-degeneracy

In order for these equations to be solvable for the highest derivatives at every possible set of initial conditions, the symmetric matrix

$$H_L(x,p) = \left(\frac{\partial^2 L}{\partial p^i \partial p^j}(x,p)\right).$$

must be invertible at every point (x,p).

Definition 4. A Lagrangian L is said to be *non-degenerate* if, relative to every local coordinate system $x: U \to \mathbb{R}^n$, the matrix H_L is invertible at every point of TU.

For example, if $L: TM \to \mathbb{R}$ restricts to each tangent space $T_m M$ to be a non-degenerate quadratic form, then L is a non-degenerate Lagrangian. In particular, when L is a Riemannian metric, L is non-degenerate.

Although Definition 4 is fairly explicit, it is certainly not coordinate free. Here is a result which may clarify the meaning of non-degenerate.

Proposition 4. *The following are equivalent for a Lagrangian* $L: TM \to \mathbb{R}$:
1. *L is a non-degenerate Lagrangian.*
2. *In local coordinates (x,p), the functions $x^1, \ldots, x^n, \partial L/\partial p^1, \ldots, \partial L/\partial p^n$ have everywhere independent differentials.*
3. *The 2-form $d\omega_L$ is non-degenerate at every point of TM, i.e., for any tangent vector $v \in T(TM)$, $v \lrcorner d\omega_L = 0$ implies that $v = 0$.*

Proof. The equivalence of (1) and (2) follows directly from the Chain Rule and is left as an exercise. The equivalence of (2) and (3) can be seen as follows: Let $v \in T_a(TM)$ be a tangent vector based at $a \in T_m M$. Choose any canonical local coordinate system (x,p) with $m \in U$ and write $q_i = \partial L/\partial p^i$ for $1 \le i \le n$. Then ω_L takes the form

$$d\omega_L = dq_i \wedge dx^i.$$

Thus,

$$v \lrcorner d\omega_L = dq_i(v)\, dx^i - dx^i(v)\, dq_i.$$

Now, suppose that the differentials $dx^1, \ldots, dx^n, dq_i, \ldots, dq_n$ are linearly independent at a and hence span $T_a^*(TM)$. Then, if $v \lrcorner d\omega_L = 0$, we must have $dq_i(v) = dx^i(v) = 0$, which, because the given $2n$ differentials form a spanning set, implies that $v = 0$ Thus, $d\omega_L$ is non-degenerate at a.

On the other hand, suppose that that the differentials $dx^1, \ldots, dx^n, dq_1, \ldots, dq_n$ are linearly dependent at a. Then, by linear algebra, there exists a non-zero vector $v \in T_a^*(TM)$ so that $dq_i(v) = dx^i(v) = 0$. However, it is then clear that $v \lrcorner d\omega_L = 0$ for such a v, so that $d\omega_L$ will be degenerate at a.

For physical reasons, the function q_i is usually called the *conjugate momentum* to the coordinate x^i. Before exploring the geometric meaning of the coordinate system (x, q), we want to give the following description of the L-critical curves of a non-degenerate Lagrangian.

Proposition 5. *If $L: TM \to \mathbb{R}$ is a non-degenerate Lagrangian, then there exists a unique vector field Y on TM so that, for every L-critical curve $\gamma: [a, b] \to M$, the associated curve $\dot\gamma: [a, b] \to TM$ is an integral curve of Y. Conversely, for any integral curve $\varphi: [a, b] \to TM$ of Y, the composition $\phi = \pi \circ \varphi: [a, b] \to M$ is an L-critical curve in M.*

Proof. It is clear that we should take Y to be the unique vector field on TM which satisfies $Y \lrcorner d\omega_L = -dE_L$. (There is only one since, by Proposition 4, $d\omega_L$ is non-degenerate.) Proposition 3 then says that for every L-critical curve, its lift $\dot\gamma$ satisfies $\ddot\gamma(t) = Y(\dot\gamma(t))$ for all t, i.e., that $\dot\gamma$ is indeed an integral curve of Y.

The details of the converse will be left to the reader. First, one must check that, with ϕ defined as above, we have $\dot\phi = \varphi$. This is best done in local coordinates. Second, one must check that ϕ is indeed L-critical, even though it may not lie entirely within a coordinate neighborhood. This may be done by computing the variation of ϕ restricted to appropriate subintervals and taking account of the boundary terms introduced by integration by parts when the endpoints are not fixed. Details are in the Exercises.

The canonical vector field Y on TM is just the coordinate free way of expressing the fact that, for non-degenerate Lagrangians, the Euler-Lagrangian equations are simply a non-singular system of second order ODE for maps $\gamma: [a, b] \to M$

Unfortunately, the expression for Y in canonical (x, p)-coordinates on TM is not very nice; it involves the inverse of the matrix H_L. However, in the (x, q)-coordinates, it is a completely different story. In these coordinates, everything takes a remarkably simple form, a fact which is the cornerstone of symplectic geometry and the calculus of variations.

The Hamiltonian

Before taking up the geometric interpretation of these new coordinates, let us do a few calculations. We have already seen that, in these coordinates, the canonical 1-form ω_L takes the simple form $\omega_L = q_i \, dx^i$. We can also express E_L as a function of (x, q). It is traditional to denote this expression by $H(x, q)$ and call it the *Hamiltonian* of the variational problem (even though, in a certain sense, it is the same function as E_L). The equation determining the vector field Y is expressed in these coordinates as

$$Y \lrcorner d\omega_L = Y \lrcorner (dq_i \wedge dx^i) = dq_i(Y)\, dx^i - dx^i(Y)\, dq_i$$
$$= -dH = -\frac{\partial H}{\partial x^i}\, dx^i - \frac{\partial H}{\partial q_i}\, dq_i,$$

so the expression for Y in these coordinates is

$$Y = \frac{\partial H}{\partial q_i}\frac{\partial}{\partial x^i} - \frac{\partial H}{\partial x^i}\frac{\partial}{\partial q_i}.$$

In particular, the flow of Y takes the form

$$\dot{x}^i = \frac{\partial H}{\partial q_i} \quad \text{and} \quad \dot{q}_i = -\frac{\partial H}{\partial x^i}.$$

These equations are known as *Hamilton's Equations* or, sometimes, as the *Hamiltonian form* of the Euler-Lagrange equations.

Part of the reason for the importance of the (x,q) coordinates is the symmetric way they treat the positions and momenta. Another reason comes from the form the infinitesimal symmetries take in these coordinates: If X is an infinitesimal symmetry of L and X' is the induced vector field on TM with conserved quantity $G = \omega_L(X')$, then, since $\mathcal{L}_{X'}\omega_L = 0$,

$$X' \lrcorner \omega_L = -d(\omega_L(X')) = -dG.$$

Thus, by the same analysis as above, the ODE represented by X' in the (x,q) coordinates becomes

$$\dot{x}^i = \frac{\partial G}{\partial q_i} \quad \text{and} \quad \dot{q}_i = -\frac{\partial G}{\partial x^i}.$$

In other words, in the (x,q) coordinates, the flow of a symmetry X' has the same Hamiltonian form as the flow of the vector field Y which gives the solutions of the Euler-Lagrange equations! This method of putting the symmetries of a Lagrangian and the solutions of the Lagrangian on a sort of equal footing will be seen to have powerful consequences.

The cotangent bundle

Canonical coordinates

Early on in this lecture, we introduced, for each coordinate chart $x: U \to \mathbb{R}^n$, a canonical extension $(x,p): TU \to \mathbb{R}^n \times \mathbb{R}^n$ and characterized it by a geometric property. There is also a canonical extension $(x,\xi): T^*U \to \mathbb{R}^n \times \mathbb{R}^n$ where $\xi = (\xi_i): T^*U \to \mathbb{R}^n$ is characterized by the condition that, if $f: U \to \mathbb{R}$ is any smooth function on U, then, regarding its exterior derivative df as a section $df: U \to T^*U$, we have

$$\xi_i \circ df = \frac{\partial f}{\partial x^i}.$$

I will leave to the reader the task of showing that (x,ξ) is indeed a coordinate system on T^*U.

It is a remarkable fact that the cotangent bundle $\pi: T^*M \to M$ of any smooth manifold carries a canonical 1-form ω defined by the following property: For each

$\alpha \in T_x^*M$, we define the linear function $\omega_\alpha: T_\alpha(T^*M) \to \mathbb{R}$ by the rule $\omega_\alpha(v) = \alpha(\pi'(\alpha)(v))$. I leave to the reader the task of showing that, in canonical coordinates $(x,\xi): T^*U \to \mathbb{R}^n \times \mathbb{R}^n$, this canonical 1-form has the expression

$$\omega = \xi_i \, dx^i.$$

The Legendre transformation

Now consider a smooth Lagrangian $L: TM \to \mathbb{R}$ as before. We can use L to construct a smooth mapping $\tau_L: TM \to T^*M$ as follows: At each $v \in TM$, the 1-form $\omega_L(v)$ is semi-basic, i.e., there exists a (necessarily unique) 1-form $\tau_L(v) \in T^*_{\pi(v)}M$ so that $\omega_L(v) = \pi^*(\tau_L(v))$. This mapping is known as the *Legendre transformation* associated to the Lagrangian L.

This definition is rather abstract, but, in local coordinates, it takes a simple form. The reader can easily check that in canonical coordinates associated to a coordinate chart $x: U \to \mathbb{R}^n$, we have

$$(x,\xi) \circ \tau_L = (x,q) = \left(x^i, \frac{\partial L}{\partial p^i}\right).$$

In other words, the (x,q) coordinates are just the canonical coordinates on the cotangent bundle composed with the Legendre transformation! It is now immediate that τ_L is a local diffeomorphism if and only if L is a non-degenerate Lagrangian. Moreover, we clearly have $\omega_L = \tau_L^*(\omega)$, so the 1-form ω_L is also expressible in terms of the canonical 1-form ω and the Legendre transform.

What about the function E_L on TM? Let us put the following condition on the Lagrangian L: Let us assume that $\tau_L: TM \to T^*M$ is a (one-to-one) diffeomorphism onto its image $\tau_L(TM) \subset T^*M$. (Note that this implies that L is non-degenerate, but is stronger than this.) Then there clearly exists a function on $\tau_L(TM)$ which pulls back to TM to be E_L. In fact, as the reader can easily verify, this is none other than the Hamiltonian function H constructed above.

The fact that the Hamiltonian H naturally lives on T^*M (or at least an open subset thereof) rather than on TM justifies it being regarded as distinct from the function E_L. There is another reason for moving over to the cotangent bundle when one can: The vector field Y on TM corresponds under the Legendre transformation to a vector field Z on $\tau_L(TM)$ which is characterized by the simple rule $Z \lrcorner \omega = -dH$. Thus, just knowing the Hamiltonian H on an open set in T^*M determines the vector field which sweeps out the solution curves! We will see that this is a very useful observation in what follows.

Poincaré recurrence

To conclude this lecture, I want to give an application of the geometry of the form ω_L to understanding the global behavior of the L-critical curves when L is a non-degenerate Lagrangian. First, I make the following observation:

LECTURE 4. SYMMETRIES AND CONSERVATION LAWS

Proposition 6. *Let $L: TM \to \mathbb{R}$ be a non-degenerate Lagrangian. Then the $2n$-form $\mu_L = (d\omega_L)^n$ is a volume form on TM (i.e., it is nowhere vanishing). Moreover the (local) flow of the vector field Y preserves this volume form.*

Proof. To see that μ_L is a volume form, just look in local (x,q)-coordinates:

$$\mu_L = (d\omega_L)^n = (dq_i \wedge dx^i)^n$$
$$= n!\, dq_1 \wedge dx^1 \wedge dq_2 \wedge dx^2 \wedge \cdots \wedge dq_n \wedge dx^n.$$

By Proposition 4, this latter form is not zero.

Finally, since

$$\mathcal{L}_Y(d\omega_L) = d(Y \lrcorner\, d\omega_L) = -d(dE_L) = 0,$$

it follows that the (local) flow of Y preserves $d\omega_L$ and hence preserves μ_L.

This volume form μ_L can be used to define a measure on TM which is known as the *Liouville measure*. Now we shall give an application of Proposition 6. This is the famous Poincaré Recurrence Theorem.

Theorem 2. *Let $L: TM \to \mathbb{R}$ be a non-degenerate Lagrangian and suppose that E_L is a proper function on TM. Then the vector field Y is complete, with flow $\Phi: \mathbb{R} \times TM \to TM$. Moreover, this flow is recurrent in the following sense: For any point $v \in TM$, any open neighborhood U of v, and any positive time interval $T > 0$, there exists an integer $m > 0$ so that $\Phi(mT, U) \cap U \neq \emptyset$.*

Proof. The completeness of the flow of Y follows immediately from the fact that the integral curve of Y which passes through $v \in TM$ must stay in the compact set $E_L^{-1}(E_L(v))$. (Recall that E_L is constant on all of the integral curves of Y.) Details are left to the reader.

I now turn to the proof of the recurrence property. Let $E_0 = E_L(v)$. By hypothesis, the set $C = E_L^{-1}([E_0 - 1, E_0 + 1])$ is compact, so the μ_L-volume of the open set $W = E_L^{-1}((E_0 - 1, E_0 + 1))$ (which lies inside C) is finite. It clearly suffices to prove the recurrence property for any open neighborhood U of v which lies inside W, so let us assume that $U \subset W$.

Let $\phi: W \to W$ be the diffeomorphism $\phi(w) = \Phi(T, w)$. This diffeomorphism is clearly invertible and preserves the μ_L-volume of open sets in W. Consider the open sets $U^k = \phi^k(U)$ for $k > 0$ (integers). These open sets all have the same μ_L-volume and hence cannot be all disjoint since then their union (which lies in W) would have infinite μ_L-volume. Let $0 < j < k$ be two integers so that $U^j \cap U^k \neq \emptyset$. Then, since

$$U^j \cap U^k = \phi^j(U) \cap \phi^k(U) = \phi^j(U \cap \phi^{k-j}(U)),$$

it follows that $U \cap \phi^{k-j}(U) \neq \emptyset$, as we wished to show.

This theorem has the amazing consequence that, whenever one has a non-degenerate Lagrangian with a proper energy function, the corresponding mechanical system "recurs" in the sense that "arbitrarily near any given initial condition, there is another initial condition so that the evolution brings this initial condition back arbitrarily close to the first initial condition". I realize that this statement is somewhat vague and subject to misinterpretation, but the precise statement has already been given, so there seems not to be much harm in giving the paraphrase.

Exercises

1. Show that two Lagrangians $L_1, L_2 \colon TM \to \mathbb{R}$ satisfy

$$E_{L_1} = E_{L_2} \quad \text{and} \quad d\omega_{L_1} = d\omega_{L_2}$$

if and only if there is a smooth function f on M so that $L_1 = L_2 + df$. Such Lagrangians are said to differ by a "divergence term." Show that such Lagrangians share the same critical curves and that one is non-degenerate if and only if the other is.

2. What does Conservation of Energy mean for the case where L defines a Riemannian metric on M?

3. Show that the equations for geodesics of a rotationally invariant metric of the form

$$I = E(r)\, dr^2 + 2\, F(r)\, dr\, d\theta + G(r)\, d\theta^2$$

can be integrated by separation of variables and quadratures. (Hint: Start with the conservation laws we already know:

$$E(r)\,\dot{r}^2 + 2\,F(r)\,\dot{r}\,\dot{\theta} + G(r)\,\dot{\theta}^2 = v_0^2$$

$$F(r)\,\dot{r} + G(r)\,\dot{\theta} = u_0$$

where v_0 and u_0 are constants. Then eliminate $\dot{\theta}$ and go on from there.)

4. The definition of ω_L given in the text might be regarded as somewhat unsatisfactory since it is given in coordinates and not "invariantly". Show that the following invariant description of ω_L is valid: The manifold TM inherits some extra structure by virtue of being the tangent bundle of another manifold M. Let $\pi \colon TM \to M$ be the basepoint projection. Then π is a submersion: For every $a \in TM$,

$$\pi'(a) \colon T_a TM \to T_{\pi(a)} M$$

is a surjection and the fiber at $\pi(a)$ is equal to

$$\pi^{-1}(\pi(a)) = T_{\pi(a)} M.$$

It follows that the kernel of $\pi'(a)$ (i.e., the "vertical space" of the bundle $\pi \colon TM \to M$ at a) is naturally isomorphic to $T_{\pi(a)}M$. Call this isomorphism $\alpha \colon T_{\pi(a)} M \xrightarrow{\sim} \ker(\pi'(a))$. Then the 1-form ω_L is defined by

$$\omega_L(v) = dL(\alpha \circ \pi'(v)) \quad \text{for } v \in T(TM).$$

Hint: Show that, in local canonical coordinates, the map $\alpha \circ \pi'$ satisfies

$$\alpha \circ \pi'\left(a^i \frac{\partial}{\partial x^i} + b^i \frac{\partial}{\partial p^i}\right) = a^i \frac{\partial}{\partial p^i}.$$

5. For any vector field X on M, let the associated vector field on TM be denoted X'. Show that if X has the form

$$X = a^i \frac{\partial}{\partial x^i}$$

in some local coordinate system, then, in the associated canonical (x,p) coordinates, X' has the form

$$X' = a^i \frac{\partial}{\partial x^i} + p^j \frac{\partial a^i}{\partial x^j} \frac{\partial}{\partial p^i}.$$

6. Show that conservation of angular momenta in the motion of a point mass in a central force field implies Kepler's Law that "equal areas are swept out over equal time intervals." Show also that, in the $n=2$ case, employing the conservation of energy and angular momentum allows one to integrate the equations of motion by quadratures. (Hint: For the second part of the problem, introduce polar coordinates: $(x^1, x^2) = (r\cos\theta, r\sin\theta)$.)

7. In the example of the motion of a rigid body, show that the Lagrangian on G is always non-negative and is non-degenerate (so that L defines a left-invariant metric on G) if and only if the matrix μ has at most one zero eigenvalue. Show that L is degenerate if and only if the rigid body lies in a subspace of dimension at most $n-2$.

8. Supply the details in the proof of Proposition 5. You will want to go back to the integration-by-parts derivation of the Euler-Lagrange equations and show that, even if the variation Γ induced by h does not have fixed endpoints, we still get a local coordinate formula of the form

$$\mathcal{F}'_{L,\Gamma}(0) = \frac{\partial L}{\partial p^k}(y(b), \dot{y}(b)) h^k(b) - \frac{\partial L}{\partial p^k}(y(a), \dot{y}(a)) h^k(a)$$

for any variation of a solution of the Euler-Lagrange equations. Give these "boundary terms" an invariant geometric meaning and show that they cancel out when we sum over a partition of a (fixed-endpoint) variation of an L-critical curve γ into subcurves which lie in coordinate neighborhoods.)

9. (Alternate to Exercise 8.) Here is another approach to proving Proposition 5. Instead of dividing the curve up into sub-curves, show that for any variation Γ of a curve $\gamma\colon [a,b] \to M$ (not necessarily with fixed endpoints), we have the formula

$$\mathcal{F}'_{L,\Gamma}(0) = \omega_L(V(b)) - \omega_L(V(a)) - \int_a^b d\omega_L(\dot{\gamma}(t), V(t)) + dE_L(V(t))\, dt$$

where $V(t) = \dot{\Gamma}'(t,0)(\partial/\partial s)$ is the "variation vector field" at $s=0$ of the lifted variation $\dot{\Gamma}$ in TM. Conclude that, whether L is non-degenerate or not, the condition $\ddot{\gamma}\lrcorner d\omega_L + dE_L(\dot{\gamma}(t)) = 0$ is the necessary and sufficient condition that γ be L-critical.

LECTURE 4. SYMMETRIES AND CONSERVATION LAWS

10. THE TWO BODY PROBLEM. Consider a pair of point masses (with masses m_1 and m_2) which move freely subject to a force between them which depends only on the distance between the two bodies and is directed along the line joining the two bodies. This is what is classically known as the Two Body Problem. It is represented by a Lagrangian on the manifold $M = \mathbb{R}^n \times \mathbb{R}^n$ with position coordinates $x_1, x_2 \colon M \to \mathbb{R}^n$ of the form

$$L(x_1, x_2, p_1, p_2) = \frac{m_1}{2}|p_1|^2 + \frac{m_2}{2}|p_2|^2 - V(|x_1 - x_2|^2).$$

(Here, (p_1, p_2) are the canonical fiber (velocity) coordinates on TM associated to the coordinate system (x_1, x_2).) Notice that L has the form "kinetic minus potential". Show that rotations and translations in $\mathbb{R}^n$ generate a group of symmetries of this Lagrangian and compute the conserved quantities. What is the interpretation of the conservation law associated to the translations?

11. THE SLIDING PARTICLE. Suppose that a particle of unit weight and mass slides without friction on a smooth hypersurface $x^{n+1} = F(x^1, \ldots, x^n)$ subject only to the force of gravity (which is directed downward along the x^{n+1}-axis). Show that the "kinetic-minus-potential" Lagrangian for this motion in the x-coordinates is

$$L = \tfrac{1}{2}\left((p^1)^2 + \cdots + (p^n)^2 + \left(\frac{\partial F}{\partial x^i}p^i\right)^2\right) - F(x^1, \ldots, x^n).$$

Show that this is a non-degenerate Lagrangian and that its energy E_L is proper if and only if $F^{-1}((-\infty, a])$ is compact for all $a \in \mathbb{R}$.

Suppose that F is invariant under rotation, i.e., that

$$F(x^1, \ldots, x^n) = f\big((x^1)^2 + \cdots + (x^n)^2\big)$$

for some smooth function f. Show that the "shadow" of the particle in $\mathbb{R}^n$ stays in a fixed 2-plane. Show that the equations of motion can be integrated by quadrature.

Remark: This Lagrangian is also used to model a small ball of unit mass and weight "rolling without friction in a cup". Of course, in this formulation, the kinetic energy stored in the ball by its spinning is ignored. If you want to take this "spinning" energy into account, then you must study quite a different Lagrangian, especially if you assume that the ball rolls without slipping. This goes into the very interesting theory of "non-holonomic systems", into which we (unfortunately) do not have time to go.

12. Let L be a Lagrangian which restricts to each fiber T_xM to be a non-degenerate (though not necessarily positive definite) quadratic form. Show that L is non-degenerate as a Lagrangian and that the Legendre mapping $\tau_L \colon TM \to T^*M$ is an isomorphism of vector bundles. Show that, if L is, in addition, a positive definite quadratic form on each fiber, then the new Lagrangian defined by

$$\tilde{L} = (L+1)^{\frac{1}{2}}$$

is also a non-degenerate Lagrangian, but that the map $\tau_{\tilde{L}}:TM \to T^*M$, though one-to-one, is not onto.

LECTURE 5
Symplectic Manifolds, I

In Lecture 4, I associated a non-degenerate 2-form $d\omega_L$ on TM to every non-degenerate Lagrangian $L: TM \to \mathbb{R}$. In this section, I want to begin a more systematic study of the geometry of manifolds on which there is specified a closed, non-degenerate 2-form.

Symplectic Algebra

First, I will develop the algebraic precursors of the manifold concepts which are to follow. For simplicity, all of these constructions will be carried out on vector spaces over the reals, but they could equally well have been carried out over any field of characteristic not equal to 2.

Symplectic vector spaces

A bilinear pairing $B: V \times V \to \mathbb{R}$ is said to be *skew-symmetric* (or *alternating*) if $B(x, y) = -B(y, x)$ for all x and y in V. The space of skew-symmetric bilinear pairings on V will be denoted by $A^2(V)$. The set $A^2(V)$ is a vector space under the obvious addition and scalar multiplication and is naturally identified with $\Lambda^2(V^*)$, the space of exterior 2-forms on V. The elements of $A^2(V)$ are often called skew-symmetric bilinear *forms* on V. A pairing $B \in A^2(V)$ is said to be *non-degenerate* if, for every non-zero $v \in V$, there is a $w \in V$ for which $B(v, w) \neq 0$.

Definition 1. A *symplectic space* is a pair (V, B) where V is a vector space and B is a non-degenerate, skew-symmetric, bilinear pairing on V.

Example. Let $V = \mathbb{R}^{2n}$ and let J_n be the $2n$-by-$2n$ matrix

$$J_n = \begin{pmatrix} 0_n & I_n \\ -I_n & 0_n \end{pmatrix}.$$

For vectors $v, w \in \mathbb{R}^{2n}$, define

$$B_0(x, y) = {}^t x \, J_n \, y.$$

Then it is clear that B_0 is bilinear and skew-symmetric. Moreover, in components

$$B_0(x,y) = x^1 y^{n+1} + \cdots + x^n y^{2n} - x^{n+1} y^1 - \cdots - x^{2n} y^n$$

so it is clear that if $B_0(x,y) = 0$ for all $y \in \mathbb{R}^{2n}$ then $x = 0$. Hence, B_0 is non-degenerate.

Generally, in order for $B(x,y) = {}^t x \, A \, y$ to define a skew-symmetric bilinear form on $\mathbb{R}^n$, it is only necessary that A be a skew-symmetric n-by-n matrix. Conversely, every skew-symmetric bilinear form B on $\mathbb{R}^n$ can be written in this form for some unique skew-symmetric n-by-n matrix A. In order that this B be non-degenerate, it is necessary and sufficient that A be invertible. (See the Exercises.)

The symplectic group

Now, a linear transformation $R \colon \mathbb{R}^{2n} \to \mathbb{R}^{2n}$ preserves B_0, i.e., satisfies $B_0(Rx, Ry) = B_0(x,y)$ for all $x, y \in \mathbb{R}^{2n}$, if and only if ${}^t R \, J_n \, R = J_n$. This motivates the following definition:

Definition. The subgroup of $\mathrm{GL}(2n, \mathbb{R})$ defined by

$$\mathrm{Sp}(n, \mathbb{R}) = \{ R \in \mathrm{GL}(2n, \mathbb{R}) \mid {}^t R \, J_n \, R = J_n \}$$

is called the *symplectic group of rank n*.

It is clear that $\mathrm{Sp}(n, \mathbb{R})$ is a (closed) subgroup of $\mathrm{GL}(2n, \mathbb{R})$. In the Exercises, you are asked to prove that $\mathrm{Sp}(n, \mathbb{R})$ is a Lie group of dimension $2n^2 + n$ and to derive other of its properties.

Symplectic normal form

The following proposition shows that there is a normal form for finite dimensional symplectic spaces.

Proposition 1. *If (V, B) is a finite dimensional symplectic space, then there exists a basis $e_1, \ldots, e_n, f^1, \ldots, f^n$ of V so that, for all $1 \leq i, j \leq n$,*

$$B(e_i, e_j) = 0, \qquad B(e_i, f^j) = \delta_i^j, \quad \text{and} \quad B(f^i, f^j) = 0.$$

Proof. The desired basis will be constructed in two steps. Let $m = \dim(V)$.

Suppose that for some $n \geq 0$, we have found a sequence of linearly independent vectors $e_1, \ldots, e_n$ so that $B(e_i, e_j) = 0$ for all $1 \leq i, j \leq n$. Consider the vector space $W_n \subset V$ which consists of all vectors $w \in V$ so that $B(e_i, w) = 0$ for all $1 \leq i \leq n$. Since the e_i are linearly independent and since B is non-degenerate, it follows that W_n has dimension $m - n$. We must have $n \leq m - n$ since all of the vectors $e_1, \ldots, e_n$ clearly lie in W_n.

If $n < m - n$, then there exists a vector $e_{n+1} \in W_n$ which is linearly independent from $e_1, \ldots, e_n$. It follows that the sequence $e_1, \ldots, e_{n+1}$ satisfies $B(e_i, e_j) = 0$ for all $1 \leq i, j \leq n+1$. (Since B is skew-symmetric, $B(e_{n+1}, e_{n+1}) = 0$ is automatic.) This extension process can be repeated until we reach a stage where $n = m - n$, i.e.,

LECTURE 5. SYMPLECTIC MANIFOLDS, I

$m = 2n$. At that point, we will have a sequence $e_1, \ldots, e_n$ for which $B(e_i, e_j) = 0$ for all $1 \leq i, j \leq n$.

Next, we construct the sequence $f^1, \ldots, f^n$. For each j in the range $1 \leq j \leq n$, consider the set of n linear equations

$$B(e_i, w) = \delta_i^j, \qquad 1 \leq i \leq n.$$

We know that these n equations are linearly independent, so there exists a solution f_0^j. Of course, once one particular solution is found, any other solution is of the form $f^j = f_0^j + a^{ji} e_i$ for some n^2 numbers a^{ji}. Thus, we have found the general solutions f^j to the equations $B(e_i, f^j) = \delta_i^j$.

We now show that we can choose the a^{ij} so as to satisfy the last remaining set of conditions, $B(f^i, f^j) = 0$. If we set $b^{ij} = B(f_0^i, f_0^j) = -b^{ji}$, then we can compute

$$B(f^i, f^j) = B(f_0^i, f_0^j) + B(a^{ik} e_k, f_0^j) + B(f_0^i, a^{jl} e_l) + B(a^{ik} e_k, a^{jl} e_l)$$
$$= b^{ij} + a^{ij} - a^{ji} + 0.$$

Thus, it suffices to set $a^{ij} = -b^{ij}/2$. (This is where the hypothesis that the characteristic of $\mathbb{R}$ is not 2 is used.)

Finally, it remains to show that the vectors $e_1, \ldots, e_n, f^1 \ldots, f^n$ form a basis of V. Since we already know that $\dim(V) = 2n$, it is enough to show that these vectors are linearly independent. However, any linear relation of the form

$$a^i e_i + b_j f^j = 0,$$

implies $b_k = B(e_k, a^i e_i + b_j f^j) = 0$ and $a^k = -B(f^k, a^i e_i + b_j f^j) = 0$.

We often say that a basis of the form found in Proposition 1 is a *symplectic* or *standard* basis of the symplectic space (V, B).

Symplectic reduction of vector spaces

If $B: V \times V \to \mathbb{R}$ is a skew-symmetric bilinear form which is not necessarily non-degenerate, then we define the *null space* of B to be the subspace

$$N_B = \{v \in V \mid B(v, w) = 0 \text{ for all } w \in V\}.$$

On the quotient vector space $\overline{V} = V/N_B$, there is a well-defined skew-symmetric bilinear form $\overline{B}: \overline{V} \times \overline{V} \to \mathbb{R}$ given by

$$\overline{B}(\overline{x}, \overline{y}) = B(x, y)$$

where $\overline{x}$ and $\overline{y}$ are the cosets in $\overline{V}$ of x and y in V. It is easy to see that $(\overline{V}, \overline{B})$ is a symplectic space.

Definition 2. If B is a skew-symmetric bilinear form on a vector space V, then the symplectic space $(\overline{V}, \overline{B})$ is called the *symplectic reduction* of (V, B).

Here is an application of the symplectic reduction idea: Using the identification of $A^2(V)$ with $\Lambda^2(V^*)$ mentioned earlier, Proposition 1 allows us to write down a normal form for any alternating 2-form on any finite dimensional vector space.

Proposition 2. *For any non-zero $\beta \in \Lambda^2(V^*)$, there exist an integer $n \leq \frac{1}{2} \dim(V)$ and linearly independent 1-forms $\omega^1, \omega^2, \ldots, \omega^{2n} \in V^*$ for which*

$$\beta = \omega^1 \wedge \omega^2 + \omega^3 \wedge \omega^4 \ldots + \omega^{2n-1} \wedge \omega^{2n}.$$

Thus, n is the largest integer so that $\beta^n \neq 0$.

Proof. Regard β as a skew-symmetric bilinear form B on V in the usual way. Let $(\overline{V}, \overline{B})$ be the symplectic reduction of (V, B). Since $B \neq 0$, we known that $\overline{V} \neq \{0\}$. Let $\dim(\overline{V}) = 2n \geq 2$ and let $e_1, \ldots, e_n, f^1 \ldots, f^n$ be elements of V so that $\overline{e}_1, \ldots, \overline{e}_n, \overline{f}^1 \ldots, \overline{f}^n$ forms a symplectic basis of $\overline{V}$ with respect to $\overline{B}$. Let $p = \dim(V) - 2n$, and let $b_1, \ldots, b_p$ be a basis of N_B.

It is easy to see that

$$\mathbf{b} = \begin{pmatrix} e_1 & f^1 & e_2 & f^2 & \cdots & e_n & f^n & b_1 & \cdots & b_p \end{pmatrix}$$

forms a basis of V. Let

$$\omega^1 \quad \cdots \quad \omega^{2n+p}$$

denote the dual basis of V^*. Then, as the reader can easily check, the 2-form

$$\Omega = \omega^1 \wedge \omega^2 + \omega^3 \wedge \omega^4 \ldots + \omega^{2n-1} \wedge \omega^{2n}$$

has the same values as β does on all pairs of elements of $\mathbf{b}$. Of course this implies that $\beta = \Omega$. The rest of the Proposition also follows easily since, for example, we have

$$\beta^n = n! \, \omega^1 \wedge \cdots \wedge \omega^{2n} \neq 0,$$

although β^{n+1} clearly vanishes.

If we regard β as an element of $A^2(V)$, then n is one-half the dimension of $\overline{V}$. Some sources call the integer n the half-rank of β and others call n the rank. I use 'half-rank'.

Note that, unlike the case of *symmetric* bilinear forms, there is no notion of signature type or "positive definiteness" for skew-symmetric forms.

It follows from Proposition 2 that for β in $A^2(V)$, where V is finite dimensional, the pair (V, β) is a symplectic space if and only if V has dimension $2n$ for some n and $\beta^n \neq 0$.

Subspaces of symplectic vector spaces

Let Ω be a symplectic form on a vector space V. For any subspace $W \subset V$, we define the Ω-complement to W to be the subspace

$$W^\perp = \{v \in V \,|\, \Omega(v, w) = 0 \text{ for all } w \in W\}.$$

The Ω-complement of a subspace W is sometimes called its *skew-complement*. It is an exercise for the reader to check that, because Ω is non-degenerate, $(W^\perp)^\perp = W$ and that, when V is finite-dimensional,

$$\dim W + \dim W^\perp = \dim V.$$

However, unlike the case of an orthogonal with respect to a positive definite inner product, the intersection $W \cap W^\perp$ does not have to be the zero subspace. For example, in an Ω-standard basis for V, the vectors $e_1, \ldots, e_n$ obviously span a subspace L which satisfies $L^\perp = L$.

If V is finite dimensional, it turns out (see the Exercises) that, up to symplectic linear transformations of V, a subspace $W \subset V$ is characterized by the numbers $d = \dim W$ and $\nu = \dim(W \cap W^\perp) \leq d$. If $\nu = 0$ we say that W is a *symplectic subspace* of V. This corresponds to the case that Ω restricts to W to define a symplectic structure on W. At the other extreme is when $\nu = d$, for then we have $W \cap W^\perp = W$. Such a subspace is called *Lagrangian*.

Symplectic Manifolds

We are now ready to return to the study of manifolds.

Definition 3. A *symplectic structure* on a smooth manifold M is a non-degenerate, closed 2-form $\Omega \in \mathcal{A}^2(M)$. The pair (M, Ω) is called a *symplectic manifold*. If Ω is a symplectic structure on M and Υ is a symplectic structure on N, then a smooth map $\phi \colon M \to N$ satisfying $\phi^*(\Upsilon) = \Omega$ is called a *symplectic map*. If, in addition, ϕ is a diffeomorphism, we say that ϕ is a *symplectomorphism*.

Examples

Before developing any of the theory, it is helpful to see a few examples.

Surfaces with area forms. If S is an orientable smooth surface, then there exists a volume form μ on S. By definition, μ is a non-degenerate closed 2-form on S and hence defines a symplectic structure on S.

Lagrangian structures on TM. From Lecture 4, any non-degenerate Lagrangian $L \colon TM \to \mathbb{R}$ defines the 2-form $d\omega_L$, which is a symplectic structure on TM.

The "standard" structure on $\mathbb{R}^{2n}$. Think of $\mathbb{R}^{2n}$ as a smooth manifold and let Ω be the 2-form with constant coefficients

$$\Omega = \tfrac{1}{2}{}^t dx \, J_n \, dx = dx^1 \wedge dx^{n+1} + \cdots + dx^n \wedge dx^{2n}.$$

Symplectic submanifolds. Let (M^{2m}, Ω) be a symplectic manifold. Suppose that $P^{2p} \subset M^{2m}$ be any submanifold to which the form Ω pulls back to be a non-degenerate 2-form Ω_P. Then (P, Ω_P) is a symplectic manifold. We say that P is a *symplectic submanifold* of M.

It is not obvious just how to find symplectic submanifolds of M. Even though being a symplectic submanifold is an "open" condition on submanifolds of M, is is not "dense". One cannot hope to perturb an arbitrary even dimensional submanifold of M slightly so as to make it symplectic. There are even restrictions on the topology of the submanifolds of M on which a symplectic form restricts to be non-degenerate.

For example, no symplectic submanifold of $\mathbb{R}^{2n}$ (with any symplectic structure on $\mathbb{R}^{2n}$) could be compact for the following simple reason: Since $\mathbb{R}^{2n}$ is contractible, its second deRham cohomology group vanishes. In particular, for any symplectic form Ω on $\mathbb{R}^{2n}$, there must be a 1-form ω so that $\Omega = d\omega$ which implies that $\Omega^m = d(\omega \wedge \Omega^{m-1})$. Thus, for all $m > 0$, the $2m$-form Ω^m is exact on $\mathbb{R}^{2n}$ (and every submanifold of $\mathbb{R}^{2n}$). By Proposition 2, if M^{2m} were a submanifold of $\mathbb{R}^{2n}$ on which Ω restricted to be non-degenerate, then Ω^m would be a volume form on M. However, on a compact manifold the volume form is never exact (just apply Stokes' Theorem).

Complex submanifolds. Nevertheless, there are many symplectic submanifolds of $\mathbb{R}^{2n}$. One way to construct them is to regard $\mathbb{R}^{2n}$ as $\mathbb{C}^n$ in such a way that the linear map $J: \mathbb{R}^{2n} \to \mathbb{R}^{2n}$ represented by J_n becomes complex multiplication. (For example, just define the complex coordinates by $z^k = x^k + ix^{k+n}$.) Then, for any non-zero vector $v \in \mathbb{R}^{2n}$, we have $\Omega(v, Jv) = -|v|^2 \neq 0$. In particular, Ω is non-degenerate on every complex subspace $S \subset \mathbb{C}^n$. Thus, if $M^{2m} \subset \mathbb{C}^n$ is any complex submanifold (i.e., all of its tangent spaces are m-dimensional complex subspaces of $\mathbb{C}^m$), then Ω restricts to be non-degenerate on M.

The cotangent bundle. Let M be any smooth manifold and let T^*M be its cotangent bundle. As we saw in Lecture 4, there is a canonical 2-form on T^*M which can be defined as follows: Let $\pi: T^*M \to M$ be the basepoint projection. Then, for every $v \in T_\alpha(T^*M)$, define

$$\omega(v) = \alpha(\pi'(v)).$$

I claim that ω is a smooth 1-form on T^*M and that $\Omega = d\omega$ is a symplectic form on T^*M.

To see this, let us compute ω in local coordinates. Let $x: U \to \mathbb{R}^n$ be a local coordinate chart. Since the 1-forms $dx^1, \ldots, dx^n$ are linearly independent at every point of U, it follows that there are unique functions ξ_i on T^*U so that, for $\alpha \in T_a^*U$,

$$\alpha = \xi_1(\alpha) \, dx^1|_a + \cdots + \xi_n(\alpha) \, dx^n|_a.$$

The functions $x^1,\ldots,x^n,\xi_1,\ldots,\xi_n$ then form a smooth coordinate system on T^*U in which the projection mapping π is given by

$$\pi(x,p) = x.$$

It is then straightforward to compute that, in this coordinate system,

$$\omega = \xi_i\,dx^i.$$

Hence, $\Omega = d\xi_i \wedge dx^i$ and so is non-degenerate.

Symplectic products. If (M,Ω) and (N,Υ) are symplectic manifolds, then $M \times N$ carries a natural symplectic structure, called the product symplectic structure $\Omega \oplus \Upsilon$, defined by

$$\Omega \oplus \Upsilon = \pi_1^*(\Omega) + \pi_2^*(\Upsilon).$$

Thus, for example, n-fold products of compact surfaces endowed with area forms give examples of compact symplectic $2n$-manifolds.

Coadjoint orbits. Let $\mathrm{Ad}^*\colon G \to \mathrm{GL}(\mathfrak{g}^*)$ denote the coadjoint representation of G. This is the so-called "contragredient" representation to the adjoint representation. Thus, for any $a \in G$ and $\xi \in \mathfrak{g}^*$, the element $\mathrm{Ad}^*(a)(\xi) \in \mathfrak{g}^*$ is determined by the rule

$$\mathrm{Ad}^*(a)(\xi)(x) = \xi\bigl(\mathrm{Ad}(a^{-1})(x)\bigr) \qquad \text{for all } x \in \mathfrak{g}.$$

One must be careful not to confuse $\mathrm{Ad}^*(a)$ with $\bigl(\mathrm{Ad}(a)\bigr)^*$. Instead, as our definition shows, $\mathrm{Ad}^*(a) = \bigl(\mathrm{Ad}(a^{-1})\bigr)^*$.

Note that the induced homomorphism of Lie algebras, $\mathrm{ad}^*\colon \mathfrak{g} \to \mathfrak{gl}(\mathfrak{g}^*)$ is given by

$$\mathrm{ad}^*(x)(\xi)(y) = -\xi\bigl([x,y]\bigr)$$

The orbits $G \cdot \xi$ in $\mathfrak{g}^*$ are called the *coadjoint orbits*. Each of them carries a natural symplectic structure. To see how this is defined, let $\xi \in \mathfrak{g}^*$ be fixed, and let $\phi\colon G \to G \cdot \xi$ be the usual submersion induced by the Ad^*-action, $\phi(a) = \mathrm{Ad}^*(a)(\xi) = a \cdot \xi$. Now let ω_ξ be the left-invariant 1-form on G whose value at e is ξ. Thus, $\omega_\xi = \xi(\omega)$ where ω is the canonical $\mathfrak{g}$-valued 1-form on G.

Proposition 3. *There is a unique symplectic form Ω_ξ on the orbit $G \cdot \xi = G/G_\xi$ satisfying $\phi^*(\Omega_\xi) = d\omega_\xi$.*

Proof. If Proposition 3 is to be true, then Ω_ξ must satisfy the rule

$$\Omega_\xi\bigl(\phi'(v),\phi'(w)\bigr) = d\omega_\xi(v,w) \quad \text{for all } v,w \in T_aG.$$

What we must do is show that this rule actually does define a symplectic 2-form on $G \cdot \xi$.

First, note that, for $x, y \in \mathfrak{g} = T_e G$, we may compute via the structure equations that
$$d\omega_\xi(x, y) = \xi\big(d\omega(x,y)\big) = \xi(-[x,y]) = ad^*(x)(\xi)(y).$$

In particular, $ad^*(x)(\xi) = 0$, if and only if x lies in the null space of the 2-form $d\omega_\xi(e)$. In other words, the null space of $d\omega_\xi(e)$ is $\mathfrak{g}_\xi$, the Lie algebra of G_ξ. Since $d\omega_\xi$ is left-invariant, it follows that the null space of $d\omega_\xi(a)$ is $L'_a(\mathfrak{g}_\xi) \subset T_a G$. Of course, this is precisely the tangent space at a to the left coset aG_ξ. Thus, for each $a \in G$,
$$N_{d\omega_\xi(a)} = \ker \phi'(a),$$

It follows that, $T_{a\cdot\xi}(G \cdot \xi) = \phi'(a)(T_a G)$ is naturally isomorphic to the symplectic quotient space $(T_a G)/(L'_a(\mathfrak{g}_\xi))$ for each $a \in G$. Thus, there is a unique, non-degenerate 2-form Ω_a on $T_{a\cdot\xi}(G \cdot \xi)$ so that $\big(\phi'(a)\big)^*(\Omega_a) = d\omega_\xi(a)$.

It remains to show that $\Omega_a = \Omega_b$ if $a \cdot \xi = b \cdot \xi$. However, this latter case occurs only if $a = bh$ where $h \in G_\xi$. Now, for any $h \in G_\xi$, we have
$$R_h^*(\omega_\xi) = \xi\big(R_h^*(\omega)\big) = \xi\big(\mathrm{Ad}(h^{-1})(\omega)\big) = \mathrm{Ad}^*(h)(\xi)(\omega) = \xi(\omega) = \omega_\xi.$$

Thus, $R_h^*(d\omega_\xi) = d\omega_\xi$. Since the following square commutes, it follows that $\Omega_a = \Omega_b$.

$$\begin{array}{ccc} T_a G & \xrightarrow{R'_h} & T_b G \\ \phi'(a) \downarrow & & \downarrow \phi'(b) \\ T_{a\cdot\xi}(G\cdot\xi) & \xrightarrow{id} & T_{b\cdot\xi}(G\cdot\xi) \end{array}$$

All this shows that there is a well-defined, non-degenerate 2-form Ω_ξ on $G \cdot \xi$ which satisfies $\phi^*(\Omega_\xi) = d\omega_\xi$. Since ϕ is a smooth submersion, the equation
$$\phi^*(d\Omega_\xi) = d(d\omega_\xi) = 0$$
implies that $d\Omega_\xi = 0$, as promised.

Note that a consequence of Proposition 3 is that all of the coadjoint orbits are actually even dimensional. As we shall see when we take up the subject of reduction, the coadjoint orbits are particularly interesting symplectic manifolds.

Examples of coadjoint orbits. Let $G = O(n)$, with Lie algebra $\mathfrak{so}(n)$, the space of skew-symmetric n-by-n matrices. Now there is an $O(n)$-equivariant positive definite pairing of $\mathfrak{so}(n)$ with itself $\langle,\rangle$ given by
$$\langle x, y \rangle = -\mathrm{tr}(xy).$$

Thus, we can identify $\mathfrak{so}(n)^*$ with $\mathfrak{so}(n)$ by this pairing. The reader can check that, in this case, the coadjoint action is isomorphic to the adjoint action
$$\mathrm{Ad}(a)(x) = axa^{-1}.$$

If ξ is the rank 2 matrix
$$\xi = \begin{pmatrix} 0 & -1 & 0 \\ 1 & 0 & \\ 0 & & 0 \end{pmatrix},$$
then it is easy to check that the stabilizer G_ξ is just the set of matrices of the form
$$\begin{pmatrix} a & 0 \\ 0 & A \end{pmatrix}$$
where $a \in \mathrm{SO}(2)$ and $A \in \mathrm{O}(n-2)$. The quotient $\mathrm{O}(n)/(\mathrm{SO}(2) \times \mathrm{O}(n-2))$ thus has a symplectic structure. It is not difficult to see that this homogeneous space can be identified with the space of oriented 2-planes in $\mathbb{E}^n$.

As another example, if $n = 2m$, then J_m lies in $\mathfrak{so}(2m)$, and its stabilizer is $\mathrm{U}(m) \subset \mathrm{SO}(2m)$. It follows that the quotient space $\mathrm{SO}(2m)/\mathrm{U}(m)$, which is identifiable as the set of orthogonal complex structures on $\mathbb{E}^{2m}$, is a symplectic space.

Finally, if $G = \mathrm{U}(n)$, then, again, we can identify $\mathfrak{u}(n)^*$ with $\mathfrak{u}(n)$ via the $\mathrm{U}(n)$-invariant, positive definite pairing
$$\langle x, y \rangle = -\mathrm{Re}\bigl(\mathrm{tr}(xy)\bigr).$$
Again, under this identification, the coadjoint action agrees with the adjoint action. For $0 < p < n$, the stabilizer of the element
$$\xi_p = \begin{pmatrix} iI_p & 0 \\ 0 & -iI_{n-p} \end{pmatrix}$$
is easily seen to be $\mathrm{U}(p) \times \mathrm{U}(n-p)$. The orbit of ξ_p is identifiable with the space $\mathrm{Gr}_p(\mathbb{C}^n)$, i.e., the Grassmannian of (complex) p-planes in $\mathbb{C}^n$, and, by Proposition 3, carries a canonical, $\mathrm{U}(n)$-invariant symplectic structure.

Darboux' theorem

There is a manifold analogue of Proposition 1 which says that symplectic manifolds of a given dimension are all locally "isomorphic". This fundamental result is known as *Darboux' Theorem*. I will give the classical proof (due to Darboux) here, deferring the more modern proof (due to Weinstein) to the next lecture.

Theorem 1 (Darboux' Theorem). *If Ω is a closed 2-form on a manifold M^{2n} which satisfies the condition that Ω^n be nowhere vanishing, then for every $p \in M$, there is a neighborhood U of p and a coordinate system $x_1, x_2, \ldots, x_n, y^1, y^2, \ldots, y^n$ on U so that*
$$\Omega|_U = dx_1 \wedge dy^1 + dx_2 \wedge dy^2 + \cdots + dx_n \wedge dy^n.$$

Proof. We will proceed by induction on n. Assume that we know the theorem for $n-1 \geq 0$. We will prove it for n. Fix p, and let y^1 be a smooth function on M for which dy^1 does not vanish at p. Now let X be the unique (smooth) vector field which satisfies

$$X \lrcorner \Omega = dy^1.$$

This vector field does not vanish at p, so there is a function x_1 on a neighborhood U of p which satisfies $X(x_1) = 1$. Now let Y be the vector field on U which satisfies

$$Y \lrcorner \Omega = -dx_1.$$

Since $d\Omega = 0$, the Cartan formula, now gives

$$\mathcal{L}_X \Omega = \mathcal{L}_Y \Omega = 0.$$

We now compute

$$[X,Y] \lrcorner \Omega = \mathcal{L}_X Y \lrcorner \Omega = \mathcal{L}_X (Y \lrcorner \Omega) - Y \lrcorner (\mathcal{L}_X \Omega)$$
$$= \mathcal{L}_X(-dx_1) = -d\left(X(x^1)\right) = -d(1) = 0.$$

Since Ω has maximal rank, this implies $[X,Y] = 0$. By the simultaneous flow-box theorem, it follows that there exist local coordinates $x_1, y^1, z^1, z^2, \ldots, z^{2n-2}$ on some neighborhood $U_1 \subset U$ of p so that

$$X = \frac{\partial}{\partial x_1} \quad \text{and} \quad Y = \frac{\partial}{\partial y^1}.$$

Now consider the form $\Omega' = \Omega - dx_1 \wedge dy^1$. Clearly $d\Omega' = 0$. Moreover,

$$X \lrcorner \Omega' = \mathcal{L}_X \Omega' = Y \lrcorner \Omega' = \mathcal{L}_Y \Omega' = 0.$$

It follows that Ω' can be expressed as a 2-form in the variables $z^1, z^2, \ldots, z^{2n-2}$ alone. Hence, in particular, $(\Omega')^{n+1} \equiv 0$. On the other hand, by the binomial theorem, then

$$0 \neq \Omega^n = n \, dx_1 \wedge dy^1 \wedge (\Omega')^{n-1}.$$

It follows that Ω' may be regarded as a closed 2-form of maximal half-rank $n-1$ on an open set in $\mathbb{R}^{2n-2}$. Now apply the inductive hypothesis to Ω'.

Darboux' Theorem has a generalization which covers the case of closed 2-forms of constant (though not necessarily maximal) rank. It is the analogue for manifolds of the symplectic reduction of a vector space.

Theorem 2 (Darboux' Reduction Theorem). *Suppose that Ω is a closed 2-form of constant half-rank n on a manifold M^{2n+k}. Then the "null bundle"*

$$N_\Omega = \{v \in TM \,|\, \Omega(v,w) = 0 \text{ for all } w \in T_{\pi(v)}M\}$$

is integrable and of constant rank k. Moreover, any point of M has a neighborhood U on which there exist local coordinates $x_1, \ldots, x_n, y^1, \ldots, y^n, z^1, \ldots z^k$ in which

$$\Omega|_U = dx_1 \wedge dy^1 + dx_2 \wedge dy^2 + \cdots + dx_n \wedge dy^n.$$

Proof. Note that a vector field X on M is a section of N_Ω if and only if $X \lrcorner \Omega = 0$. In particular, since Ω is closed, the Cartan formula implies that $\mathcal{L}_X \Omega = 0$ for all such X.

If X and Y are two sections of N_Ω, then

$$[X,Y] \lrcorner \Omega = \mathcal{L}_X(Y \lrcorner \Omega) - Y(\mathcal{L}_X \Omega) = 0 - 0 = 0,$$

so it follows that $[X,Y]$ is a section of N_Ω as well. Thus, N_Ω is integrable.

Now apply the Frobenius Theorem. For any point $p \in M$, there exists a neighborhood U on which there exist local coordinates $z^1 \ldots, z^{2n+k}$ so that N_Ω restricted to U is spanned by the vector fields $Z_i = \partial/\partial z^i$ for $1 \leq i \leq k$. Since $Z_i \lrcorner \Omega = \mathcal{L}_{Z_i} \Omega = 0$ for $1 \leq i \leq k$, it follows that Ω can be expressed on U in terms of the variables $z^{k+1}, \ldots, z^{2n+k}$ alone. In particular, Ω restricted to U may be regarded as a non-degenerate closed 2-form on an open set in $\mathbb{R}^{2n}$. The stated result now follows from Darboux' Theorem.

Symplectic and Hamiltonian vector fields

I now want to examine some of the special vector fields which are defined on symplectic manifolds. Let M^{2n} be manifold and let Ω be a symplectic form on M.

Symplectic vector fields. Let $\text{Sp}(\Omega) \subset \text{Diff}(M)$ denote the subgroup of symplectomorphisms of (M,Ω). We would like to follow Lie in regarding $\text{Sp}(\Omega)$ as an "infinite dimensional Lie group". In that case, the Lie algebra of $\text{Sp}(\Omega)$ should be the space of vector fields whose flows preserve Ω. Of course, Ω will be invariant under the flow of a vector field X if and only if $\mathcal{L}_X \Omega = 0$. This motivates the following definition:

Definition 4. A vector field X on M is said to be *symplectic* if $\mathcal{L}_X \Omega = 0$. The space of symplectic vector fields on M will be denoted $\text{sp}(\Omega)$.

It turns out that there is a very simple characterization of the symplectic vector fields on M: Since $d\Omega = 0$, it follows that for any vector field X on M,

$$\mathcal{L}_X \Omega = d(X \lrcorner \Omega).$$

Thus, X is a symplectic vector field if and only if $X \lrcorner \Omega$ is closed.

Now, since Ω is non-degenerate, for any vector field X on M, the 1-form $\flat(X) = -X \lrcorner \Omega$ vanishes only where X does. Since TM and T^*M have the same rank, it follows that the mapping $\flat \colon \mathfrak{X}(M) \to \mathcal{A}^1(M)$ is an isomorphism of $C^\infty(M)$-modules.

In particular, $\flat$ has an inverse, $\sharp\colon \mathcal{A}^1(M) \to \mathfrak{X}(M)$. With this notation, we can write $\mathrm{sp}(\Omega) = \sharp\bigl(\mathcal{Z}^1(M)\bigr)$ where $\mathcal{Z}^1(M)$ denotes the vector space of closed 1-forms on M.

Hamiltonian vector fields. Now, $\mathcal{Z}^1(M)$ contains, as a subspace, $\mathcal{B}^1(M) = d\bigl(C^\infty(M)\bigr)$, the space of exact 1-forms on M. This subspace is of particular interest; we encountered it already in Lecture 4.

Definition 5. For each $f \in C^\infty(M)$, the vector field $X_f = \sharp(df)$ is called the *Hamiltonian vector field* associated to f. The set of all Hamiltonian vector fields on M is denoted $\mathrm{h}(\Omega)$.

Thus, by definition, $\mathrm{h}(\Omega) = \sharp\bigl(\mathcal{B}^1(M)\bigr)$. For this reason, Hamiltonian vector fields are often called *exact*. Note that a Hamiltonian vector field is one whose equations, written in symplectic coordinates, represent an ODE in Hamiltonian form.

The Poisson bracket. The following formula shows that, not only is $\mathrm{sp}(\Omega)$ a Lie algebra of vector fields, but that $\mathrm{h}(\Omega)$ is an ideal in $\mathrm{sp}(\Omega)$, i.e., that $[\mathrm{sp}(\Omega), \mathrm{sp}(\Omega)] \subset \mathrm{h}(\Omega)$.

Proposition 4. *For all $Y, Z \in \mathrm{sp}(\Omega)$, we have $[Y, Z] = X_{\Omega(Y,Z)}$. In particular, $[X_f, X_g] = X_{\{f,g\}}$ where, by definition, $\{f, g\} = \Omega(X_f, X_g)$.*

Proof. We use the fact that, for any vector field X, the operator $\mathcal{L}_X$ is a derivation with respect to any natural pairing between tensors on M:

$$[X, Y] \lrcorner \,\Omega = \bigl(\mathcal{L}_X Y\bigr) \lrcorner \,\Omega = \mathcal{L}_X\bigl(Y \lrcorner \,\Omega\bigr) - Y \lrcorner \bigl(\mathcal{L}_X \Omega\bigr)$$
$$= d\bigl(X \lrcorner (Y \lrcorner \,\Omega)\bigr) + X \lrcorner \, d(Y \lrcorner \,\Omega) + 0 = d\bigl(\Omega(Y, X)\bigr) + 0$$
$$= -d\bigl(\Omega(X, Y)\bigr) = \bigl(X_{\Omega(X,Y)}\bigr) \lrcorner \,\Omega.$$

This proves our first equation. The remaining equation follows immediately.

The definition $\{f, g\} = \Omega(X_f, X_g)$ is an important one. The bracket $(f, g) \mapsto \{f, g\}$ is called the *Poisson bracket* of the functions f and g. Proposition 4 implies that the Poisson bracket gives the functions on M the structure of a Lie algebra. The Poisson bracket is slightly more subtle than the pairing $(X_f, X_g) \mapsto X_{\{f,g\}}$ since the mapping $f \mapsto X_f$ has a non-trivial kernel, namely, the locally constant functions.

Thus, if M is connected, then we get an exact sequence of Lie algebras

$$0 \longrightarrow \mathbb{R} \longrightarrow C^\infty(M) \longrightarrow \mathrm{h}(\Omega) \longrightarrow 0$$

which is not, in general, split (see the Exercises). Since $\{1, f\} = 0$ for all functions f on M, it follows that the Poisson bracket on $C^\infty(M)$ makes it into a central extension of the algebra of Hamiltonian vector fields. The geometry of this central extension plays an important role in quantization theories on symplectic manifolds (see [GS 2] or [We]).

Also of great interest is the exact sequence

$$0 \longrightarrow \mathrm{h}(\Omega) \longrightarrow \mathrm{sp}(\Omega) \longrightarrow H^1_{dR}(M, \mathbb{R}) \longrightarrow 0,$$

where the right hand arrow is just the map described by $X \mapsto [X \lrcorner \Omega]$. Since the bracket of two elements in $\mathrm{sp}(\Omega)$ lies in $\mathrm{h}(\Omega)$, it follows that this linear map is actually a Lie algebra homomorphism when $H^1_{dR}(M,\mathbb{R})$ is given the abelian Lie algebra structure. This sequence also may or may not split (see the Exercises), and the properties of this extension have a great deal to do with the study of groups of symplectomorphisms of M. See the Exercises for further developments.

Involution

I now want to make some remarks about the meaning of the Poisson bracket and its applications.

Definition 5. Let (M, Ω) be a symplectic manifold. Two functions f and g are said to be *in involution* (with respect to Ω) if they satisfy the condition $\{f, g\} = 0$.

Note that, since $\{f, g\} = dg(X_f) = -df(X_g)$, it follows that two functions f and g are in involution if and only if each is constant on the integral curves of the other's Hamiltonian vector field.

Now, if one is trying to describe the integral curves of a Hamiltonian vector field, X_f, the more independent functions on M that one can find which are constant on the integral curves of X_f, the more accurately one can describe those integral curves. If one were able find, in addition to f itself, $2n-2$ additional independent functions on M which are constant on the integral curves of X_f, then one could describe the integral curves of X_f implicitly by setting those functions equal to a constant.

It turns out, however, that this is too much to hope for in general. It can happen that a Hamiltonian vector field X_f has no functions in involution with it except for functions of the form $F(f)$.

Complete integrability. Nevertheless, in many cases which arise in practice, we can find several functions in involution with a given function $f = f_1$ and, moreover, in involution with each other. In case one can find $n-1$ such independent functions, $f_2, \ldots, f_n$, we have the following theorem of Liouville which says that the remaining $n-1$ required functions can be found (at least locally) by quadrature alone. In the classical language, a vector field X_f for which such functions are known is said to be 'completely integrable by quadratures', or, more simply as 'completely integrable'.

Theorem 3. *Let $f^1, f^2, \ldots, f^n$ be n functions in involution on a symplectic manifold (M^{2n}, Ω). Suppose that the functions f^i are independent in the sense that the differentials $df^1, \ldots, df^n$ are linearly independent at every point of M. Then each point of M has an open neighborhood U on which there are functions $a_1, \ldots, a_n$ on U so that*

$$\Omega = df^1 \wedge da_1 + \cdots + df^n \wedge da_n.$$

Moreover, the functions a_i can be found by "finite" operations and quadrature.

Proof. By hypothesis, the forms $df^1, \ldots, df^n$ are linearly independent at every point of M, so it follows that the Hamiltonian vector fields $X_{f^1}, \ldots, X_{f^n}$ are also linearly independent at every point of M. Also by hypothesis, the functions f^i are in involution, so it follows that $df^i(X_{f^j}) = 0$ for all i and j.

The vector fields X_{f^i} are linearly independent on M, so by 'finite' operations, we can construct 1-forms $\bar{\beta}_1, \ldots, \bar{\beta}_n$ which satisfy the conditions

$$\bar{\beta}_i(X_{f^j}) = \delta_{ij} \qquad \text{(Kronecker delta)}.$$

Any other set of forms β_i which satisfy these conditions are given by expressions:

$$\beta_i = \bar{\beta}_i + g_{ij}\, df^j.$$

for some functions g_{ij} on M. Let us regard the functions g_{ij} as unknowns for a moment. Let $Y_1, \ldots, Y_n$ be the vector fields which satisfy

$$Y_i \lrcorner\, \Omega = \beta_i,$$

with $\bar{Y}_i$ denoting the corresponding quantities when the g_{ij} are set to zero. Then it is easy to see that

$$Y_i = \bar{Y}_i - g_{ij}\, X_{f^j}.$$

Now, by construction,

$$\Omega(X_{f^i}, X_{f^j}) = 0 \quad \text{and} \quad \Omega(Y_i, X_{f^j}) = \delta_{ij}.$$

Moreover, as is easy to compute,

$$\Omega(Y_i, Y_j) = \Omega(\bar{Y}_i, \bar{Y}_j) - g_{ji} + g_{ij}.$$

Thus, choosing the functions g_{ij} appropriately, say $g_{ij} = -\frac{1}{2}\Omega(\bar{Y}_i, \bar{Y}_j)$, we may assume that $\Omega(Y_i, Y_j) = 0$. It follows that the sequence of 1-forms $df^1, \ldots, df^n, \beta_1, \ldots, \beta_n$ is the dual basis to the sequence of vector fields $Y_1, \ldots, Y_n, X_{f^1}, \ldots, X_{f^n}$. In particular, we see that

$$\Omega = df^1 \wedge \beta_1 + \cdots + df^n \wedge \beta_n,$$

since the 2-forms on either side of this equation have the same values on all pairs of vector fields drawn from this basis.

Now, since Ω is closed, we have

$$d\Omega = df^1 \wedge d\beta_1 + \cdots + df^n \wedge d\beta_n = 0.$$

If, for example, we wedge both sides of this equation with $df^2, \ldots, df^n$, we see that

$$df^1 \wedge df^2 \wedge \ldots \wedge df^n \wedge d\beta_1 = 0.$$

Hence, it follows that $d\beta_1$ lies in the ideal generated by the forms $df^1, \ldots, df^n$. Of course, there was nothing special about the first term, so we clearly have

$$d\beta_i \equiv 0 \bmod df^1, \ldots, df^n \qquad \text{for all } 1 \leq i \leq n.$$

In particular, it follows that, if we pull back the 1-forms β_i to any n-dimensional level set $M_c \subset M$ defined by equations $f^i = c^i$ where the c^i are constants, then each β_i becomes closed.

Let $m \in M$ be fixed and choose functions $g_1, \ldots, g_n$ on a neighborhood U of m in M so that $g_i(m) = 0$ and so that the functions $g_1, \ldots, g_n, f^1, \ldots, f^n$ form a coordinate chart on U. By shrinking U if necessary, we may assume that the image of this coordinate chart in $\mathbb{R}^n \times \mathbb{R}^n$ is an open set of the form $B_1 \times B_2$, where B_1 and B_2 are open balls in $\mathbb{R}^n$ (with B_1 centered on 0). In this coordinate chart, the β_i can be expressed in the form

$$\beta_i = B_i^j(g, f)\, dg_j + C_{ij}(g, f)\, df^j.$$

Define new functions h_i on $B^1 \times B^2$ by the rule

$$h_i(g, f) = \int_0^1 B_i^j(tg, f) g_j\, dt.$$

(This is just the Poincaré homotopy formula with the f's held fixed. It is also the first place where we use "quadrature".) Since setting the f's equals to constants makes β_i a closed 1-form, it follows easily that

$$\beta_i = dh_i + A_{ij}(g, f)\, df^j$$

for some functions A_{ij} on $B^1 \times B^2$. Thus, on U, the form Ω has the expression

$$\Omega = df^i \wedge dh_i + A_{ij}\, df^i \wedge df^j.$$

It follows that the 2-form $A = A_{ij}\, df^i \wedge df^j$ is closed on (the contractible open set) $B^1 \times B^2$. Thus, the functions A_{ij} do not depend on the g-coordinates at all. Hence, by employing quadrature once more (i.e., the second time) in the Poincaré homotopy formula, we can write $A = -d(s_i\, df^i)$ for some functions s_i of the f's alone. Setting $a_i = h_i + s_i$, we have the desired local normal form $\Omega = df^i \wedge da_i$.

In many useful situations, one does not need to restrict to a local neighborhood U to define the functions a_i (at least up to additive constants) and the 1-forms da_i can be defined globally on M (or, at least away from some small subset in M where degeneracies occur). In this case, the construction above is often called the construction of 'action-angle' coordinates. We will discuss this further in Lecture 6.

Exercises

1. Show that the bilinear form on $\mathbb{R}^n$ defined in the text by the rule $B(x,y) = {}^t\!x\,A\,y$ (where A is a skew-symmetric n-by-n matrix) is non-degenerate if and only if A is invertible. Show directly (i.e., without using Proposition 1) that a skew-symmetric, n-by-n matrix A cannot be invertible if n is odd. (Hint: For the last part, compute $\det(A)$ two ways.)

2. Let (V, B) be a symplectic space and let $\mathbf{b} = (b_1, b_2, \ldots, b_m)$ be a basis of B. Define the m-by-m skew-symmetric matrix $A_{\mathbf{b}}$ whose ij-entry is $B(b_i, b_j)$. Show that if $\mathbf{b}' = \mathbf{b}R$ is any other basis of V (where $R \in \mathrm{GL}(m, \mathbb{R})$), then

$$A_{\mathbf{b}'} = {}^t\!R\, A_{\mathbf{b}}\, R.$$

Use Proposition 1 and Exercise 1 to conclude that, if A is an invertible, skew-symmetric $2n$-by-$2n$ matrix, then there exists a matrix $R \in \mathrm{GL}(2n, \mathbb{R})$ so that

$$A = {}^t\!R \begin{pmatrix} 0_n & I_n \\ -I_n & 0_n \end{pmatrix} R.$$

In other words, the $\mathrm{GL}(2n, \mathbb{R})$-orbit of the matrix J_n defined in the text (under the "standard" (right) action of $\mathrm{GL}(2n, \mathbb{R})$ on the skew-symmetric $2n$-by-$2n$ matrices) is the open set of all invertible skew-symmetric $2n$-by-$2n$ matrices.

3. Show that $\mathrm{Sp}(n, \mathbb{R})$, as defined in the text, is indeed a Lie subgroup of $\mathrm{GL}(2n, \mathbb{R})$ and has dimension $2n^2 + n$. Compute its Lie algebra $\mathfrak{sp}(n, \mathbb{R})$. Show that $\mathrm{Sp}(1, \mathbb{R}) = \mathrm{SL}(2, \mathbb{R})$.

4. In Lecture 2, we defined the groups $\mathrm{GL}(n, \mathbb{C}) = \{R \in \mathrm{GL}(2n, \mathbb{R}) \mid J_n R = R J_n\}$ and $\mathrm{O}(2n) = \{R \in \mathrm{GL}(2n, \mathbb{R}) \mid {}^t\!R\, R = I_{2n}\}$. Show that

$$\mathrm{GL}(n, \mathbb{C}) \cap \mathrm{Sp}(n, \mathbb{R}) = \mathrm{O}(2n) \cap \mathrm{Sp}(n, \mathbb{R}) = \mathrm{GL}(n, \mathbb{C}) \cap \mathrm{O}(2n) = \mathrm{U}(n).$$

5. Let Ω be a symplectic form on a vector space V of dimension $2n$. Let $W \subset V$ be a subspace which satisfies $\dim W = d$ and $\dim(W \cap W^\perp) = \nu$. Show that there exists an Ω-standard basis of V so that W is spanned by the vectors

$$e_1, \ldots, e_{\nu+m}, f_1, \ldots, f_m$$

where $d - \nu = 2m$. In this basis of V, what is a basis for $W^\perp$?

6. **THE PFAFFIAN.** Let V be a vector space of dimension $2n$. Fix a basis $\mathbf{b} = (b_1, \ldots, b_{2n})$. For any skew-symmetric $2n$-by-$2n$ matrix $F = (f^{ij})$, define the 2-vector

$$\Phi_F = \tfrac{1}{2} f^{ij} b_i \wedge b_j = \tfrac{1}{2} \mathbf{b} \wedge F \wedge {}^t\mathbf{b}.$$

Then there is a unique polynomial function Pf, homogeneous of degree n, on the space of skew-symmetric $2n$-by-$2n$ matrices for which

$$(\Phi_F)^n = n!\,\mathrm{Pf}(F)\, b_1 \wedge \ldots \wedge b_{2n}.$$

Show that

$$\mathrm{Pf}(F) = f^{12} \qquad \text{when } n=1,$$
$$\mathrm{Pf}(F) = f^{12}f^{34} + f^{13}f^{42} + f^{14}f^{23} \qquad \text{when } n=2.$$

Show also that

$$\mathrm{Pf}(A F\,{}^t\!A) = \det(A)\,\mathrm{Pf}(F)$$

for all $A \in \mathrm{GL}(2n,\mathbb{R})$. (Hint: Examine the effect of a change of basis $\mathbf{b} = \mathbf{b}'A$. Compare Problem 2.) Use this to conclude that $\mathrm{Sp}(n,\mathbb{R})$ is a subgroup of $\mathrm{SL}(2n,\mathbb{R})$. Finally, show that $(\mathrm{Pf}(F))^2 = \det(F)$. (Hint: Show that the left and right hand sides are polynomial functions which agree on a certain open set in the space of skew-symmetric $2n$-by-$2n$ matrices.)

The polynomial function Pf is called the *Pfaffian*. It plays an important role in differential geometry.

7. Verify that, for any $B \in A^2(V)$, the symplectic reduction $(\overline{V}, \overline{B})$ is a well-defined symplectic space.
8. Show that if there is a G-invariant non-degenerate pairing $(\,,\,)\colon \mathfrak{g} \times \mathfrak{g} \to \mathbb{R}$, then $\mathfrak{g}$ and $\mathfrak{g}^*$ are isomorphic as G-representations.
9. Compute the adjoint and coadjoint representations for

$$G = \left\{ \begin{pmatrix} a & b \\ 0 & 1 \end{pmatrix} \,\bigg|\, a \in \mathbb{R}^+, b \in \mathbb{R} \right\}$$

Show that $\mathfrak{g}$ and $\mathfrak{g}^*$ are *not* isomorphic as G-representations! (For a general G, the Ad-orbits of G in $\mathfrak{g}$ are not generally of even dimension, so they can't be symplectic manifolds.)

10. For any Lie group G and any $\xi \in \mathfrak{g}^*$, show that the symplectic structures Ω_ξ and $\Omega_{a\cdot\xi}$ on $G \cdot \xi$ are the same for any $a \in G$.
11. For any Lie group G show that, for the 'generic' $\xi \in \mathfrak{g}^*$, the identity component of the stabilizer subgroup G_ξ is abelian.

    (Hint: For each $\xi \in \mathfrak{g}^*$, define $\delta\xi \in A^2(\mathfrak{g})$ by the rule $\delta\xi(x,y) = -\xi([x,y])$. Show that the Lie algebra of G_ξ is $N_{\delta\xi}$, i.e., the null space of $\delta\xi$. Let $\mathcal{O} \subset \mathfrak{g}^*$ denote the subset consisting of those $\xi \in \mathfrak{g}^*$ for which $\dim N_{\delta\xi}$ has its minimum dimension, say $r \le \dim \mathfrak{g}$.[2] Show that $\mathcal{O}$ is both open and dense in $\mathfrak{g}^*$. Show, moreover, that for $\xi \in \mathcal{O}$, the subspace $N_{\delta\xi} = \mathfrak{g}_\xi$ is abelian as follows: If $x, y \in \mathfrak{g}_\xi$ satisfy $[x,y] \ne 0$, then let $\eta \in \mathfrak{g}^*$ satisfy $\eta([x,y]) \ne 0$ and, that for $t \ne 0$ sufficiently small, the element $\xi' = \xi + t\eta$ will have $N_{\delta\xi'}$ of dimension strictly smaller than r, a contradiction.)

    This result will turn out to have important implications for the solvability of ODE by the method of reduction (See Lecture 7).

12. This exercise concerns the splitting properties of the two Lie algebras sequences associated to any symplectic structure Ω on a connected manifold

---

[2] This number r is frequently called the *rank* of $\mathfrak{g}$, especially when $\mathfrak{g}$ is reductive, i.e., a sum of simple subalgebras plus a center.

M:
$$0 \longrightarrow \mathbb{R} \longrightarrow C^\infty(M) \longrightarrow \mathsf{h}(\Omega) \longrightarrow 0$$

and
$$0 \longrightarrow \mathsf{h}(\Omega) \longrightarrow \mathsf{sp}(\Omega) \longrightarrow H^1_{dR}(M,\mathbb{R}) \longrightarrow 0.$$

Define the "divided powers" of Ω by the rule $\Omega^{[k]} = (1/k!)\,\Omega^k$, for each $0 \leq k \leq n$.

(i) Show that, for any vector fields X and Y on M,
$$\Omega(X,Y)\,\Omega^{[n]} = -(X \lrcorner\, \Omega) \wedge (Y \lrcorner\, \Omega) \wedge \Omega^{[n-1]}.$$

Conclude that the first of the above two sequences splits if M is compact. (Hint: For the latter statement, show that the set of functions f on M for which $\int_M f\,\Omega^{[n]} = 0$ forms a Poisson subalgebra of $C^\infty(M)$.)

(ii) On the other hand, show that for $\mathbb{R}^2$ with the symplectic structure $\Omega = dx \wedge dy$, the first sequence does *not* split. (Hint: Show that every smooth function on $\mathbb{R}^2$ is of the form $\{x,g\}$ for some $g \in C^\infty(\mathbb{R}^2)$. Why does this help?)

(iii) Suppose that M is compact. Define a skew-symmetric pairing
$$\beta_\Omega \colon H^1_{dR}(M,\mathbb{R}) \times H^1_{dR}(M,\mathbb{R}) \to \mathbb{R}$$

by the formula
$$\beta_\Omega(a,b) = \int_M \tilde{a} \wedge \tilde{b} \wedge \Omega^{[n-1]},$$

where $\tilde{a}$ and $\tilde{b}$ are closed 1-forms representing the cohomology classes a and b respectively. Show that if there is a Lie algebra splitting $\sigma \colon H^1_{dR}(M,\mathbb{R}) \to \mathsf{sp}(\Omega)$ then
$$\Omega\bigl(\sigma(a), \sigma(b)\bigr) = -\frac{\beta_\Omega(a,b)}{vol(M, \Omega^{[n]})}$$

for all $a,b \in H^1_{dR}(M,\mathbb{R})$. (Remember that the Lie algebra structure on $H^1_{dR}(M,\mathbb{R})$ is the abelian one.) Use this to conclude that the second sequence does split for a symplectic structure on the standard 1-holed torus, but does *not* split for any symplectic structure on the 2-holed torus. (Hint: To show the non-splitting result, use the fact that any tangent vector field on the 2-holed torus must have a zero.)

13. THE FLUX HOMOMORPHISM. The object of this exercise is to try to identify the subgroup of $Sp(\Omega)$ whose Lie algebra is $\mathsf{h}(\Omega)$, following some ideas of Calabi [Ca]. Thus, let (M,Ω) be a symplectic manifold.

First, I remind you how the construction of the (smooth) universal cover of the identity component of $Sp(\Omega)$ goes. Let $p\colon [0,1] \times M \to M$ be a smooth map with the property that the map $p_t \colon M \to M$ defined by $p_t(m) = p(t,m)$ is a symplectomorphism for all $0 \leq t \leq 1$. Such a p is called a (smooth) path in $Sp(M)$. We say that p is *based at the identity map* $e\colon M \to M$ if $p_0 = e$.

The set of smooth paths in $Sp(M)$ which are based at e will be denoted by $P_e(Sp(\Omega))$.

Two paths p and p' in $P_e(Sp(\Omega))$ satisfying $p_1 = p'_1$ are said to be *homotopic* if there is a smooth map $P\colon [0,1] \times [0,1] \times M \to M$ which satisfies the following conditions: First, $P(s, 0, m) = m$ for all s and m. Second, $P(s, 1, m) = p_1(m) = p'_1(m)$ for all s and m. Third, $P(0, t, m) = p(t, m)$ and $P(1, t, m) = p'(t, m)$ for all t and m. The set of homotopy classes of elements of $P_e(Sp(\Omega))$ is then denoted by $\widetilde{Sp}^0(\Omega)$. In any reasonable topology on $Sp(\Omega)$, this should be the universal covering space of the identity component of $Sp(\Omega)$. There is a natural group structure on $\widetilde{Sp}^0(\Omega)$ in which $\tilde{e}$, the homotopy class of the constant path at e, is the identity element (cf. the covering spaces exercise in Exercise Set 2).

We are now going to construct a homomorphism $\Phi\colon \widetilde{Sp}^0(\Omega) \to H^1(M, \mathbb{R})$, called the *flux homomorphism*. Let $p \in P_e(Sp(\Omega))$ be chosen, and let $\gamma\colon S^1 \to M$ be a closed curve representing an element of $H_1(M, \mathbb{Z})$. Then we can define

$$F(p, \gamma) = \int_{[0,1] \times S^1} (p \cdot \gamma)^*(\Omega)$$

where $(p \cdot \gamma)\colon [0,1] \times S^1 \to M$ is defined by $(p \cdot \gamma)(t, \theta) = p(t, \gamma(\theta))$. The number $F(p, \gamma)$ is called the *flux* of p through γ.

(i) Show that $F(p, \gamma) = F(p', \gamma')$ if p is homotopic to p' and γ is homologous to γ'. (Hint: Use Stokes' Theorem several times.)

Thus, F is actually well defined as a map $F\colon \widetilde{Sp}^0(\Omega) \times H_1(M, \mathbb{R}) \to \mathbb{R}$.

(ii) Show that $F\colon \widetilde{Sp}^0(\Omega) \times H_1(M, \mathbb{R}) \to \mathbb{R}$ is linear in its second variable and that, under the obvious multiplication, we have

$$F(pp', \gamma) = F(p, \gamma) + F(p', \gamma).$$

(Hint: Use Stokes' Theorem again.)

Thus, F may be transposed to become a homomorphism

$$\Phi\colon \widetilde{Sp}^0(\Omega) \to H^1(M, \mathbb{R}).$$

Show (by direct computation) that if ζ is a closed 1-form on M for which the symplectic vector field $Z = \sharp\zeta$ is complete on M, then the path p in $Sp(M)$ defined by the flow of Z from $t = 0$ to $t = 1$ satisfies $\Phi(p) = [\zeta] \in H^1(M, \mathbb{R})$. Conclude that the flux homomorphism Φ is always surjective and that its derivative $\Phi'(\tilde{e})\colon sp(\Omega) \to H^1(M, \mathbb{R})$ is just the operation of taking cohomology classes. (Recall that we identify $sp(\Omega)$ with $\mathcal{Z}^1(M)$.)

(iii) Show that if M is a compact surface of genus $g > 1$, then the flux homomorphism is actually well defined as a map from $Sp(\Omega)$ to $H^1(M, \mathbb{R})$. Would the same result be true if M were of genus 1? How could you modify the map so as to make it well-defined on $Sp(\Omega)$ in the case of the torus? (Hint: Show that if you have two paths p and p' with the same endpoint, then you can express the difference of their fluxes across a circle γ as an integral of the form

$$\int_{S^1 \times S^1} \Psi^*(\Omega)$$

where $\Psi: S^1 \times S^1 \to M$ is a certain piecewise smooth map from the torus into M. Now use the fact that, for any piecewise smooth map $\Psi: S^1 \times S^1 \to M$, the induced map $\Psi^*: H^2(M, \mathbb{R}) \to H^2(S^1 \times S^1, \mathbb{R})$ on cohomology is zero. (Why does this follow from the assumption that the genus of M is greater than 1?))

In any case, the subgroup $\ker \Phi$ (or its image under the natural projection from $\widetilde{\mathrm{Sp}}^0(\Omega)$ to $\mathrm{Sp}(\Omega)$) is known as the group $\mathrm{H}(\Omega)$ of *exact* or *Hamiltonian* symplectomorphisms. Note that, at least formally, its Lie algebra is $\mathrm{h}(\Omega)$.

14. In the case of the geodesic flow on a surface of revolution (see Lecture 4), show that the energy $f^1 = E_L$ and the conserved quantity $f^2 = F(r)\dot{r} + G(r)\dot{\theta}$ are in involution. Use the algorithm described in Theorem 3 to compute the functions a_1 and a_2, thus verifying that the geodesic equations on a surface of revolution are integrable by quadrature.

LECTURE 6
Symplectic Manifolds, II

I want to turn now to the problem of describing the symplectic structures a manifold M can have. This is a surprisingly delicate problem and is currently a subject of research.

Obstructions

Of course, one fundamental question is whether a given manifold has any symplectic structures at all. I want to begin this lecture with a discussion of the two known obstructions for a manifold to have a symplectic structure.

The cohomology ring condition

If $\Omega \in \mathcal{A}^2(M^{2n})$ is a symplectic structure on a compact manifold M, then the cohomology class $[\Omega] \in H^2_{dR}(M, \mathbb{R})$ is non-zero. In fact, $[\Omega]^n = [\Omega^n]$, but the class $[\Omega^n]$ cannot vanish in $H^{2n}_{dR}(M)$ because the integral of Ω^n over M is clearly non-zero. Thus, we have

Proposition 1. *If M^{2n} is compact and has a symplectic structure, there must exist an element $u \in H^2(M, \mathbb{R})$ so that $u^n \neq 0 \in H^{2n}_{dR}(M)$.*

Example. This immediately rules out the existence of a symplectic structure on S^{2n} for all $n > 1$. One consequence of this, as you are asked to show in the Exercises, is that there cannot be any simple notion of connected sum in the category of symplectic manifolds (except in dimension 2).

The bundle obstruction

If M admits a symplectic structure Ω, then, in particular, this defines a symplectic structure on each of the tangent spaces $T_m M$ which varies continuously with m. In other words, TM must carry the structure of a symplectic vector bundle. There are topological obstructions to the existence of such a structure on the tangent bundle of a general manifold. As a simple example, if M has a symplectic structure, then TM must be orientable.

There are more subtle obstructions than orientation. Unfortunately, a description of these obstructions requires some acquaintance with the theory of characteristic classes. However, part of the following discussion will be useful even to those who aren't familiar with characteristic class theory, so I will give it now, even though the concepts will only reveal their importance in later Lectures.

Definition 1. An *almost symplectic structure* on a manifold M^{2n} is a smooth 2-form Ω defined on M which is non-degenerate but not necessarily closed. An *almost complex structure* on M^{2n} is a smooth bundle map $J: TM \to TM$ which satisfies $J^2 v = -v$ for all v in TM.

The reason that I have introduced both of these concepts at the same time is that they are intimately related. The really deep aspects of this relationship will only become apparent in the Lecture 9, but we can, at least, give the following result now.

Proposition 2. *A manifold M^{2n} has an almost symplectic structure if and only if it has an almost complex structure.*

Proof. First, suppose that M has an almost complex structure J. Let g_0 be any Riemannian metric on M. (Thus, $g_0: TM \to \mathbb{R}$ is a smooth function which restricts to each $T_m M$ to be a positive definite quadratic form.) Now define a new Riemannian metric by the formula

$$g(v) = g_0(v) + g_0(Jv).$$

Then g has the property that $g(Jv) = g(v)$ for all $v \in TM$ since

$$g(Jv) = g_0(Jv) + g_0(J^2 v) = g_0(Jv) + g_0(-v) = g(v).$$

Now let $\langle,\rangle$ denote the (symmetric) inner product associated with g. Thus, $\langle v, v \rangle = g(v)$, so we have $\langle Jx, Jy \rangle = \langle x, y \rangle$ when x and y are tangent vector with the same base point. For $x, y \in T_m M$ define $\Omega(x, y) = \langle Jx, y \rangle$. I claim that Ω is a non-degenerate 2-form on M. To see this, first note that

$$\Omega(x, y) = \langle Jx, y \rangle = -\langle Jx, J^2 y \rangle = -\langle J^2 y, Jx \rangle = -\langle Jy, x \rangle = -\Omega(y, x),$$

so Ω is a 2-form. Moreover, if x is a non-zero tangent vector, then $\Omega(x, Jx) = \langle Jx, Jx \rangle = g(x) > 0$, so it follows that $x \lrcorner \Omega \neq 0$. Thus Ω is non-degenerate.

To go the other way is a little more delicate. Suppose that Ω is given and fix a Riemannian metric g on M with associated inner product $\langle,\rangle$. Then, by linear algebra there exists a unique bundle mapping $A: TM \to TM$ so that $\Omega(x, y) = \langle Ax, y \rangle$. Since Ω is skew-symmetric and non-degenerate, it follows that A must be skew-symmetric relative to $\langle,\rangle$ and must be invertible. It follows that $-A^2$ must be symmetric and positive definite relative to $\langle,\rangle$. Now, standard results from linear algebra imply that there is a unique smooth bundle map $B: TM \to TM$ which positive definite and symmetric with respect to $\langle,\rangle$ and which satisfies $B^2 = -A^2$. Moreover, this linear mapping B must commute with A. (See the Exercises if you are not familiar with this fact). Thus, the mapping $J = AB^{-1}$ satisfies $J^2 = -I$, as desired.

It is not hard to show that the mappings $(J, g_0) \mapsto \Omega$ and $(\Omega, g) \mapsto J$ constructed in the proof of Proposition 1 depend continuously (in fact, smoothly) on their arguments. Since the set of Riemannian metrics on M is contractible, it follows that the set of homotopy classes of almost complex structures on M is in natural one-to-one correspondence with the set of homotopy classes of almost symplectic structures.

(The reader who is familiar with the theory of principal bundles knows that at the heart of Proposition 1 is the fact that $\mathrm{Sp}(n, \mathbb{R})$ and $\mathrm{GL}(n, \mathbb{C})$ have the same maximal compact subgroup, namely $\mathrm{U}(n)$.)

Characteristic classes as obstructions. Suppose that M has a symplectic structure Ω and let J be any one of the almost complex structures on M we constructed above. Then the tangent bundle of M can be regarded as a complex bundle, which we will denote by T^J and hence has a total Chern class

$$c(T^J) = \big(1 + c_1(J) + c_2(J) + \cdots + c_n(J)\big)$$

where $c_i(J) \in H^{2i}(M, \mathbb{Z})$. Now, by the properties of Chern classes, $c_n(J) = e(TM)$, where $e(TM)$ is the Euler class of the tangent bundle given the orientation determined by the volume form Ω^n.

These classes are related to the Pontrijagin classes of TM by the Whitney sum formula (see [MS]):

$$\begin{aligned} p(TM) &= 1 - p_1(TM) + p_2(TM) - \cdots + (-1)^{[n/2]} p_{[n/2]}(TM) \\ &= c(T^J \oplus T^{-J}) \\ &= \big(1 + c_1(J) + c_2(J) + \cdots + c_n(J)\big)\big(1 - c_1(J) + c_2(J) - \cdots + (-1)^n c_n(J)\big) \end{aligned}$$

Since $p(TM)$ depends only on the diffeomorphism class of M, this gives quadratic equations for the $c_i(J)$,

$$p_k(T) = \big(c_k(J)\big)^2 - 2c_{k-1}(J)c_{k+1}(J) + \cdots + (-1)^k 2c_0(J)c_{2k}(J),$$

to which any manifold with an almost complex structure must have solutions. Since not every $2n$-manifold has cohomology classes $c_i(J)$ satisfying these equations, it follows that some $2n$-manifolds have no almost complex structure and hence, by Proposition 2, no almost symplectic structure either.

Examples. Here are two examples in dimension 4 to show that the cohomology ring condition and the bundle obstruction are independent.

- $M = S^1 \times S^3$ does not have a symplectic structure because $H^2(M, \mathbb{R}) = 0$. However the bundle obstruction vanishes because M is parallelizable (why?). Thus M does have an almost symplectic structure.
- $M = \mathbb{CP}^2 \# \mathbb{CP}^2$. The cohomology ring of M in this case is generated over $\mathbb{Z}$ by two generators u_1 and u_2 in $H^2(M, \mathbb{Z})$ which are subject to the relations

$u_1 u_2 = 0$ and $u_1^2 = u_2^2 = v$ where v generates $H^4(M,\mathbb{Z})$. For any non-zero class $u = n_1 u_1 + n_2 u_2$, we have $u^2 = (n_1^2 + n_2^2)v \neq 0$. Thus the cohomology ring condition is satisfied.

However, M has no almost symplectic structure: If it did, then $T = TM$ would have a complex structure J, with total Chern class $c(J)$ and the equations above would give $p_1(T) = \bigl(c_1(J)\bigr)^2 - 2c_2(J)$. Moreover, we would have $e(T) = c_2(J)$. Thus, we would have to have

$$\bigl(c_1(J)\bigr)^2 = p_1(T) + 2e(T).$$

For any compact, simply-connected, oriented 4-manifold M with orientation class $\mu \in H^4(M,\mathbb{Z})$, the Hirzebruch Signature Theorem (see [MS]) implies $p_1(T) = 3(b_2^+ - b_2^-)\mu$, where $b_2^\pm$ are the number of positive and negative eigenvalues respectively of the intersection pairing $H^2(M,\mathbb{Z}) \times H^2(M,\mathbb{Z}) \to \mathbb{Z}$. In addition, $e(T) = (2 + b_2^+ + b_2^-)\mu$. Substituting these into the above formula, we would have $\bigl(c_1(J)\bigr)^2 = (4 + 5b_2^+ - b_2^-)\mu$ for any complex structure J on the tangent bundle of M.

In particular, if $M = \mathbb{CP}^2 \# \mathbb{CP}^2$ had an almost complex structure J, then $\bigl(c_1(J)\bigr)^2$ would be either $14v$ (if $\mu = v$, since then $b_2^+ = 2$ and $b_2^- = 0$) or $-2v$ (if $\mu = -v$, since then $b_2^+ = 0$ and $b_2^- = 2$). However, by our previous calculations, neither $14v$ nor $-2v$ is the square of a cohomology class in $H^2(\mathbb{CP}^2 \# \mathbb{CP}^2, \mathbb{Z})$.

This example shows that, in general, one cannot hope to have a connected sum operation for symplectic manifolds.

The actual conditions for a manifold to have an almost symplectic structure can be expressed in terms of characteristic classes, so, in principle, this can always be determined once the manifold is given explicitly. In Lecture 9 we will describe more fully the following remarkable result of Gromov:

> *If M^{2n} has no compact components and has an almost symplectic structure Υ, then there exists a symplectic structure Ω on M which is homotopic to Υ through almost symplectic structures.*

Thus, the problem of determining which manifolds have symplectic structures is now reduced to the compact case. In this case, no obstruction beyond what I have already described is known. Thus, I can state the following basic open problem *If a compact manifold M^{2n} satisfies the cohomology ring condition and has an almost symplectic structure, does it have a symplectic structure?*

Even (perhaps especially) for 4-manifolds, this problem is extremely interesting and very poorly understood.

Deformations of symplectic structures

We will now turn to some of the features of the space of symplectic structures on a given manifold which does admit symplectic structures. We will examine the "deformation problem". The following theorem due to Moser (see [We]) shows that symplectic structures determining a fixed cohomology class in H^2 on a compact manifold are "rigid".

Theorem 1. *If M^{2n} is a compact manifold and Ω_t for $t \in [0,1]$ is a continuous 1-parameter family of smooth symplectic structures on M which has the property that the cohomology classes $[\Omega_t]$ in $H^2_{dR}(M, \mathbb{R})$ are independent of t, then for each $t \in [0,1]$, there exists a diffeomorphism ϕ_t so that $\phi_t^*(\Omega_t) = \Omega_0$.*

Proof. We will start by proving a special case and then deduce the general case from it. Suppose that Ω_0 is a symplectic structure on M and that $\varphi \in \mathcal{A}^1(M)$ is a 1-form so that, for all $s \in (-1, 1)$, the 2-form

$$\Omega_s = \Omega_0 + s\,d\varphi$$

is a symplectic form on M as well. (This is true for all sufficiently "small" 1-forms on M since M is compact.) Now consider the 2-form on $(-1,1) \times M$ defined by the formula

$$\Omega = \Omega_0 + s\,d\varphi - \varphi \wedge ds.$$

(Here, we are using s as the coordinate on the first factor $(-1,1)$ and, as usual, we write Ω_0 and φ instead of $\pi_2^*(\Omega_0)$ and $\pi_2^*(\varphi)$ where $\pi_2\colon (-1,1) \times M \to M$ is the projection on the second factor.)

The reader can check that Ω is closed on $(-1,1) \times M$. Moreover, since Ω pulls back to each slice $\{s_0\} \times M$ to be the non-degenerate form Ω_{s_0} it follows that Ω has half-rank n everywhere. Thus, the kernel N_Ω is 1-dimensional and is transverse to each of the slices $\{t\} \times M$. Hence there is a unique vector field X which spans N_Ω and satisfies $ds(X) = 1$.

Now because M is compact, it is not difficult to see that each integral curve of X projects by $s = \pi_1$ diffeomorphically onto $(-1,1)$. Moreover, it follows that there is a smooth map $\phi\colon (-1,1) \times M \to M$ so that, for each m, the curve $t \mapsto \phi(t, m)$ is the integral curve of X which passes through $(0, m)$.

It follows that the map $\Phi\colon (-1,1) \times M \to (-1,1) \times M$ defined by

$$\Phi(t, m) = \bigl(t, \phi(t, m)\bigr)$$

carries the vector field $\partial/\partial s$ to the vector field X. Moreover, since Ω_0 and Ω have the same value when pulled back to the slice $\{0\} \times M$ and since

$$\begin{aligned}\mathcal{L}_{\partial/\partial s}\Omega_0 &= 0 \\ \partial/\partial s \lrcorner\, \Omega_0 &= 0\end{aligned} \quad \text{and} \quad \begin{aligned}\mathcal{L}_X \Omega &= 0 \\ X \lrcorner\, \Omega &= 0,\end{aligned}$$

it follows easily that $\Phi^*(\Omega) = \Omega_0$. In particular, $\phi_t^*(\Omega_t) = \Omega_0$ where ϕ_t is the diffeomorphism of M given by $\phi_t(m) = \phi(t, m)$.

Now let us turn to the general case. If Ω_t for $0 \le t \le 1$ is any continuous family of smooth closed 2-forms for which the cohomology classes $[\Omega_t]$ are all equal to $[\Omega_0]$, then for any two values t_1 and t_2 in the unit interval, consider the 1-parameter family of 2-forms

$$\Upsilon_s = (1-s)\Omega_{t_1} + s\Omega_{t_2}.$$

Using the compactness of M, it is not difficult to show that for t_2 sufficiently close to t_1, the family Υ_s is a 1-parameter family of symplectic forms on M for s in some open interval containing $[0,1]$. Moreover, by hypothesis, $[\Omega_{t_2} - \Omega_{t_1}] = 0$, so there exists a 1-form φ on M so that $d\varphi = \Omega_{t_2} - \Omega_{t_1}$. Thus,

$$\Upsilon_s = \Omega_{t_1} + s\, d\varphi.$$

By the special case already treated, there exists a diffeomorphism ϕ_{t_2,t_1} of M so that $\phi_{t_2,t_1}^*(\Omega_{t_2}) = \Omega_{t_1}$.

Finally, using the compactness of the interval $[0,t]$ for any $t \in [0,1]$, we can subdivide this interval into a finite number of intervals $[t_1, t_2]$ on which the above argument works. Then, by composing diffeomorphisms, we can construct a diffeomorphism ϕ_t of M so that $\phi_t^*(\Omega_t) = \Omega_0$.

The reader may have wanted the family of diffeomorphisms ϕ_t to depend continuously on t and smoothly on t if the family Ω_t is smooth in t. This can, in fact, be arranged. However, it involves showing that there is a smooth family of 1-forms φ_t on M so that $\frac{d}{dt}\Omega_t = d\varphi_t$, i.e., smoothly solving the d-equation. This can be done, but requires some delicacy or use of elliptic machinery (e.g., Hodge-deRham theory).

Warning: Theorem 1 does not hold without the hypothesis of compactness. For example, if Ω is the restriction of the standard structure on $\mathbb{R}^{2n}$ to the unit ball B^{2n}, then for the family $\Omega_t = e^t \Omega$ there cannot be any family of diffeomorphisms of the ball ϕ_t so that $\phi_t^*(\Omega_t) = \Omega$ since the integrals over B of the volume forms $(\Omega_t)^n = e^{nt}\Omega^n$ are all different.

Intuitively, Theorem 1 says that the "connected components" of the space of symplectic structures on a manifold are orbits of the group $\mathrm{Diff}^0(M)$ of diffeomorphisms isotopic to the identity. (The reason this is only intuitive is that we have not actually defined a topology on the space of symplectic structures on M.)

It is an interesting question as to how many "connected components" the space of symplectic structures on M has. The work of Gromov has yielded methods to attack this problem and I will have more to say about this in Lecture 9.

Submanifolds of symplectic manifolds

We will now pass on to the study of the geometry of submanifolds of a symplectic manifold. The following result describes the behaviour of symplectic structures near closed submanifolds. This theorem, due to Weinstein (see [Weinstein]), can be regarded as a generalization of Darboux' Theorem. The reader will note that the proof is quite similar to the proof of Theorem 1.

Theorem 2. *Let $P \subset M$ be a closed submanifold and let Ω_0 and Ω_1 be symplectic structures on M which have the property that $\Omega_0(p) = \Omega_1(p)$ for all $p \in P$. Then there exist open neighborhoods U_0 and U_1 of P and a diffeomorphism $\phi: U_0 \to U_1$ satisfying $\phi^*(\Omega_1) = \Omega_0$ and which moreover fixes P pointwise and satisfies $\phi'(p) = \mathrm{id}_p: T_p M \to T_p M$ for all $p \in P$.*

Proof. Consider the linear family of 2-forms

$$\Omega_t = (1-t)\Omega_0 + t\Omega_1$$

which "interpolates" between the forms Ω_0 and Ω_1. Since $[0,1]$ is compact and since, by hypothesis, $\Omega_0(p) = \Omega_1(p)$ for all $p \in P$, it easily follows that there is an open neighborhood U of P in M so that Ω_t is a symplectic structure on U for all t in some open interval $I = (-\varepsilon, 1+\varepsilon)$ containing $[0,1]$.

We may even suppose that U is a "tubular neighborhood" of P which has a smooth retraction $R: [0,1] \times U \to U$ into P. Since $\Phi = \Omega_1 - \Omega_0$ vanishes on P, it follows without too much difficultly (see the Exercises) that there is a 1-form φ on U which vanishes on P and which satisfies $d\varphi = \Phi$.

Now, on $I \times U$, consider the 2-form

$$\Omega = \Omega_0 + s\, d\varphi - \varphi \wedge ds.$$

This is a closed 2-form of half-rank n on $I \times U$. Just as in the previous theorem, it follows that there exists a unique vector field X on $I \times U$ so that $ds(X) = 1$ and $X \lrcorner \Omega = 0$.

Since φ and $d\varphi$ vanish on P, the vector field X has the property that $X(s,p) = \partial/\partial s$ for all $p \in P$ and $s \in I$. In particular, the set $\{0\} \times P$ lies in the domain of the time 1 flow of X. Since this domain is an open set, it follows that there is an open neighborhood U_0 of P in U so that $\{0\} \times U_0$ lies in the domain of the time 1 flow of X. The image of $\{0\} \times U_0$ under the time 1 flow of X is of the form $\{1\} \times U_1$ where U_1 is another open neighborhood of P in U.

Thus, the time 1 flow of X generates a diffeomorphism $\phi: U_0 \to U_1$. By the arguments of the previous theorem, it follows that $\phi^*(\Omega_1) = \Omega_0$. I leave it to the reader to check that ϕ fixes P in the desired fashion.

Theorem 2 has a useful corollary:

Corollary. *Let Ω be a symplectic structure on M and let f_0 and f_1 be smooth embeddings of a manifold P into M so that $f_0^*(\Omega) = f_1^*(\Omega)$ and so that there exists a smooth bundle isomorphism $\tau: f_0^*(TM) \to f_1^*(TM)$ which extends the identity map on the subbundle $TP \subset f_i^*(TM)$ and which identifies the symplectic structures on $f_i^*(TM)$. Then there exist open neighborhoods U_i of $f_i(P)$ in M and a diffeomorphism $\phi: U_0 \to U_1$ which satisfies $\phi^*(\Omega) = \Omega$ and, moreover, $\phi \circ f_0 = f_1$.*

Proof. It is an elementary result in differential topology that, under the hypotheses of the Corollary, there exists an open neighborhood W_0 of $f_0(P)$ in M and a smooth diffeomorphic embedding $\psi: W_0 \to M$ so that $\psi \circ f_0 = f_1$ and $\psi'(f_0(p)): T_{f_0(p)}(M) \to T_{f_1(p)}(M)$ is equal to $\tau(p)$. It follows that $\psi^*(\Omega)$ is a symplectic form on W_0 which agrees with Ω along $f_0(P)$. By Theorem 2, it follows that there is a neighborhood U_0 of $f_0(P)$ which lies in W_0 and a smooth map $\nu: U_0 \to W_0$ which is a diffeomorphism onto its image, fixes $f_0(P)$ pointwise, satisfies $\nu'(f_0(p)) = id_{f_0(p)}$ for all $p \in P$, and also satisfies $\nu^*(\psi^*(\Omega)) = \Omega$. Now just take $\phi = \psi \circ \nu$.

I will now give two particularly important applications of this result:

Symplectic submanifolds. If $P \subset M$ is a symplectic submanifold, then by using Ω, one can define a normal bundle for P as follows:

$$\nu(P) = \{(p,v) \in P \times TM \mid v \in T_pM,\ \Omega(v,w) = 0 \text{ for all } w \in T_pP\}.$$

The bundle $\nu(P)$ has a natural symplectic structure on each of its fibers (see the Exercises), and hence is a symplectic vector bundle. The following proposition shows that, up to local diffeomorphism, this normal bundle determines the symplectic structure Ω on a neighborhood of P.

Proposition 3. *Let (P, Υ) be a symplectic manifold and let $f_0, f_1 \colon P \to M$ be two symplectic embeddings of P as submanifolds of M so that the normal bundles $\nu_0(P)$ and $\nu_1(P)$ are isomorphic as symplectic vector bundles. Then there are open neighborhoods U_i of $f_i(P)$ in M and a symplectic diffeomorphism $\phi \colon U_0 \to U_1$ which satisfies $f_1 = \phi \circ f_0$.*

Proof. It suffices to construct the map τ required by the hypotheses of Theorem 2. Now, we have a symplectic bundle decomposition $f_i^*(TM) = TP \oplus \nu_i(P)$ for $i = 1, 2$. If $\alpha \colon \nu_0(P) \to \nu_1(P)$ is a symplectic bundle isomorphism, we then define $\tau = id \oplus \alpha$ in the obvious way and we are done.

Lagrangian submanifolds. At the other extreme, we want to consider submanifolds of M to which the form Ω pulls back to be as degenerate as possible.

Definition 2. If Ω is a symplectic structure on M^{2n}, an immersion $f \colon P \to M$ is said to be *isotropic* if $f^*(\Omega) = 0$. If the dimension of P is n, we say that f is a *Lagrangian* immersion. If in addition, f is one-to-one, then we say that $f(P)$ is a *Lagrangian submanifold* of M.

Note that the dimension of an isotropic submanifold of M^{2n} is at most n, so the Lagrangian submanifolds of M have maximal dimension among all isotropic submanifolds.

Example: graphs of symplectic mappings. If $f \colon M \to N$ is a symplectic mapping where Ω and Υ are the symplectic forms on M and N respectively, then the graph of f in $M \times N$ is an isotropic submanifold of $M \times N$ endowed with the symplectic structure $(-\Omega) \oplus \Upsilon = \pi_1^*(-\Omega) + \pi_2^*(\Upsilon)$. If M and N have the same dimension, then the graph of f in $M \times N$ is a Lagrangian submanifold.

Example: closed 1-forms. If α is a 1-form on M, then the graph of α in T^*M is a Lagrangian submanifold of T^*M if and only if $d\alpha = 0$. This follows because Ω on T^*M has the "reproducing property" that $\alpha^*(\Omega) = d\alpha$ for any 1-form on M.

Proposition 4. *Let Ω be a symplectic structure on M and let P be a closed Lagrangian submanifold of M. Then there exists an open neighborhood U of the zero section in T^*P and a smooth map $\phi \colon U \to M$ satisfying $\phi(0_p) = p$ which is a diffeomorphism onto an open neighborhood of P in M, and which pulls back Ω to be the standard symplectic structure on U.*

Proof. From the earlier proofs, the reader probably can guess what I will do. Let $\iota\colon P \to M$ be the inclusion mapping and let $\zeta\colon P \to T^*P$ be the zero section of T^*P. I leave as an exercise for the reader to show that $\zeta^*(T(T^*P)) = TP \oplus T^*P$, and that the induced symplectic structure Υ on this sum is simply the natural one on the sum of a bundle and its dual:

$$\Upsilon\big((v_1,\xi_1),(v_2,\xi_2)\big) = \xi_1(v_2) - \xi_2(v_1)$$

I will show that there is a bundle isomorphism $\tau\colon TP \oplus T^*P \to \iota^*(TM)$ which restricts to the subbundle TP to be $\iota'\colon TP \to \iota^*(TM)$.

First, select an n-dimensional subbundle $L \subset \iota^*(TM)$ which is complementary to $\iota'(TP) \subset \iota^*(TM)$. It is not difficult to show (and it is left as an exercise for the reader) that it is possible to choose L so that it is a Lagrangian subbundle of $\iota^*(TM)$ so that there is an isomorphism $\alpha\colon T^*P \to L$ so that $\tau\colon TP \oplus T^*P \to \iota'(TP) \oplus L$ defined by $\tau = \iota' \oplus \alpha$ is a symplectic bundle isomorphism.

Now apply the Corollary to Theorem 2.

Fixed points of symplectic mappings. Proposition 4 shows that the symplectic structure on a manifold M in a neighborhood of a closed Lagrangian submanifold P is completely determined by the diffeomorphism type of P. This fact has several interesting applications. We will only give one of them here.

Proposition 5. *Let (M,Ω) be a compact symplectic manifold with $H^1_{dR}(M,\mathbb{R}) = 0$. Then in $\mathrm{Diff}(M)$ endowed with the C^1 topology, there exists an open neighborhood $\mathcal{U}$ of the identity map so that any symplectomorphism $\phi\colon M \to M$ which lies in $\mathcal{U}$ has at least two fixed points.*

Proof. Consider the manifold $M \times M$ endowed with the symplectic structure $\Omega \oplus (-\Omega)$. The diagonal $\Delta \subset M \times M$ is a Lagrangian submanifold. Proposition 4 implies that there exists an open neighborhood U of the zero section in T^*M and a symplectic map $\psi\colon U \to M \times M$ which is a diffeomorphism onto its image so that $\psi(0_p) = (p,p)$.

Now, there is an open neighborhood $\mathcal{U}_0$ of the identity map on M in $\mathrm{Diff}(M)$ endowed with the C^0 topology which is characterized by the condition that ϕ belongs to $\mathcal{U}_0$ if and only if the graph of ϕ in $M \times M$, namely $id \times \phi$ lies in the open set $\psi(U) \subset M \times M$. Moreover, there is an open neighborhood $\mathcal{U} \subset \mathcal{U}_0$ of the identity map on M in $\mathrm{Diff}(M)$ endowed with the C^1 topology which is characterized by the condition that ϕ belongs to $\mathcal{U}$ if and only if $\psi^{-1} \circ (id \times \phi)\colon M \to T^*M$ is the graph of a 1-form α_ϕ.

Now suppose that $\phi \in \mathcal{U}$ is a symplectomorphism. By our previous discussion, it follows that the graph of ϕ in $M \times M$ is Lagrangian. This implies that the graph of α_ϕ is Lagrangian in T^*M which, by our second example, implies that α_ϕ is closed. Since $H^1_{dR}(M,\mathbb{R}) = 0$, this, in turn, implies that $\alpha_\phi = df_\phi$ for some smooth function f on M.

Since M is compact, it follows that f_ϕ must have at least two critical points. However, these critical points are zeros of the 1-form $df_\phi = \alpha_\phi$. It is a consequence of our construction that these points must then be places where the graph of ϕ intersects the diagonal Δ. In other words, they are fixed points of ϕ.

This theorem can be generalized considerably. According to a theorem of Hamilton [Ha], if M is compact, then there is an open neighborhood $\mathcal{U}$ of the identity map id in $\mathrm{Sp}(\Omega)$ (with the C^1 topology) so that every $\phi \in \mathcal{U}$ is the time-one flow of a symplectic vector field $X_\phi \in \mathrm{sp}(\Omega)$. If X_ϕ is actually Hamiltonian (which would, of course, follow if $H^1_{dR}(M, \mathbb{R}) = 0$), then $-X_\phi \lrcorner \Omega = df_\phi$, so X_ϕ will vanish at the critical points of f_ϕ and these will be fixed points of ϕ.

Appendix: Lie's transformation groups, II

Lie's theory of transformations

The reader who is learning symplectic geometry for the first time may be astonished by the richness of the subject and, at the same time, be wondering "Are there other geometries like symplectic geometry which remain to be explored?" The point of this appendix is to give one possible answer to this very vague question.

When Lie began his study of transformation groups in n variables, he modeled his attack on the known study of the finite groups. Thus, his idea was that he would find all of the "simple groups" first and then assemble them (by solving the extension problem) to classify the general group. Thus, if one "group" G had a homomorphism onto another "group" H

$$1 \longrightarrow K \longrightarrow G \longrightarrow H \longrightarrow 1$$

then one could regard G as a semi-direct product of H with the kernel subgroup K.

Guided by this idea, Lie decided that the first task was to classify the transitive transformation groups G, i.e., the ones which acted transitively on $\mathbb{R}^n$ (at least locally). The reason for this was that, if G had an orbit S of dimension $0 < k < n$, then the restriction of the action of G to S would give a non-trivial homomorphism of G into a transformation group in fewer variables.

Second, Lie decided that he needed to classify first the "groups" which, in his language, "did not preserve any subset of the variables." The example he had in mind was the group of diffeomorphisms of $\mathbb{R}^2$ of the form

$$\phi(x, y) = \big(f(x), g(x, y)\big).$$

Clearly the assignment $\phi \mapsto f$ provides a homomorphism of this group into the group of diffeomorphisms in one variable. Lie called groups which "did not preserve any subset of the variables" *primitive*. In modern language, *primitive* is taken to mean that G does not preserve any foliation on $\mathbb{R}^n$ (coordinates on the leaf space would furnish a "proper subset of the variables" which was preserved by G).

Thus, the fundamental problem was to classify the "primitive transitive continuous transformation groups".

The finite dimensional simple groups

When the algebra of infinitesimal generators of G was finite dimensional, Lie and his coworkers made good progress. Their work culminated in the work of Cartan and Killing, classifying the finite dimensional simple Lie groups. (Interestingly enough, they did not then go on to solve the extension problem and so classify all Lie groups. Perhaps they regarded this as a problem of lesser order. Or, more likely, the classification turned out to be messy, uninteresting, and ultimately intractable.)

They found that the simple groups fell into two types. Besides the special linear groups, such as $\mathrm{SL}(n,\mathbb{R})$, $\mathrm{SL}(n,\mathbb{C})$ and other complex analogs; orthogonal groups, such as $\mathrm{SO}(p,q)$ and its complex analogs; and symplectic groups, such as $\mathrm{Sp}(n,\mathbb{R})$ and its complex analogs (which became known as the classical groups), there were five "exceptional" types. This story is quite long, but very interesting. The "finite dimensional Lie groups" went on to become an essential part of the foundation of modern differential geometry. A complete account of this classification (along with very interesting historical notes) can be found in [He].

The infinite dimensional primitive groups

However, when the algebra of infinitesimal generators of G was *infinite* dimensional, the story was not so complete. Lie himself identified four classes of these "infinite dimensional primitive transitive transformation groups". They were

- In every dimension n, the full diffeomorphism group, $\mathrm{Diff}(\mathbb{R}^n)$.
- In every dimension n, the group of diffeomorphisms which preserve a fixed volume form μ, denoted by $\mathsf{SDiff}(\mu)$.
- In every even dimension $2n$, the group of diffeomorphisms which preserve the standard symplectic form

$$\Omega_n = dx_1 \wedge dy^1 + \cdots + dx_n \wedge dy^n,$$

denoted by $\mathsf{Sp}(\Omega_n)$.

- In every odd dimension $2n+1$, the group of diffeomorphisms which preserve, *up to a scalar function multiple*, the 1-form

$$\omega_n = dz + x_1\, dy^1 + \cdots + x_n\, dy^n.$$

This "group" was known as the *contact group* and I will denote it by $\mathsf{Ct}(\omega_n)$.

However, Lie and his coworkers were never able to discover any others, though they searched diligently. (By the way, Lie was aware that there were also holomorphic analogs acting in $\mathbb{C}^n$, but, at that time, the distinction between real and complex was not generally made explicit. Apparently, an educated reader was supposed to know or be able to guess what the generalizations to the complex category were.)

Cartan's work. In a series of four papers spanning from 1902 to 1910, Élie Cartan reformulated Lie's problem in terms of systems of partial differential equations and, under the hypothesis of analyticity (real and complex were not carefully distinguished), he proved that Lie's classes were essentially all of the infinite dimensional primitive transitive transformation groups. The slight extension was that $\mathsf{SDiff}(\mu)$ had a companion extension to $\mathbb{R}\cdot\mathsf{SDiff}(\mu)$, the diffeomorphisms which preserve μ up to a constant multiple and that $\mathsf{Sp}(\Omega_n)$ had a companion extension to $\mathbb{R}\cdot\mathsf{Sp}(\Omega_n)$, the diffeomorphisms which preserve Ω_n up to a constant multiple. Of course, there were also the holomorphic analogues of these. Notice the remarkable fact that there are no "exceptional infinite dimensional primitive transitive transformation groups".

These papers are remarkable, not only for their results, but for the wealth of concepts which Cartan introduced in order to solve his problem. In these papers, Cartan introduces the notion of G-structures (of all orders), principal bundles and their connections, jet bundles, prolongation (both of group actions and exterior differential systems), and a host of other ideas which were only appreciated much later. Perhaps because of its originality, Cartan's work in this area was essentially ignored for many years.

Lie pseudo-groups

In the 1950's, when algebraic varieties were being explored and developed as complex manifolds, it began to be understood that complex manifolds were to be thought of as manifolds with an atlas of coordinate charts whose "overlaps" were holomorphic. Generalizing this example, it became clear that, for any collection Γ of local diffeomorphisms of $\mathbb{R}^n$ which satisfied the following definition, one could define a category of Γ-manifolds as manifolds endowed with an atlas $\mathcal{A}$ of coordinate charts whose overlaps lay in $\mathcal{A}$.

Definition 3. A *local diffeomorphism* of $\mathbb{R}^n$ is a pair (U, ϕ) where $U \subset \mathbb{R}^n$ is an open set and $\phi: U \to \mathbb{R}^n$ is a one-to-one diffeomorphism onto its image. A set Γ of local diffeomorphisms of $\mathbb{R}^n$ is said to form a *pseudo-group* on $\mathbb{R}^n$ if it satisfies the following three properties:
  1. (Composition and Inverses) If (U, ϕ) and (V, ψ) are in Γ, then $(\phi^{-1}(V), \psi \circ \phi)$ and $(\phi(U), \phi^{-1})$ also belong to Γ.
  2. (Localization and Globalization) If (U, ϕ) is in Γ, and $W \subset U$ is open, then $(W, \phi_{|W})$ is also in Γ. Moreover, if (U, ϕ) is a local diffeomorphism of $\mathbb{R}^n$ such that U can be written as the union of open subsets W_α for which $(W_\alpha, \phi_{|W_\alpha})$ is in Γ for all α, then (U, ϕ) is in Γ.
  3. (Non-triviality) $(\mathbb{R}^n, id)$ is in Γ.

As it turned out, the pseudo-groups Γ of interest in geometry were exactly the ones which could be characterized as the (local) solutions of a system of partial differential equations, i.e., they were Lie's transformation groups. This caused a revival of interest in Cartan's work. Consequently, much of Cartan's work has now been redone in modern language. In particular, Cartan's classification was redone according to modern standards of rigor and a very readable account of this theory can be found in [SS].

In any case, symplectic geometry, seen in this light, is one of a small handful of "natural" geometries that one can impose on manifolds.

Exercises

1. Assume $n > 1$. Show that if $A_{r,R} \subset \mathbb{R}^{2n}$ (with its standard symplectic structure) is the annulus described by the relations $r < |\mathbf{x}| < R$, then there cannot be a symplectic diffeomorphism $\phi \colon A_{r,R} \to A_{s,S}$ which "exchanges the boundaries". (Hint: Show that if ϕ existed one would be able to construct a symplectic structure on S^{2n}.) Conclude that one cannot naïvely define connected sum in the category of symplectic manifolds. (The "naïve" definition would be to try to take two symplectic manifolds M_1 and M_2 of the same dimension, choose an open ball in each one, cut out a sub-ball of each and identify the resulting annuli by an appropriate diffeomorphism which was chosen to be a symplectomorphism.)

2. This exercise completes the proof of Proposition 1.
   (i) Let $\mathcal{S}_n^+$ denote the space of n-by-n positive definite symmetric matrices. Show that the map $\sigma \colon \mathcal{S}_n^+ \to \mathcal{S}_n^+$ defined by $\sigma(s) = s^2$ is a one-to-one diffeomorphism of $\mathcal{S}_n^+$ onto itself. Conclude that every element of $\mathcal{S}_n^+$ has a unique positive definite square root and that the map $s \mapsto \sqrt{s}$ is a smooth mapping. Show also that, for any $r \in O(n)$, we have $\sqrt{{}^t r a r} = {}^t r \sqrt{a}\, r$, so that the square root function is $O(n)$-equivariant.
   (ii) Let $\mathcal{A}_n^{\bullet}$ denote the space of n-by-n invertible anti-symmetric matrices. Show that, for $a \in \mathcal{A}_n^{\bullet}$, the matrix $-a^2$ is symmetric and positive definite. Show that the matrix $b = \sqrt{-a^2}$ is the unique symmetric positive definite matrix which satisfies $b^2 = -a^2$ and moreover that b commutes with a. Check also that the mapping $a \mapsto \sqrt{-a^2}$ is $O(n)$-equivariant.
   (iii) Now verify the claim made in the proof of Proposition 1 that, for any smooth vector bundle E over a manifold M endowed with a smooth inner product on the fibers and any smooth, invertible skew-symmetric bundle mapping $A \colon E \to E$, there exists a unique smooth positive definite symmetric bundle mapping $B \colon E \to E$ which satisfies $B^2 = -A^2$ and which commutes with A.

3. This exercise requires that you know something about characteristic classes.
   (i) Show that S^{4n} has no almost complex structure for any n. (Hint: What could the total Chern and Pontrjagin class of the tangent bundle be?)
      (Using the Bott Periodicity Theorem, it can be shown that the characteristic class c_n of any complex bundle over S^{2n} must be an integer multiple of $(n-1)!\, v$ where $v \in H^{2n}(S^{2n})$ is a generator. It follows that, among the spheres, only S^2 and S^6 could have almost complex structures and, in fact, they both do. It is a long standing problem whether or not S^6 has a complex structure.)
   (ii) Using the formulas for 4-manifolds developed in the Lecture, determine how many possibilities there are for the first Chern class $c_1(J)$ of an almost complex structure J on M where M a connected sum of 3 or 4 copies of $\mathbb{CP}^2$.

4. Show that, if Ω_0 is a symplectic structure on a compact manifold M, then there is an open neighborhood U in $H^2(M, \mathbb{R})$ of $[\Omega_0]$, such that, for all $u \in U$, there is a symplectic structure Ω_u on M with $[\Omega_u] = u$. (Hint: Since M is

compact, for any closed 2-form Υ, the 2-form $\Omega + t\Upsilon$ is non-degenerate for all sufficiently small t.)

5. Mimic the proof of Theorem 1 to prove another theorem of Moser: For any compact, connected, oriented manifold M, two volume forms μ_0 and μ_1 differ by an oriented diffeomorphism (i.e., there exists an orientation preserving diffeomorphism $\phi: M \to M$ which satisfies $\phi^*(\mu_1) = \mu_0$) if and only if

$$\int_M \mu_0 = \int_M \mu_1.$$

(This theorem is also true without the hypothesis of compactness, but the proof is slightly more delicate.)

6. Let M be a connected, smooth oriented 4-manifold and let $\mu \in \mathcal{A}^4(M)$ be a volume form which satisfies $\int_M \mu = 1$. (By the previous problem, any two such forms differ by an oriented diffeomorphism of M.) For any (smooth) $\Omega \in \mathcal{A}^2(M)$, define $*(\Omega^2) \in C^\infty(M)$ by the equation

$$\Omega^2 = *(\Omega^2)\,\mu.$$

Now, fix a cohomology class $u \in H^2_{dR}(M)$ satisfying $u^2 = r[\mu]$ where $r \neq 0$. Define the functional $\mathcal{F}: u \to \mathbb{R}$

$$\mathcal{F}(\Omega) = \int_M *(\Omega^2)\,\Omega^2 \qquad \text{for } \Omega \in u.$$

Show that any $\mathcal{F}$-critical 2-form $\Omega \in u$ is a symplectic form satisfying $*(\Omega^2) = r$ and that $\mathcal{F}$ has no critical values other than r^2. Show also that $\mathcal{F}(\Omega) \geq r^2$ for all $\Omega \in u$.

This motivates defining an invariant of the class u by

$$\mathcal{I}(u) = \inf_{\Omega \in u} \mathcal{F}(\Omega).$$

Gromov has suggested (private communication) that perhaps $\mathcal{I}(u) = r^2$ for all u, even when the infimum is not attained.

7. Let $P \subset M$ be a closed submanifold and let $U \subset M$ be an open neighborhood of P in M which can be retracted onto P, i.e., there exists a smooth map $R: U \times [0,1] \to U$ so that $R(u,1) = u$ for all $u \in U$, $R(p,t) = p$ for all $p \in P$ and $t \in [0,1]$, and $R(u,0)$ lies in P for all $u \in U$. (Every closed submanifold of M has such a neighborhood.)

Show that if Φ is a closed k-form on U which vanishes at every point of P, then there exists a $(k-1)$-form ϕ on U which vanishes on P and satisfies $d\phi = \Phi$. (Hint: Mimic Poincaré's Homotopy Argument: Let $\Upsilon = R^*(\Phi)$ and set $v = \frac{\partial}{\partial t} \lrcorner \Upsilon$. Then, using the fact that $v(u,t)$ can be regarded as a $(k-1)$-form at u for all t, define

$$\phi(u) = \int_0^1 v(u,t)\,dt.$$

Now verify that ϕ has the desired properties.)
8. Show that Theorem 2 implies Darboux' Theorem. (Hint: Take P to be a point in a symplectic manifold M.)
9. This exercise assumes that you have done Exercise 5.13. Let (M, Ω) be a symplectic manifold. Show that the following description of the flux homomorphism is valid. Let p be an e-based path in $\mathrm{Sp}(\Omega)$. Thus, $p \colon [0,1] \times M \to M$ satisfies $p_t^*(\Omega) = \Omega$ for all $0 \leq t \leq 1$. Show that $p^*(\Omega) = \Omega + \varphi \wedge dt$ for some 1-form φ on $[0,1] \times M$. Let $\iota_t \colon M \to [0,1] \times M$ be the "t-slice inclusion": $\iota_t(m) = (t, m)$, and set $\varphi_t = \iota_t^*(\varphi)$.

Show that φ_t is closed for all $0 \leq t \leq 1$. Show that if we set

$$\tilde{\Phi}(p) = \int_0^1 \varphi_t \, dt,$$

then the cohomology class $[\tilde{\Phi}(p)] \in H^1_{dR}(M, \mathbb{R})$ depends only on the homotopy class of p and hence defines a map $\Phi \colon \widetilde{\mathrm{Sp}^0}(\Omega) \to H^1_{dR}(M, \mathbb{R})$. Verify that this map is the same as the flux homomorphism defined in Exercise 5.13.

Use this description to show that if p is in the kernel of Φ, then p is homotopic to a path p' for which the forms φ_t' are all exact. This shows that the kernel of Φ is actually connected.

10. The point of this exercise is to show that any symplectic vector bundle over a symplectic manifold (M, Ω) can occur as the symplectic normal bundle for some symplectic embedding M into some other symplectic manifold.

Let (M, Ω) be a symplectic manifold and let $\pi \colon E \to M$ be a symplectic vector bundle over M of rank $2n$. (I.e., E comes equipped with a section B of $\Lambda^2(E^*)$ which restricts to each fiber E_m to be a symplectic structure B_m.) Show that there exists a symplectic structure Ψ on an open neighborhood in E of the zero section of E which satisfies the condition that $\Psi_{0_m} = \Omega_m + B_m$ under the natural identification $T_{0_m} E = T_m M \oplus E_m$.

(Hint: Choose a locally finite open cover $\mathcal{U} = \{U_\alpha \,|\, \alpha \in A\}$ of M so that, if we define $E_\alpha = \pi^{-1}(U_\alpha)$, then there exists a symplectic trivialization $\tau_\alpha \colon E_\alpha \to \mathbb{R}^{2n}$ (where $\mathbb{R}^{2n}$ is given its standard symplectic structure $\Omega_0 = dx_i \wedge dy^i$). Now let $\{\lambda_\alpha \,|\, \alpha \in A\}$ be a partition of unity subordinate to the cover $\mathcal{U}$. Show that the form

$$\Psi = \pi^*(\Omega) + \sum_\alpha d\bigl(\lambda_\alpha \, \tau_\alpha^*(x_i \, dy^i)\bigr)$$

has the desired properties.)

11. Show that if E is a symplectic vector bundle over M and $L \subset E$ is a Lagrangian subbundle, then E is isomorphic to $L \oplus L^*$ as a symplectic bundle. (The symplectic bundle structure Υ on $L \oplus L^*$ is the one which, on each fiber satisfies

$$\Upsilon\bigl((v, \alpha), (w, \beta)\bigr) = \alpha(w) - \beta(v). \;)$$

(Hint: First choose a complementary subbundle $F \subset E$ so that $E = L \oplus F$. Show that F is naturally isomorphic to L^* abstractly by using the fact that the symplectic structure on E is non-degenerate. Then show that there exists a bundle map $A: F \to L$ so that

$$\tilde{F} = \{v + Av \mid v \in F\}$$

is also a Lagrangian subbundle of E which is complementary to L and isomorphic to L^* via some bundle map $\alpha: L^* \to \tilde{F}$. Now show that $id \oplus \alpha: L \oplus L^* \to L \oplus \tilde{F} \simeq E$ is a symplectic bundle isomorphism.)

12. ACTION-ANGLE COORDINATES. Proposition 4 can be used to prove a result of Arnol'd about the existence of so-called action angle coordinates in the neighborhood of a compact level set of a completely integrable Hamiltonian system. (See Lecture 5). Here is how this goes: Let (M^{2n}, Ω) be a symplectic manifold and let $f = (f^1, \ldots, f^n): M \to \mathbb{R}^n$ be a smooth submersion with the property that the coordinate functions f^i are in involution, i.e., $\{f^i, f^j\} = 0$. Suppose that, for some $c \in \mathbb{R}^n$, the f-level set $M_c = f^{-1}(c)$ is compact. Replacing f by $f - c$, we may assume that $c = 0$, which we do from now on.

Show that $M_0 \subset M$ is a closed Lagrangian submanifold of M.

Use Proposition 4 to show that there is an open neighborhood B of $0 \in \mathbb{R}^n$ so that $(f^{-1}(B), \Omega)$ is symplectomorphic to a neighborhood U of the zero section in T^*M_0 (endowed with its standard symplectic structure) in such a way that, for each $b \in B$, the submanifold $M_b = f^{-1}(b)$ is identified with the graph of a closed 1-form ω_b on M_0. Show that it is possible to choose $b_1, \ldots, b_n$ in B so that the corresponding closed 1-forms $\omega_1, \ldots, \omega_n$ are linearly independent at every point of M_0.

Conclude that M_0 is diffeomorphic to a torus $T = \mathbb{R}^n/\Lambda$ where $\Lambda \subset \mathbb{R}^n$ is a lattice, in such a way that the forms ω_i become identified with $d\theta_i$ where θ_i are the corresponding linear coordinates on $\mathbb{R}^n$.

Now prove that for any $b \in B$, the 1-form ω_b must be a linear combination of the ω_i with constant coefficients. Thus, there are functions a^i on B so that $\omega_b = a^i(b)\omega_i$. (Hint: Show that the coefficients must be invariant under the flows of the vector fields dual to the ω_i.)

Conclude that, under the symplectic map identifying M_B with U, the form Ω gets identified with $da^i \wedge d\theta_i$. The functions a^i and θ_i are the so-called "action-angle coordinates".

Extra Credit: Trace through the methods used to prove Proposition 4 and show that, in fact, the action-angle coordinates can be constructed using quadrature and "finite" operations.

LECTURE 7
Classical Reduction

In this section, we return to the study of group actions. This time, however, we will concentrate on group actions on symplectic manifolds which preserve the symplectic structure. Such actions happen to have quite interesting properties and moreover, turn out to have a wide variety of applications.

Symplectic group actions

First, the basic definition.

Definition 1. Let (M, Ω) be a symplectic manifold and let G be a Lie group. A left action $\lambda: G \times M \to M$ of G on M is a *symplectic* action if $\lambda_a^*(\Omega) = \Omega$ for all $a \in G$.

We have already encountered several examples:

Example: Lagrangian symmetries. If G acts on a manifold M is such a way that it preserves a non-degenerate Lagrangian $L: TM \to \mathbb{R}$, then, by construction, it preserves the symplectic 2-form $d\omega_L$.

Example: cotangent actions. A left G-action $\lambda: G \times M \to M$, induces an action $\tilde{\lambda}$ of G on T^*M. Namely, for each $a \in G$, the diffeomorphism $\lambda_a: M \to M$ induces a diffeomorphism $\tilde{\lambda}_a: T^*M \to T^*M$. Since the natural symplectic structure on T^*M is invariant under diffeomorphisms, it follows that $\tilde{\lambda}$ is a symplectic action.

Example: coadjoint orbits. As we saw in Lecture 5, for every $\xi \in \mathfrak{g}^*$, the coadjoint orbit $G \cdot \xi$ carries a natural G-invariant symplectic structure Ω_ξ. Thus, the left action of G on $G \cdot \xi$ is symplectic.

Example: circle actions on $\mathbb{C}^n$. Let $z^1, \ldots, z^n$ be linear complex coordinates on $\mathbb{C}^n$ and let this vector space be endowed with the symplectic structure

$$\Omega = \tfrac{i}{2}(dz^1 \wedge d\bar{z}^1 + \cdots + dz^n \wedge d\bar{z}^n)$$
$$= dx^1 \wedge dy^1 + \cdots + dx^n \wedge dy^n$$

where $z^k = x^k + iy^k$. Then for any integers $(k_1, \ldots, k_n)$, we can define an action of S^1 on $\mathbb{C}^n$ by the formula

$$e^{i\theta} \cdot \begin{pmatrix} z^1 \\ \vdots \\ z^n \end{pmatrix} = \begin{pmatrix} e^{ik_1\theta} z^1 \\ \vdots \\ e^{ik_n\theta} z^n \end{pmatrix}$$

The reader can easily check that this defines a symplectic circle action on $\mathbb{C}^n$.

Generally we will be interested in the following situation: Y will be a Hamiltonian vector field on a symplectic manifold (M, Ω) and G will act symplectically on M as a group of symmetries of the flow of Y. We want to understand how to use the action of G to "reduce" the problem of integrating the flow of Y.

In Lecture 3, we saw that when Y was the Euler-Lagrange vector field associated to a non-degenerate Lagrangian L, then the infinitesimal generators of symmetries of L could be used to generate conserved quantities for the flow of Y. We want to extend this process (as far as is reasonable) to the general case.

For the rest of the lecture, I will assume that G is a Lie group with a symplectic action λ on a connected symplectic manifold (M, Ω).

Hamiltonian actions

Since λ is symplectic, it follows that the mapping $\lambda_*: \mathfrak{g} \to \mathfrak{X}(M)$ actually has image in $\mathrm{sp}(\Omega)$, the algebra of symplectic vector fields on M. As we saw in Lecture 3, λ_* is an anti-homomorphism, i.e., $\lambda_*([x,y]) = -[\lambda_*(x), \lambda_*(y)]$. Since, as we saw in Lecture 5, $[\mathrm{sp}(\Omega), \mathrm{sp}(\Omega)] \subset \mathrm{h}(\Omega)$, it follows that $\lambda_*([\mathfrak{g},\mathfrak{g}]) \subset \mathrm{h}(\Omega)$. Thus, $H_\lambda: \mathfrak{g} \to H^1_{dR}(M, \mathbb{R})$ defined by $H_\lambda(x) = [\lambda_*(x) \lrcorner \Omega]$ is a homomorphism of Lie algebras with kernel containing the commutator subalgebra $[\mathfrak{g}, \mathfrak{g}]$.

The map H_λ is the obstruction to finding a Hamiltonian function associated to each infinitesimal symmetry $\lambda_*(x)$ since $H_\lambda(x) = 0$ if and only if $\lambda_*(x) \lrcorner \Omega = -df$ for some $f \in C^\infty(M)$.

Definition 2. A symplectic action $\lambda: G \times M \to M$ is said to be *Hamiltonian* if $H_\lambda = 0$, i.e., if $\lambda_*(\mathfrak{g}) \subset \mathrm{h}(\Omega)$.

There are a few particularly interesting cases where the obstruction H_λ must vanish:

1. • If $H^1_{dR}(M, \mathbb{R}) = 0$. In particular, if M is simply connected.
2. • If $\mathfrak{g}$ is perfect, i.e., $[\mathfrak{g}, \mathfrak{g}] = \mathfrak{g}$. For example this happens whenever the Killing form on $\mathfrak{g}$ is non-degenerate (this is the first Whitehead Lemma, see Exercise 3). However, this is not the only case: For example, if G is the group of rigid motions in $\mathbb{R}^n$ for $n \geq 3$, then $\mathfrak{g}$ is perfect, even though its Killing form is degenerate.
3. • If there exists a 1-form ω on M which is invariant under G and satisfies $\Omega = d\omega$. (This is the case for symmetries of a Lagrangian.) To see this, note that if X is a vector field on M which preserves ω, then

$$0 = \mathcal{L}_X \omega = d(X \lrcorner \omega) + X \lrcorner \Omega,$$

so $X \lrcorner \Omega$ is exact.

Poisson actions. For a Hamiltonian action λ, every infinitesimal symmetry $\lambda_*(x)$ has a Hamiltonian function $f_x \in C^\infty$. However, the choice of f_x is not unique since we can add any constant to f_x without changing its Hamiltonian vector field. This non-uniqueness causes some problems in the theory we wish to develop.

To see why, suppose that we choose a (linear) lifting $\rho\colon \mathfrak{g} \to C^\infty(M)$ of $-\lambda_*\colon \mathfrak{g} \to h(\Omega)$. (The choice of $-\lambda_*$ instead of λ_* was made to get rid of the annoying sign in the formula for the bracket.)

$$\begin{array}{ccccccccc} & & & & \mathfrak{g} & & & & \\ & & & \rho\downarrow & & \searrow^{-\lambda_*} & & & \\ 0 & \to & \mathbb{R} & \to & C^\infty(M) & \to & h(\Omega) & \to & 0 \end{array}$$

Thus, for every $x \in \mathfrak{g}$, we have $\lambda_*(x) \lrcorner \Omega = d(\rho(x))$. A short calculation (see the Exercises) now shows that $\{\rho(x), \rho(y)\}$ is a Hamiltonian function for $-\lambda_*([x,y])$, i.e., that

$$\lambda_*([x,y]) \lrcorner \Omega = d(\{\rho(x), \rho(y)\}).$$

In particular, it follows (since M is connected) that there must be a skew-symmetric bilinear map $c_\rho\colon \mathfrak{g} \times \mathfrak{g} \to \mathbb{R}$ so that

$$\{\rho(x), \rho(y)\} = \rho([x,y]) + c_\rho(x,y).$$

An application of the Jacobi identity implies that the map c_ρ satisfies the condition

$$c_\rho([x,y], z) + c_\rho([y,z], x) + c_\rho([z,x], y) = 0 \quad \text{for all } x, y, z \in \mathfrak{g}.$$

This condition is known as the *2-cocycle condition* for c_ρ regarded as an element of $A^2(\mathfrak{g}) = \Lambda^2(\mathfrak{g}^*)$. (See Exercise 3 for an explanation of this terminology.)

For purposes of simplicity, it would be nice if we could choose ρ so that c_ρ were identically zero. In order to see whether this is possible, let us choose another linear map $\tilde{\rho}\colon \mathfrak{g} \to C^\infty(M)$ which satisfies $\tilde{\rho}(x) = \rho(x) + \xi(x)$ where $\xi\colon \mathfrak{g} \to \mathbb{R}$ is any linear map. Every possible lifting of $-\lambda_*$ is clearly of this form for some ξ. Now we compute that

$$\{\tilde{\rho}(x), \tilde{\rho}(y)\} = \{\rho(x), \rho(y)\} = \rho([x,y]) + c_\rho(x,y)$$
$$= \tilde{\rho}([x,y]) + c_\rho(x,y) - \xi([x,y]).$$

Thus, $c_{\tilde{\rho}}(x,y) = c_\rho(x,y) - \xi([x,y])$. Thus, in order to be able to choose $\tilde{\rho}$ so that $c_{\tilde{\rho}} = 0$, we see that there must exist a $\xi \in \mathfrak{g}^*$ so that $c_\rho = -\delta\xi$ where $\delta\xi$ is the skew-symmetric bilinear map on $\mathfrak{g}$ which satisfies $\delta\xi(x,y) = -\xi([x,y])$ (see Exercise 3 for an explanation of this notation). This is known as the *2-coboundary condition*.

There are several important cases where we can assure that c_ρ can be written in the form $-\delta\xi$. Among them are:

1. • If M is compact, then the sequence
$$0 \to \mathbb{R} \to C^\infty(M) \to \mathsf{H}(\Omega) \to 0$$
splits: If we let $C_0^\infty(M,\Omega) \subset C^\infty(M)$ denote the space of functions f for which $\int_M f\,\Omega^n = 0$, then these functions are closed under Poisson bracket (see Exercise 5.6 for a hint as to why this is true) and we have a splitting of Lie algebras $C^\infty(M) = \mathbb{R} \oplus C_0^\infty(M,\Omega)$. Now just choose the unique ρ so that it takes values in $C_0^\infty(M,\Omega)$. This will clearly have $c_\rho = 0$.

2. • If $\mathfrak{g}$ has the property that every 2-cocycle for $\mathfrak{g}$ is actually a 2-coboundary. This happens, for example, if the Killing form of $\mathfrak{g}$ is non-degenerate (this is the second Whitehead Lemma, see Exercise 3), though it can also happen for other Lie algebras. For example, for the non-abelian Lie algebra of dimension 2, it is easy to see that every 2-cocycle is a 2-coboundary.

3. • If there is a 1-form ω on M which is preserved by the G action and satisfies $d\omega = \Omega$. (This is true in the case of symmetries of a Lagrangian.) In this case, we can merely take $\rho(x) = -\omega(\lambda_*(x))$. I leave as an exercise for the reader to check that this works.

Definition 3. A Hamiltonian action $\lambda\colon G \times M \to M$ is said to be a *Poisson* action if there exists a lifting ρ with $c_\rho = 0$.

Henceforth in this Lecture, I am only going to consider Poisson actions. By my previous remarks, this case includes all of the Lagrangians with symmetries, but it also includes others. I will assume that, in addition to having the action $\lambda\colon G \times M \to M$ specified, we have chosen a lifting $\rho\colon \mathfrak{g} \to C^\infty(M)$ of $-\lambda_*$ which satisfies $\{\rho(x),\rho(y)\} = \rho([x,y])$ for all $x,y \in \mathfrak{g}$.

The moment map

We are now ready to make one of the most important constructions in the theory.

Definition 4. The *moment map* associated to λ and ρ is the mapping $\mu\colon M \to \mathfrak{g}^*$ which satisfies
$$\mu(m)(y) = \rho(y)(m).$$

Note that, for fixed $m \in M$, the assignment $y \mapsto \rho(y)(m)$ is a linear map from $\mathfrak{g}$ to $\mathbb{R}$, so the definition makes sense.

It is worth pausing to consider why this mapping is called the moment map. The reader should calculate this mapping in the case of a free particle or a rigid body moving in space. In either case, the Lagrangian is invariant under the action of the group G of rigid motions of space. If $y \in \mathfrak{g}$ corresponds to a translation, then $\rho(y)$ gives the function on $T\mathbb{R}^3$ which evaluates at each point (i.e., each position-plus-velocity) to be the linear momentum in the direction of translation. If y corresponds to rotation about a fixed axis, then $\rho(y)$ turns out to be the angular momentum of the body about that axis.

LECTURE 7. CLASSICAL REDUCTION

One important reason for studying the moment map is the following generalization of the classical conservation of momentum theorems:

Proposition 1. *If Y is a symplectic vector field on M which is invariant under the action of G, then μ is constant on the integral curves of Y.*

Thus, the moment map gives us conserved quantities for any G-invariant vector field.

The main result about the moment map is the following one. (I remind the reader that I am assuming that all actions are Poisson.)

Theorem 1. *If G is connected, then the moment map $\mu\colon M \to \mathfrak{g}^*$ is G-equivariant.*

Proof. Recall that the coadjoint action of G on $\mathfrak{g}^*$ is defined by $\mathrm{Ad}^*(g)(\xi)(x) = \xi(\mathrm{Ad}(g^{-1})x)$. That μ be G-equivariant, i.e., that $\mu(g \cdot m) = \mathrm{Ad}^*(g)(\mu(m))$ for all $m \in M$ and $g \in G$, is thus seen to be equivalent to the condition that

$$\rho\big(\mathrm{Ad}(g^{-1})y\big)(m) = \rho(y)(g \cdot m)$$

for all $m \in M$, $g \in G$, and $y \in \mathfrak{g}$. This is the identity I shall prove.

Since G is connected and since each side of the above equation represents a G-action, if we prove that the above formula holds for g of the form $g = e^{tx}$ for any $x \in \mathfrak{g}$ and any $t \in \mathbb{R}$, the formula for general g will follow. Thus, we want to prove that

$$\rho\big(\mathrm{Ad}(e^{-tx})y\big)(m) = \rho(y)(e^{tx} \cdot m)$$

for all t. Since this latter equation holds at $t = 0$, it is enough to show that both sides have the same derivative with respect to t.

Now the derivative of the right hand side of the formula is

$$\begin{aligned}
d\big(\rho(y)\big)\big(\lambda_*(x)(e^{tx} \cdot m)\big) &= \Omega\big(\lambda_*(y)(e^{tx} \cdot m), \lambda_*(x)(e^{tx} \cdot m)\big) \\
&= \Omega\big(\lambda_*(\mathrm{Ad}(e^{-tx})y)(m), \lambda_*(\mathrm{Ad}(e^{-tx})x)(m)\big) \\
&= \Omega\big(\lambda_*(\mathrm{Ad}(e^{-tx})y)(m), \lambda_*(x)(m)\big)
\end{aligned}$$

where, to verify the second equality we have used the identity

$$\lambda'_a\big(\lambda_*(y)(m)\big) = \lambda_*\big(\mathrm{Ad}(a)y\big)(a \cdot m)$$

and the fact that Ω is G-invariant.

On the other hand, the derivative of the left hand side of the formula is clearly

$$\begin{aligned}
\rho\big([-x, \mathrm{Ad}(e^{-tx})y]\big)(m) &= -\big\{\rho(x), \rho\big(\mathrm{Ad}(e^{-tx})y\big)\big\}(m) \\
&= \Omega\big(\lambda_*(\mathrm{Ad}(e^{-tx})y)(m), \lambda_*(x)(m)\big)
\end{aligned}$$

so we are done. (Note that I have used my assumption that $c_\rho = 0$!)

Example: left-invariant metrics on Lie groups. Let G be a Lie group and let $Q: \mathfrak{g} \to \mathbb{R}$ be a non-degenerate quadratic form with associated inner product $\langle , \rangle_Q$. Let $L: TG \to \mathbb{R}$ be the Lagrangian

$$L = \tfrac{1}{2} Q(\omega)$$

where $\omega: TG \to \mathfrak{g}$ is, as usual, the canonical left-invariant form on G. Then, using the basepoint map $\pi: TG \to G$, we compute that

$$\omega_L = \langle \omega, \pi^*(\omega) \rangle_Q.$$

As we saw in Lecture 3, the assumption that Q is non-degenerate implies that $d\omega_L$ is a symplectic form on TG. Now, since the flow of a right-invariant vector field Y_x is multiplication on the left by e^{tx}, it follows that, for this action, we may define

$$\rho(x) = -\omega_L(Y'_x) = -\langle \omega, \omega(Y_x) \rangle_Q = -\langle \omega, \mathrm{Ad}(g^{-1})x \rangle_Q$$

(where $g: TG \to G$ is merely a more descriptive name for the base point map than π).

Now, there is an isomorphism $\tau_Q: \mathfrak{g} \to \mathfrak{g}^*$, called *transpose with respect to* Q which satisfies $\tau_Q(x)(y) = \langle x, y \rangle_Q$ for all $x, y \in \mathfrak{g}$. In terms of τ_Q, we can express the moment map as

$$\mu(v) = -\mathrm{Ad}^*(g)\bigl(\tau_Q(\omega(v))\bigr)$$

for all $v \in TG$. Note that μ is G-equivariant, as promised by the theorem.

According to the Proposition 1, the function μ is a conserved quantity for the solutions of the Euler-Lagrange equations. In one of the Exercises, you are asked to show how this information can be used to help solve the Euler-Lagrange equations for the L-critical curves.

Reduction

I now want to discuss a method of taking quotients by group actions in the symplectic category. Now, when a Lie group G acts symplectically on the left on a symplectic manifold M, it is not generally true that the space of orbits $G\backslash M$ can be given a symplectic structure, even when this orbit space can be given the structure of a smooth manifold (for example, the quotient need not be even dimensional).

However, when the action is Poisson, there is a natural method of breaking the orbit space $G\backslash M$ into a union of symplectic submanifolds provided that certain regularity criteria are met. The procedure I will describe is known as *symplectic reduction*. It is due, in its modern form, to Marsden and Weinstein (see [GS 2]).

The idea is simple: If $\mu: M \to \mathfrak{g}^*$ is the moment map, then the G-equivariance of μ implies that there is a well-defined set map

$$\bar{\mu}: G\backslash M \to G\backslash \mathfrak{g}^*.$$

The theorem we are about to prove asserts that, provided certain regularity criteria are met, the subsets $M_\xi = \bar\mu^{-1}(\bar\xi) \subset G\backslash M$ are symplectic manifolds in a natural way.

Cleanliness. To facilitate the discussion, it is helpful to introduce some terminology: Let $f: X \to Y$ be a smooth map. We say that $y \in Y$ is a *clean* value of f if the set $f^{-1}(y) \subset X$ is a smooth manifold in X and, moreover, for each $x \in f^{-1}(y)$, we have $T_x f^{-1}(y) = \ker f'(x)$. Note that every regular value of f is clean, but not every clean value of f need be regular. The concept of cleanliness is very frequently encountered in the reduction theory we are about to develop.

Theorem 2. *Let $\lambda: G \times M \to M$ be a Poisson action on the symplectic manifold M. Let $\mu: M \to \mathfrak{g}^*$ be a moment map for λ. Suppose that, $\xi \in \mathfrak{g}^*$ is a clean value of μ. Then G_ξ acts smoothly on $\mu^{-1}(\xi)$. Suppose further that the space of G_ξ-orbits on $\mu^{-1}(\xi)$, say $M_\xi = G_\xi \backslash (\mu^{-1}(\xi))$ can be given the structure of a smooth manifold in such a way that the quotient mapping $\pi_\xi : \mu^{-1}(\xi) \to M_\xi$ is a smooth submersion. Then there exists a symplectic structure Ω_ξ on M_ξ which is defined by the condition that $\pi_\xi^*(\Omega_\xi)$ be the pullback of Ω to $\mu^{-1}(\xi)$.*

Proof. Since ξ is a clean value of μ, we know that $\mu^{-1}(\xi)$ is a smooth submanifold of M. Because of the G-equivariance of the moment map, it follows that G_ξ, the stabilizer of ξ, acts on M in such a way that it carries $\mu^{-1}(\xi)$ into itself. It is clear that this action will be smooth.

Now, I claim that for each $m \in \mu^{-1}(\xi)$, the Ω-complementary subspace to $T_m(\mu^{-1}(\xi))$ is the space $T_m(G \cdot m)$, i.e., the tangent to the G-orbit through m. To see this, first note that the space $T_m(G \cdot m)$ is spanned by the values at m assumed by the vector fields $\lambda_*(x)$ where $x \in \mathfrak{g}$. Thus, a vector $v \in T_m M$ lies in the Ω-complementary space of $T_m(G \cdot m)$ if and only if v satisfies $\Omega(\lambda_*(x)(m), v) = 0$ for all $x \in \mathfrak{g}$. Since, by definition, $\Omega(\lambda_*(x)(m), v) = d(\rho(x))(v)$, it follows that this is equivalent to the condition that v lies in $\ker \mu'(m) = T_m(\mu^{-1}(\xi))$ (since ξ is a clean value of μ).

Now, by the equivariance of the moment map, it is clear that, for all $m \in \mu^{-1}(\xi)$, we have $\mu^{-1}(\xi) \cap (G \cdot m) = G_\xi \cdot m$. It is also clear that

$$T_m(\mu^{-1}(\xi)) \cap T_m(G \cdot m) = \ker \mu'(m) \cap T_m(G \cdot m) = T_m(G_\xi \cdot m).$$

Thus, since the Ω-complementary spaces $T_m(\mu^{-1}(\xi))$ and $T_m(G \cdot m)$ intersect in the tangents to the G_ξ-orbits, it follows that if $\tilde\Omega_\xi$ denotes the pullback of Ω to $\mu^{-1}(\xi)$, then the null space of $\tilde\Omega_\xi$ at m is precisely $T_m(G_\xi \cdot m)$.

Finally, let us assume, as in the theorem, that there is a smooth manifold structure on the orbit space $M_\xi = G_\xi \backslash \mu^{-1}(\xi)$ so that the orbit space projection $\pi_\xi : \mu^{-1}(\xi) \to M_\xi$ is a smooth submersion. Since $\tilde\Omega_\xi$ is clearly G_ξ invariant and closed and moreover, since its null space at each point of M is precisely the tangent space to the fibers of π_ξ, it follows that there exists a unique "push down" 2-form Ω_ξ on M_ξ as described in the theorem. That Ω_ξ is closed and non-degenerate is now immediate.

The point of Theorem 2 is that, even though the quotient of a symplectic manifold by a symplectic group action is not, in general, a symplectic manifold, there *is* a way to produce a family of symplectic quotients parametrized by the elements of the space $\mathfrak{g}^*$. The quotients M_ξ often turn out to be quite interesting, even though the original symplectic manifold M may not be too interesting.

Remarks

Before I pass on to the examples, let me make a few comments about the hypotheses in Theorem 2.

First, there will always be clean values ξ of μ for which $\mu^{-1}(\xi)$ is not empty (even when there are no such regular values). This follows because, if we look at the closed subset $D_\mu \subset M$ consisting of points m where $\mu'(m)$ does not reach its maximum rank, then $\mu(D_\mu)$ can be shown (by a sort of Sard's Theorem argument) to be a proper subset of $\mu(M)$. The values $\mu(M) \setminus \mu(D_\mu)$ are then all clean.

Second, it quite frequently does happen that the G_ξ-orbit space M_ξ has a manifold structure for which π_ξ is a submersion. This can be guaranteed by various hypotheses which are often met with in practice. For example, if G_ξ is compact and acts freely on $\mu^{-1}(\xi)$, then M_ξ will be a manifold.

Weaker hypotheses also work. Basically, one needs to know that, at every point m of $\mu^{-1}(\xi)$, there is a smooth "slice" to the action of G_ξ, i.e., a smoothly embedded disk D in $\mu^{-1}(\xi)$ passing through m and intersecting each G_ξ-orbit exactly once. (Compare the construction of a smooth structure on each G-orbit in Theorem 1 of Lecture 3.)

In any case, even when there is not a slice around each point of $\mu^{-1}(\xi)$, there is very often a "near-slice", i.e., a smoothly embedded disk D in $\mu^{-1}(\xi)$ passing through m and intersecting each G_ξ-orbit only a finite number of times. In this case, the quotient space M_ξ inherits the structure of a symplectic *orbifold*, and these "generalized manifolds" have turned out to be quite useful.

Finally, it is worth remarking on the dimension of M_ξ when it does turn out to be a manifold. Let $G_m \subset G$ be the stabilizer of $m \in \mu^{-1}(\xi)$. I leave as an exercise for the reader to check that

$$\dim M_\xi = \dim M - \dim G - \dim G_\xi + 2\dim G_m$$
$$= \dim M - 2\dim G/G_m + \dim G/G_\xi.$$

Some examples

Since we will see so many examples in the Exercises and the next lecture, I will content myself with only mentioning two here:

Example: The cotangent bundle of G. Let $M = T^*G$ and let G act on T^*G on the left in the obvious way. Then the reader can easily check that, for each $x \in \mathfrak{g}$, we have $\rho(x)(\alpha) = \alpha(Y_x)$ for all $\alpha \in T^*G$ where, as usual, Y_x denotes the right invariant vector field on G whose value at e is $x \in \mathfrak{g}$. From this, it follows without too much trouble that $\mu: T^*G \to \mathfrak{g}^*$ is given by $\mu(\alpha) = R^*_{\pi(\alpha)}(\alpha)$.

Thus, it follows that $\mu^{-1}(\xi) \subset T^*G$ is merely the graph in T^*G of the left-invariant 1-form ω_ξ (i.e., the left-invariant 1-form whose value at e is $\xi \in \mathfrak{g}^*$).

Thus, we can use ξ as a section of T^*G to pull back Ω to get the 2-form $d\omega_\xi$ on G. As we already saw in Lecture 5, and is now borne out by Theorem 2, the null space of $d\omega_\xi$ on G is the coset space of G_ξ, the quotient is merely the coadjoint orbit G/G_ξ, and the symplectic structure Ω_ξ is just the one we already constructed.

Note, by the way, that every value of μ is clean in this example (in fact, they are all regular), even though the dimensions of the quotients G/G_ξ vary with ξ.

Example: Space rotations on the cotangent bundle of space. Let $G = \mathrm{SO}(3)$ act on $\mathbb{R}^6 = T^*\mathbb{R}^3$ by the extension of rotation about the origin in $\mathbb{R}^3$. Then, in standard coordinates (x, y) (where $x, y \in \mathbb{R}^3$), the action is simply $g \cdot (x, y) = (gx, gy)$, and the symplectic form is $\Omega = dx \cdot dy = {}^t dx \wedge dy$.

We can identify $\mathfrak{so}(3)^*$ with $\mathfrak{so}(3)$ itself by interpreting $a \in \mathfrak{so}(3)$ as the linear functional $b \mapsto -\mathrm{tr}(ab)$. It is easy to see that the co-adjoint action in this case gets identified with the adjoint action.

We compute that $\rho(a)(x, y) = -{}^t x a y$, so it follows without too much difficulty that, with respect to our identification of $\mathfrak{so}(3)^*$ with $\mathfrak{so}(3)$, we have $\mu(x, y) = x\,{}^t y - y\,{}^t x$.

The reader can check that all of the values of μ are clean except for $0 \in \mathfrak{so}(3)$. Even this value would be clean if, instead of using all of $\mathbb{R}^6$, we removed the origin.

I leave it to the reader to check that the G-invariant map $P\colon \mathbb{R}^6 \to \mathbb{R}^3$ defined by
$$P(x, y) = (x \cdot x,\ x \cdot y,\ y \cdot y)$$
maps the set $\mu^{-1}(0)$ onto the "cone" consisting of those points $(a, b, c) \in \mathbb{R}^3$ with $a, c > 0$ and $b^2 = ac$ and its fibers are the G_0 orbits of the points in $\mu^{-1}(0)$. Also, for $\xi \neq 0$, the P-image of the set $\mu^{-1}(\xi)$ is one nappe of the hyperboloid of two sheets described as $ac - b^2 = -\mathrm{tr}(\xi^2)$. The reader should compute the area forms Ω_ξ on these sheets.

Exercises

1. Let M be the torus $\mathbb{R}^2/\mathbb{Z}^2$ and let dx and dy be the standard 1-forms on M. Let $\Omega = dx \wedge dy$. Show that the "translation action" $(a,b) \cdot [x,y] = [x+a, y+b]$ of $\mathbb{R}^2$ on M is symplectic but not Hamiltonian.

2. Let (M, Ω) be a connected symplectic manifold and let $\lambda : G \times M \to M$ be a Hamiltonian group action.
   (i) Prove that, if $\rho : \mathfrak{g} \to C^\infty(M)$ is a linear mapping which satisfies $\lambda_*(x) \lrcorner \Omega = d(\rho(x))$, then $\lambda_*([x,y]) \lrcorner \Omega = d(\{\rho(x), \rho(y)\})$.
   (ii) Show that the associated linear mapping $c_\rho : \mathfrak{g} \times \mathfrak{g} \to \mathbb{R}$ defined in the text does indeed satisfy $c_\rho([x,y], z) + c_\rho([y,z], x) + c_\rho([z,x], y) = 0$ for all $x, y, z \in \mathfrak{g}$. (Hint: Use the fact the Poisson bracket satisfies the Jacobi identity and that the Poisson bracket of a constant function with any other function is zero.)

3. LIE ALGEBRA COHOMOLOGY. The purpose of this exercise is to acquaint the reader with the rudiments of Lie algebra cohomology.

    The Lie bracket of a Lie algebra $\mathfrak{g}$ can be regarded as a linear map $\partial : \Lambda^2(\mathfrak{g}) \to \mathfrak{g}$. The dual of this linear map is a map $-\delta : \mathfrak{g}^* \to \Lambda^2(\mathfrak{g}^*)$. (Thus, for $\xi \in \mathfrak{g}^*$, we have $\delta\xi(x,y) = -\xi([x,y])$.) This map δ can be extended uniquely to a graded, derivation of degree one $\delta : \Lambda^*(\mathfrak{g}^*) \to \Lambda^*(\mathfrak{g}^*)$.
   (i) For any $c \in \Lambda^2(\mathfrak{g}^*)$, show that $\delta c(x,y,z) = -c([x,y],z) - c([y,z],x) - c([z,x],y)$. (Hint: Every $c \in \Lambda^2(\mathfrak{g}^*)$ is a sum of wedge products $\xi \wedge \eta$ where $\xi, \eta \in \mathfrak{g}^*$.) Conclude that the Jacobi identity in $\mathfrak{g}$ is equivalent to the condition that $\delta^2 = 0$ on all of $\Lambda^*(\mathfrak{g}^*)$.

   Thus, for any Lie algebra $\mathfrak{g}$, we can define the k'th *cohomology group* of $\mathfrak{g}$, denoted $H^k(\mathfrak{g})$, as the kernel of δ in $\Lambda^k(\mathfrak{g}^*)$ modulo the subspace $\delta(\Lambda^{k-1}(\mathfrak{g}^*))$.

   (ii) Let G be a Lie group whose Lie algebra is $\mathfrak{g}$. For each $\Phi \in \Lambda^k(\mathfrak{g}^*)$, define ω_Φ to be the left-invariant k-form on G whose value at the identity is Φ. Show that $d\omega_\Phi = \omega_{\delta\Phi}$. (Hint: the space of left-invariant forms on G is clearly closed under exterior derivative and is generated over $\mathbb{R}$ by the left-invariant 1-forms. Thus, it suffices to prove this formula for Φ of degree 1. Why?)

   Thus, the cohomology groups $H^k(\mathfrak{g})$ measure "closed-mod-exact" in the space of left-invariant forms on G. If G is compact, then these cohomology groups are isomorphic to the corresponding deRham cohomology groups of the manifold G.

   (iii) (The Whitehead Lemmas) Show that if the Killing form of $\mathfrak{g}$ is non-degenerate, then $H^1(\mathfrak{g}) = H^2(\mathfrak{g}) = 0$. (Hint: You should have already shown that if κ is non-degenerate, then $[\mathfrak{g}, \mathfrak{g}] = \mathfrak{g}$. Show that this implies that $H^1(\mathfrak{g}) = 0$. Next show that for $\Phi \in \Lambda^2(\mathfrak{g}^*)$, we can write $\Phi(x,y) = \kappa(Lx, y)$ where $L : \mathfrak{g} \to \mathfrak{g}$ is skew-symmetric. Then show that if $\delta\Phi = 0$, then L is a derivation of $\mathfrak{g}$. Now see Exercise 3.3, part (iv).)

4. HOMOGENEOUS SYMPLECTIC MANIFOLDS. Suppose that (M, Ω) is a symplectic manifold and suppose that there exists a transitive symplectic action $\lambda : G \times M \to M$ where G is a group whose Lie algebra satisfies $H^1(\mathfrak{g}) = H^2(\mathfrak{g}) = 0$. Show that there is a G-equivariant symplectic covering map

$\pi\colon M \to G/G_\xi$ for some $\xi \in \mathfrak{g}^*$. Thus, up to passing to covers, the only symplectic homogeneous spaces of a Lie group satisfying $H^1(\mathfrak{g}) = H^2(\mathfrak{g}) = 0$ are the coadjoint orbits. This result is usually associated with the names Kostant, Souriau, and Symes.

(Hint: Since G acts homogeneously on M, it follows that, as G-spaces, $M = G/H$ for some closed subgroup $H \subset G$ which is the stabilizer of a point m of M. Let $\phi\colon G \to M$ be $\phi(g) = g \cdot m$. Now consider the left-invariant 2-form $\phi^*(\Omega)$ on G in light of the previous Exercise. Why do we also need the hypothesis that $H^1(\mathfrak{g}) = 0$?)

I warn the reader that this characterization of homogeneous symplectic spaces is sometimes misquoted. Either the covering ambiguity is overlooked or else, instead of hypotheses about the cohomology groups, sometimes compactness is assumed, either for M or G. The example of $S^1 \times S^1$ acting on itself and preserving the bi-invariant area form shows that compactness is not generally helpful. Here is an example which shows that you must allow for the covering possibility: Let $H \subset \mathrm{SL}(2,\mathbb{R})$ be the subgroup of diagonal matrices with *positive* entries on the diagonal. Then $\mathrm{SL}(2,\mathbb{R})/H$ has an $\mathrm{SL}(2,\mathbb{R})$-invariant area form, but it double covers the associated coadjoint orbit.

5. Verify the claim made in the text that, if there exists a G-invariant 1-form ω on M so that $d\omega = \Omega$, then the formula $\rho(x) = -\omega(\lambda_*(x))$ yields a lifting ρ for which $c_\rho = 0$.

6. Show that if $\mathbb{R}^2$ acts on itself by translation then, with respect to the standard area form $\Omega = dx \wedge dy$, this action is Hamiltonian but not Poisson.

7. Verify the claim made in the proof of Theorem 1 that the following identity holds for all $a \in G$, all $y \in \mathfrak{g}$, and all $m \in M$:

$$\lambda'_a\bigl(\lambda_*(y)(m)\bigr) = \lambda_*\bigl(\mathrm{Ad}(a)y\bigr)(a \cdot m).$$

8. MATRIX CALCULATIONS. The purpose of this exercise is to let you get some practice in a case where everything can be written out in coordinates.

Let $G = \mathrm{GL}(n,\mathbb{R})$ and let $Q\colon \mathfrak{gl}(n,\mathbb{R}) \to \mathbb{R}$ be a non-degenerate quadratic form. Show that if we use the inclusion mapping $x\colon \mathrm{GL}(n,\mathbb{R}) \to \mathcal{M}_{n \times n}$ as a coordinate chart, then, in the associated canonical coordinates (x,p), the Lagrangian L takes the form $L = \frac{1}{2}\langle x^{-1}p, x^{-1}p\rangle_Q$. Show also that $\omega_L = \langle x^{-1}p, x^{-1}dx\rangle_Q$.

Now compute the expression for the momentum mapping μ and the Euler-Lagrange equations for motion under the Lagrangian L. Show directly that μ is constant on the solutions of the Euler-Lagrange equations.

Suppose that Q is Ad-invariant, i.e., $Q(\mathrm{Ad}(g)(x)) = Q(x)$ for all $g \in G$ and $x \in \mathfrak{g}$. Show that the constancy of μ is equivalent to the assertion that $p\,x^{-1}$ is constant on the solutions of the Euler-Lagrange equations. Show that, in this case, the L-critical curves in G are just the curves $\gamma(t) = \gamma_0\, e^{tv}$ where $\gamma_0 \in G$ and $v \in \mathfrak{g}$ are arbitrary.

Finally, repeat all of these constructions for the general Lie group G, translating everything into invariant notation (as opposed to matrix notation).

9. EULER'S EQUATION. Look back over the example given in the Lecture of left-invariant metrics on Lie groups. Suppose that $\gamma: \mathbb{R} \to G$ is an L-critical curve. Define $\xi(t) = \tau_Q(\omega(\dot\gamma(t)))$. Thus, $\xi: \mathbb{R} \to \mathfrak{g}^*$. Show that the image of ξ lies on a single coadjoint orbit. Moreover, show that ξ satisfies *Euler's Equation*:
$$\dot\xi + \mathrm{ad}^*\left(\tau_Q^{-1}(\xi)\right)(\xi) = 0.$$

The reason Euler's Equation is so remarkable is that it only involves "half of the variables" of the curve $\dot\gamma$ in TG.

Once a solution to Euler's Equation is found, the equation for finding the original curve γ is just $\dot\gamma = L'_\gamma(\tau_Q^{-1}(\xi))$, which is a Lie equation for γ and hence is amenable to Lie's method of reduction.

Actually more is true. Show that, if we set $\xi(0) = \xi_0$, then the equation $\mathrm{Ad}^*(\gamma)(\xi) = \xi_0$ determines the solution γ of the Lie equation with initial condition $\gamma(0) = e$ up to right multiplication by a curve in the stabilizer subgroup G_{ξ_0}. Thus, we are reduced to solving a Lie equation for a curve in G_{ξ_0}. (It may be of some interest to note that the stabilizer of the generic element $\eta \in \mathfrak{g}^*$ is an abelian group (see Exercise 5.11). Of course, for such η, the corresponding Lie equation can be solved by quadratures.)

10. PROJECT: ANALYSIS OF THE RIGID BODY IN $\mathbb{R}^3$. Go back to the example of the motion of a rigid body in $\mathbb{R}^3$ presented in Lecture 4. Use the information provided in the previous two Exercises to show that the equations of motion for a free rigid body are integrable by quadratures. You will want to first compute the coadjoint action and describe the coadjoint orbits and their stabilizers.

11. Verify that, under the hypotheses of Theorem 2, the dimension of the reduced space M_ξ is given by the formula
$$\dim M_\xi = \dim M - \dim G - \dim G_\xi + 2\dim G_m$$
where G_m is the stabilizer of any $m \in \mu^{-1}(\xi)$.
    (Hint: Show that for any $m \in \mu^{-1}(\xi)$, we have
$$\dim T_m \mu^{-1}(\xi) + \dim T_m(G \cdot m) = \dim M$$
and then do some arithmetic.)

12. In the reduction process, what is the relationship between M_ξ and $M_{\mathrm{Ad}^*(g)(\xi)}$?

13. Suppose that $\lambda: G \times M \to M$ is a Poisson action and that Y is a symplectic vector field on M which is G-invariant. Then according to Proposition 1, Y is tangent to each of the submanifolds $\mu^{-1}(\xi)$ (when ξ is a clean value of μ). Show that, when the symplectic quotient M_ξ exists, then there exists a unique vector field Y_ξ on M_ξ which satisfies $Y_\xi(\pi_\xi(m)) = \pi'_\xi(Y(m))$. Show also that Y_ξ is symplectic. Finally show that, given an integral curve $\gamma: \mathbb{R} \to M_\xi$ of Y_ξ, then the problem of lifting this to an integral curve of Y is reducible by "finite" operations to solving a Lie equation for G_ξ.

This procedure is extremely helpful for two reasons: First, since M_ξ is generally quite a bit smaller than M, it should, in principle, be easier to find integral curves of Y_ξ than integral curves of Y. For example, if M_ξ is two dimensional, then Y_ξ can be integrated by quadratures (Why?). Second, it very frequently happens that G_ξ is a solvable or even abelian group (see Exercise 5.11). As we have already seen, when this happens the "lifting problem" can be integrated by (a sequence of) quadratures.

LECTURE 8
Recent Applications of Reduction

In this Lecture, we will see some examples of symplectic reduction and its generalizations in somewhat non-classical settings.

In many cases, we will be concerned with extra structure on M which can be carried along in the reduction process to produce extra structure on M_ξ. Often this extra structure takes the form of a Riemannian metric with special holonomy, so we begin with a short review of this topic.

Riemannian holonomy

Let M^n be a connected and simply connected n-manifold, and let g be a Riemannian metric on M. Associated to g is the notion of *parallel transport* along curves. Thus, for each (piecewise C^1) curve $\gamma\colon [0,1] \to M$, there is associated a linear mapping $P_\gamma\colon T_{\gamma(0)}M \to T_{\gamma(1)}M$, called parallel transport along γ, which is an isometry of vector spaces and which satisfies the conditions $P_{\bar\gamma} = P_\gamma^{-1}$ and $P_{\gamma_2\gamma_1} = P_{\gamma_2} \circ P_{\gamma_1}$ where $\bar\gamma$ is the path defined by $\bar\gamma(t) = \gamma(1-t)$ and $\gamma_2\gamma_1$ is defined only when $\gamma_1(1) = \gamma_2(0)$ and, in this case, is given by the formula

$$\gamma_2\gamma_1(t) = \begin{cases} \gamma_1(2t) & \text{for } 0 \le t \le \tfrac{1}{2}, \\ \gamma_2(2t-1) & \text{for } \tfrac{1}{2} \le t \le 1. \end{cases}$$

These properties imply that, for any $x \in M$, the set of linear transformations of the form P_γ where $\gamma(0) = \gamma(1) = x$ is a subgroup $H_x \subset O(T_xM)$ and that, for any other point $y \in M$, we have $H_y = P_\gamma H_x P_{\bar\gamma}$ where $\gamma\colon [0,1] \to M$ satisfies $\gamma(0) = x$ and $\gamma(1) = y$. Because we are assuming that M is simply connected, it is easy to show that H_x is actually connected and hence is a subgroup of $\mathrm{SO}(T_xM)$.

Élie Cartan was the first to define and study H_x. He called it the *holonomy* of g at x. He assumed that H_x was always a closed Lie subgroup of $\mathrm{SO}(T_xM)$, a result which was only later proved by Borel and Lichnerowicz (see [KN]).

Georges deRham, a student of Cartan, proved that, if there is a splitting $T_xM = V_1 \oplus V_2$ which remains invariant under all the action of H_x, then, in fact, the metric g is locally a product metric in the following sense: The metric g can be written as a sum of the form $g = g_1 + g_2$ in such a way that, for every point $y \in M$ there exists a neighborhood U of y, a coordinate chart $(x_1, x_2): U \to \mathbb{R}^{d_1} \times \mathbb{R}^{d_2}$, and metrics $\bar{g}_i$ on $\mathbb{R}^{d_i}$ so that $g_i = x_i^*(\bar{g}_i)$.

He also showed that in this reducible case the holonomy group H_x is a direct product of the form $H_x^1 \times H_x^2$ where $H_x^i \subset \mathrm{SO}(V_i)$. Moreover, it turns out (although this is not obvious) that, for each of the factor groups H_x^i, there is a submanifold $M_i \subset M$ so that $T_x M_i = V_i$ and so that H_x^i is the holonomy of the Riemannian metric g_i on M_i.

From this discussion it follows that, in order to know which subgroups of $\mathrm{SO}(n)$ can occur as holonomy groups of simply connected Riemannian manifolds, it is enough to find the ones which, in addition, act irreducibly on $\mathbb{R}^n$. Using a great deal of machinery from the theory of representations of Lie groups, M. Berger [Ber] determined a relatively short list of possibilities for irreducible Riemannian holonomy groups. This list was slightly reduced a few years later, independently by Alexseevski and by Brown and Gray. The result of their work can be stated as follows:

Theorem 1. *Suppose that g is a Riemannian metric on a connected and simply connected n-manifold M and that the holonomy H_x acts irreducibly on $T_x M$ for some (and hence every) $x \in M$. Then either (M, g) is locally isometric to an irreducible Riemannian symmetric space or else there is an isometry $\iota: T_x M \to \mathbb{R}^n$ so that $H = \iota H_x \iota^{-1}$ is one of the subgroups of $\mathrm{SO}(n)$ in the following table.*

Subgroup	Conditions	Geometrical Type
$\mathrm{SO}(n)$	any n	generic metric
$\mathrm{U}(m)$	$n = 2m > 2$	Kähler
$\mathrm{SU}(m)$	$n = 2m > 2$	Ricci-flat Kähler
$\mathrm{Sp}(m)\mathrm{Sp}(1)$	$n = 4m > 4$	Quaternionic Kähler
$\mathrm{Sp}(m)$	$n = 4m > 4$	Hyperkähler
G_2	$n = 7$	Associative
$\mathrm{Spin}(7)$	$n = 8$	Cayley

A few words of explanation and comment about Theorem 1 are in order.

First, a Riemannian symmetric space is a Riemannian manifold diffeomorphic to a homogeneous space G/H where $H \subset G$ is essentially the fixed subgroup of an involutory homomorphism $\sigma: G \to G$ which is endowed with a G-invariant metric g which is also invariant under the involution $\iota: G/H \to G/H$ defined by $\iota(aH) = \sigma(a)H$. The classification of the Riemannian symmetric spaces reduces to a classification problem in the theory of Lie algebras and was solved by Cartan. Thus, the Riemannian symmetric spaces may be regarded as known.

Second, among the holonomies of non-symmetric metrics listed in the table, the ranges for n have been restricted so as to avoid repetition or triviality. Thus, $\mathrm{U}(1) = \mathrm{SO}(2)$ and $\mathrm{SU}(1) = \{e\}$ while $\mathrm{Sp}(1) = \mathrm{SU}(2)$, and $\mathrm{Sp}(1)\mathrm{Sp}(1) = \mathrm{SO}(4)$.

Third, according to S. T. Yau's celebrated proof of the Calabi Conjecture, any compact complex manifold for which the canonical bundle is trivial and which has

LECTURE 8. RECENT APPLICATIONS OF REDUCTION

a Kähler metric also has a Ricci-flat Kähler metric (see [Bes]). For this reason, metrics with holonomy SU(m) are often referred to as Calabi-Yau metrics.

Finally, I will not attempt to discuss the proof of Theorem 1 in these notes. Even with modern methods, the proof of this result is non-trivial and, in any case, would take us far from our present interests. Instead, I will content myself with the remark that it is now known that every one of these groups does, in fact, occur as the holonomy of a Riemannian metric on a manifold of the appropriate dimension. I refer the reader to [Bes] for a complete discussion.

We will be particularly interested in the Kähler and Hyperkähler cases since these cases can be characterized by the condition that the holonomy of g leaves invariant certain closed non-degenerate 2-forms. Hence these cases represent symplectic manifolds with "extra structure", namely a compatible metric.

The basic result will be that, for a manifold M which carries one of these two structures, there is a reduction process which can be applied to suitable group actions on M which preserve the structure.

Kähler manifolds and algebraic geometry

In this section, we give a very brief study of Kähler manifolds. These are symplectic manifolds which are also complex manifolds in such a way that the complex structure is "maximally compatible" with the symplectic structure. These manifolds arise with great frequency in Algebraic Geometry, and it is beyond the scope of these Lectures to do more than make an introduction to their uses here.

Hermitian linear algebra

As usual, we begin with some linear algebra. Let $H\colon \mathbb{C}^n \times \mathbb{C}^n \to \mathbb{C}$ be the hermitian inner product given by

$$H(z,w) = {}^t\bar{z}w = \bar{z}^1 w^1 + \cdots + \bar{z}^n w^n.$$

Then U(n) $\subset$ GL(n, $\mathbb{C}$) is the group of complex linear transformations of $\mathbb{C}^n$ which preserve H: $H(Az, Aw) = H(z, w)$ for all $z, w \in \mathbb{C}^n$ if and only if ${}^t\bar{A}A = I_n$. Note that H can be split into real and imaginary parts as

$$H(z,w) = \langle z, w \rangle + \imath\, \Omega(z,w).$$

It is clear from the relation $H(z,w) = \overline{H(w,z)}$ that $\langle,\rangle$ is symmetric and Ω is skew-symmetric. I leave it to the reader to show that $\langle,\rangle$ is positive definite and that Ω is non-degenerate.

Moreover, since $H(z, \imath w) = \imath H(z,w)$, it also follows that $\Omega(z,w) = \langle \imath z, w \rangle$ and $\langle z, w \rangle = \Omega(z, \imath w)$. It easily follows from these equations that, if we let $J\colon \mathbb{C}^n \to \mathbb{C}^n$ denote multiplication by $\imath$, then knowing any two of the three objects $\langle,\rangle$, Ω, or J on $\mathbb{R}^{2n}$ determines the third.

Definition 1. Let V be a vector space over $\mathbb{R}$. A non-degenerate 2-form Ω and a complex structure $J\colon V \to V$ are said to be *compatible* if $\Omega(x, Jy) = \Omega(y, Jx)$ for all $x, y \in T_m M$. If the pair (Ω, J) is compatible, then we say that the pair forms an

Hermitian structure on V if, in addition, $\Omega(x, Jx) > 0$ for all non-zero $x \in V$. The positive definite quadratic form $g(x,x) = \Omega(x, Jx)$ is called the associated metric on V.

I leave to the reader the task of showing that any two Hermitian structures on V are isomorphic via some invertible endomorphism of V (see the Exercises). Since, as is easily verified, $\Omega(v,w) = g(Jv, w)$, it follows that knowing g and either J or Ω determines the remaining element of the triple. In an extension of the notion of compatibility, we define a metric g to be compatible with a non-degenerate 2-form Ω if the linear map $J: TM \to TM$ defined by the relation $\Omega(v,w) = g(Jv, w)$ satisfies $J^2 = -1$. Also, we define a metric g to be compatible with a complex structure J on V if $g(Jv,w) = -g(Jw,v)$, so that $\Omega(v,w) = g(Jv,w)$ defines a 2-form on V.

Almost Hermitian manifolds

Since our main interest is in symplectic and complex structures, I will introduce the notion of an almost Hermitian structure on a manifold in terms of its almost complex and almost symplectic structures:

Definition 2. Let M^{2n} be a manifold. A 2-form Ω and an almost complex structure J define an *almost Hermitian structure* on M if, for each $m \in M$, the pair (Ω_m, J_m) defines a Hermitian structure on $T_m M$.

When (Ω, J) defines an almost Hermitian structure on M, the Riemannian metric g on M defined by $g(v) = \Omega(v, Jv)$ is called the *associated metric*.

Just as one must place conditions on an almost symplectic structure in order to get a symplectic structure, there are conditions which an almost complex structure must satisfy in order to be a complex structure.

Definition 3. An almost complex structure J on M^{2n} is *integrable* if every point of M has a neighborhood U on which there exists a coordinate chart $z: U \to \mathbb{C}^n$ so that for all $v \in TU$, we have $z'(Jv) = \imath\, z'(v)$. Such a coordinate chart is said to be *J-holomorphic*.

According to the Korn-Lichtenstein theorem, when $n = 1$ all almost complex structures are integrable. However, for $n \geq 2$, one can easily write down examples of almost complex structures J which are not integrable. (See the Exercises.)

When J is an integrable almost complex structure on M, the set

$$\mathcal{U}_J = \{(U, z) \mid z: U \to \mathbb{C}^n \text{ is } J\text{-holomorphic}\}$$

forms an atlas of charts which are holomorphic on overlaps. Thus, $\mathcal{U}_J$ defines a complex structure on M.

The reader may be wondering just how one determines whether an almost complex structure is integrable or not. In the Exercises, you are asked to show that, for an integrable almost complex structure J, the identity $\mathcal{L}_{JX} J - J \circ \mathcal{L}_X J = 0$ must hold for all vector fields X on M. It is a remarkable result, due to Newlander and Nirenberg, that this condition is sufficient for J to be integrable.*

---

*The reason that I mention this condition is that it shows that integrability is determined by J and its first derivatives in any local coordinate system. This condition can be rephrased as the

LECTURE 8. RECENT APPLICATIONS OF REDUCTION

We are now ready to name the various integrability conditions that can be defined for an almost Hermitian manifold.

Definition 4. We call an almost Hermitian pair (Ω, J) on a manifold M *almost Kähler* if Ω is closed, *Hermitian* if J is integrable, and *Kähler* if Ω is closed and J is integrable.

We already saw in Lecture 6 that a manifold has an almost complex structure if and only if it has an almost symplectic structure. However, this relationship does not, in general, hold between complex structures and symplectic structures.

Example. Here is a complex manifold which has no symplectic structure. Let $\mathbb{Z}$ act on $M = \mathbb{C}^2 \setminus \{0\}$ by $n \cdot z = 2^n z$. This free action preserves the standard complex structure on M. Let $N = \mathbb{Z} \setminus \tilde{M}$, then, via the quotient mapping, N inherits the structure of a complex manifold.

However, N is diffeomorphic to $S^1 \times S^3$ as a smooth manifold. Thus N is a compact manifold satisfying $H^2_{dR}(N, \mathbb{R}) = 0$. In particular, by the cohomology ring obstruction discussed in Lecture 6, we see that M cannot be given a symplectic structure.

Example. Here is an example due to Thurston, of a compact 4-manifold which has a complex structure and has a symplectic structure, but has no Kähler structure.

Let $H_3 \subset \mathrm{GL}(3, \mathbb{R})$ be the Heisenberg group, defined in Lecture 2 as the set of matrices of the form
$$g = \begin{pmatrix} 1 & x & z + \tfrac{1}{2}xy \\ 0 & 1 & y \\ 0 & 0 & 1 \end{pmatrix}.$$

The left invariant forms and their structure equations on H_3 are easily computed in these coordinates as

$$\omega_1 = dx \qquad\qquad d\omega_1 = 0$$
$$\omega_2 = dy \qquad\qquad d\omega_2 = 0$$
$$\omega_3 = dz - \tfrac{1}{2}(x\,dy - y\,dx) \qquad d\omega_3 = -\omega_1 \wedge \omega_2$$

Now, let $\Gamma = H_3 \cap \mathrm{GL}(3, \mathbb{Z})$ be the subgroup of H_3 consisting of those elements of H_3 all of whose entries are integers. Let $X = \Gamma \backslash H_3$ be the space of right cosets of Γ. Since the forms ω_i are left-invariant, it follows that they are well-defined on X and form a basis for the 1-forms on X.

Now let $M = X \times S^1$ and let $\omega_4 = d\theta$ be the standard 1-form on S^1. Then the forms ω_i for $1 \leq i \leq 4$ form a basis for the 1-forms on M. Since $d\omega_4 = 0$, it follows that the 2-form

$$\Omega = \omega_1 \wedge \omega_3 + \omega_2 \wedge \omega_4$$

is closed and non-degenerate on M. Thus, M has a symplectic structure.

---

condition that the vanishing of a certain tensor N_J, called the Nijnhuis tensor of J and constructed out of the first order jet of J at each point, is necessary and sufficient for the integrability of J.

Next, I want to construct a complex structure on M. In order to do this, I will produce the appropriate local holomorphic coordinates on M. Let $\tilde{M} = H_3 \times \mathbb{R}$ be the simply connected cover of M with coordinates (x, y, z, θ). We regard $\tilde{M}$ as a Lie group. Define the functions $w^1 = x + \imath y$ and $w^2 = z + \imath\left(\theta + \frac{1}{4}(x^2 + y^2)\right)$ on $\tilde{M}$. Then I leave to the reader to check that, if g_0 is the element of $\tilde{M}$ with coordinates $(x_0, y_0, z_0, \theta_0)$, then

$$L_{g_0}^*(w^1) = w^1 + w_0^1 \qquad \text{and} \qquad L_{g_0}^*(w^2) = w^2 + (\imath/2)\bar{w}_0^1\, w^1 + w_0^2.$$

Thus, the coordinates w^1 and w^2 define a left-invariant complex structure on $\tilde{M}$. Since M is obtained from $\tilde{M}$ by dividing by the obvious left action of $\Gamma \times \mathbb{Z}$, it follows that there is a unique complex structure on M for which the covering projection is holomorphic.

Finally, we show that M cannot carry a Kähler structure. Since Γ is a discrete subgroup of H_3, the projection $H_3 \to X$ is a covering map. Since $H_3 = \mathbb{R}^3$ as manifolds, it follows that $\pi_1(X) = \Gamma$. Moreover, X is compact since it is the image under the projection of the cube in H_3 consisting of those elements whose entries lie in the closed interval $[0, 1]$. On the other hand, since $[\Gamma, \Gamma] \simeq \mathbb{Z}$, it follows that $\Gamma/[\Gamma, \Gamma] \simeq \mathbb{Z}^2$. Thus, $H^1(M, \mathbb{Z}) = H^1(X \times S^1, \mathbb{Z}) = \mathbb{Z}^2 \oplus \mathbb{Z}$. From this, we get that $H_{dR}^1(M, \mathbb{R}) = \mathbb{R}^3$. In particular, the first Betti number of M is 3. Now, it is a standard result in Kähler geometry that the odd degree Betti numbers of a compact Kähler manifold must be even (for example, see [Ch]). Hence, M cannot carry any Kähler metric.

Example. Because of the classification of compact complex surfaces due to Kodaira, we know exactly which compact 4-manifolds can carry complex structures. Fernandez, Gotay, and Gray [FGG] have constructed a compact, symplectic 4-manifold M whose underlying manifold is not on Kodaira's list, thus, providing an example of a compact symplectic 4-manifold which carries no complex structure.

The fundamental theorem relating the two "integrability conditions" to the idea of holonomy is the following one. We only give the idea of the proof because a complete proof would require the development of considerable machinery.

Theorem 2. *An almost Hermitian structure (Ω, J) on a manifold M is Kähler if and only if the form Ω is parallel with respect to the parallel transport of the associated metric g.*

Proof. (IDEA) Once the formulas are developed, it is not difficult to see that the covariant derivatives of Ω with respect to the Levi-Civita connection of g are expressible in terms of the exterior derivative of Ω and the Nijnhuis tensor of J. Conversely, the exterior derivative of Ω and the Nijnhuis tensor of J can be expressed in terms of the covariant derivative of Ω with respect to the Levi-Civita connection of g. Thus, Ω is covariant constant (i.e., invariant under parallel translation with respect to g) if and only it is closed and J is integrable.

It is worth remarking that J is invariant under parallel transport with respect to g if and only if Ω is. The reason for this is that J is determined from and

LECTURE 8. RECENT APPLICATIONS OF REDUCTION

determines Ω once g is fixed. The observation now follows, since g is invariant under parallel transport with respect to its own Levi-Civita connection.

Kähler reduction

We are now ready to state the first of the reduction theorems we will discuss in this Lecture.

For simplicity of notation, I will say that a value $\xi \in \mathfrak{g}^*$ is *good* if, in addition to being clean, it has the property that the quotient space $M_\xi = G_\xi \backslash \mu^{-1}(\xi)$ can be given the structure of a smooth manifold in such a way that the orbit projection $\pi_\xi : \mu^{-1}(\xi) \to M_\xi$ is a smooth submersion.

Theorem 3 (KÄHLER REDUCTION). *Let (Ω, g) be a Kähler structure on M^{2n}. Let $\lambda : G \times M \to M$ be a left action which is Poisson with respect to Ω and preserves the metric g. Let $\mu : M \to \mathfrak{g}^*$ be the associated moment map. Then, for every good value $\xi \in \mathfrak{g}^*$ of μ, there is a unique Kähler structure (Ω_ξ, g_ξ) on M_ξ defined by the conditions that $\pi_\xi^*(\Omega_\xi)$ be equal to the pullback of Ω to $\mu^{-1}(\xi) \subset M$ and that $\pi_\xi : \mu^{-1}(\xi) \to M_\xi$ be a Riemannian submersion.*

Proof. Let $\tilde{g}_\xi$ and $\tilde{\Omega}_\xi$ be the pullbacks of g and Ω respectively to $\mu^{-1}(\xi)$. By the very hypotheses of the theorem, $\tilde{g}_\xi$ and $\tilde{\Omega}_\xi$ are invariant under the action of G_ξ.

From Theorem 2 of Lecture 7, we already know that there exists a unique symplectic structure Ω_ξ on M_ξ for which $\pi_\xi^*(\Omega_\xi) = \tilde{\Omega}_\xi$.

Here is how we construct g_ξ. There is a well defined $\tilde{g}_\xi$-orthogonal splitting

$$T_m \mu^{-1}(\xi) = T_m(G_\xi \cdot m) \oplus H_m$$

which is clearly G_ξ-invariant. Since, by hypothesis, $\pi_\xi : \mu^{-1}(\xi) \to M_\xi$ is a submersion, it easily follows that $\pi'_\xi(m) : H_m \to T_{\pi_\xi(m)} M_\xi$ is an isomorphism of vector spaces. Moreover, the G_ξ-invariance of $\tilde{g}_\xi$ shows that there is a well-defined quadratic form $g_\xi(m)$ on $T_{\pi_\xi(m)} M_\xi$ which corresponds to the restriction of $\tilde{g}_\xi$ to H_m under this isomorphism. By the very definition of Riemannian submersion, it follows that g_ξ is a Riemannian metric on M_ξ for which π_ξ is a Riemannian submersion.

It remains to show that (Ω_ξ, g_ξ) defines a Kähler structure on M_ξ. First, we show that it is an almost Kähler structure, i.e., that Ω_ξ and g_ξ are actually compatible. Since $\pi'_\xi(m) : H_m \to T_{\pi_\xi(m)} M_\xi$ is an isomorphism of vector spaces which identifies (Ω_ξ, g_ξ) with the restriction of (Ω, g) to H_m, it suffices to show that H_m is invariant under the action of J.

Here is how we do this. Tracing back through the definitions, we see that $x \in T_m M$ lies in the subspace H_m if and only if x satisfies both of the conditions $\Omega(x, y) = 0$ for all $y \in T_m(G \cdot m)$ and $g(x, y) = 0$ for all $y \in T_m(G_\xi \cdot m)$. However, since $\Omega(x, y) = g(Jx, y)$ for all y, it follows that the necessary and sufficient conditions that x lie in H_m can also be expressed as the two conditions $g(Jx, y) = 0$ for all $y \in T_m(G \cdot m)$ and $\Omega(Jx, y) = 0$ for all $y \in T_m(G_\xi \cdot m)$. Of course, these conditions are exactly the conditions that Jx lie in H_m. Thus, $x \in H_m$ implies that $Jx \in H_m$, as we wished to show.

Example: Kähler reduction in algebraic geometry. By far the most common examples of Kähler manifolds arise in Algebraic Geometry. Here is a sample of what Kähler reduction yields:

Let $M = \mathbb{C}^{n+1}$ with complex coordinates $z^0, z^1, \ldots, z^n$. We let $z^k = x^k + \imath y^k$ define real coordinates on M. Let $G = S^1$ act on M by the rule

$$e^{\imath\theta} \cdot z = e^{\imath\theta} z.$$

Then G clearly preserves the Kähler structure defined by the natural complex structure on M and the symplectic form

$$\Omega = \frac{\imath}{2}{}^t dz \wedge d\bar{z} = dx^1 \wedge dy^1 + \cdots + dx^n \wedge dy^n.$$

The associated metric is easily seen to be just

$$g = {}^t dz \circ d\bar{z} = \left(dx^1\right)^2 + \left(dy^1\right)^2 + \cdots + \left(dx^n\right)^2 + \left(dy^n\right)^2.$$

Now, setting $X = \frac{\partial}{\partial \theta}$, we can compute that

$$\lambda_*(X) = x^k \frac{\partial}{\partial y^k} - y^k \frac{\partial}{\partial x^k}.$$

Thus, it follows that

$$d\rho(X) = \lambda_*(X) \lrcorner \Omega = -x^k\, dx^k - y^k\, dy^k = d\left(-\tfrac{1}{2}|z|^2\right).$$

Thus, identifying $\mathfrak{g}^*$ with $\mathbb{R}$, we have that $\mu: \mathbb{C}^n \to \mathbb{R}$ is merely $\mu(z) = -\frac{1}{2}|z|^2$.

It follows that every negative number is a non-trivial clean value for μ. For example, $S^{2n+1} = \mu^{-1}(-\tfrac{1}{2})$. Clearly $G = S^1$ itself is the stabilizer subgroup of all of the values of μ. Thus, $M_{-\frac{1}{2}}$ is the quotient of the unit sphere by the action of S^1. Since each G-orbit is merely the intersection of S^{2n+1} with a (unique) complex line through the origin, it is clear that $M_{-\frac{1}{2}}$ is diffeomorphic to $\mathbb{CP}^n$.

It is instructive to compute what the Kähler structure looks like in local coordinates. Let $\mathbb{A}_0 \subset \mathbb{CP}^n$ be the subset consisting of those points $[z^0, \ldots, z^n]$ for which $z^0 \neq 0$. Then $\mathbb{A}_0$ can be parametrized by $\phi: \mathbb{C}^n \to \mathbb{A}_0$ by $\phi(w) = [1, w]$. Now, over $\mathbb{A}_0$, we can choose a section $\sigma: \mathbb{A}_0 \to S^{2n+1}$ by the rule

$$\sigma \circ \phi(w) = \frac{(1, w)}{W}$$

where $W^2 = 1 + |w^1|^2 + \cdots + |w^n|^2 > 0$. It follows that

$$\phi^*(\Omega_{-\frac{1}{2}}) = (\sigma \circ \phi)^*(\Omega)$$
$$= \frac{\imath}{2}\left(d\left(\frac{w^k}{W}\right) \wedge d\left(\frac{\bar{w}^k}{W}\right)\right) = \frac{\imath}{2}\left(\frac{dw^k \wedge d\bar{w}^k}{W^2} + (w^k\, d\bar{w}^k - \bar{w}^k\, dw^k) \wedge \frac{dW}{W^3}\right).$$

Up to a normalizing constant, this is the usual formula for the Fubini-Study Kähler structure on $\mathbb{CP}^n$ in an affine chart.

Of course, the Fubini-Study metric induces a Kähler structure on every complex submanifold of $\mathbb{CP}^n$. However, we can just as easily see how this arises from the reduction procedure: If $P(z^0, \ldots, z^n)$ is a non-zero homogeneous polynomial of degree d, then the set $\tilde{M}_P = P^{-1}(0) \subset \mathbb{C}^{n+1}$ is a complex subvariety of $\mathbb{C}^{n+1}$ which is invariant under the S^1 action since, by homogeneity, we have

$$P(e^{\imath\theta} \cdot z) = e^{\imath d\theta} P(z).$$

It is easy to show that if the variety $\tilde{M}_P$ has no singularity other than $0 \in \mathbb{C}^{n+1}$, then the Kähler reduction of the Kähler structure that it inherits from the standard structure on $\mathbb{C}^{n+1}$ is just the Kähler structure on the corresponding projectivized variety $M_P \subset \mathbb{CP}^n$ which is induced by restriction of the Fubini-Study structure.

'Example': flat bundles over compact Riemann surfaces. The following is not really an example of the theory as we have developed it since it will deal with 'infinite dimensional manifolds', however it is suggestive and the formal calculations yield an interesting result. (For a review of the terminology used in this and the next example, see the Appendix.)

Let G be a Lie group with Lie algebra $\mathfrak{g}$, and let $\langle,\rangle$ be a positive definite, Ad-invariant inner product on $\mathfrak{g}$. (For example, if $G = \mathrm{SU}(n)$, we could take $\langle x, y \rangle = -\mathrm{tr}(xy)$.)

Let Σ be a connected compact Riemann surface. Then there is a star operation $*\colon \mathcal{A}^1(\Sigma) \to \mathcal{A}^1(\Sigma)$ which satisfies $*^2 = -id$, and $\alpha \wedge *\alpha \geq 0$ for all 1-forms α on Σ.

Let P be a principal right G-bundle over Σ, and let $\mathrm{Ad}(P) = P \times_{\mathrm{Ad}} \mathfrak{g}$ denote the vector bundle over M associated to the adjoint representation $\mathrm{Ad}\colon G \to \mathrm{Aut}(\mathfrak{g})$. Let $\mathrm{Aut}(P)$ denote the group of automorphisms of P, also known as the *gauge group* of P.

Let $\mathfrak{A}(P)$ denote the space of connections on P. Then it is well known that $\mathfrak{A}(P)$ is an affine space modeled on the vector space $\mathcal{A}^1(\mathrm{Ad}(P))$, which consists of the 1-forms on M with values in $\mathrm{Ad}(P)$. Thus, in particular, for every $A \in \mathfrak{A}(P)$, we have a natural isomorphism

$$T_A \mathfrak{A}(P) = \mathcal{A}^1(\mathrm{Ad}(P)).$$

I now want to define a "Kähler" structure on $\mathfrak{A}(P)$. In order to do this, I will define the metric $\mathfrak{g}$ and the 2-form Ω.

First, for $\alpha \in T_A \mathfrak{A}(P)$, I define

$$\mathbf{g}(\alpha) = \int_\Sigma \langle \alpha, *\alpha \rangle.$$

(I extend the operator $*$ in the obvious way to $\mathcal{A}^1(\mathrm{Ad}(P))$.) It is clear that $\mathbf{g}(\alpha) \geq 0$ with equality if and only if $\alpha = 0$. Thus, $\mathbf{g}$ defines a "Riemannian metric" on $\mathfrak{A}(P)$. Since $\mathbf{g}$ is "translation invariant", it "follows" that $\mathbf{g}$ is "flat".

Second, I define $\mathbf{\Omega}$ by the rule:

$$\mathbf{\Omega}(\alpha, \beta) = \int_\Sigma \langle \alpha, \beta \rangle.$$

Since $\mathbf{\Omega}(\alpha, \beta) = \mathbf{g}(\alpha, *\beta)$, it follows that $\mathbf{\Omega}$ is actually non-degenerate. Moreover, because $\mathbf{\Omega}$ too is "translation invariant", it "must" be "parallel" with respect to $\mathbf{g}$.

Thus, $(\mathbf{\Omega}, \mathbf{g})$ is a "flat Kähler" structure on $\mathfrak{A}(P)$. Now, I claim that both $\mathbf{\Omega}$ and $\mathbf{g}$ are invariant under the natural right action of $\mathrm{Aut}(P)$ on $\mathfrak{A}(P)$. To see this, note that an element $\phi \in \mathrm{Aut}(P)$ determines a map $\varphi: P \to G$ by the rule $p \cdot \varphi(p) = \phi(p)$ and that this φ satisfies the identity $\varphi(p \cdot g) = g^{-1}\varphi(p)g$. In terms of φ, the action of $\mathrm{Aut}(P)$ on $\mathfrak{A}(P)$ is given by the classical formula

$$A \cdot \phi = \phi^*(A) = \varphi^*(\omega_G) + \mathrm{Ad}(\varphi^{-1})(A).$$

From this, it follows easily that $\mathbf{\Omega}$ and $\mathbf{g}$ are $\mathrm{Aut}(P)$-invariant.

Now, I want to compute the moment map $\boldsymbol{\mu}$. The Lie algebra of $\mathrm{Aut}(P)$, namely $\mathfrak{aut}(P)$, can be naturally identified with $\mathcal{A}^0(\mathrm{Ad}(P))$, the space of sections of the bundle $\mathrm{Ad}(P)$. I leave to the reader the task of showing that the induced map from $\mathfrak{aut}(P)$ to vector fields on $\mathfrak{A}(P)$ is given by $d_A: \mathcal{A}^0(\mathrm{Ad}(P)) \to \mathcal{A}^1(\mathrm{Ad}(P))$. Thus, in order to construct the moment map, we must find, for each $f \in \mathcal{A}^0(\mathrm{Ad}(P))$, a function $\boldsymbol{\rho}(f)$ on $\mathfrak{A}$ so that the 1-form $d\boldsymbol{\rho}(f)$ is given by

$$d\boldsymbol{\rho}(f)(\alpha) = d_A f \lrcorner \mathbf{\Omega}(\alpha) = \int_\Sigma \langle d_A f, \alpha \rangle = -\int_\Sigma \langle f, d_A \alpha \rangle.$$

However, this is easy. We just set

$$\boldsymbol{\rho}(f)(A) = -\int_\Sigma \langle f, F_A \rangle$$

and the reader can easily check that

$$\left.\frac{d}{dt}\right|_{t=0} (\boldsymbol{\rho}(f)(A + t\alpha)) = -\int_\Sigma \langle f, d_A \alpha \rangle$$

as desired. Finally, using the natural isomorphism

$$(\mathfrak{aut}(P))^* = (\mathcal{A}^0(\mathrm{Ad}(P)))^* = \mathcal{A}^2(\mathrm{Ad}(P)),$$

LECTURE 8. RECENT APPLICATIONS OF REDUCTION

we see that (up to sign) the formula for the moment map simply becomes

$$\mu(A) = F_A = dA + \tfrac{1}{2}[A, A].$$

Now, can we do reduction? What we need is a clean value of μ. As a reasonable first guess, let's try 0. Thus, $\mu^{-1}(0)$ consists exactly of the flat connections on P and the reduced space $\mathfrak{M}_0$ should be the flat connections modulo gauge equivalence, i.e., $\mu^{-1}(0)/\mathrm{Aut}(P)$.

How can we tell whether 0 is a clean value? One way to know this would be to know that 0 is a regular value. We have already seen that $\mu'(A)(\alpha) = d_A \alpha$, so we are asking whether the map $d_A \colon \mathcal{A}^1(\mathrm{Ad}(P)) \to \mathcal{A}^2(\mathrm{Ad}(P))$ is surjective for any flat connection A. Now, because A is flat, the sequence

$$0 \longrightarrow \mathcal{A}^0(\mathrm{Ad}(P)) \xrightarrow{d_A} \mathcal{A}^1(\mathrm{Ad}(P)) \xrightarrow{d_A} \mathcal{A}^2(\mathrm{Ad}(P)) \longrightarrow 0$$

forms a complex and the usual Hodge theory pairing shows that $H^2(\Sigma, d_A)$ is the dual space of $H^0(\Sigma, d_A)$. Thus, $\mu'(A)$ is surjective if and only if $H^0(\Sigma, d_A) = 0$. Now, an element $f \in \mathcal{A}^0(\mathrm{Ad}(P))$ which satisfies $d_A f = 0$ exponentiates to a 1-parameter family of automorphisms of P which commute with the parallel transport of A. I leave to the reader to show that $H^0(\Sigma, d_A) = 0$ is equivalent to the condition that the holonomy group $H_A(p) \subset G$ has a centralizer of positive dimension in G for some (and hence every) point of P. For example, for $G = \mathrm{SU}(2)$, this would be equivalent to saying that the holonomy groups $H_A(p)$ were each contained in an $S^1 \subset G$.

Let us let $\widetilde{\mathfrak{M}}^* \subset \mu^{-1}(0)$ denote the (open) subset consisting of those flat connections whose holonomy groups have at most discrete centralizers in G. If G is compact, of course, this implies that these centralizers are finite. Then it "follows" that $\mathfrak{M}_0^* = \widetilde{\mathfrak{M}}^*/\mathrm{Aut}(P)$ is a Kähler manifold wherever it is a manifold. (In general, at the connections where the centralizer of the holonomy is trivial, one expects the quotient to be a manifold.)

Since the space of flat connections on P modulo gauge equivalence is well-known to be identifiable as the space $R(\pi_1(\Sigma, s), G) = \mathrm{Hom}(\pi_1(\Sigma, s), G)/G$ of equivalence classes of representation of $\pi_1(\Sigma)$ into G, our discussion leads us to believe that this space (which is finite dimensional) should have a natural Kähler structure on it. This is indeed the case, and the geometry of this Kähler metric is the subject of current interest.

Hyperkähler manifolds

In this section, we will generalize the Kähler reduction procedure to the case of manifolds with holonomy $\mathrm{Sp}(m)$, the so-called Hyperkähler case.

Quaternion Hermitian linear algebra

We begin with some linear algebra over the ring $\mathbb{H}$ of *quaternions*. For our purposes, $\mathbb{H}$ can be identified with the vector space of dimension 4 over $\mathbb{R}$ of matrices of the

form
$$x = \begin{pmatrix} x^0 + \imath x^1 & x^2 + \imath x^3 \\ -x^2 + \imath x^3 & x^0 - \imath x^1 \end{pmatrix} \stackrel{\text{def}}{=} x^0 1 + x^1 i + x^2 j + x^3 k.$$

(We are identifying the 2-by-2 identity matrix with 1 in this representation.) It is easy to see that $\mathbb{H}$ is closed under matrix multiplication. If we define $\bar{x} = x^0 - x^1 i - x^2 j - x^3 k$, then we easily get $\overline{xy} = \bar{y}\bar{x}$ and

$$x\bar{x} = \left((x^0)^2 + (x^1)^2 + (x^2)^2 + (x^3)^2\right) 1 = \det(x) 1 \stackrel{\text{def}}{=} |x|^2 1.$$

It follows that every non-zero element of $\mathbb{H}$ has a multiplicative inverse. Note that the space of quaternions of unit norm, S^3 defined by $|x| = 1$, is simply SU(2).

Much of the linear algebra which works for the complex numbers can be generalized to the quaternions. However, some care must be taken since $\mathbb{H}$ is not commutative. In the following exposition, it turns out to be most convenient to define vector spaces over $\mathbb{H}$ as *right* vector spaces instead of left vector spaces. Thus, the standard $\mathbb{H}$-vector space of $\mathbb{H}$-dimension n is $\mathbb{H}^n$ (thought of as columns of quaternions of height n) where the action of the scalars *on the right* is given by

$$\begin{pmatrix} x^1 \\ \vdots \\ x^n \end{pmatrix} \cdot q = \begin{pmatrix} x^1 q \\ \vdots \\ x^n q \end{pmatrix}.$$

With this convention, a quaternion linear map $A: \mathbb{H}^n \to \mathbb{H}^m$, i.e., an additive map satisfying $A(v q) = A(v) q$, can be represented by an m-by-n matrix of quaternions acting via matrix multiplication on the *left*.

Let $H: \mathbb{H}^n \times \mathbb{H}^n \to \mathbb{H}$ be the "quaternion Hermitian" inner product given by

$$H(z,w) = {}^t\bar{z}w = \bar{z}^1 w^1 + \cdots + \bar{z}^n w^n.$$

Then by our conventions, we have

$$H(z q, w) = \bar{q} H(z, w) \quad \text{and} \quad H(z, w q) = H(z, w) q.$$

We also have $H(z, w) = \overline{H(w, z)}$, just as before.

We define $\operatorname{Sp}(n) \subset \operatorname{GL}(n, \mathbb{H})$ to be the group of $\mathbb{H}$-linear transformations of $\mathbb{H}^n$ which preserve H, i.e., $H(Az, Aw) = H(z, w)$ for all $z, w \in \mathbb{H}^n$. It is easy to see that

$$\operatorname{Sp}(n) = \left\{ A \in \operatorname{GL}(n, \mathbb{H}) \mid {}^t\bar{A}A = I_n \right\}.$$

I leave as an exercise for the reader to show that $\operatorname{Sp}(n)$ is a compact Lie group of dimension $2n^2 + n$. Also, it is not difficult to show that $\operatorname{Sp}(n)$ is connected and acts irreducibly on $\mathbb{H}^n$. (see the Exercises)

Now H can be split into one real and three imaginary parts as

$$H(z, w) = \langle z, w \rangle + \Omega_1(z, w) i + \Omega_2(z, w) j + \Omega_3(z, w) k.$$

It is clear from the relations above that $\langle,\rangle$ is symmetric and positive definite and that each of the Ω_a is skew-symmetric. Moreover, we have the following identities:

$$\langle z,w\rangle = \Omega_1(z,w\,i) = \Omega_2(z,w\,j) = \Omega_3(z,w\,k)$$

and
$$\Omega_2(z,w\,i) = \Omega_3(z,w)$$
$$\Omega_3(z,w\,j) = \Omega_1(z,w)$$
$$\Omega_1(z,w\,k) = \Omega_2(z,w).$$

Proposition 1. *The subgroup of $GL(4n,\mathbb{R})$ which fixes the three 2-forms $(\Omega_1,\Omega_2,\Omega_3)$ is equal to $Sp(n)$.*

Proof. Let $G \subset GL(4n,\mathbb{R})$ be the subgroup which fixes each of the Ω_a. Clearly we have $Sp(n) \subset G$.

Now, from the first of the identities above, it follows that each of the forms Ω_a is non-degenerate. Then, from the second set of these identities, it follows that the subgroup G must also fix the linear transformations of $\mathbb{R}^{4n}$ which represent multiplication on the right by i, j, and k. Of course, this, by definition, implies that G is a subgroup of $GL(n,\mathbb{H})$. Returning to the first of the identities, it also follows that G must preserve the inner product defined by $\langle,\rangle$. Finally, since we have now seen that G must preserve all of the components of H, it follows that G must preserve H as well. However, this was the very definition of $Sp(n)$.

Proposition 1 motivates the way we will want to define HyperKähler structures on manifolds: as triples of 2-forms which satisfy certain conditions. Here is the linear algebra definition on which the manifold definition will be based.

Definition 5. Let V be a vector space over $\mathbb{R}$. A *Hyperkähler structure* on V is a choice of a triple of non-degenerate 2-forms $(\Omega_1,\Omega_2,\Omega_3)$ which satisfy the following properties: First, that the linear maps R_i, R_j which are defined by the equations

$$\Omega_2(v,R_i\,w) = \Omega_3(v,w) \qquad \Omega_1(v,R_j\,w) = -\Omega_3(v,w)$$

should satisfy $(R_i)^2 = (R_j)^2 = -id$ and should skew-commute, i.e., $R_i R_j = -R_j R_i$. Second, if we set $R_k = -R_i R_j$, then we should have

$$\Omega_1(v,R_i\,w) = \Omega_2(v,R_j\,w) = \Omega_3(v,R_k\,w) = \langle v,w\rangle$$

where $\langle,\rangle$ (defined by these equations) is a positive definite symmetric bilinear form on V. The inner product $\langle,\rangle$ is called the *associated metric* on V.

This may seem to be a rather cumbersome definition (and I admit that it is), but it is sufficient to prove the following Proposition (which I leave as an exercise for the reader).

Proposition 2. *If $(\Omega_1,\Omega_2,\Omega_3)$ is a Hyperkähler structure on a real vector space V, then $\dim(V) = 4n$ for some n and, moreover, there is an $\mathbb{R}$-linear isomorphism* of V with $\mathbb{H}^n$ which identifies the Hyperkähler structure on V with the standard one on $\mathbb{H}^n$.

Hyperkähler structures

We are now ready for the analogs of Definitions 3 and 4:

Definition 6. If M is a manifold of dimension $4n$, an *almost Hyperkähler structure* on M is a triple $(\Omega_1, \Omega_2, \Omega_3)$ of 2-forms on M which have the property that they induce a Hyperkähler structure on each tangent space $T_m M$.

Definition 7. An almost Hyperkähler structure $(\Omega_1, \Omega_2, \Omega_3)$ on a manifold M^{4n} is a Hyperkähler structure on M if each of the forms Ω_a is closed.

At first glance, Definition 6 may seem surprising, after all, no mention has been made of the almost complex structures R_i, R_j, and R_k which are defined on M by any almost Hyperkähler structure on M and one would surely want these to be integrable if the analogy with Kähler geometry is to be kept up. The nice result is that the integrability of these structures comes for free:

Theorem 4. *For an almost Hyperkähler structure $(\Omega_1, \Omega_2, \Omega_3)$ on a manifold M^{4n}, the following are equivalent:*
1. $d\Omega_1 = d\Omega_2 = d\Omega_3 = 0$.
2. *Each of the 2-forms Ω_a is parallel with respect to the Levi-Civita connection of the associated metric.*
3. *Each of the almost complex structures R_i, R_j, and R_k are integrable.*

Proof. (IDEA) The proof of Theorem 4 is much like the proof of Theorem 2. One shows by local calculations in Gauss normal coordinates at any point on M that the covariant derivatives of the forms Ω_a with respect to the Levi-Civita connection of the associated metric can be expressed in terms of the coefficients of their exterior derivatives and vice-versa. Similarly, one shows that the formulas for the Nijnhuis tensors of the three almost complex structures on M can be expressed in terms of the covariant derivatives of the three 2-forms and vice-versa. This is a rather formidable linear algebra problem, but it is nothing else. I will not do the computation here.

Note that Theorem 4 implies that the holonomy H of the associated metric of a Hyperkähler structure on M^{4n} must be a subgroup of $\mathrm{Sp}(n)$. If H is a proper subgroup of $\mathrm{Sp}(n)$, then by Theorem 1, the associated metric must be locally a product metric. Now, as is easy to verify, the only products from Berger's List which can appear as subgroups of $\mathrm{Sp}(n)$ are products of the form

$$\{e\}_{n_0} \times \mathrm{Sp}(n_1) \times \cdots \times \mathrm{Sp}(n_k)$$

where $\{e\}_{n_0} \subset \mathrm{Sp}(n_0)$ is just the identity subgroup and $n = n_0 + \cdots + n_k$. Thus, it follows that a Hyperkähler structure can be decomposed locally into a product of the "flat" example with Hyperkähler structures whose holonomy is the full $\mathrm{Sp}(n_i)$. (If M is simply connected and the associated metric is complete, then the deRham Splitting Theorem implies that M can be globally written as a product of such metrics.) This motivates our calling a Hyperkähler structure on M^{4n} *irreducible* if its holonomy is equal to $\mathrm{Sp}(n)$.

The reader may be wondering just how common these Hyperkähler structures are (aside from the flat ones of course). The answer is that they are not so easy to come by. The first known non-flat example was the Eguchi-Hanson metric (often called a "gravitational instanton") on $T^*\mathbb{CP}^1$. The first known irreducible example in dimensions greater than 4 was discovered by Eugenio Calabi, who, working independently from Eguchi and Hanson, constructed an irreducible Hyperkähler structure on $T^*\mathbb{CP}^n$ for each n which happened to agree with the Eguchi-Hanson metric for $n = 1$. (We will see Calabi's examples a little further on.)

The first known compact example was furnished by Yau's solution of the Calabi Conjecture:

Example: K3 surfaces. A $K3$ surface is a compact simply connected 2-dimensional complex manifold S with trivial canonical bundle. What this latter condition means is that there is nowhere-vanishing holomorphic 2-form Υ on S. An example of such a surface is a smooth algebraic surface of degree 4 in $\mathbb{CP}^3$.

A fundamental result of Siu [Si] is that every $K3$ surface is Kähler, i.e., that there exists a 2-form Ω on S so that the hermitian structure (Ω, J) on S is actually Kähler. Moreover, Yau's solution of the Calabi Conjecture implies that Ω can be chosen so that Υ is parallel with respect to the Levi-Civita connection of the associated metric.

Multiplying Υ by an appropriate constant, we can arrange that $2\Omega^2 = \Upsilon \wedge \overline{\Upsilon}$. Since $\Omega \wedge \Upsilon = 0$ and $\Upsilon \wedge \Upsilon = 0$, it easily follows (see the Exercises) that if we write $\Omega = \Omega_1$ and $\Upsilon = \Omega_2 - \imath \Omega_3$, then the triple $(\Omega_1, \Omega_2, \Omega_3)$ defines a Hyperkähler structure on S.

For a long time, the $K3$ surfaces were the only known compact manifolds with Hyperkähler structures. In fact, a "proof" was published showing that there were no other compact ones. However, this turned out not to be correct.

Example: holomorphic symplectic manifolds. Let M^{4n} be a simply connected, compact complex manifold (of complex dimension $2n$) with a *holomorphic symplectic* form Υ. Then Υ^n is a non-vanishing holomorphic volume form, and hence the canonical bundle of M is trivial. If M has a Kähler structure which is compatible with its complex structure, then, by Yau's solution of the Calabi Conjecture, there is a Kähler metric g on M for which the volume form Υ^n is parallel. This implies that the holonomy of g is a subgroup of $\mathrm{SU}(2n)$. However, this in turn implies that g is Ricci-flat and hence, by a Bochner vanishing argument, that every holomorphic form on M is parallel with respect to g. Thus, Υ is also parallel with respect to g and hence the holonomy is a subgroup of $\mathrm{Sp}(n)$. If M can be constructed in such a way that it cannot be written as a non-trivial product of complex submanifolds, then the holonomy of g must act irreducibly on $\mathbb{C}^{2n}$ and hence must equal $\mathrm{Sp}(n)$.

Fujita was the first to construct a simply connected, compact complex 4-manifold which carried a holomorphic 2-form and which could not be written non-trivially as a product. This work is written up in detail in a survey article by [Bea].

Hyperkähler reduction

I am now ready to describe another method of constructing Hyperkähler structures, known as *Hyperkähler reduction*. This method first appeared in a famous paper by Hitchin, Karlhede, Lindström, and Roček [HKLR].

Theorem 5. *Suppose that $(\Omega_1, \Omega_2, \Omega_3)$ is a Hyperkähler structure on M and that there is a left action $\lambda \colon G \times M \to M$ which is Poisson with respect to each of the three symplectic forms Ω_a. Let*

$$\mu = (\mu_1, \mu_2, \mu_3) \colon M \to \mathfrak{g}^* \oplus \mathfrak{g}^* \oplus \mathfrak{g}^*$$

be a G-equivariant moment map. Suppose that $\xi \in \mathfrak{g}^* \oplus \mathfrak{g}^* \oplus \mathfrak{g}^*$ is a clean value for μ and that the quotient $M_\xi = G_\xi \backslash \mu^{-1}(\xi)$ has a smooth structure for which the projection $\pi_\xi \colon \mu^{-1}(\xi) \to M_\xi$ is a smooth submersion. Then there is a unique Hyperkähler structure $(\Omega_1^\xi, \Omega_2^\xi, \Omega_3^\xi)$ on M_ξ with the property that $\pi_\xi^*(\Omega_a^\xi)$ is the pull back of Ω_a to $\mu^{-1}(\xi) \subset M$ for each $a = 1, 2,$ or 3.

Proof. Assume the hypotheses of the Theorem. Let $\tilde{\Omega}_a^\xi$ be the pullback of Ω_a to $\mu^{-1}(\xi) \subset M$. It is clear that each of the forms $\tilde{\Omega}_a^\xi$ is a closed, G_ξ-invariant 2-form on $\mu^{-1}(\xi)$.

I first want to show that each of these can be written as a pullback of a 2-form on M_ξ, i.e., that each is semi-basic for π_ξ. To do this, I need to characterize $T_m \mu^{-1}(\xi)$ in an appropriate fashion. Now, the assumption that ξ is a clean value for μ is equivalent to the assumption that $v \in T_m M$ lies in $T_\mu^{-1}(\xi)$ if and only if $\Omega_a(v, \lambda_*(x)(m)) = 0$ for all $x \in \mathfrak{g}$ and all three values of a. Thus,

$$T_\mu^{-1}(\xi) = \{v \in T_m M \mid \langle v, w\, i \rangle = \langle v, w\, j \rangle = \langle v, w\, k \rangle = 0, \text{ for all } w \in T_m(G \cdot m)\}.$$

Since $G_\xi \cdot m = \mu^{-1}(\xi) \cap (G \cdot m)$ and since $T_m(G_\xi \cdot m) = T_m \mu^{-1}(\xi) \cap T_m(G \cdot m)$, it follows that $v \in T_M(G_\xi \cdot m)$ implies that v is in the null space of each of the forms $\tilde{\Omega}_a^\xi$. Thus, each of the forms $\tilde{\Omega}_a^\xi$ is semi-basic for π_ξ, as we wished to show. This, combined with G_ξ-equivariance, implies that there exist unique forms Ω_a^ξ on M_ξ which satisfy $\pi_\xi^*(\Omega_a^\xi) = \tilde{\Omega}_a^\xi$. Since π_ξ is a submersion, it automatically follows that these three 2-forms are closed.

To complete the proof, it suffices to show that the triple $(\Omega_1^\xi, \Omega_2^\xi, \Omega_3^\xi)$ actually defines an almost Hyperkähler structure on M_ξ, for then we can apply Theorem 4.

We do this as follows: Use the associated metric $\langle, \rangle$ to define an orthogonal splitting

$$T_m \mu^{-1}(\xi) = T_m(G_\xi \cdot m) \oplus H_m.$$

It is clear that, for each $m \in \mu^{-1}(\xi)$, the differential $\pi_\xi'(m)$ induces an isomorphism from H_m to $T_{\pi_\xi(m)} M_\xi$ in such a way that the restriction of the form $\tilde{\Omega}_a^\xi$ to H_m gets identified with Ω_a^ξ. Thus, it suffices to show that the forms $(\tilde{\Omega}_1^\xi, \tilde{\Omega}_2^\xi, \tilde{\Omega}_3^\xi)$ define a Hyperkähler structure when restricted to H_m. Now, by Proposition 2, to do this, it would suffice to show that H_m is stable under the action of R_i, R_j, and R_k.

However, the very definition of H_m is that it is the subspace of T_mM which is orthogonal to the $\mathbb{H}$-linear subspace $(T_m(G \cdot m)) \cdot \mathbb{H} \subset T_mM$. Since the orthogonal complement of an $\mathbb{H}$-linear subspace of T_mM is also an $\mathbb{H}$-linear subspace, we are done.

I leave as an exercise for the reader to determine a formula for the dimension of M_ξ analogous to the formula that we developed for symplectic reduction.

Example: reductions of $\mathbb{H}^n$. One of the simplest things that one could do is take $M = \mathbb{H}^n$ and let $G \subset \mathrm{Sp}(n)$ be any closed subgroup. This yields an astonishing variety of examples. Here, I will present just one.

Let $S^1 \subset \mathrm{Sp}(n)$ act diagonally on $\mathbb{H}^n$ by the action

$$e^{i\theta} \cdot \begin{pmatrix} q^1 \\ \vdots \\ q^n \end{pmatrix} = \begin{pmatrix} e^{i\theta} q^1 \\ \vdots \\ e^{i\theta} q^n \end{pmatrix}.$$

Then it is not difficult to see that the moment map can be identified with the map

$$\mu(q) = {}^t\bar{q}\, i\, q.$$

The reduced space M_p for any $p \neq 0$ is easily seen to be complex analytically equivalent to $T^*\mathbb{CP}^{n-1}$, and the induced Hyperkähler structure is the one found by Calabi. In particular, for $n = 2$, one recovers the Eguchi-Hansen metric.

In the Exercises, there are other examples for you to try.

The method of Hyperkähler reduction has a wide variety of applications. Many of the interesting moduli spaces for Yang-Mills theory turn out to have Hyperkähler structures because of this reduction procedure. For example, as Atiyah and Hitchin [AH] show, the space of magnetic monopoles of "charge" k on $\mathbb{R}^3$ turns out to have a natural Hyperkähler structure which is derived by methods extremely similar to the example presented earlier of a Kähler structure on the moduli space of flat connections over a Riemann surface.

Peter Kronheimer [Kr] has used the method of Hyperkähler reduction to construct, for each quotient manifold Σ of S^3, an asymptotically locally Euclidean (ALE) Ricci-flat self-dual Einstein metric on a 4-manifold M_Σ whose boundary at infinity is Σ. He then went on to prove that all such metrics on 4-manifolds arise in this way.

Finally, it should also be mentioned that the case of metrics on manifolds M^{4n} with holonomy $\mathrm{Sp}(n) \cdot \mathrm{Sp}(1)$ can also be treated by the method of reduction. I don't have time to go into this here, but the reader can find a complete account in [GL].

Exercises

1. Show that the following two definitions of compatibility between an almost complex structure J and a metric g on M^{2n} are equivalent
   (i) (g, J) are compatible if $g(v) = g(Jv)$ for all $v \in TM$.
   (ii) (g, J) are compatible if $\Omega(v, w) = \langle Jv, w \rangle$ defines a (skew-symmetric) 2-form on M.

2. A NON-INTEGRABLE ALMOST COMPLEX STRUCTURE. Let J be an almost complex structure on M. Let $\mathcal{A}^{1,0} \subset \mathbb{C} \otimes \mathcal{A}^1(M)$ denote the space of $\mathbb{C}$-valued 1-forms on M which satisfy $\alpha(Jv) = \imath \alpha(v)$ for all $v \in TM$.
   (i) Show that if we define $\mathcal{A}^{0,1}(M) \subset \mathbb{C} \otimes \mathcal{A}^1(M)$ to be the space of $\mathbb{C}$-valued 1-forms on M which satisfy $\alpha(Jv) = -\imath \alpha(v)$ for all $v \in TM$, then $\mathcal{A}^{0,1}(M) = \overline{\mathcal{A}^{1,0}(M)}$ and that $\mathcal{A}^{1,0}(M) \cap \mathcal{A}^{0,1}(M) = \{0\}$.
   (ii) Show that if J is an integrable almost complex structure, then, for any $\alpha \in \mathcal{A}^{1,0}(M)$, the 2-form $d\alpha$ is (at least locally) in the ideal generated by $\mathcal{A}^{1,0}(M)$. (Hint: Show that, if $z : U \to \mathbb{C}^n$ is a holomorphic coordinate chart, then, on U, the space $\mathcal{A}^{1,0}(U)$ is spanned by the forms $dz^1, \ldots, dz^n$. Now consider the exterior derivative of any linear combination of the dz^i.) It is a celebrated result of Newlander and Nirenberg that this condition is sufficient for J to be integrable.
   (iii) Show that there is an almost complex structure on $\mathbb{C}^2$ for which $\mathcal{A}^{1,0}(\mathbb{C}^2)$ is spanned by the 1-forms
   $$\omega^1 = dz^1 - \bar{z}^1 \, d\bar{z}^2$$
   $$\omega^2 = dz^2$$
   and that this almost complex structure is not integrable.

3. Let $M = \mathbb{C}^{n_1} \oplus \mathbb{C}^{n_2} \setminus \{(0,0)\}$. Let $G = S^1$ act on M by the action
   $$e^{\imath\theta} \cdot (z_1, z_2) = \left(e^{\imath d_1 \theta} z_1, e^{\imath d_2 \theta} z_2\right)$$
   where d_1 and d_2 are integers. Let M have the standard flat Kähler structure. Compute the moment map μ and the Kähler structures on the reduced spaces. How do the relative signs of d_1 and d_2 affect the answer? What interpretation can you give to these spaces?

4. Go back to the the example of the 'Kähler structure' on the space $\mathfrak{A}(P)$ of connections on a principal right G-bundle P over a connected compact Riemann surface Σ. Assume that $G = S^1$ and identify $\mathfrak{g}$ with $\mathbb{R}$ in the natural way. Thus, F_A is a well-defined 2-form on Σ and the cohomology class $[F_A] \in H^2_{dR}(\Sigma, \mathbb{R})$ is independent of the choice of A. Assume that $[F_A] \neq 0$. Then, in this case, $\mu^{-1}(0)$ is empty so the construction we made in the example in the Lecture is vacuous. Here is how we can still get some information.

   Fix any non-vanishing 2-form Ψ on Σ so that $[\Psi] = [F_A]$. Show that even though μ has no regular values, Ψ is a non-trivial clean value of μ. Show also that, for any $A \in \mu^{-1}(\Psi)$, the stabilizer $G_A \subset \text{Aut}(P)$ is a discrete (and hence finite) subgroup of S^1. Describe, as fully as you can, the reduced space $\mathfrak{M}_\Psi$ and its Kähler structure.

LECTURE 8. RECENT APPLICATIONS OF REDUCTION

5. Verify that $\mathrm{Sp}(n)$ is a connected Lie group of dimension $2n^2+n$. (Hint: You will probably want to study the function $f(A) = {}^t\bar{A}A = I_n$.) Show that $\mathrm{Sp}(n)$ acts transitively on the unit sphere $S^{4n-1} \subset \mathbb{H}^n$ defined by the relation $H(x,x) = 1$. (Hint: First show that, by acting by diagonal matrices in $\mathrm{Sp}(n)$, you can move any element of $\mathbb{H}^n$ into the subspace $\mathbb{R}^n$. Then note that $\mathrm{Sp}(n)$ contains $\mathrm{SO}(n)$.) By analysing the stabilizer subgroup in $\mathrm{Sp}(n)$ of an element of S^{4n-1}, show that there is a fibration

$$\mathrm{Sp}(n-1) \to \mathrm{Sp}(n)$$
$$\downarrow$$
$$S^{4n-1}$$

and use this to conclude by induction that $\mathrm{Sp}(n)$ is connected and simply connected for all n.

6. Prove Proposition 2. (Hint: First show how the maps R_i, R_j, and R_k define the structure of a right $\mathbb{H}$-module on V. Then show that V has a basis $b_1,\ldots,b_n$ over $\mathbb{H}$ and use this to construct an $\mathbb{H}$-linear isomorphism of V with $\mathbb{H}^n$. If you pick the basis b_a carefully, you will be done at this point. Warning: You must use the positive definiteness of $\langle,\rangle$!)

7. Determine a formula for the dimension of the Hyperkähler reduced space M_ξ in terms of the dimensions of M, G, G_ξ and G_m (where m lies in $\mu^{-1}(\xi)$).

8. Apply the Hyperkähler reduction procedure to $\mathbb{H}^2$ with $\mathbb{R}$ acting by the rule

$$\theta \cdot \begin{pmatrix} q^1 \\ q^2 \end{pmatrix} = \begin{pmatrix} e^{i\theta} q^1 \\ q^2 + \theta \end{pmatrix}.$$

Determine which values of μ are clean and describe the resulting complex surfaces and their Hyperkähler structures.

LECTURE 9
The Gromov School of Symplectic Geometry

In this Lecture, I want to describe some of the remarkable new information we have about symplectic manifolds owing to the influence of the ideas of Mikhail Gromov. The basic reference for much of this material is Gromov's remarkable book *Partial Differential Relations*.

The fundamental idea of studying complex structures "tamed by" a given symplectic structure was developed by Gromov in a remarkable paper *Pseudo-holomorphic Curves on Almost Complex Manifolds* and has proved extraordinarily fruitful. In the latter part of this lecture, I will try to introduce the reader to this theory.

Soft techniques in symplectic manifolds

Symplectic immersions and embeddings

Classical immersion theory. Before beginning on the topic of symplectic immersions, let me recall how the theory of immersions in the ordinary sense goes.

Recall that the Whitney Immersion Theorem (in the weak form) asserts that any smooth n-manifold M has an immersion into $\mathbb{R}^{2n}$. This result is proved by first immersing M into some $\mathbb{R}^N$ for $N \gg 0$ and then using Sard's Theorem to show that if $N > 2n$, one can find a vector $u \in \mathbb{R}^N$ so that u is not tangent to $f(M)$ at any point. Then the projection of $f(M)$ onto a hyperplane orthogonal to u is still an immersion, but now into $\mathbb{R}^{N-1}$.

This result is not the best possible. Whitney himself showed that one could always immerse M^n into $\mathbb{R}^{2n-1}$, although "general position" arguments are not sufficient to do this. This raises the question of determining what the best possible immersion or embedding dimension is.

One topological obstruction to immersing M^n into $\mathbb{R}^{n+k}$ can be described as follows: If $f\colon M \to \mathbb{R}^{n+k}$ is an immersion, then the trivial bundle $f^*(T\mathbb{R}^{n+k}) = M \times \mathbb{R}^{n+k}$ can be split into a direct sum $f^*(T\mathbb{R}^{n+k}) = TM \oplus \nu^f$ where ν^f is the normal bundle of the immersion f. Thus, if there is no bundle ν of rank k over M so that $TM \oplus \nu$ is trivial, then there can be no immersion of M into $\mathbb{R}^{n+k}$.

The remarkable fact is that this topological necessary condition is almost sufficient. In fact, we have the following result of Hirsch and Smale for the general immersion problem.

Theorem 1. *Let M and N be connected smooth manifolds and suppose either that M is non-compact or else that $\dim(M) < \dim(N)$. Let $f: M \to N$ be a continuous map, and suppose that there is a vector bundle ν over M so that $f^*(TN) = TM \oplus \nu$. Then f is homotopic to an immersion of M into N.*

The h-Principle. Theorem 1 can be interpreted as an example of what Gromov calls the *h-principle*, which I now want to describe.

Let $\pi: V \to X$ be a surjective submersion. A *section* of π is, by definition, a map $\sigma: X \to V$ which satisfies $\pi \circ \sigma = id_X$. Let $J^k(X, V)$ denote the space of k-jets of sections of V, and let $\pi^k: J^k(X, V) \to X$ denote the basepoint or "source" projection. Given any section s of π, there is an associated section $j^k(s)$ of π^k which is defined by letting $j^k(s)(x)$ be the k-jet of s at $x \in X$. A section σ of π^k is said to be *holonomic* if $\sigma = j^k(s)$ for some section s of π.

A *partial differential relation* of order k for π is a subset $R \subset J^k(X, V)$. A section s of π is said to *satisfy R* if $j^k(s)(X) \subset R$. We can now make the following definition:

Definition 1. A partial differential relation $R \subset J^k(X, V)$ satisfies the *h-principle* if, for every section σ of π^k which satisfies $\sigma(X) \subset R$, there is a one-parameter family of sections σ_t ($0 \leq t \leq 1$) of π^k which satisfy the conditions that $\sigma_t(X) \subset R$ for all t, that $\sigma_0 = \sigma$, and that σ_1 is holonomic.

Very roughly speaking, a partial differential relation satisfies the h-principle if, whenever the "topological" conditions for a solution to exist are satisfied, then a solution exists.

For example, if $X = M$ and $V = M \times N$, where $\dim(N) \geq \dim(M)$, then there is an (open) subset $R = \mathrm{Imm}(M, N) \subset J^1(M, M \times N)$ which consists of the 1-jets of graphs of (local) immersions of M into N. What the Hirsch-Smale immersion theory says is that $\mathrm{Imm}(M, N)$ satisfies the h-principle if either $\dim M = \dim N$ and M has no compact component or else $\dim M < \dim N$.

Of course, the h-principle does not hold for every relation R. The real question is how to determine when the h-principle holds for a given R. Gromov has developed several extremely general methods for proving that the h-principle holds for various partial differential relations R which arise in geometry. These methods include his theory of topological sheaves and techniques like his method of convex integration. They generally work in situations where the local solutions of a given partial differential relation R are easy to come by and it is mainly a question of "patching together" local solutions which are fairly "flexible".

Gromov calls this collection of techniques "soft" to distinguish them from the "hard" techniques, such as elliptic theory, which come from analysis and deal with situations where the local solutions are somewhat "rigid".

Here is a sample of some of the results which Gromov obtains by these methods:

Theorem 2. *Let X^{2n} be a smooth manifold and let $V \subset \Lambda^2(T^*(M))$ denote the open subbundle consisting of non-degenerate 2-forms $\omega \in T_xX$. Let $Z^1(X, V) \subset J^1(X, V)$ denote the space of 1-jets of closed non-degenerate 2-forms on X. Then, if X has no compact component, $Z^1(X, V)$ satisfies the h-principle.*

In particular, Theorem 2 implies that a non-compact, connected X has a symplectic structure if and only if it has an almost symplectic structure.

Note that this result is definitely *not* true for compact manifolds. We have already seen several examples, e.g., $S^1 \times S^3$, which have almost symplectic structures but no symplectic structures because they do not satisfy the cohomology ring obstruction. Gromov has asked the following:

Question : *If X^{2n} is compact and connected and satisfies the condition that there exists an element $u \in H^2_{dR}(X, \mathbb{R})$ which satisfies $u^n \neq 0$, does $Z^1(X,V)$ satisfy the h-principle?*

The next result I want to describe is Gromov's theorem on symplectic immersions. This theorem is an example of a sort of "restricted h-principle" in that it is only required to apply to sections σ which satisfy specified cohomological conditions.

First, let me make a few definitions: Let (X, Ξ) and (Y, Ψ) be two connected symplectic manifolds. Let $\mathcal{S}(X,Y) \subset J^1(X, X \times Y)$ denote the space of 1-jets of graphs of (local) symplectic maps $f: X \to Y$. i.e., (local) maps $f: X \to Y$ which satisfy $f^*(\Psi) = \Xi$. Let $\tau: \mathcal{S}(X,Y) \to Y$ be the obvious "target projection".

Theorem 3. *If either X is non-compact, or $\dim(X) < \dim(Y)$, then any section σ of $\mathcal{S}(X,Y)$ for which the induced map $s = \tau \circ \sigma: X \to Y$ satisfies the cohomological condition $s^*([\Psi]) = [\Xi]$ is homotopic to a holonomic section of $\mathcal{S}(X,Y)$.*

This result can be also stated as follows: Suppose that either X is non-compact or else that $\dim(X) < \dim(Y)$. Let $\phi: X \to Y$ be a smooth map which satisfies the cohomological condition $\phi^*([\Psi]) = [\Xi]$. Suppose that there exists a bundle map $f: TX \to \phi^*(TY)$ which is symplectic in the obvious sense. Then ϕ is homotopic to a symplectic immersion.

Theorem 3 implies the following result of Narasimham and Ramanan.

Corollary. *Any compact symplectic manifold (M, Ω) for which the cohomology class $[\Omega]$ is integral admits a symplectic immersion into $(\mathbb{CP}^N, \Omega_N)$ for some N.*

Proof. Since the cohomology class $[\Omega]$ is integral, there exists a smooth map $\phi: M \to \mathbb{CP}^N$ for some N sufficiently large so that $\phi^*([\Omega_N]) = [\Omega]$. Then, choosing $N \gg n$, we can arrange that there also exists a symplectic bundle map $f: TM \to f^*(T\mathbb{CP}^N)$ (see the Exercises). Now apply Theorem 3.

As a final example along these lines, let me state Gromov's embedding result. Here, the reader should be thinking of the difference between the Whitney Immersion Theorems and the Whitney Embedding Theorems: One needs slightly more room to embed than to immerse.

Theorem 4. *Suppose that (X, Ξ) and (Y, Ψ) are connected symplectic manifolds and that either X is non-compact and $\dim(X) < \dim(Y)$ or else that $\dim(X) < \dim(Y) - 2$. Suppose that there exists a smooth embedding $\phi: X \to Y$ and that the induced map on bundles $\phi': TX \to \phi^*(TY)$ is homotopic through a 1-parameter family of injective bundle maps $\varphi_t: TX \to \phi^*(TY)$ (with $\varphi_0 = \phi'$) to a symplectic bundle map $\varphi_1: TX \to \phi^*(TY)$. Then ϕ is isotopic to a symplectic embedding $\varphi: X \to Y$.*

This result is actually the best possible, for, as Gromov has shown using 'hard' techniques (see below), there are counterexamples if one leaves out the dimensional restrictions. Note by the way that, because Theorem 4 deals with embeddings rather than immersions, it not straightforward to place it in the framework of the h-principle.

Blowing up in the symplectic category

We have already seen in Lecture 6 that certain operations on smooth manifolds cannot be carried out in the symplectic category. For example, one cannot form connected sums in the symplectic category.

However, certain of the operations from the geometry of complex manifolds *can* be carried out. Gromov has shown how to define the operation of "blowing up" in the symplectic category.

Recall how one "blows up" the origin in $\mathbb{C}^n$. To avoid triviality, let me assume that $n > 1$. Consider the subvariety

$$X = \{(v, [w]) \in \mathbb{C}^n \times \mathbb{CP}^{n-1} \mid v \in [w]\} \subset \mathbb{C}^n \times \mathbb{CP}^{n-1}.$$

It is easy to see that X is a smooth embedded submanifold of the product and that the projection $\pi: X \to \mathbb{C}^n$ is a biholomorphism away from the "exceptional point" $0 \in \mathbb{C}^n$. Moreover, if Ω_0 and Φ are the standard Kähler 2-forms on $\mathbb{C}^n$ and $\mathbb{CP}^{n-1}$ respectively, then, for each $\epsilon > 0$, the 2-form $\Omega_\epsilon = \Omega_0 + \epsilon \Phi$ is a Kähler 2-form on X.

Now, Gromov realized that this can be generalized to a "blow up" construction for any point p on any symplectic manifold (M^{2n}, Ω). Here is how this goes:

First, choose a neighborhood U of p on which there exists a local chart $z: U \to \mathbb{C}^n$ which is symplectic, i.e., satisfies $z^*(\Omega_0) = \Omega$, and satisfies $z(p) = 0$. Suppose that the ball $B_{2\delta}(0)$ in $\mathbb{C}^n$ of radius 2δ centered on 0 lies inside $z(U)$. Since $\pi: \pi^{-1}\bigl(B^*_{2\delta}(0)\bigr) \to B^*_{2\delta}(0)$ is a diffeomorphism, there exists a closed 2-form $\tilde{\Phi}$ on $B^*_{2\delta}(0)$ so that $\pi^*(\tilde{\Phi}) = \Phi$. Since $H^2_{dR}\bigl(B^*_{2\delta}(0)\bigr) = 0$, there exists a 1-form φ on $B^*_{2\delta}(0)$ so that $d\varphi = \tilde{\Phi}$.

Now consider the family of symplectic forms $\Omega_0 + \epsilon \, d\varphi$ on $B^*_{2\delta}(0)$. By using a homotopy argument exactly like the one used to Prove Theorem 1 in Lecture 6, it easily follows that for all $t > 0$ sufficiently small, there exists an open annulus $A(\delta - \varepsilon, \delta + \varepsilon)$ and a one-parameter family of diffeomorphisms $\phi_t: A(\delta - \varepsilon, \delta + \varepsilon) \to B^*_{2\delta}(0)$ so that

$$\phi_t^*(\Omega_0) = \Omega_0 + t \, d\varphi.$$

It follows that we can set

$$\hat{M} = \pi^{-1}\bigl(B^*_{\delta + \varepsilon}(0)\bigr) \cup_{\psi_t} M \setminus z^{-1}\bigl(\phi_t\bigl(B^*_{\delta - \varepsilon}(0)\bigr)\bigr)$$

where

$$\psi_t: \pi^{-1}\bigl(A(\delta - \varepsilon, \delta + \varepsilon)\bigr) \to M \setminus z^{-1}\bigl(\phi_t\bigl(A(\delta - \varepsilon, \delta + \varepsilon)\bigr)\bigr)$$

is given by $\psi_t = z^{-1} \circ \phi_t \circ \pi$. Since ψ_t identifies the symplectic structure Ω_t with Ω_0 on the annulus "overlap", it follows that $\hat{M}$ is symplectic. This is Gromov's symplectic blow up procedure. Note that it can be effected in such a way that the symplectic structure on $M \setminus \{p\}$ is not disturbed outside of an arbitrarily small ball around p. Note also that there is a parameter involved, and that the symplectic structure is certainly not unique.

This is only describes a simple case. However, Gromov has shown how any compact symplectic submanifold S^{2k} of M^{2n} can be blown up to become a symplectic "hypersurface" $\hat{S}$ in a new symplectic manifold $\hat{M}$ which has the property that $M \setminus S$ is diffeomorphic to $\hat{M} \setminus \hat{S}$.

The basic idea is the same as what we have already done: First, one mimics the topological operations which would be performed if one were blowing up a complex submanifold of a complex manifold. Thus, the submanifold S gets "replaced" by the complex projectivization $\hat{S} = \mathbb{P}N_S$ of a complex normal bundle. Second, one shows how to define a symplectic structure on the resulting smooth manifold which can be made to agree with the old structure outside of an arbitrarily small neighborhood of the blow up.

The details in the general case are somewhat more complicated than the case of blowing up a single point, and Dusa McDuff has written out a careful construction [M 2]. She has also used the method of blow ups to produce an example of a simply connected compact symplectic manifold which has no Kähler structure.

Hard techniques in symplectic manifolds

(Pseudo-) holomorphic curves

I begin with a fundamental definition.

Definition 2. Let M^{2n} be a smooth manifold and let $J: TM \to TM$ be an almost complex structure on M. For any Riemann surface Σ, we say that a map $f: \Sigma \to M$ is J-holomorphic if $f'(\imath v) = J f'(v)$ for all $v \in T\Sigma$.

Often, when J is clear from context, I will simply say "f is holomorphic". Several authors use the terminology "almost holomorphic" or "pseudo-holomorphic" for this concept, reserving the word "holomorphic" for use only when the almost complex structure on M is integrable to an actual complex structure. This distinction does not seem to be particularly useful, so I will not maintain it.

It is instructive to see what this looks like in local coordinates. Let $z = x + \imath y$ be a local holomorphic coordinate on Σ and let $w: U \to \mathbb{R}^{2n}$ be a local coordinate on M. Then there exists a matrix $\mathrm{J}(w)$ of functions on $w(U) \subset \mathbb{R}^{2n}$ which satisfies $w'(Jv) = \mathrm{J}(w(p))w'(v)$ for all $v \in T_p U$. This matrix of functions satisfies the relation $\mathrm{J}^2 = -I_{2n}$. Now, if $f: \Sigma \to M$ is holomorphic and carries the domain of the z-coordinate into U, then $F = w \circ f$ is easily seen to satisfy the first order system of partial differential equations

$$\frac{\partial F}{\partial y} = \mathrm{J}(F)\frac{\partial F}{\partial x}.$$

Since $\mathrm{J}^2 = -I_{2n}$, it follows that this is a first-order, elliptic, determined system of partial differential equations for F. In fact, the principal symbol of these equations is the same as that for the Cauchy-Riemann equations. Assuming that J is sufficiently regular (C^∞ is sufficient and we will always have this) there are plenty of local solutions. What is at issue is the nature of the global solutions.

A *parametrized holomorphic curve* in M is a holomorphic map $f: \Sigma \to M$. Sometimes we will want to consider *unparametrized holomorphic curves* in M, namely equivalence classes $[\Sigma, f]$ of holomorphic curves in M where (Σ_1, f_1) is equivalent to (Σ_2, f_2) if there exists a holomorphic map $\phi: \Sigma_1 \to \Sigma_2$ satisfying $f_1 = f_2 \circ \phi$.

We are going to be particularly interested in the space of holomorphic curves in M. Here are some properties that hold in the case of holomorphic curves in actual complex manifolds and it would be nice to know if they also hold for holomorphic curves in almost complex manifolds.

LOCAL FINITE DIMENSIONALITY. If Σ is a compact Riemann surface, and $f: \Sigma \to M$ is a holomorphic curve, it is reasonable to ask what the space of "nearby" holomorphic curves looks like. Because the equations which determine these mappings are elliptic and because Σ is compact, it follows without too much difficulty that the space of nearby holomorphic curves is finite dimensional. (We do *not*, in general know that it is a smooth manifold!)

INTERSECTIONS. A pair of distinct complex curves in a complex surface always intersect at isolated points and with positive "multiplicity." This follows from complex analytic geometry. This result is extremely useful because it allows us to derive information about actual numbers of intersection points of holomorphic curves by applying topological intersection formulas. (Usually, these topological intersection formulas only count the number of signed intersections, but if the surfaces can only intersect positively, then the topological intersection numbers (counted with multiplicity) are the actual intersection numbers.)

KÄHLER AREA BOUNDS. If M happens to be a Kähler manifold, with Kähler form Ω, then the area of the image of a holomorphic curve $f: \Sigma \to M$ is given by the formula

$$\text{Area}(f(\Sigma)) = \int_\Sigma f^*(\Omega).$$

In particular, since Ω is closed, the right hand side of this equation depends only on the homotopy class of f as a map into M. Thus, if (Σ_t, f_t) is a continuous one-parameter family of closed holomorphic curves in a Kähler manifold, then they all have the same area. This is a powerful constraint on how the images can behave, as we shall see.

Now the first two of these properties go through without change in the case of almost complex manifolds.

In the case of local finiteness, this is purely an elliptic theory result. Studying the linearization of the equations at a solution will even allow one to predict, using the Atiyah-Singer Index Theorem, an upper bound for the local dimension of the moduli space and, in some cases, will allow us to conclude that the moduli space near a given closed curve is actually a smooth manifold (see below).

As for pairs of complex curves in an almost complex surface, Gromov has shown that they do indeed only intersect in isolated points and with positive multiplicity

LECTURE 9. THE GROMOV SCHOOL OF SYMPLECTIC GEOMETRY

(unless they have a common component, of course). Both Gromov and Dusa McDuff have used this fact to study the geometry of symplectic 4-manifolds.

The third property is only valid for Kähler manifolds, but it is highly desirable. The behaviour of holomorphic curves in compact Kähler manifolds is well understood in a large part because of this area bound. This motivated Gromov to investigate ways of generalizing this property.

Symplectic tamings

Following Gromov, we make the following definition.

Definition 3. A symplectic form Ω on M *tames* an almost complex structure J if it is J-positive, i.e., if it satisfies $\Omega(v, Jv) > 0$ for all non-zero tangent vectors $v \in TM$.

The reader should be thinking of Kähler geometry. In that case, the symplectic form Ω and the complex structure J satisfy $\Omega(v, Jv) = \langle v, v \rangle > 0$. Of course, this generalizes to the case of an arbitrary almost Kähler structure.

Now, if M is compact and Ω tames J, then for any Riemannian metric g on M (not necessarily compatible with either J or Ω) there is a constant $C > 0$ so that

$$|v \wedge Jv| \leq C\,\Omega(v, Jv)$$

where $|v \wedge Jv|$ represents the area in T_pM of the parallelogram spanned by v and Jv in T_pM (see the Exercises). In particular, it follows that, for any holomorphic curve $f: \Sigma \to M$, we have the inequality

$$\text{Area}(f(\Sigma)) \leq C \int_\Sigma f^*(\Omega).$$

Just as in the Kähler case, the integral on the right hand side depends only on the homotopy class of f. Thus, if an almost complex structure can be tamed, it follows that, in any metric on M, there is a uniform upper bound on the areas of the curves in any continuous family of compact holomorphic curves in M.

Example: an untameable complex structure. Let $N_3^{\mathbb{C}}$ denote the complex Heisenberg group. Thus, $N_3^{\mathbb{C}}$ is the complex Lie group of matrices of the form

$$g = \begin{pmatrix} 1 & x & z \\ 0 & 1 & y \\ 0 & 0 & 1 \end{pmatrix}.$$

Let $\Gamma \subset N_3^{\mathbb{C}}$ be the subgroup all of whose entries belong to the ring of Gaussian integers $\mathbb{Z}[\imath]$.

Let $M = N_3^{\mathbb{C}}/\Gamma$. Then M is a compact complex 3-manifold. I claim that the complex structure on M cannot be tamed by any symplectic form.

To see this, consider the right-invariant 1-form

$$dg\, g^{-1} = \begin{pmatrix} 0 & \omega_1 & \omega_3 \\ 0 & 0 & \omega_2 \\ 0 & 0 & 0 \end{pmatrix}.$$

Since they are right-invariant, it follows that the complex 1-forms $\omega_1, \omega_2, \omega_3$ are also well-defined on M. Define the metric G on M to be the quadratic form

$$G = \omega_1 \circ \overline{\omega_1} + \omega_2 \circ \overline{\omega_2} + \omega_3 \circ \overline{\omega_3}.$$

Now consider the holomorphic curve $Y: \mathbb{C} \to N_3^{\mathbb{C}}$ defined by

$$Y(y) = \begin{pmatrix} 1 & 0 & 0 \\ 0 & 1 & y \\ 0 & 0 & 1 \end{pmatrix}.$$

Let $\psi: \mathbb{C} \to M$ be the composition. It is clear that ψ is doubly periodic and hence defines an embedding of a complex torus into M. It is clear that the G-area of this torus is 1.

Now $N_3^{\mathbb{C}}$ acts holomorphically on M on the left (not by G-isometries, of course). We can consider what happens to the area of the torus $\psi(\mathbb{C})$ under the action of this group. Specifically, for $x \in \mathbb{C}$, let ψ_x denote ψ acted on by left multiplication by the matrix

$$\begin{pmatrix} 1 & x & 0 \\ 0 & 1 & 0 \\ 0 & 0 & 1 \end{pmatrix}.$$

Then, as the reader can easily check,

$$\psi_x^*(G) = (1 + |x|^2)|dy|^2.$$

Thus, the G-area of the torus $\psi_x(\mathbb{C})$ goes off to infinity as x tends to infinity. Obviously, there can be no taming of the complex manifold M. (In particular, M cannot carry a Kähler structure compatible with its complex structure.)

Gromov's compactness theorem

In this section, I want to discuss Gromov's approach to compactifying the connected components of the space of unparametrized holomorphic curves in M.

Example: Plane algebraic curves. Before looking at the general case, let us look at what happens in a very familiar case: The case of algebraic curves in $\mathbb{CP}^2$ with its standard Fubini-Study metric and symplectic form Ω (normalized so as to give the lines in $\mathbb{CP}^2$ an area of 1).

Since this is a Kähler metric, we know that the area of a connected one-parameter family of holomorphic curves (Σ_t, f_t) in $\mathbb{CP}^2$ is constant and is equal to an integer $d = \int_{\Sigma_t} f_t^*(\Omega)$, called the *degree*. To make matters as simple as possible, let me consider the curves degree by degree.

$d = 0$. In this case, the "curves" are just the constant maps and (in the unparametrized case) clearly constitute a copy of $\mathbb{CP}^2$ itself. Note that this is already compact.

$d = 1$. In this case, the only possibility is that each Σ_t is just $\mathbb{CP}^1$ and the holomorphic map f_t must be just a biholomorphism onto a line in $\mathbb{CP}^2$. Of course, the space of lines in $\mathbb{CP}^2$ is compact, just being a copy of the dual $\mathbb{CP}^2$. Thus, the space $\mathfrak{M}_1$ of unparametrized holomorphic curves in $\mathbb{CP}^2$ is compact. Note, however, that the space $\mathfrak{H}_1$ of holomorphic maps $f\colon \mathbb{CP}^1 \to \mathbb{CP}^2$ of degree 1 is not compact. In fact, the fibers of the natural map $\mathfrak{H}_1 \to \mathfrak{M}_1$ are copies of $\mathrm{Aut}(\mathbb{CP}^1) = \mathrm{PSL}(2,\mathbb{C})$.

$d = 2$. This is the first really interesting case. Here again, degree 2 (connected, parametrized) curves in $\mathbb{CP}^2$ consist of rational curves, and the images $f\colon \mathbb{CP}^1 \to \mathbb{CP}^2$ are of two kinds: the smooth conics and the double covers of lines. However, not only is this space not compact, the corresponding space of unparametrized curves is not compact either, for it is fairly clear that one can approach a pair of intersecting lines as closely as one wishes.

In fact, the reader may want to contemplate the one-parameter family of hyperbolas $xy = \lambda^2$ as $\lambda \to 0$. If we choose the parametrization

$$f_\lambda(t) = [t, \lambda t^2, \lambda] = [1, x, y],$$

then the pullback $\Phi_\lambda = f_\lambda^*(\Omega)$ is an area form on $\mathbb{CP}^1$ whose total integral is 2, but (and the reader should check this), as $\lambda \to 0$, the form Φ_λ accumulates equally at the points $t = 0$ and $t = \infty$ and goes to zero everywhere else. (See the Exercises for a further discussion.)

Now, if we go ahead and add in the pairs of lines, then this "completed" moduli space is indeed compact. It is just the space of non-zero quadratic forms in three variables (irreducible or not) up to constant multiples. It is well known that this forms a $\mathbb{CP}^5$.

In fact, a further analysis of low degree mappings indicates that the following phenomena are typical: If one takes a sequence (Σ_k, f_k) of smooth holomorphic curves in $\mathbb{CP}^2$, then after reparametrizing and passing to a subsequence, one can arrange that the holomorphic curves have the property that, at a finite number of points $p_k^\alpha \in \Sigma$, the induced metric $f_k^*(\Omega)$ on the surface goes to infinity and the integral of the induced area form on a neighborhood of each of these points approaches an integer while along a finite number of loops γ_i, the induced metric goes to zero.

The first type of phenomenon is called "bubbling", for what is happening is that a small 2-sphere is inflating and "breaking off" from the surface and covering a line in $\mathbb{CP}^2$. The second type of phenomenon is called "vanishing cycles", a loop in the surface is literally contracting to a point. It turns out that the limiting object in $\mathbb{CP}^2$ is a union of algebraic curves whose total degree is the same as that of the members of the varying family.

Thus, for $\mathbb{CP}^2$, the moduli space $\mathfrak{M}_d$ of unparametrized curves of degree d has a compactification $\overline{\mathfrak{M}_d}$ where the extra points represent decomposable or degenerate curves with "cusps".

Other instances of this "bubbling" phenomena have been discovered. Sacks and Uhlenbeck showed that when one wants to study the question of representing elements of $\pi_2(X)$ (where X is a Riemannian manifold) by harmonic or minimal surfaces, one has to deal with the possibility of pieces of the surface "bubbling off" in exactly the fashion described above.

More recently, this sort of phenomenon has been used in "reverse" by Taubes to construct solutions to the (anti-)self dual Yang-Mills equations over compact 4-manifolds.

With all of this evidence of good compactifications of moduli spaces in other problems, Gromov had the idea of trying to compactify the connected components of the "moduli space" $\mathfrak{M}$ of holomorphic curves in a general almost complex manifold M. Since one would certainly expect the area function to be continuous on each compactified component, it follows that there is not much hope of finding a good compactification of the components of $\mathfrak{M}$ in a case where the area function is not bounded on the components of $\mathfrak{M}$ (as in the case of the Heisenberg example above).

However, it is still possible that one might be able to produce such a compactification if one *can* get an area bound on the curves in each component. Gromov's insight was that having the area bound was enough to furnish *a priori* estimates on the derivatives of curves with an area bound, at least away from a finite number of points.

With all of this in mind, I can now very roughly state Gromov's Compactification Theorem:

Theorem 5. *Let M be a compact almost complex manifold with almost complex structure J and suppose that Ω tames J. Then every component $\mathfrak{M}_\alpha$ of the moduli space $\mathfrak{M}$ of connected unparametrized holomorphic curves in M can be compactified to a space $\overline{\mathfrak{M}}_\alpha$ by adding a set of "cusp" curves, where a cusp curve is essentially a finite union of (possibly) singular holomorphic curves in M which is obtained as a limit of a sequence of connected curves in $\mathfrak{M}_\alpha$ by "pinching loops" and "bubbling".*

For the precise definition of "cusp curve" consult [Gr 1], [Wo], or [Pa]. The method that Gromov uses to prove his compactness theorem is basically a generalization of the Schwarz Lemma. This allows him to get control of the sup-norm of the first derivatives of a holomorphic curve in M terms of the L_1^2-norm (i.e., area norm) at least in regions where the area form stays bounded.

Unfortunately, although the ideas are intuitively compelling, the actual details are non-trivial. However, there are, by now, several good sources, from different viewpoints, for proofs of Gromov's Compactification Theorem. The articles [M 4] and [Wo] listed in the Bibliography are very readable accounts and are highly recommended. I hear that the (unpublished) [Pa] is also an excellent account which is closer in spirit to Gromov's original ideas of how the proof should go. Finally, there is the quite recent [Ye], which generalizes this compactification theorem to the case of curves with boundary.

Actually, the most fruitful applications of these ideas have been in the situation when, for various reasons, it turns out that there cannot be any cusp curves, so that, by Gromov's compactness theorem, the moduli space is already compact. Here is a case where this happens.

Proposition 1. *Suppose that (M, J) is a compact, almost complex manifold and that Ω is a 2-form which tames J. Suppose that there exists a non-constant holomorphic curve $f \colon S^2 \to M$, and suppose that there is a number $A > 0$ so that $\int_{S^2} f^*(\Omega) \geq A$ for all non-constant holomorphic maps $f \colon S^2 \to M$. Then for any $B < 2A$, the set $\mathfrak{M}_B$ of unparametrized holomorphic curves $f(S^2) \subset M$ which satisfy $\int_{S^2} f^*(\Omega) = B$ is compact.*

Proof. (IDEA) If the space $\mathfrak{M}_B$ were not compact, then a point of the compactification would would correspond a union of cusp curves which would contain at least two distinct non-constant holomorphic maps of S^2 into M. Of course, this would imply that the limiting value of the integral of Ω over this curve would be at least $2A > B$, a contradiction.

An example of this phenomenon is when the taming form Ω represents an integral class in cohomology. Then the presence of any holomorphic rational curves at all implies that there is a compact moduli space at some level.

Applications

It is reasonable to ask how the Compactness Theorem can be applied in symplectic geometry.

To do this, what one typically does is first fix a symplectic manifold (M, Ω) and then considers the space $\mathcal{J}(\Omega)$ of almost complex structures on M which Ω tames. We already know from Lecture 5 that $\mathcal{J}(\Omega)$ is not empty. We even know that the space $\mathcal{K}(\Omega) \subset \mathcal{J}(\Omega)$ of Ω-compatible almost complex structures is non-empty. Moreover, it is not hard to show that these spaces are contractible (see the Exercises).

▶ Thus, any invariant of the almost complex structures $J \in \mathcal{J}(\Omega)$ or of the almost Kähler structures $J \in \mathcal{K}(\Omega)$ which is constant under homotopy through such structures is an invariant of the underlying symplectic manifold (M, Ω).

This idea is extremely powerful. Gromov has used it to construct many new invariants of symplectic manifolds. He has then gone on to use these invariants to detect features of symplectic manifolds which are not presently accessible by any other means.

Here is a sample of some of the applications of Gromov's work on holomorphic curves. Unfortunately, I will not have time to discuss the proofs of any of these results.

Theorem 6 (GROMOV). *If there is a symplectic embedding of $B^{2n}(r) \subset \mathbb{R}^{2n}$ into $B^2(R) \times \mathbb{R}^{2n-2}$, then $r \leq R$.*

One corollary of Theorem 6 is that any diffeomorphism of a symplectic manifold which is a C^0-limit of symplectomorphisms is itself a symplectomorphism.

Theorem 7 (GROMOV). *If Ω is a symplectic structure on $\mathbb{CP}^2$ and there exists an embedded Ω-symplectic sphere $S \subset \mathbb{CP}^2$, then Ω is equivalent to the standard symplectic structure.*

The next two theorems depend on the notion of asymptotic flatness: We say that a non-compact symplectic manifold M^{2n} is *asymptotically flat* if there is a compact set $K_1 \subset M^{2n}$ and a compact set $K_2 \subset \mathbb{R}^{2n}$ so that $M \setminus K_1$ is symplectomorphic to $\mathbb{R}^{2n} \setminus K_2$ (with the standard symplectic structure on $\mathbb{R}^{2n}$).

Theorem 8 (MCDUFF). *Suppose that M^4 is a non-compact symplectic manifold which is asymptotically flat. Then M^4 is symplectomorphic to $\mathbb{R}^4$ with a finite number of points blown up.*

Theorem 9 (MCDUFF, FLOER, ELIASHBERG). *Suppose that M^{2n} is asymptotically flat and contains no symplectic 2-spheres. Then M^{2n} is diffeomorphic to $\mathbb{R}^{2n}$.*

It is not known whether one might replace "diffeomorphic" with "symplectomorphic" in this theorem for $n > 2$.

Epilogue

I hope that this Lecture has intrigued you as to the possibilities of applying the ideas of Gromov in modern geometry. Let me close by quoting from Gromov's survey paper on symplectic geometry in the Proceedings of the 1986 ICM:

> Differential forms (of any degree) taming partial differential equations provide a major (if not the only) source of integro-differential inequalities needed for a priori estimates and vanishing theorems. These forms are defined on spaces of jets (of solutions of equations) and they are often (e.g., in Bochner-Weitzenbock formulas) exact and invariant under pertinent (infinitesimal) symmetry groups. Similarly, convex (in an appropriate sense) functions on spaces of jets are responsible for the maximum principles. A great part of hard analysis of PDE will become redundant when the algebraic and geometric structure of taming forms and corresponding convex functions is clarified. (From the PDE point of view, symplectic geometry appears as a taming device on the space of 0-jets of solutions of the Cauchy-Riemann equation.)

Exercises

1. Use the fact that an orientable 3-manifold M^3 is parallelizable (i.e., its tangent bundle is trivial) and Theorem 1 to show that a compact 3-manifold can always be immersed in $\mathbb{R}^4$ and a 3-manifold with no compact component can always be immersed in $\mathbb{R}^3$.

2. Show that Theorem 2 does, in fact imply that any connected non-compact symplectic manifold which has an almost complex structure has a symplectic structure. (Hint: Show that the natural projection $Z^1(X,V) \to V$ has contractible fibers (in fact, $Z^1(X,V)$ is an affine bundle over V, and then use this to show that a non-degenerate 2-form on X can be homotoped to a closed non-degenerate 2-form on X.)

3. Show that the hypothesis in Theorem 3 that X either be non-compact or that $\dim(X) < \dim(Y)$ is essential.

4. Show that if E is a symplectic bundle over a compact manifold M^{2n} whose rank is $2n + 2k$ for some $k > 0$, then there exists a symplectic splitting $E = F \oplus T$ where T is a trivial symplectic bundle over M of rank k. (Hint: Use transversality to pick a non-vanishing section of E. Now what?)

    Show also that, if E is a symplectic bundle over a compact manifold M^{2n}, then there exists another symplectic bundle E' over M so that $E \oplus E'$ is trivial. (Hint: Mimic the proof for complex bundles.)

    Finally, use these results to complete the proof of the Corollary to Theorem 3.

5. Show that if a symplectic manifold M is simply connected, then the symplectic blow up $\hat{M}$ of M along a symplectic submanifold S of M is also simply connected. (Hint: Any loop in $\hat{M}$ can be deformed into a loop which misses $\hat{S}$. Now what?)

6. Prove, as stated in the text, that if M is compact and Ω tames J, then for any Riemannian metric g on M (not necessarily compatible with either J or Ω) there is a constant $C > 0$ so that

$$|v \wedge Jv| \leq C\,\Omega(v, Jv)$$

where $|v \wedge Jv|$ represents the area in T_pM of the parallelogram spanned by v and Jv in T_pM.

7. FIRST ORDER EQUATIONS AND HOLOMORPHIC CURVES. The point of this problem is to show how elliptic quasi-linear determined PDE for two functions of two unknowns can be reformulated as a problem in holomorphic curves in an almost complex manifold.

    Suppose that $\pi\colon V^4 \to X^2$ is a smooth submersion from a 4-manifold onto a 2-manifold. Suppose also that $R \subset J^1(X,V)$ is smooth submanifold of dimension 6 which has the property that it locally represents an elliptic, quasi-linear pair of first order PDE for sections s of π. Show that there exists a unique almost complex structure on V so that a section s of π is a solution of R if and only if its graph in V is an (unparametrized) holomorphic curve in V.

(Hint: The hypotheses on the relation R are equivalent to the following conditions. For every point $v \in V$, there are coordinates x, y, f, g on a neighborhood of v in V with the property that x and y are local coordinates on a neighborhood of $\pi(v)$ and so that a local section s of the form $f = F(x,y)$, $g = G(x,y)$ is a solution of R if and only if they satisfy a pair of equations of the form

$$A_1 f_x + B_1 f_y + C_1 g_x + D_1 g_y + E_1 = 0$$
$$A_2 f_x + B_2 f_y + C_2 g_x + D_2 g_y + E_2 = 0$$

where the $A_1, \ldots, E_2$ are specific functions of (x, y, f, g). The ellipticity condition is equivalent to the assumption that

$$\det \begin{pmatrix} A_1 \xi + B_1 \eta & C_1 \xi + D_1 \eta \\ A_2 \xi + B_2 \eta & C_2 \xi + D_2 \eta \end{pmatrix} > 0$$

for all $(\xi, \eta) \neq (0, 0)$.)

Show that the problem of isometrically embedding a metric g of positive Gauss curvature on a surface Σ into $\mathbb{R}^3$ can be turned into a problem of finding a holomorphic section of an almost complex bundle $\pi: V \to \Sigma$. Do this by showing that the bundle V whose sections are the quadratic forms which have positive g-trace and which satisfy the algebraic condition imposed by the Gauss equation on quadratic forms which are second fundamental forms for isometric embeddings of g is a smooth rank 2 disk bundle over Σ and that the Codazzi equations then reduce to a pair of elliptic first order quasi-linear PDE for sections of this bundle.

Show that, if Σ is topologically S^2, then the topological self-intersection number of a global section of V is -4. Conclude, using the fact that distinct holomorphic curves in V must have positive intersection number, that (up to sign) there cannot be more that one second fundamental form on Σ which satisfies both the Gauss and Codazzi equations. Thus, conclude that a closed surface of positive Gauss curvature in $\mathbb{R}^3$ is rigid.

This approach to isometric embedding of surfaces has been extensively studied by Labourie [La].

8. Prove, as claimed in the text that, for the map $f_\lambda: \mathbb{CP}^1 \to \mathbb{CP}^2$ given by the rule

$$f_\lambda(t) = [t, \lambda t^2, \lambda] = [1, x, y],$$

the pull-back of the Fubini-Study metric accumulates at the points $t = 0$ and $t = \infty$ and goes to zero everywhere else. What would have happened if, instead we had used the map

$$g_\lambda(t) = [t, t^2, \lambda^2] = [1, x, y]?$$

Is there a contradiction here?

9. Verify the claim made in the text that, for a symplectic manifold (M, Ω), the spaces $\mathcal{K}(\Omega)$ and $\mathcal{J}(\Omega)$ of Ω-compatible and Ω-tame almost complex structures on M are indeed contractible. (Hint: Fix an element $J_0 \in \mathcal{K}(\Omega)$, with

associated inner product $\langle,\rangle_0$ and show that, for any $J \in \mathcal{J}(\Omega)$, we can write $J = J_0(S + A)$ where S is symmetric and positive definite with respect to $\langle,\rangle_0$ and A is anti-symmetric. Now what?)

10. The point of this exercise is to get a look at the pseudo-holomorphic curves of a non-integrable almost complex structure. Let $X^4 = \mathbb{C} \times \Delta = \{(w,z) \in \mathbb{C}^2 \,|\, |z| < 1\}$, and give X^4 the almost complex structure for which the complex valued 1-forms $\alpha = dw + \bar{z}\, d\bar{w}$ and $\beta = dz$ are a basis for the (1,0)-forms. Verify that this does indeed define a non-integrable almost complex structure on the 4-manifold X. Show that the pseudo-holomorphic curves in X can be described explicitly as follows: If M is a Riemann surface and $\phi \colon M^2 \to X$ is a pseudo-holomorphic mapping, then one of the following is true: Either $\phi^*(\beta) = 0$ and there exists a holomorphic function h on M and a constant z_0 so that

$$\phi = \left(h - \bar{z}_0\, \bar{h},\ z_0\right),$$

or else there exists a non-constant holomorphic function g on M which satisfies $|g| < 1$, a meromorphic function f on M so that $f\, dg$ and $\overline{fg\, dg}$ are holomorphic 1-forms without periods on M, and a constant w_0 so that

$$\phi(p) = \left(w_0 + \int_{p_0}^{p}(f\, dg - \overline{fg\, dg}),\ g(p)\right)$$

where the integral is taken to be taken over any path from some basepoint p_0 to p in M.

(Hint: It is obvious that you must take $g = \phi^*(z)$, but it is not completely obvious where f will be found. However, if g is not a constant function, then it will clearly be holomorphic, now consider the "function"

$$f = \frac{\phi^*(\alpha)}{(1 - |g|^2)\, dg}$$

and show that it must be meromorphic, with poles at worst along the zeroes of dg.)

BIBLIOGRAPHY

[A]   V. I. Arnol'd,, *Mathematical Methods of Classical Mechanics*, Second Edition, Springer-Verlag, Berlin, Heidelberg, New York, 1989.

[AH] M. F. Atiyah and N. Hitchin, *The Geometry and Dynamics of Magnetic Monopoles*, Princeton University Press, Princeton, 1988..

[Bea] A. Beauville, *Variétés Kähleriennes dont la première classe de Chern est nulle*, J. Differential Geom. **18** (1983), 755–782.

[Ber] M. Berger, *Sur les groupes d'holonomie homogène des variétés a connexion affine et des variétés Riemanniennes*, Bull. Soc. Math. France **83** (1955), 279–330.

[Bes] A. Besse, *Einstein Manifolds*, Springer-Verlag, Berlin, Heidelberg, New York, 1987.

[Ca] E. Calabi, *Problems in Analysis*, Princeton University Press, Princeton, 1970, pp. 1–26.

[Ch] S.-S. Chern, *On a generalization of Kähler geometry*, Algebraic Geometry and Topology: A Symposium in Honor of S. Lefshetz (1957), Princeton University Press, 103–121..

[FGG] M. Fernandez, M. Gotay, and A. Gray, *Four-dimensional parallelizable symplectic and complex manifolds*, Proc. Amer. Math. Soc. **103** (1988), 1209–1212.

[Fo] A. T. Fomenko, *Symplectic Geometry*, Gordon and Breach, New York, 1988.

[GL] K. Galicki and H.B. Lawson, *Quaternionic Reduction and quaternionic orbifolds*, Math. Ann. **282** (1988), 1–21.

[GG] M. Golubitsky and V. Guillemin, *Stable Mappings and their Singularities*, Graduate Texts in Mathematics 14, Springer-Verlag, Berlin, Heidelberg, New York, 1973.

[Gr 1] M. Gromov, *Pseudo-holomorphic curves on almost complex manifolds*, Inventiones Mathematicae **82** (1985), 307–347.

[Gr 2] M. Gromov, *Partial Differential Relations*, Ergebnisse der Math., Springer-Verlag, Berlin, Heidelberg, New York, 1986.

[Gr 3] M. Gromov, *Soft and hard symplectic geometry*, Proceedings of the ICM at Berkeley **1** (1986, 1987), Amer. Math. Soc., Providence, R.I., 81–98.

[GS 1] V. Guillemin and S. Sternberg, *Geometric Asymptotics*, Mathematical Surveys **14** (1977), Amer. Math. Soc., Providence, R.I..

[GS 2] V. Guillemin and S. Sternberg, *Symplectic Techniques in Physics*, Cambridge University, Cambridge and New York, 1984.

[GS 3] V. Guillemin and S. Sternberg, *Variations on a Theme of Kepler*, Colloquium Publications **42** (1990), Amer. Math. Soc., Providence, R.I..

[Ha] R. Hamilton, *The inverse function theorem of Nash and Moser*, Bulletin of the AMS **7** (1982), 65–222.

[He] S. Helgason, *Differential Geometry, Lie Groups, and Symmetric Spaces*, Academic Press, Princeton, 1978.

[Hi] N. Hitchin, *Metrics on moduli spaces*, Contemporary Mathematics **58** (1986), Amer. Math. Soc., Providence, R.I., 157–178.

[HKLR] N. Hitchin, A. Karlhede, U. Lindström, and M. Roček, *HyperKähler metrics and supersymmetry*, Comm. Math. Phys. **108** (1987), 535–589.

[Ki] F. Kirwan, *Cohomology of Quotients in Symplectic and Algebraic Geometry*, Mathematical Notes **31** (1984), Princeton University Press, Princeton.

[KN] S. Kobayashi and K. Nomizu, *Foundations of Differential Geometry*, vols. I and II, John Wiley & Sons, New York, 1963.

[Kr] P. Kronheimer, *The construction of ALE spaces as hyper-Kähler quotients*, J. Differential Geometry **29** (1989), 665–683.

[La] F. Labourie, *Immersions isométriques elliptiques et courbes pseudo-holomorphes*, J. Differential Geometry **30** (1989), 395–424.

[M 1] D. McDuff, *Symplectic diffeomorphisms and the flux homomorphism*, Inventiones Mathematicae **77**, 1984 353–366.

[M 2] D. McDuff, *Examples of simply-connected symplectic non-Kählerian manifolds*, J. Differential Geom **20** (1984), 267–277.

[M 3] D. McDuff, *Examples of symplectic structures*, Inventiones Mathematicae **89** (1987), 13–36.

[M 4] D. McDuff, *Symplectic 4-manifolds*, Proceedings of the ICM at Kyoto (1990).

[M 5] D. McDuff, *Elliptic methods in symplectic geometry*, Bull. Amer. Math. Soc. **23** (1990), 311–358.

[MS] J. Milnor and J. Stasheff, *Characteristic Classes*, Annals of Math. Studies **76** (1974), Princeton University Press, Princeton.

[Mo] J. Moser, *On the volume element on manifolds*, Trans. Amer. Math. Soc. **120** (1965), 280–296.

[Pa] P. Pansu, *Pseudo-holomorphic curves in symplectic manifolds*, preprint, École Polytechnique, Palaiseau, 1986.

[Po] I. Postnikov, *Groupes et algèbras de Lie*, French translation of the Russian original, Éditions Mir, Moscow, 1985.

[Sa] S. Salamon, *Riemannian geometry and holonomy groups*, Pitman Research Notes in Math. no. 201, (1989), Longman scientific & Technical, Essex.

[Si] Y. T. Siu, *Every K3 surface is Kähler*, Inventiones Math. **73** (1983), 139–150.

[SS] I. Singer and S. Sternberg, *The infinite groups of Lie and Cartan, I (The transitive groups)*, Journal d'Analyse **15** (1965).
[Va] V. S. Varadarajan, *Lie Groups, Lie Algebras, and their Representations*, Springer-Verlag, Berlin, Heidelberg, New York, 1984.
[Wa] F. Warner, *Foundations of Differentiable Manifolds and Lie Groups*, Springer-Verlag, Berlin, Heidelberg, New York, 1983.
[We] A. Weinstein, *Lectures on Symplectic Manifolds*, Regional Conference Series in Mathematics **29** (1976), Amer. Math. Soc., Providence, R.I.
[Wo] J. Wolfson, *Gromov's compactness of pseudo-holomorphic curves and symplectic geometry*, J. Differential Geom **28** (1988), 383–405.
[Ye] R. Ye, *Gromov's compactness theorem for pseudo-holomorphic curves*, Transactions of the AMS.

Introduction to Quantum Field Theory for Mathematicians

Jeffrey M. Rabin

Introduction to Quantum Field Theory for Mathematicians

Jeffrey M. Rabin

Introduction

The aim of these lectures is to introduce mathematicians to the basic physical concepts and mathematical structures of classical mechanics, quantum mechanics, and quantum field theory. From the physicist's perspective the presentation will be very standard, being a highly compressed version (minus the detailed examples and computational techniques) of the standard graduate courses in these subjects. I have deliberately avoided translating the material completely into the language and notation of modern mathematics, preferring to present the standard physics notation and terminology (but with a running mathematical commentary). My goal is to help the student learn to read the physics literature and converse with its authors on their own terms, as well as to introduce the specific subject matter of quantum field theory (QFT). I have used the standard notation of each field of physics I discuss, but this unavoidably leads to some conflicts when the standard notations of different fields happen to coincide. I have tried to point out such ambiguities when they arise. An extensive set of exercises appears at the end of these lecture notes.

I do not assume prior knowledge of physics beyond $F = ma$, although some previous exposure to quantum mechanics would cushion the shock of absorbing it all at once. The mathematical background of a first-year graduate student should be more than adequate, with perhaps some review of the following topics: the elementary methods of the calculus of variations needed for the derivation of the Euler-Lagrange equation; linear algebra, especially the spectral properties of Hermitian operators in finite-dimensional spaces and the faith that unbounded self-adjoint operators in Hilbert spaces behave similarly; what is meant by a representation of a Lie group or Lie algebra; and some idea of how to work with distributions and Fourier transforms.

These lectures may also be viewed as an introduction for mathematicians to the alien culture of the physics community. One important cultural difference to keep

---

[1] Department of Mathematics University of California at San Diego, La Jolla, CA 92093.
E-mail address: jrabin@ucsd.edu.

© 1995 American Mathematical Society

in mind is that the notation used in physics is not designed to emphasize the logical relations of the concepts involved, but rather to facilitate explicit calculations. It is often intentionally ambiguous so that a given formula can be reinterpreted in whatever way makes calculation easiest and most natural, and it discourages one from getting bogged down in questions of rigor. Another difference, in fact, is that these questions of rigor are not addressed from the start, at the stage of definitions, but later, after the heuristic calculations have been done on the basis of provisional definitions. The philosophy is to proceed formally until difficulties are encountered, and only then resolve them. The difficulties often arise from an over-idealized mathematical model of a given physical situation, and only after the calculation is completed can one understand their physical origin and base the correct resolution on the true behavior of the physical system under consideration.

It is not easy to recommend references on the subjects of these lectures for a mathematical audience. Standard graduate physics texts are written for readers whose physical intuition has been developed through exposure to many examples and computations in classical and quantum mechanics and electromagnetism, and who are thoroughly steeped in the alien culture described above. Nevertheless the following suggestions should be useful. For classical mechanics, V. Arnold, *Mathematical Methods of Classical Mechanics* [1], or the standard graduate physics text, H. Goldstein, *Classical Mechanics* [2]. Among the many textbooks on elementary quantum mechanics, A.Z. Capri, *Nonrelativistic Quantum Mechanics* [3] is more careful than most concerning unbounded operators, and R. Shankar, *Principles of Quantum Mechanics* [4] contains a discussion of path integrals. For the path integral formulation of quantum mechanics direct from the source, see R.P. Feynman and A.R. Hibbs, *Quantum Mechanics and Path Integrals* [5]. Two books on QFT which mathematicians have found palatable are P. Ramond, *Field Theory: A Modern Primer* [6], and L.H. Ryder, *Quantum Field Theory* [7]. See also R.J. Rivers, *Path Integral Methods in Quantum Field Theory* [8]. A comprehensive treatment of all aspects of renormalization theory can be found in J. Collins, *Renormalization* [9]. Finally, I recommend two books on the rigorous foundations of QFT with the warning that such books are somewhat removed from the actual practice of field theory by physicists, and from the geometric and topological subject matter of "physical mathematics". They serve to reassure mathematicians who find themselves unable to suspend their disbelief that QFT is not built on sand. These are J. Glimm and R. Jaffe, *Quantum Physics* [10], and N.N. Bogolubov, A.A. Logunov, and R.T. Todorov, *Introduction to Axiomatic Quantum Field Theory* [11].

Acknowledgements

I thank the organizers of the Regional Geometry Institute for the opportunity to deliver this course of lectures. I am indebted to Bruce Driver for acting as my assistant during the course, for proofreading these notes, and for contributing the first two groups of homework exercises.

LECTURE 1
Classical Mechanics

We begin with the elementary formulation of classical mechanics for a single particle of mass m moving on a path $\mathbf{x}(t)$ in $\mathbf{R}^3$, as described by Newton's second law of motion,

$$(1.1) \qquad m\frac{d^2\mathbf{x}}{dt^2} = \mathbf{F}(\mathbf{x}, t),$$

where $\mathbf{F}$ is the force acting on the particle. We restrict ourselves to the case of a conservative force, namely one which is the gradient of some (time-independent) potential energy function V on $\mathbf{R}^3$, $\mathbf{F} = -\nabla V(\mathbf{x})$. Then there is an energy function which is conserved along the particle's trajectory,

$$(1.2) \qquad E = \frac{1}{2}m\left|\frac{d\mathbf{x}}{dt}\right|^2 + V(\mathbf{x}),$$

$$(1.3) \qquad \frac{dE}{dt} = m\frac{d\mathbf{x}}{dt} \cdot \frac{d^2\mathbf{x}}{dt^2} + \frac{d\mathbf{x}}{dt} \cdot \nabla V(\mathbf{x}) = 0.$$

Although Newton's law of motion is in a sense a complete statement of classical mechanics, there are several reasons to desire another formulation. Newton's law is a vector equation, and it is unpleasant to transform components of vectors into exotic coordinate systems. It requires that all forces acting on a mechanical system be known explicitly, whereas in practice one often describes a force implicitly by its effects, e.g. constraining the trajectory to lie on the surface of a sphere. And it describes the motion locally, in terms of the local acceleration caused by the local force. Newton's law gives almost no insight into the mathematical structure of classical mechanics, nor is it a suitable starting point for quantization of a classical system. And it cannot describe classical systems which do not consist of particles, such as the electromagnetic field. The Lagrangian formulation of mechanics to be discussed now gives a global and coordinate-free characterization of the classical trajectory. Then we will derive the Hamiltonian formulation which reveals symplectic geometry as the mathematical structure of mechanics. Both formulations are general enough to describe fields as well as particles, and both are suitable for the transition to quantum mechanics. The Hamiltonian approach leads to the canonical formulation of quantum mechanics in terms of self-adjoint operators in a Hilbert space, while the Lagrangian approach leads to the path integral formulation.

Let us generalize to a system of many particles having coordinates $q_i(t)$ and velocities $\dot{q}_i(t) = dq_i/dt$. (Custom dictates Newton's notation in which a dot denotes the time derivative.) For N particles moving in $\mathbf{R}^3$, i runs from 1 to $3N$, and the motion is described by a path in the configuration space $\mathbf{R}^{3N}$. However, we do not assume that the q_i are necessarily rectangular coordinates. Any set of $3N$ variables labeling the particles' locations will do. Given the points in configuration space representing the system at initial time t_1 and final time t_2, we wish to characterize the actual path followed by the system, among all others joining the same points, as the one which extremizes some functional S of the path, called the action. So we introduce

$$(1.4) \qquad S = \int_{t_1}^{t_2} L(q_i, \dot{q}_i) dt.$$

The function L is called the Lagrangian and will be specified later. It depends on two independent sets of $3N$ variables, denoted q_i and $\dot{q}_i$ for mnemonic reasons, but the latter are not the derivatives of the former. This is an example of intentionally ambiguous notation. The mnemonic reminds us how to evaluate S on a given path: substitute $q_i(t)$ and $\dot{q}_i(t)$ for the arguments of L and integrate the resulting function of t.

Let $q_i(t)$ be a path for which S is stationary. That is, under the change $q_i(t) \to q_i(t) + \delta q_i(t)$, with $\delta q_i(t)$ an arbitrary (smooth) function vanishing at the endpoints t_1 and t_2, the change δS in the action should be zero to linear order in $\delta q_i(t)$. We compute to linear order

$$(1.5) \qquad \delta S = \int_{t_1}^{t_2} \left(\delta q_i \frac{\partial L}{\partial q_i} + \delta \dot{q}_i \frac{\partial L}{\partial \dot{q}_i} \right) dt$$

$$(1.6) \qquad = \int_{t_1}^{t_2} \delta q_i \left(\frac{\partial L}{\partial q_i} - \frac{d}{dt} \frac{\partial L}{\partial \dot{q}_i} \right) dt + \left[\delta q_i \frac{\partial L}{\partial \dot{q}_i} \right]_{t_1}^{t_2},$$

where we have integrated by parts. The Einstein summation convention is in effect throughout these lectures: a sum is implied over any index such as i which is repeated in a given term. Since $\delta q_i = 0$ at the endpoints, the boundary contribution vanishes and we obtain the Euler-Lagrange equations for the extremal path,

$$(1.7) \qquad \frac{\partial L}{\partial q_i} - \frac{d}{dt} \frac{\partial L}{\partial \dot{q}_i} = 0.$$

To recover Newton's equations for a particle we can take

$$(1.8) \qquad L(x_i, \dot{x}_i) = \frac{1}{2} m \dot{x}_i \dot{x}_i - V(x_i),$$

leading to the Euler-Lagrange equations

$$-\frac{\partial V}{\partial x_i} - \frac{d}{dt}(m\dot{x}_i) = 0, \tag{1.9}$$

which indeed agree with Newton's law. In general the Euler-Lagrange equations are second-order ODEs which can be expected to have a unique solution given the initial positions and velocities of the particles.

A great advantage of the Lagrangian formulation of mechanics, and a running theme of these lectures, is the beautiful connection it reveals between the symmetries of a physical system and its conservation laws. The first example of a conservation law is evident from Eq. (1.7), namely, if the Lagrangian is independent of a particular coordinate q_i then the quantity $p_i \equiv \partial L/\partial \dot{q}_i$ is conserved during the motion. It is called the momentum conjugate to the coordinate q_i. For example, for a free particle with $V = 0$, L is independent of all coordinates and the three conserved quantities $p_i = m\dot{x}_i$ are just the components of the usual momentum vector of elementary mechanics.

More generally, suppose there is a continuous symmetry (a 1-parameter Lie group action on the space of paths) which leaves the action unchanged regardless of the limits of integration t_1, t_2. Infinitesimally this is a variation $q_i(t) \to q_i(t) + \delta q_i(t)$ for which we no longer assume $\delta q_i(t_1) = \delta q_i(t_2) = 0$. Noether's theorem states that such a symmetry implies the existence of a conserved quantity. To see this, we reverse our viewpoint on Eq. (1.6): the boundary term no longer vanishes automatically, but the first term is zero along the extremal path because that path obeys the Euler-Lagrange equations. Then we have

$$0 = \delta S = \left[\delta q_i \frac{\partial L}{\partial \dot{q}_i}\right]_{t_1}^{t_2}. \tag{1.10}$$

Because t_1 and t_2 can be arbitrary times, we see that $p_i \delta q_i$ is conserved along the extremal trajectory. The simplest example is again the free particle. The fact that the Lagrangian is independent of the coordinates can be recast as the invariance of the Lagrangian under the three infinitesimal translations $q_i \to q_i + \epsilon_i$, which are of the above form with $\delta q_i = \epsilon_i$. By Noether's theorem, $p_i \delta q_i = \epsilon_i p_i$ is conserved. Since the ϵ_i are arbitrary, we recover the conservation of the three components of momentum. In general there is a conserved quantity corresponding to each generator of the Lie algebra of a Lie group of symmetries. However, discrete groups of symmetries do not lead to conservation laws: it was crucial to the entire argument that the symmetries could be taken to lie in an infinitesimal neighborhood of the identity.

The Lagrangian approach to mechanics gives a global and coordinate-free description of the path in configuration space, as well as a deep understanding of the origin of conservation laws. As we will see later, the same framework can describe physical systems such as fields which do not consist of moving particles, simply by choosing a different Lagrangian function. Papers in the physics literature always specify the system under consideration by writing down the Lagrangian. From our present viewpoint, perhaps the only complaint about this formulation of mechanics that comes to mind is that the Euler-Lagrange equations are second order ODEs, whereas it is first order ODEs which have a nice geometric interpretation as flows along vector fields. We motivate the Hamiltonian formulation of mechanics by the desire to interpret the time evolution of a physical system as such a flow.

We define the Hamiltonian of a system as the function of $6N$ suggestively-named variables $H(q_i, p_i) \equiv p_i \dot{q}_i - L(q_i, \dot{q}_i)$. The meaning of this equation is as follows. The right side is a function of q_i and $\dot{q}_i$. We assume that the defining equations for the momenta, $p_i = \partial L/\partial \dot{q}_i$, can be inverted so as to solve for the $\dot{q}_i$ in terms of q_i and p_i, and we use this to eliminate the $\dot{q}_i$ and reexpress the Hamiltonian in terms of the q_i and p_i only. This procedure is called a Legendre transformation. By the implicit function theorem, we can carry it out provided that

$$(1.11) \qquad \frac{\partial p_i}{\partial \dot{q}_j} = \frac{\partial^2 L}{\partial \dot{q}_i \partial \dot{q}_j}$$

is a nonsingular matrix. This is always true in simple examples, and we can give a physical argument that it is true in all particle mechanics problems. Recall that L normally contains a kinetic energy term which is quadratic in the velocities $\dot{q}_i$. The matrix will be nonsingular provided that each velocity component contributes to the kinetic energy, which is certainly expected. However, as we will see, it is not true in gauge theories, and this is the origin of all the difficulties in quantizing gauge theories.

Computing the differential of H, we find

$$(1.12) \qquad \begin{aligned} dH &= p_i d\dot{q}_i + \dot{q}_i dp_i - \frac{\partial L}{\partial q_i} dq_i - \frac{\partial L}{\partial \dot{q}_i} d\dot{q}_i \\ &= \dot{q}_i dp_i - \frac{\partial L}{\partial q_i} dq_i, \end{aligned}$$

where the first and last terms cancel by definition of the momenta. Thus dH is automatically expressed in terms of the differentials of the variables on which H depends. This in fact motivates the Legendre transformation, and it allows us to read off the partial derivatives of H as

$$(1.13) \qquad \frac{\partial H}{\partial p_i} = \dot{q}_i, \qquad \frac{\partial H}{\partial q_i} = -\frac{\partial L}{\partial q_i}.$$

All this follows purely from the definition of H. Now, however, suppose that we are on the trajectory obeying the Euler-Lagrange equations,

$$(1.14) \qquad \frac{\partial L}{\partial q_i} = \frac{d}{dt}\frac{\partial L}{\partial \dot{q}_i} = \dot{p}_i.$$

Combining this with Eq. (1.13) gives the alternative characterization of the extremal trajectory in terms of Hamilton's equations of motion,

$$(1.15) \qquad \dot{q}_i = \frac{\partial H}{\partial p_i}, \qquad \dot{p}_i = -\frac{\partial H}{\partial q_i},$$

which are first order equations as promised. Furthermore, along the extremal path we have $dH = \dot{q}_i dp_i - \dot{p}_i dq_i$, or, dividing by dt, $dH/dt = 0$, and H itself is a conserved quantity.

For an example of all this formalism we return to the particle in a potential $V(x_i)$, for which

$$\begin{aligned} H &= (m\dot{x}_i)\dot{x}_i - \frac{1}{2}m\dot{x}_i\dot{x}_i + V(x_i) \\ &= \frac{1}{2}m(\dot{x}_i)^2 + V(x_i) \\ &= \frac{p_i^2}{2m} + V(x_i). \end{aligned}$$
(1.16)

(Note the summation convention even in p_i^2!) Here the Hamiltonian is conserved because it is the total energy. Hamilton's equations of motion become $\dot{x}_i = p_i/m$ and $\dot{p}_i = -\nabla_i V$, equivalent to Newton's laws.

Now let us compute the time derivative, along the extremal path of course, of an arbitrary function $A(q, p)$ of the q's and p's.

$$\begin{aligned} \frac{dA}{dt} &= \frac{\partial A}{\partial q_i}\dot{q}_i + \frac{\partial A}{\partial p_i}\dot{p}_i \\ &= \frac{\partial A}{\partial q_i}\frac{\partial H}{\partial p_i} - \frac{\partial A}{\partial p_i}\frac{\partial H}{\partial q_i} \\ &\equiv [A, H]. \end{aligned}$$
(1.17)

The expression $[A, B]$ defined for any two functions of the variables q, p is called their Poisson bracket, and has the usual properties of a Lie bracket:

(1.18) $$[A, B] = -[B, A],$$

(1.19) $$[aA + bB, C] = a[A, C] + b[B, C],$$

where a, b are constants;

(1.20) $$[AB, C] = A[B, C] + [A, C]B,$$

(1.21) $$[A, [B, C]] + [B, [C, A]] + [C, [A, B]] = 0,$$

as well as the fundamental or canonical Poisson bracket relations

(1.22) $$[q_i, q_j] = [p_i, p_j] = 0, \quad [q_i, p_j] = \delta_{ij}.$$

The equation $dA/dt = [A, H]$ reduces to Hamilton's equations themselves if $A = q_i$ or p_i, and may be considered the basic dynamical equation of Hamiltonian mechanics. It says that the Hamiltonian is the infinitesimal generator of time evolution via the Poisson bracket, and makes it clear that the conserved quantities are precisely those whose bracket with the Hamiltonian vanishes.

The Poisson bracket allows us to expose an even deeper connection between symmetries and the associated conserved quantities. Consider a system admitting an infinitesimal symmetry $q_i \to q_i + \epsilon f_i(q)$, with ϵ an "infinitesimal parameter" (a physicist's reminder that only linearized variations are important) and the f_i some functions. The corresponding conserved quantity from Noether's theorem is $Q \equiv p_i f_i(q)$. We compute the Poisson bracket,

$$(1.23) \qquad [q_j, Q] = \frac{\partial q_j}{\partial q_i}\frac{\partial Q}{\partial p_i} - \frac{\partial q_j}{\partial p_i}\frac{\partial Q}{\partial q_i} = \delta_{ij} f_i(q) = f_j(q),$$

and therefore $\delta q_j = \epsilon [q_j, Q]$, showing that the conserved quantity is the infinitesimal generator of the symmetry via the Poisson bracket. The status of the Hamiltonian as the generator of translations in time is a special case of this general phenomenon. [Actually, before asserting that Q generates the symmetry we should verify that $\delta p_j = \epsilon [p_j, Q]$ as well. It would take us somewhat afield to discuss the effect of symmetries on the momenta p_j; this would lead us into the geometry of canonical transformations, or symplectic diffeomorphisms. We would quickly discover that if there is a Lie group G of symmetries, of dimension k, then the Poisson bracket algebra of the k conserved quantities Q_n coincides (up to a possible central extension) with the Lie algebra of G.]

The Geometry of Hamiltonian Mechanics

Although we will not use it in these lectures, I want to give a translation dictionary between the concepts of Hamiltonian mechanics developed above and the coordinate-free language of symplectic geometry. This should help mathematicians to better understand the physicists' coordinate-dependent notation, as well as connecting the treatment given here to that in Robert Bryant's lectures in this volume.

The variables denoted q_i were global coordinates on the configuration space $\mathbf{R}^{3N}$; more generally they will be local coordinates on a configuration space M which may be any manifold whose points correspond to the possible configurations of the system under discussion. The variables $\dot{q}_i$ should be viewed as fiber coordinates for the tangent bundle $T(M)$ inasmuch as they describe all possible tangent vectors (velocities) to all possible paths in M. The Lagrangian $L(q, \dot{q})$ is a globally defined function on $T(M)$. Given a smooth path in M, one can lift it to a path in $T(M)$ by lifting it along the fiber $T_p(M)$ at each point p to its tangent vector there. Then one can restrict the Lagrangian to the lifted path and integrate this restriction to obtain the action S.

The momenta p_i are fiber coordinates for the cotangent bundle $T^*(M)$ which is known as the phase space; it follows from the definition $p_i = \partial L/\partial \dot{q}_i$ that under a change of coordinates p_i and $\dot{q}_i$ transform by inverse matrices, corresponding to dual bundles. Then the change of variables from $(q, \dot{q})$ to (q, p) is supposed to be the expression in coordinates of a diffeomorphism from $T(M)$ to $T^*(M)$ which is the identity on M. In the usual case in which L is quadratic in the $\dot{q}_i$ it is a linear isomorphism on the fibers, but in general it may not be linear or even surjective. Recall that there is not even a canonical identification of the fibers $T_p(M)$ and $T_p^*(M)$ unless some additional data such as a metric is given; in the usual case the kinetic energy term in L is a quadratic form on $T_p(M)$ which serves as this metric. The Hamiltonian $H(q,p)$ is a function on $T^*(M)$. The cotangent bundle of any manifold always has a canonical symplectic structure given by the closed, nondegenerate 2-form $\omega = dq_i dp_i = -d(p_i dq_i)$. This form is globally defined despite being specified in local coordinates: the fact that p_i are fiber coordinates for the cotangent bundle means precisely that $p_i dq_i$ is an invariantly defined 1-form. To any function $A(q,p)$ on $T^*(M)$ we associate a vector field $\tilde{A}$ which is its symplectic gradient; namely the vector field which is dual to the 1-form dA in the sense that $\omega(\tilde{A}, \cdot) = dA$. In coordinates we have

$$(1.24) \qquad \tilde{A} = \left(\frac{\partial A}{\partial p_i}, -\frac{\partial A}{\partial q_i} \right) = \frac{\partial A}{\partial p_i} \frac{\partial}{\partial q_i} - \frac{\partial A}{\partial q_i} \frac{\partial}{\partial p_i}.$$

Then Hamiltonian dynamics is simply the flow along the vector field $\tilde{H}$, and the Poisson bracket is nothing but $[A, B] = \omega(\tilde{A}, \tilde{B})$. It has the properties of a Lie bracket because the correspondence between functions and vector fields associates the Poisson bracket $[A, B]$ to the Lie bracket $[\tilde{B}, \tilde{A}]$. Finally, a conserved quantity Q gives rise to a vector field $\tilde{Q}$ whose flow commutes with that of $\tilde{H}$ and represents the action of the symmetry on the points of phase space; the Lie bracket algebra of the vector fields generating a Lie group of symmetries should reproduce the Lie algebra of that group.

LECTURE 2
Classical Field Theory

We have stressed that the Lagrangian and Hamiltonian formulations of mechanics are applicable to systems such as the electromagnetic and gravitational fields, which (classically) do not consist of particles.[2] These systems are described by PDEs such as Maxwell's equations rather than the ODEs which describe the motion of particles. We consider the mathematically simplest example of a real-valued field $\phi(x)$ on $\mathbf{R}^4$ which satisfies the Klein-Gordon equation,

(2.1) $$(\Delta + m^2)\phi(x) = 0.$$

Here $x = x^\mu = (x^0, x^1, x^2, x^3) = (t, x, y, z)$ is a point in $\mathbf{R}^4$ specifying a location $\mathbf{x} = x^i$ in space and a time t at which the field is measured. Note the convention that Greek indices run from 0 to 3 while Roman indices run from 1 to 3, so that x^i are the spatial components of x^μ. $\Delta = \frac{\partial^2}{\partial t^2} - \frac{\partial^2}{\partial x^2} - \frac{\partial^2}{\partial y^2} - \frac{\partial^2}{\partial z^2}$ is the wave operator, the four-dimensional Laplacian or D'Alembertian corresponding to the Minkowski or Lorentz metric

(2.2) $$\eta^{\mu\nu} = \eta_{\mu\nu} = \begin{pmatrix} 1 & 0 & 0 & 0 \\ 0 & -1 & 0 & 0 \\ 0 & 0 & -1 & 0 \\ 0 & 0 & 0 & -1 \end{pmatrix}.$$

[Actually the four-dimensional Laplacian is more commonly denoted by a square (four sides) than by Δ (three), but this character wasn't available to my typesetting program!] We abbreviate $\partial_\mu \equiv \partial/\partial x^\mu$, so that $\Delta = \eta^{\mu\nu}\partial_\mu\partial_\nu$. We also make the

---

[2] Mathematicians sometimes bemoan the absence of a definition of the term "field" in the physics literature. But "field", like "particle", is a physical concept, not a mathematical one. An appropriate answer to the question, "What is a field?" would be, "The electromagnetic field is a field." Fields really exist and can be observed in the laboratory. Of course, within a given theory we represent fields and particles by specific mathematical structures. One theory might say that a particle is a path in $\mathbf{R}^3$ while another says it is an irreducible representation of the Lorentz group. For present purposes a field is a function from R^4 to some algebra, usually the real numbers.

convention that the indices on any tensor can be raised and lowered using the metric, e.g.

$$(2.3) \qquad \partial^\mu \equiv \eta^{\mu\nu}\partial_\nu = \left(\frac{\partial}{\partial t}, -\frac{\partial}{\partial x}, -\frac{\partial}{\partial y}, -\frac{\partial}{\partial z}\right),$$

and we write the Klein-Gordon equation as $(\partial_\mu \partial^\mu + m^2)\phi = 0$, or even as $(\partial^2 + m^2)\phi = 0$. The parameter m will ultimately turn out to be the mass of something when we reach QFT. [The contortions with indices express the fact that the metric allows us to identify cotangent vectors (those with upper indices) with tangent vectors (those with lower indices), but this geometric view is overly sophisticated when one is in Euclidean space. Think of them as merely a notational device to keep track of the minus signs arising from the indefinite metric.]

It is crucial to understand the transition from our earlier variables $q_i(t)$ to the field $\phi(x) = \phi(\mathbf{x},t)$. The field ϕ, like an electric field, is an observable quantity which can be measured at any location $\mathbf{x}$ in space at any time. Similarly q_i is an observable quantity which can be measured, for each i, at any time. Thus, ϕ corresponds to q, t is always the time, and — most importantly — $\mathbf{x}$ corresponds to the index i. Sums over i in the formulas of particle mechanics can be expected to translate into integrals over $\mathbf{x}$ in field theory. For this reason field theory is often described as the mechanics of systems having infinitely many degrees of freedom.

We can obtain the Klein-Gordon equation as the extremal condition for either of the two actions

$$(2.4) \qquad S = -\frac{1}{2}\int_V d^4x\, \phi(\partial_\mu \partial^\mu + m^2)\phi,$$

or

$$(2.5) \qquad S = \frac{1}{2}\int_V d^4x\, (\partial_\mu \phi \partial^\mu \phi - m^2\phi^2).$$

These differ by a boundary term resulting from integration by parts, but this will not change the extremal condition. The integrand for S is called the Lagrangian density $\mathcal{L}$; technically its integral over space (sum over i) would be the Lagrangian L, although commonly $\mathcal{L}$ itself is called the Lagrangian.

Let us obtain the condition for a general action to be stationary to first order under variations $\delta\phi$ which vanish at the boundary of the volume V. From

$$(2.6) \qquad S = \int_V d^4x\, \mathcal{L}(\phi, \partial_\mu \phi)$$

we obtain

$$(2.7) \qquad \delta S = \int_V d^4x \left(\delta\phi \frac{\partial \mathcal{L}}{\partial \phi} + \partial_\mu \delta\phi \frac{\partial \mathcal{L}}{\partial \partial_\mu \phi}\right)$$

$$(2.8) \qquad = \int_V d^4x\, \partial_\mu \left(\delta\phi \frac{\partial \mathcal{L}}{\partial \partial_\mu \phi}\right) + \int_V d^4x\, \delta\phi \left(\frac{\partial \mathcal{L}}{\partial \phi} - \partial_\mu \frac{\partial \mathcal{L}}{\partial \partial_\mu \phi}\right).$$

The derivatives appearing here are formal derivatives of $\mathcal{L}$ thought of as a function of the independent variables ϕ and $\partial_\mu \phi$. The first integral is zero by the divergence theorem for $\delta\phi$ vanishing on ∂V, so the Euler-Lagrange equations are

$$\text{(2.9)} \qquad \frac{\partial \mathcal{L}}{\partial \phi} - \partial_\mu \frac{\partial \mathcal{L}}{\partial \partial_\mu \phi} = 0.$$

For the Lagrangian of Eq. (2.5) they are easily seen to give the Klein-Gordon equation.

In field theory there is a much stronger, local version of Noether's theorem. Again we suppose that $\delta\phi$ is an infinitesimal symmetry, so that $\delta S = 0$ for any region V even though $\delta\phi$ no longer vanishes on the boundary. Along the path satisfying the Euler-Lagrange equations the second integral in Eq. (2.8) is now zero, whereas the vanishing of the first integral for all regions V implies that

$$\text{(2.10)} \qquad \partial_\mu j^\mu = 0, \quad j^\mu = \delta\phi \frac{\partial \mathcal{L}}{\partial \partial_\mu \phi}.$$

Separating the time and space components of the current vector $j^\mu = (j^0, \mathbf{j})$ gives the differential conservation law $\partial_0 j^0 = -\nabla \cdot \mathbf{j}$. If we define the total charge at time t by

$$\text{(2.11)} \qquad Q(t) = \int d^3x\, j^0(t, \mathbf{x}),$$

and assume that $\mathbf{j}$ vanishes fast enough at infinity, then the divergence theorem implies the conservation of charge $dQ/dt = 0$. However, we can also define the charge inside a specified region,

$$\text{(2.12)} \qquad Q_V(t) = \int_V d^3x\, j^0(t, \mathbf{x}),$$

and deduce that any change in this charge must be accounted for by a flux of the current vector $\mathbf{j}$ through the boundary of V. This local conservation law is much stronger than the overall conservation of Q throughout space, which would allow charge to simply vanish from one region and simultaneously reappear elsewhere without traversing the space between. These conserved currents play a fundamental role in QFT and particle physics. [The terminology stems from electromagnetism, where the Q corresponding to the $U(1)$ symmetry group is indeed the electric charge, and $\mathbf{j}$ is the electric current density. Physicists often refer to any conserved quantity as a charge or current.]

The above discussion gives only the simplest version of Noether's theorem, applicable to an "internal symmetry". For a symmetry such as a spatial rotation, which acts on the coordinates x^μ as well as the field ϕ, the derivation requires more care because the integration region V is different in the new coordinate system. Goldstein [2] derives the conserved current

$$\text{(2.13)} \qquad j^\mu = (\delta\phi - \delta x^\nu \partial_\nu \phi) \frac{\partial \mathcal{L}}{\partial \partial_\mu \phi} + \mathcal{L}\, \delta x^\mu$$

in this case.

I will give only a brief discussion of the Hamiltonian formulation of field theory, mainly to fix some notation and introduce the notion of functional derivative which will be useful later. Each field ϕ appearing in the Lagrangian has a conjugate momentum field defined by $\pi(x) \equiv \partial \mathcal{L}/\partial \partial_0 \phi$, and the Hamiltonian is defined by

$$(2.14) \qquad H = \int d^3x [\pi(x) \partial_0 \phi(x) - \mathcal{L}].$$

H is constructed from the fields at a fixed time t, but as usual it will be conserved and thus independent of t. For example, for the Klein-Gordon field we find $\pi(x) = \partial^0 \phi(x)$ and

$$(2.15) \qquad H = \frac{1}{2} \int d^3x \left(\pi^2 + |\nabla \phi|^2 + m^2 \phi^2 \right),$$

a positive definite functional of $\phi(x)$ which can be identified as the energy in the field. The Hamilton equations of motion can be written in nearly the same form they had in particle mechanics by introducing the functional derivative notation,

$$(2.16) \qquad \dot{\phi}(\mathbf{x}, t) = \frac{\delta H}{\delta \phi(\mathbf{x})}, \quad \dot{\pi}(\mathbf{x}, t) = -\frac{\delta H}{\delta \pi(\mathbf{x})},$$

or alternatively in the Poisson bracket form $\dot{A} = [A, H]$ with

$$(2.17) \qquad [A, B] \equiv \int d^3z \left[\frac{\delta A}{\delta \phi(\mathbf{z})} \frac{\delta B}{\delta \pi(\mathbf{z})} - \frac{\delta A}{\delta \pi(\mathbf{z})} \frac{\delta B}{\delta \phi(\mathbf{z})} \right].$$

The functional derivative appearing in these equations is merely the physicists' way of expressing the (Fréchet) derivative of a functional on an infinite-dimensional function space. To define the derivative of a functional F on a function space at a particular point ϕ, a mathematician would linearize the functional about that point, writing $F[\phi + \delta \phi] = F[\phi] + F_\phi[\delta \phi] + O(\delta \phi^2)$, and define the linear map F_ϕ to be the derivative. Physicists, however, insist on expressing such linear maps in terms of their integral kernels. For example, the simple linear functional $\phi \to \phi(0)$ is thought of in terms of the Dirac delta function, $\phi(0) = \int dx\, \phi(x) \delta(x)$. Similarly, the functional derivative is the integral kernel of F_ϕ:

$$(2.18) \qquad F_\phi[\delta \phi] = \int d^3x\, \delta \phi(\mathbf{x}) \frac{\delta F}{\delta \phi(\mathbf{x})}.$$

The integral d^3x here is appropriate for Hamiltonian mechanics, in which the fields are thought of as functions on $\mathbf{R}^3$ at each fixed time; later we will encounter functional derivatives with respect to functions on $\mathbf{R}$ or $\mathbf{R}^4$. An alternative definition,

$$(2.19) \qquad \frac{\delta F}{\delta \phi(\mathbf{y})} = \lim_{\epsilon \to 0} \frac{1}{\epsilon} \{ F[\phi(\mathbf{x}) + \epsilon \delta^3(\mathbf{x} - \mathbf{y})] - F[\phi(\mathbf{x})] \},$$

where $\delta^3(\mathbf{x} - \mathbf{y})$ is the Dirac delta distribution in $\mathbf{x}$ with support at $\mathbf{y}$, shows that the functional derivative measures the response of the functional F to a change in the function ϕ supported at a point $\mathbf{y}$; it does depend on the point $\mathbf{y}$ in general. Some basic examples (in four dimensions) which will be useful later are

$$\frac{\delta}{\delta\phi(y)} \int d^4x\, \phi^n(x) = n\phi^{n-1}(y), \tag{2.20}$$

$$\frac{\delta}{\delta\phi(y)} \exp \int d^4x\, J(x)\phi(x) = J(y) \exp \int d^4x\, J(x)\phi(x). \tag{2.21}$$

Returning to mechanics we also have $\delta\phi(\mathbf{x})/\delta\phi(\mathbf{y}) = \delta^3(\mathbf{x} - \mathbf{y})$, which leads to the canonical brackets

$$[\phi(\mathbf{x}), \pi(\mathbf{y})] = \delta^3(\mathbf{x} - \mathbf{y}). \tag{2.22}$$

The fact that the canonical Poisson bracket is a distribution rather than a smooth function reflects our choice of coordinates for field theory. The configuration space M is the space of smooth functions $\phi(\mathbf{x})$ on $\mathbf{R}^3$, perhaps with some boundary conditions at infinity. Compare the configuration space $\mathbf{R}^3$ for a single particle, which can be viewed as the space of functions on the index set $i \in \{1,2,3\}$, recalling that $\mathbf{x}$ is analogous to i. The coordinate function q_i on $\mathbf{R}^3$ assigns to any function on the index set its value on the index i. Similarly, $\phi(\mathbf{x})$ above is the coordinate function on M which assigns to any function on $\mathbf{R}^3$ its value at $\mathbf{x}$. $\pi(\mathbf{x})$ is a similar coordinate function on the fibers of $T^*(M)$. But we know that the values of functions at points are never the best coordinates on function spaces; much better coordinates are things like Fourier coefficients which are the integrals of the functions against various kernels. For each test function f on $\mathbf{R}^3$ we can introduce a better coordinate function $\phi[f] = \int d^3x\, f(\mathbf{x})\phi(\mathbf{x})$ and similarly for $\pi[f]$. Then the Poisson bracket becomes nonsingular; $[\phi[f], \pi[g]] = \int d^3x\, f(\mathbf{x})g(\mathbf{x})$. These ideas will reappear in QFT.

As an example of a physically important field theory, in fact a gauge theory, let us now discuss electromagnetism. The classical electromagnetic field in a vacuum is described by a pair of time-dependent vector-valued functions on $\mathbf{R}^3$, the electric field $\mathbf{E}(x)$ and the magnetic field $\mathbf{B}(x)$. In some system of electromagnetic units, in which the speed of light c is equal to unity, they obey Maxwell's equations,

$$\nabla \cdot \mathbf{B} = 0, \quad \nabla \times \mathbf{E} + \frac{\partial \mathbf{B}}{\partial t} = 0, \tag{2.23}$$

$$\nabla \cdot \mathbf{E} = 0, \quad \nabla \times \mathbf{B} - \frac{\partial \mathbf{E}}{\partial t} = 0. \tag{2.24}$$

These equations assume a more covariant appearance in terms of the antisymmetric electromagnetic field strength tensor,

$$(2.25) \quad F^{\mu\nu} = \begin{bmatrix} 0 & -E^1 & -E^2 & -E^3 \\ E^1 & 0 & -B^3 & B^2 \\ E^2 & B^3 & 0 & -B^1 \\ E^3 & -B^2 & B^1 & 0 \end{bmatrix},$$

or the corresponding 2-form $F = \frac{1}{2}F^{\mu\nu}dx_\mu dx_\nu$. The Hodge dual $*F$ corresponds to the dual tensor $\tilde{F}^{\mu\nu} = \frac{1}{2}\epsilon^{\mu\nu\lambda\kappa}F_{\lambda\kappa}$, where $\epsilon^{\mu\nu\lambda\kappa}$ is totally antisymmetric in its indices and $\epsilon^{0123} = +1$. Then Maxwell's equations (2.23) read $\partial_\mu \tilde{F}^{\mu\nu} = 0$ or $dF = 0$, while Eqns. (2.24) become $\partial_\mu F^{\mu\nu} = 0$ or $d*F = 0$. Since we are in Euclidean space, $dF = 0$ is solved by $F = dA$ or $F^{\mu\nu} = \partial^\mu A^\nu - \partial^\nu A^\mu$ for some 1-form called the potential. Of course A is not unique; we have the freedom to change it by any gauge transformation $A \to A + d\Lambda$ or $A^\mu \to A^\mu + \partial^\mu \Lambda$ with an arbitrary function Λ.

If we assume the representation $F = dA$, then the remaining Maxwell equations $\partial_\mu F^{\mu\nu} = 0$ can be obtained by varying A^μ in the gauge-invariant action

$$(2.26) \quad S = -\frac{1}{4}\int d^4x\, F_{\mu\nu}F^{\mu\nu} = -\frac{1}{2}\int F \wedge {}^*F.$$

Alternatively, both these equations and the relation between A and F can be obtained from

$$(2.27) \quad S = -\frac{1}{2}\int d^4x\, F_{\mu\nu}(\partial^\mu A^\nu - \partial^\nu A^\mu) + \frac{1}{4}\int d^4x\, F_{\mu\nu}F^{\mu\nu}$$

by varying A and F independently.

We see that the equations of this gauge theory coincide with the equations for a connection on a $U(1)$ principal fiber bundle over $\mathbf{R}^4$. A is the connection viewed as a $U(1)$ Lie algebra ($= \mathbf{R}$) valued 1-form on the base space, and $F = dA + \frac{1}{2}[A, A]$ is the curvature, with $[A, A] = 0$ for this Abelian algebra. $A \to A + d\Lambda$ is the change in A under a change of bundle trivialization. To see this, write an element of the gauge group as $g = e^{i\Lambda}$, so that $g^{-1}dg = id\Lambda$ and $A \to A + d\Lambda$ can be rewritten as $A \to g^{-1}Ag - ig^{-1}dg$, which looks more familiar as $iA \to g^{-1}iAg + g^{-1}dg$. So the mathematician's connection is iA; this relative factor i between mathematicians' and physicists' conventions occurs frequently. Its origin is that physicists want A to be real, or more generally Hermitian, whereas mathematicians prefer skew Hermitian.

From the action of Eq. (2.26) we compute the momenta $\pi^\mu = \partial \mathcal{L}/\partial \partial_0 A_\mu$ conjugate to A^μ as $\pi^0 = 0, \pi^i = F^{i0} = E^i = \partial^i A^0 - \partial^0 A^i$. As advertised, in this gauge theory the velocity $\partial_0 A_0$ contributes nothing to the kinetic energy, consequently $\pi^0 = 0$ and we cannot solve for the velocities in terms of the momenta to perform the Legendre transform. It is easy to understand why this phenomenon is linked with gauge invariance. If it were possible to pass to the Hamiltonian formulation, we would obtain a unique solution to the initial value problem for the equations of motion: given the initial point in phase space, simply flow along the Hamiltonian vector field. However, the initial value problem cannot have a unique solution in a gauge theory: a gauge transformation with a time-dependent function Λ can produce a different solution with the same initial data.

One way to evade this difficulty is by gauge-fixing: seeking solutions to the equations of motion which obey some additional condition which picks out one A from each gauge orbit. In general there are topological obstructions to such a choice of gauge [12], but evidently it is enough to pick out a subset of A's whose stabilizer contains only time-independent gauge transformations. In electrodynamics, either the temporal gauge $A^0 = 0$ or the Coulomb gauge $\nabla \cdot \mathbf{A} = 0$ plus $A^0 = 0$ is a satisfactory choice, but in quantized non-Abelian gauge theories a straightforward gauge-fixing is computationally intractable. The extensive machinery of BRST quantization and Faddeev-Popov ghost fields has been developed to handle such systems. For an introduction, see [**13, 14, 15**].

LECTURE 3
The Lorentz Group and Spinors

The particles which are produced and studied in accelerators have velocities comparable to the velocity of light c (which is 1 in our units!), and consequently exhibit clearly the effects predicted by the special theory of relativity. The mathematical content of special relativity is that the Lorentz group, in fact the Poincaré group, (or at least the connected component of the identity in these groups) must act on the space of solutions to the equations of motion of particles and fields.

In relativistic physics we are concerned with inertial observers, namely observers for whom free particles appear to move with constant velocity. Intuitively such observers are themselves unaccelerated (e.g. not observing the world from a roller coaster) but may be in uniform motion relative to one another. Each such observer sets up a three dimensional coordinate system as well as a clock to measure time, and assigns to each event of interest a set of coordinates in the resulting coordinate system on $\mathbf{R}^4$. The coordinate transformations relating the coordinate systems of different observers comprise the Poincaré group; the subgroup fixing the origin of $\mathbf{R}^4$ is the Lorentz group (by convention the origins of the spatial coordinate systems of the corresponding observers coincide at time $t = 0$). The Poincaré group must act on the space of solutions of the equations of motion of any physical system, so that each orbit can be interpreted as a single physical solution viewed in the coordinates of all possible inertial observers. The simplest way to ensure this in the Lagrangian formulation of mechanics is to require the action to be Poincaré invariant. By Noether's theorem, any relativistic theory will then contain conserved quantities whose Poisson bracket algebra is that of the Poincaré group; these quantities are the energy, momentum, and angular momentum of the system.

The basic result of special relativity is the identification of the Lorentz group as the group $O(1, 3)$ of linear transformations of $\mathbf{R}^4$ preserving the quadratic form $\eta_{\mu\nu} x^\mu x^\nu = t^2 - x^2 - y^2 - z^2$. The null vectors of this form satisfy $x^2 + y^2 + z^2 = t^2$, which is the equation of a sphere in $\mathbf{R}^3$ of radius t. If a pulse of light is emitted at the origin at $t = 0$ it will reach this sphere at time t, so the constancy of the speed of light for all observers implies that the Lorentz group must preserve this sphere. Some more physical input, such as the homogeneity and isotropy of space, is needed to conclude that the quadratic form and not merely its nullspace is preserved. The Poincaré group is the semidirect product of the Lorentz group with the translation group of $\mathbf{R}^4$.

Let the 4×4 matrix Λ belong to the Lorentz group, so that $x'^\mu = \Lambda^\mu{}_\nu x^\nu$ is a Lorentz transformation (humor me regarding the index locations). The condition of preserving the quadratic form implies $\Lambda^T \eta \Lambda = \eta$. Taking the determinant of this equation, we learn that $\det \Lambda = \pm 1$. Taking its 00 component gives $(\Lambda^0{}_0)^2 = 1 + \Lambda^i{}_0 \Lambda^i{}_0 \geq 1$. Therefore, the Lorentz group consists of at least — in fact, exactly — four connected components, characterized by the signs of $\det \Lambda$ and $\Lambda^0{}_0$. The identity component, for which both signs are $+$, is the proper Lorentz group. The other components contain various reflections in the space and time axes, which in fact are known not to be exact symmetries of the laws of particle physics. (The violations of parity and time-reversal invariance were among the revolutionary experimental discoveries of the 1950s and '60s in particle physics.)

To expose the important relation between the Lorentz group and $SL(2,C)$ we associate to any vector x^μ in $\mathbf{R}^4$ the Hermitian matrix

$$(3.1) \qquad X = \begin{pmatrix} x^0 + x^3 & x^1 - ix^2 \\ x^1 + ix^2 & x^0 - x^3 \end{pmatrix} = x^\mu \sigma_\mu.$$

Here the Pauli matrices σ_μ form a convenient basis for the space of 2×2 Hermitian matrices,

$$(3.2) \qquad \sigma_0 = \begin{pmatrix} 1 & 0 \\ 0 & 1 \end{pmatrix}, \quad \sigma_1 = \begin{pmatrix} 0 & 1 \\ 1 & 0 \end{pmatrix},$$

$$(3.3) \qquad \sigma_2 = \begin{pmatrix} 0 & -i \\ i & 0 \end{pmatrix}, \quad \sigma_3 = \begin{pmatrix} 1 & 0 \\ 0 & -1 \end{pmatrix}.$$

They obey $\sigma_i \sigma_j = i\epsilon_{ijk}\sigma_k$ and are orthogonal with respect to the inner product $\text{Tr}(\sigma_\mu \sigma_\nu) = 2\delta_{\mu\nu}$. This allows us to invert relations like Eq. (3.1) by dotting each side with σ_ν to obtain $x^\mu = \frac{1}{2}\text{Tr}(X\sigma_\mu)$.

We note that under the correspondence (3.1) the Lorentz norm corresponds to the determinant: $x^\mu x_\mu = \det X$. For any matrix A belonging to $SL(2,C)$, the complex 2×2 matrices of determinant 1, the transformation $X' = AXA^\dagger$ preserves the Hermiticity and the determinant of X and consequently induces a Lorentz transformation on x^μ, proper in fact because $SL(2,C)$ is connected ($A^\dagger$ is the standard physics notation for the adjoint or Hermitian conjugate of a matrix, the complex conjugate transposed). We can find this Lorentz transformation as follows:

$$(3.4) \qquad x'^\mu = \frac{1}{2}\text{Tr}(X'\sigma_\mu) = \frac{1}{2}\text{Tr}(AXA^\dagger \sigma_\mu) = \frac{1}{2}x^\nu \text{Tr}(A\sigma_\nu A^\dagger \sigma_\mu).$$

Thus $\Lambda^\mu{}_\nu = \frac{1}{2}\text{Tr}(\sigma_\mu A\sigma_\nu A^\dagger)$. It also follows that

$$(3.5) \qquad \Lambda^\mu{}_\nu \sigma_\nu = A^\dagger \sigma_\mu A,$$

a formula which we will need shortly. It is not difficult to show (exercise) that we have defined a homomorphism of $SL(2,\mathbf{C})$ onto the proper Lorentz group, with kernel exactly $\{\pm 1\}$. Therefore the proper Lorentz group is doubly connected, and $SL(2,\mathbf{C})$ is its universal cover.

Suppose that two observers O and O' each measure some physical quantities, such as the electromagnetic field components, at a point P to which they assign coordinates x^μ and $x'^\mu = \Lambda^\mu{}_\nu x^\nu$ respectively, obtaining the results $F^{\mu\nu}(x)$ and $F'^{\mu\nu}(x')$, respectively. The simplest assumption one can make is that their results are related by the matrix associated to Λ in some representation of the Lorentz group which is characteristic of the particular quantities measured.

Example 3.1. $\phi(x)$ appearing in the Klein-Gordon equation is called a scalar field because it is associated to the trivial representation: $\phi'(x') = \phi(x)$. Indeed, the Klein-Gordon equation is invariant under this representation in the sense that if $(\Delta + m^2)\phi(x) = 0$ then automatically $(\Delta' + m^2)\phi'(x') = 0$, because m is a scalar and Δ is a Lorentz-invariant differential operator.

Example 3.2. Maxwell's equations are invariant if the potential A transforms according to the fundamental vector representation of the Lorentz group, $A'^\mu(x') = \Lambda^\mu{}_\nu A^\nu(x)$, while the field strength F transforms by the tensor product of two vector representations, $F'^{\mu\nu}(x') = \Lambda^\mu{}_\sigma \Lambda^\nu{}_\kappa F^{\sigma\kappa}(x)$. We see the motivation for the fanatical index conventions: any product of tensors which is summed over a pair of upper and lower indices will transform correctly as a tensor having the remaining indices. Contracting all pairs of indices so that none remain gives a Lorentz scalar, suitable for use as a Lagrangian — the physicist's solution to the problem of invariant theory.

Example 3.3. A representation of $SL(2, \mathbf{C})$ assigns to the elements $\pm A$ either the same matrix, or a pair of matrices $\pm M$. In the former case it induces a true representation of the Lorentz group, while in the latter case we say we have a "double-valued representation" of the Lorentz group. There are in fact two inequivalent complex two-dimensional representations of $SL(2, \mathbf{C})$: left-handed spinors which obey $\psi'_L(x') = A\psi_L(x)$, and right-handed spinors obeying $\psi'_R(x') = (A^\dagger)^{-1}\psi_R(x)$. Each ψ is a two-component complex column vector. The direct sum of these two "chiral representations" is called the Dirac spinor representation, and it acts on four-component complex column vectors by $\psi'(x') = S(\Lambda)\psi(x)$, where

(3.6) $$\psi = \begin{pmatrix} \psi_L \\ \psi_R \end{pmatrix}, \quad S = \begin{pmatrix} A & 0 \\ 0 & A^{\dagger -1} \end{pmatrix}.$$

Before constructing a Lorentz-invariant wave equation for $\psi(x)$, let me clarify the physical interpretation of double-valued representations. The notation $S(\Lambda)$ is misleading because the Lorentz transformation Λ only determines S up to sign. Therefore, the relation between the values of ψ measured by observers O and O' is also ambiguous by a sign. Of course I have not yet given a physical interpretation of ψ or a procedure for measuring it, and in fact no satisfactory interpretation exists outside the context of QFT. However, part of the interpretation must be that the overall sign of ψ is an arbitrary convention and not determined by the physical measurements. Observable quantities such as the energy or momentum associated to the field ψ will turn out to be quadratic in ψ and unaffected by this sign ambiguity.

For convenience in manipulating Dirac spinors we define 4 × 4 matrices analogous to σ_μ, the Dirac γ matrices which in terms of 2 × 2 subblocks are

$$\gamma^0 = \begin{pmatrix} 0 & \sigma_0 \\ \sigma_0 & 0 \end{pmatrix}, \quad \gamma^j = \begin{pmatrix} 0 & -\sigma_j \\ \sigma_j & 0 \end{pmatrix}. \tag{3.7}$$

They satisfy the Clifford algebra $\gamma^\mu \gamma^\nu + \gamma^\nu \gamma^\mu = 2\eta^{\mu\nu} I$, where I is the 4 × 4 identity matrix which is often omitted from this and similar formulas. This is a hint of the well-known relationship between Clifford algebras and the (pseudo-)orthogonal Lie groups. Note that γ^0 is Hermitian but γ^j is skew Hermitian. The projection operators onto the two-dimensional subspaces of the left and right handed spinors are $(1 \pm \gamma^5)/2$, where

$$\gamma^5 = i\gamma^0 \gamma^1 \gamma^2 \gamma^3 = \begin{pmatrix} 1 & 0 \\ 0 & -1 \end{pmatrix}. \tag{3.8}$$

(The story behind the notation γ^5 is this. In the early days of relativity some authors used four-dimensional coordinates x^μ with $\mu = 0, 1, 2, 3$ while others preferred $\mu = 1, 2, 3, 4$. Consequently one author's γ^4 was likely to be another's γ^0, and to avoid confusion this new matrix had to be called γ^5.) Two crucial properties of the γ matrices are $S^{-1} = \gamma^0 S^\dagger \gamma^0$, which can be verified directly, and $S^{-1} \gamma^\mu S = \Lambda^\mu{}_\nu \gamma^\nu$, which is a consequence of Eq. (3.5). They imply that in a certain sense the index on γ^μ behaves like a Lorentz vector index and can be used as such to form invariants.

For example, consider the four complex functions (actually 1 × 1 matrices) $j^\mu = \bar{\psi} \gamma^\mu \psi$, where $\bar{\psi} \equiv \psi^\dagger \gamma^0$ is the "Dirac conjugate" of ψ. I claim that under Lorentz transformations they behave as the components of a vector field. Proof:

$$\begin{aligned} j'^\mu(x') &= \bar{\psi}'(x')\gamma^\mu \psi'(x') = \psi^\dagger(x) S^\dagger \gamma^0 \gamma^\mu S \psi(x) = \bar{\psi}(x) \gamma^0 S^\dagger \gamma^0 \gamma^\mu S \psi(x) \\ &= \bar{\psi}(x) S^{-1} \gamma^\mu S \psi(x) = \Lambda^\mu{}_\nu \bar{\psi}(x) \gamma^\nu \psi(x) = \Lambda^\mu{}_\nu j^\nu(x). \end{aligned} \tag{3.9}$$

A similar computation shows that the following is a Lorentz invariant action:

$$\int d^4 x\, \bar{\psi}(x)(i\gamma^\mu \partial_\mu - m)\psi(x), \tag{3.10}$$

whose variation with respect to the components of ψ yields the Dirac equation

$$(i\gamma^\mu \partial_\mu - m)\psi = 0. \tag{3.11}$$

The parenthesized expression is a 4 × 4 matrix differential operator; the constant m (which really means mI) will ultimately be the mass of some particle, such as an electron, which is somehow related to ψ. By applying the operator $i\gamma^\nu \partial_\nu + m$ to both sides of the Dirac equation, one can verify that each component of ψ individually obeys the Klein-Gordon equation. In terms of the right and left handed spinors the Dirac equation reads

$$\begin{bmatrix} i(\sigma_0 \partial_0 - \sigma_j \partial_j)\psi_R \\ i(\sigma_0 \partial_0 + \sigma_j \partial_j)\psi_L \end{bmatrix} = \begin{bmatrix} m\psi_L \\ m\psi_R \end{bmatrix}. \tag{3.12}$$

Physicists call the self-adjoint operator $i\gamma^\mu \partial_\mu - m$ the Dirac operator, whereas mathematicians would use this term for one operator of the adjoint pair $i(\sigma_0 \partial_0 \pm \sigma_j \partial_j)$ which map the chiral subspaces to each other. We note that for $m = 0$ we can write an invariant equation for each chiral spinor separately, but for $m \neq 0$ Lorentz invariance necessarily couples them. This is the motivation for introducing the reducible Dirac representation. If, as some experiments now suggest, the electron's neutrino has a nonzero mass, we would have to abandon the traditional view of the "left-handed neutrino" described by the chiral spinor ψ_L alone.

The Dirac equation is invariant under the $U(1)$ action $\psi \to e^{i\theta}\psi$, and the corresponding conserved current is precisely j^μ of Eq. (3.9). It represents the electric current carried by the moving particles (e.g. electrons) which are (somehow) described by the Dirac equation. As promised earlier it is quadratic in the components of ψ and therefore insensitive to the sign ambiguity in those components. As a last note on the Dirac equation, note that we are free to change basis in the four dimensional Dirac representation space by $\psi \to K\psi$ and $\gamma^\mu \to K\gamma^\mu K^{-1}$ with K a constant matrix. This leads to different representations of the γ matrices which are more convenient in various situations when it is not important to separate the chiral components of ψ.

Finally, a comment on the geometric interpretation of fields. In these lectures, we always take $\mathbf{R}^4$ as our model of spacetime. We have of course exploited the simple geometry of $\mathbf{R}^4$ to simplify our formulas, but this necessarily obscures the general situation. On a general spacetime manifold M we would have a principal fiber bundle P whose structure group is the full symmetry group of our physical theory, $G = \text{Lorentz} \times U(1) \times \cdots$, together with vector bundles associated to various representations of G. Some fields, such as the electromagnetic potential A, are viewed as connections on P, while others such as ψ are sections of the associated bundles. The action is some invariant integral over M which depends on the fields but not on arbitrary choices such as local trivializations. The Lorentz group factor of G acts on the frame bundle of M, which is one factor of P, rather than on M itself. In our case we have implicitly used the simple geometry of $M = \mathbf{R}^4$ in several ways. Since $\mathbf{R}^4$ is contractible, all bundles are trivial. Hence we can pick global trivializations and view connections and sections as functions on the base space. Under a change of global trivialization the structure group acts on these functions according to various representations, as we have described. Furthermore, $\mathbf{R}^4$ can be identified with its tangent space at any point. Hence we can set the frame bundle connection (gravitational field) to zero, and restrict ourselves to changes of trivialization which are uniform over M and hence preserve this condition. Then the Lorentz group can be viewed as acting on M rather than on the tangent spaces. A physicist would say it has become a global symmetry rather than a local one.

LECTURE 4
Quantum Mechanics

Very small objects, such as molecules, atoms, and subatomic particles, do not obey the laws of classical mechanics. Quantum mechanics was developed during the 1920s in order to account for their behavior. The Schrödinger equation is to quantum mechanics what Newton's second law is to classical mechanics: a simple and, in principle, complete statement of the basic physics which however is not the most powerful method for solving practical problems, and which moreover obscures the true mathematical structures of the theory. Nevertheless we will begin with this equation because of its familiarity to some readers.

A particle of mass m is described in quantum mechanics by a complex-valued "wave function" $\psi(\mathbf{x},t)$ (not to be confused with the four-component Dirac spinor despite the coincidence of the standard notations) which obeys the Schrödinger wave equation,

$$(4.1) \qquad i\hbar \frac{\partial \psi}{\partial t} = -\frac{\hbar^2}{2m}\nabla^2 \psi + V(\mathbf{x})\psi,$$

where $V(\mathbf{x})$ is the same potential energy function that appeared in classical mechanics, and $\hbar = h/2\pi = 1.055 \times 10^{-27}$ erg-sec is a fundamental constant called Planck's constant (Actually h was originally Planck's constant, but since $h/2\pi$ appears in most formulas a special notation was introduced for it). We will shortly adopt units in which $\hbar = 1$ as we did for the speed of light. The standard way to solve such a PDE is of course by separation of variables, seeking solutions of the form $\psi(\mathbf{x},t) = \psi(\mathbf{x})\exp(-iEt/\hbar)$, where E is a constant having the dimensions of energy. This leads to the eigenvalue problem

$$(4.2) \qquad -\frac{\hbar^2}{2m}\nabla^2 \psi + V\psi = E\psi,$$

known as the time-independent Schrödinger equation. The physical interpretation of the wave function, also known as the probability amplitude,[3] is that $|\psi(\mathbf{x},t)|^2 d^3x$

---

[3] The term amplitude refers to anything which must be squared in order to produce the quantity of physical significance. It derives from wave theory, where the energy of a wave is proportional to its amplitude squared.

represents the probability of finding the particle located in an infinitesimal volume element d^3x at the point $\mathbf{x}$ at time t. ψ should belong to $L^2(\mathbf{R}^3)$ so that the total probability can be normalized to unity, $\int d^3x\, |\psi|^2 = 1$, and this provides the boundary condition for the eigenvalue problem. The fundamental tenet of quantum mechanics is that no amount of knowledge concerning the initial conditions or the physical environment of the particle will enable one to predict more about its future motion than the probabilities given by ψ. For those accustomed to classical mechanics, this failure of determinism can be difficult to accept. It has been said that if you believe quantum mechanics when you first learn it, then you didn't understand it!

We will now give the Hamiltonian formulation of quantum mechanics which reveals its true mathematical context as the spectral theory of operators in Hilbert space. The setting is completely different from classical mechanics, but the central fact that time evolution is a flow generated by the Hamiltonian is preserved. The description of any particular physical system starts with a complex vector space $\mathcal{H}$ associated to that system and called "The Hilbert Space" whether it is one or not (generally it fails to be one only in that the scalar product may not be positive definite). Usually $\mathcal{H}$ is infinite-dimensional, often the space of complex L^2 functions on the classical configuration space M, sometimes tensored with another finite-dimensional vector space. The spectral properties of the Hermitian operators on $\mathcal{H}$ play a central role.

This creates a problem because the operators of interest are generally unbounded and often have no eigenfunctions in $\mathcal{H}$. For example, in $L^2(\mathbf{R})$ we will be interested in the operator $-id/dx$, whose eigenfunctions are $\exp ipx$ with eigenvalue p. Even in the best case, when p is real, these do not belong to $\mathcal{H}$. Even worse, the operator x (that is, multiplication by x) has no eigenfunctions at all unless we allow distributions like $\delta(x-a)$, an "eigenfunction" with eigenvalue a. Of the many ways to deal with this situation, I will describe only the setup known as a rigged[4] Hilbert space. This is a nested sequence $\Omega \subset \mathcal{H} \cong \mathcal{H}^* \subset \Omega^*$, where Ω is a "nuclear subspace" dense in $\mathcal{H}$, and Ω^* is its dual space. The definition of a nuclear subspace is complicated, but when $\mathcal{H}$ is a function space Ω can be the functions decreasing faster than any power of $|x|$ at infinity. Then the "eigenfunctions" encountered above can be viewed as distributions belonging to Ω^*, e.g. $\exp ipx$ is reinterpreted as the linear functional $f(x) \to \int dx f(x) \exp ipx$, namely the Fourier transform. Further, the inverse transform $f(x) = \int (dp/2\pi) \hat{f}(p) \exp -ipx$ can be viewed as the expansion of the vector $f(x)$ in $\mathcal{H}$ in terms of the "basis" of eigenfunctions of the Hermitian operator $-id/dx$. In this sense, the spectral theory needed to formulate quantum mechanics works fine even though the relevant eigenfunctions are not in $\mathcal{H}$.

Returning to quantum mechanics, each state of the physical system is supposed to be represented by a normalized (unit length) vector in $\mathcal{H}$. The state of a system is the totality of information about the system at a given time needed to solve the initial value problem for its evolution starting at that time. In classical mechanics the states were represented by the points of phase space, the values of all coordinates and momenta at a given time. Next, the observables of the system — the measurable quantities such as positions or momenta of particles, their energies or

---

[4] "Rigged" not in the sense of prearranged for dishonest purposes, as a rigged election, but in the sense of well-equipped for a voyage, as a rigged schooner.

angular momenta and so forth, which were functions on phase space in classical mechanics — are represented by Hermitian (self-adjoint) linear operators (usually unbounded) in $\mathcal{H}$. Notation: vectors in $\mathcal{H}$ are denoted by $|\psi>$, where ψ is any convenient mnemonic label; e.g. an eigenvector of some operator A with eigenvalue a might be denoted by $|a>$. The scalar product $(\phi, A\psi)$ is written $<\phi|A|\psi>$ and is called a matrix element of A; if there is an orthonormal basis of $\mathcal{H}$ containing the vectors $|\phi>$ and $|\psi>$ then this is an entry in the matrix of A relative to this basis. The symbol $<\phi|$ appearing by itself denotes the vector in V^* which is dual to $|\phi>$ via the scalar product. The notation $<\phi|A|\psi>$ is intentionally ambiguous, readable as a scalar product or as the evaluation of a dual vector on a vector.

Suppose then that the system is in a state $|\psi>$ and we measure some observable represented by the Hermitian operator A — what result will we obtain? The answer depends on the spectrum of A. It is conventional to discuss separately the cases of discrete and continuous spectrum, although a unified discussion could be given in terms of the spectral projection operators associated to A. We also assume for simplicity that the eigenvalues are all simple (nondegenerate). Thus, let A have eigenvectors $|a>$ obeying $A|a>=a|a>$ and normalized so that $<a'|a>=\delta_{aa'}$ for discrete spectrum, or $<a'|a>=\delta(a-a')$ for continuous spectrum.[5] Then the result of the measurement will be one of the eigenvalues a of A, with probability $|<a|\psi>|^2$ for the discrete spectrum. For continuous spectrum $|<a|\psi>|^2 da$ is the probability that the result lies between a and $a + da$. The total probability is 1 because of the normalization of $|\psi>$, and the eigenvalues are real because A is Hermitian.

We will frequently use the identity $\sum_a |a><a| = 1$ for the orthonormal eigenvectors of any Hermitian operator, called "inserting a complete set of states". Applying both sides to any vector gives $\sum_a |a><a|\psi>=|\psi>$, which is just the expansion of $|\psi>$ in an orthonormal basis. (It is conventional to write such formulas as if the spectrum were discrete, with the understanding that the sum means an integral over a spectral measure in general.) As an example of its use, let us calculate the average value obtained over many measurements of A in state $|\psi>$. This is the sum of the possible values weighted by their probabilities,

(4.3)
$$\sum_a a|<a|\psi>|^2 = \sum_a a <\psi|a><a|\psi> = \sum_a <\psi|A|a><a|\psi> = <\psi|A|\psi>,$$

giving a direct physical interpretation to the diagonal matrix elements of A.

Among the observables are the coordinate functions q_i and p_i on phase space themselves, which classically obeyed the Poisson bracket relations (1.22). We assume that the corresponding operators in quantum mechanics obey

(4.4) $\qquad [q_i, q_j] = [p_i, p_j] = 0, \quad [q_j, p_k] = i\delta_{jk},$

where the bracket is now the commutator, $[A, B] = AB - BA$. The factor i (which means the multiplication operator by the constant i) is necessary because the commutator of two Hermitian operators is not Hermitian but skew Hermitian. As a

---

[5] In the rigged Hilbert space framework, this means that for any two vectors in Ω, $|\psi>= \int da' f(a')|a'>$ and $|\phi>= \int da\, g(a)|a>$, we should have $<\psi|\phi>= \int da f^*(a)g(a)$.

first approximation, any classical function on phase space is assumed to go over to the same function of the operators p and q upon quantization. This prescription is ambiguous because, for example, $pq^2 = q^2p = qpq$ classically but not quantum-mechanically, but in simple physical examples the correct ordering of operators can be fixed.

To complete the postulates of quantum mechanics we must give the rule for time evolution. Among the observables of a system is the total energy; the corresponding Hermitian operator is the Hamiltonian H. We assume that if a system is initially in the state $|\psi(0)>$, then its state at time t will be

$$(4.5) \qquad |\psi(t)> = e^{-itH}|\psi(0)>,$$

so that time evolution is a one parameter group of unitary transformations generated by H. The Schrödinger equation follows immediately in the form

$$(4.6) \qquad i\frac{\partial}{\partial t}|\psi(t)> = H|\psi(t)>.$$

[This discussion assumes that H has no explicit dependence on t. If it does, the Schrödinger equation remains correct, but the resulting unitary time evolution operator $U(t)$ is not simply given by $\exp -itH$.] Any state can be expanded in the basis of eigenstates of H, which satisfy $H|E> = E|E>$ and change only by a phase under time evolution: $|E(t)> = \exp -iEt|E(0)>$. Therefore, diagonalizing H solves the time evolution problem for all states. These eigenstates of H are often called stationary states, because all probabilities $|<\psi|E(t)>|^2$ with fixed $|\psi>$ are constant in time.

Let us complete this circle of ideas by deriving the Schrödinger equation (4.2) from this general framework. The Hilbert space for a particle in $\mathbf{R}^3$ is $L^2(\mathbf{R}^3)$, so a state $|\psi>$ is a square-integrable function $\psi(\mathbf{x})$. The fundamental commutation relations for coordinates and momenta can be satisfied by choosing the operators x_i to be multiplication by x_i, and p_j to be $-id/dx_j$. Furthermore, a rough statement of the von Neumann uniqueness theorem is that this is the *only* realization of the canonical commutation relations up to unitary equivalence. The Hamiltonian operator should then be

$$(4.7) \qquad H = \frac{\mathbf{p}^2}{2m} + V(\mathbf{x}) = -\frac{1}{2m}\nabla^2 + V(\mathbf{x}),$$

and the Schrödinger equation is nothing but the eigenvalue problem for H. Recalling that $|\mathbf{y}> = \delta^3(\mathbf{x} - \mathbf{y})$ is an eigenfunction of the position operators x_i with eigenvalues y_i, we have

$$(4.8) \qquad <\mathbf{y}|\psi> = \int d^3x\, \delta^3(\mathbf{x} - \mathbf{y})\psi(\mathbf{x}) = \psi(\mathbf{y}),$$

so that the general postulates of quantum mechanics give the interpretation of $\psi(\mathbf{y})$ as the probability amplitude for finding the particle located near $\mathbf{y}$.

It will be helpful later to be familiar with the eigenfunctions of H for a free particle, the case $V = 0$. Then $H = \mathbf{p}^2/2m$, and the eigenfunctions $e^{i\mathbf{p}\cdot\mathbf{x}}$ of $\mathbf{p} =$

$-id/dx$ are also the eigenfunctions of H, with eigenvalues $\mathbf{p}^2/2m$. Note that $\mathbf{p}$ here refers to either the operator or its eigenvalues depending upon the context. We see that H has continuous spectrum, and its eigenfunctions do not belong to $L^2(\mathbf{R}^3)$ but must be viewed as elements of Ω^* in the rigged Hilbert space setup. The correctly normalized eigenstates satisfying $<\mathbf{p}|\mathbf{p}'> = \delta^3(\mathbf{p}-\mathbf{p}')$ are $|\mathbf{p}> = (2\pi)^{-3/2}e^{i\mathbf{p}\cdot\mathbf{x}}$.

Symmetries in Quantum Mechanics

When a physical system is invariant under some Lie symmetry group G, we assume that a unitary representation of G acts in the Hilbert space and commutes with the time evolution operator $\exp -iHt$. For each one-parameter subgroup of G we will have an Abelian group of unitary linear operators $U(\theta)$ which act on a state of the system $|\psi>$ to give the state of the system after the symmetry transformation, e.g. after rotation by angle θ about some axis. The unitarity guarantees that scalar products are preserved; since the observable predictions of quantum mechanics are given by various scalar products this ensures that the symmetry transformation does not change the results of experiments. Stone's theorem guarantees that each one-parameter subgroup has a Hermitian generator A such that $U(\theta) = \exp -i\theta A$, and it follows that the observable A commutes with the Hamiltonian, $[A, H] = 0$. Therefore A is conserved, in the sense that any matrix element

(4.9) $\qquad <\phi(t)|A|\psi(t)> = <\phi(0)|e^{iHt}Ae^{-iHt}|\psi(0)> = <\phi(0)|A|\psi(0)>$

is independent of time. Thus, in quantum mechanics we have a very direct proof that conserved quantities are the generators of symmetries.

Symmetries are of great practical use in diagonalizing H in concrete problems. Because $[A, H] = 0$, the eigenstates of H can be chosen to be simultaneously eigenstates of A, whose spectrum may already be known. In fact, for any element g of the symmetry group G, and any eigenstate $|E>$ of H, it is easy to see that $g|E>$ is another eigenstate having the same eigenvalue. Therefore, each eigenspace of H carries a unitary representation of G! Knowledge of the unitary representations of common symmetry groups such as $SO(3)$ is thus extremely valuable in understanding the spectrum of an invariant Hamiltonian, which is why it is taught in quantum mechanics courses under titles such as "theory of angular momentum".

Heisenberg Picture

Our entire discussion of quantum mechanics so far has been in the so-called Schrödinger picture which leads to the Schrödinger equation as the description of dynamics. In this picture the vectors representing the states of the system change with time, but the operators representing physical observables do not. This situation is reversed in the Heisenberg picture, which is more useful in QFT.

The Heisenberg picture is obtained by a time-dependent unitary map of $\mathcal{H}$ onto itself. Each state $|\psi>$ at time t is mapped to $e^{iHt}|\psi>$, and each operator A is mapped to $e^{iHt}Ae^{-iHt}$. Because the map is unitary, it preserves all scalar products and matrix elements of operators, hence all physically measurable quantities.[6]

---

[6] Recall that I also spoke of symmetries as preserving all measurable quantities, even though symmetries act unitarily on the states and leave the operators unchanged, hence do not preserve matrix elements. This is actually as it should be: a rotation of a system in space should not preserve the matrix elements of p_x, the momentum component along the x axis, unless the coordinate

Because it inverts the time evolution operator on the states, the state of a system in the Heisenberg picture never changes. Instead, the operator representing a given observable will change with time according to $A(t) = e^{iHt} A e^{-iHt}$, or, infinitesimally,

$$\frac{dA}{dt} = -i[A, H]. \tag{4.10}$$

This replaces the Schrödinger equation as the description of dynamics. Note that it is identical to the classical Hamilton equation of motion (1.17) with the Poisson bracket replaced by $-i$ times the commutator. This makes it especially clear that the conserved quantities are those which commute with the Hamiltonian. Finally note that the canonical commutation relations $[q_j, p_k] = i\delta_{jk}$ hold in the Heisenberg picture only if the operators involved are evaluated at equal times. The commutator $[q_j(t), p_k(t')]$ is more complicated and depends on the specific form of H.

---

axes are rotated along with the system. That second rotation, not the first, would produce a new operator having the same matrix elements as the original p_x.

LECTURE 5
The Simple Harmonic Oscillator

The simple harmonic oscillator is the most important exactly solvable system in quantum mechanics. It has been called the only exactly solvable system in the sense that virtually all others can be reduced to it by a change of variables. It is the basis for QFT as well as a first approximation to the description of any oscillating system, from atoms in a solid to gravitational radiation detectors.

At the classical level we have a particle of mass m moving along the x axis and attached to the origin by a spring of force constant k, so that the force acting on the particle is $F = -kx$. Newton's equation of motion $m\ddot{x} + kx = 0$ has the solution $x(t) = A\sin(\omega t + \phi)$, a sinusoidal oscillation with frequency $\omega^2 = k/m$. The Hamiltonian for the system is

$$(5.1) \qquad H = \frac{p^2}{2m} + \frac{1}{2}kx^2 = \frac{p^2}{2m} + \frac{1}{2}m\omega^2 x^2.$$

The Schrödinger equation for this Hamiltonian can be solved directly in terms of special functions (Hermite polynomials), but much more insight results from diagonalizing H using methods of Lie algebra representation theory.

We introduce two new operators in the Hilbert space $L^2(\mathbf{R})$,

$$(5.2) \qquad a = x\sqrt{\frac{m\omega}{2}} + ip\sqrt{\frac{1}{2m\omega}},$$

$$(5.3) \qquad a^\dagger = x\sqrt{\frac{m\omega}{2}} - ip\sqrt{\frac{1}{2m\omega}}.$$

They are indeed adjoint operators, and they obey $[a, a^\dagger] = 1$ as a consequence of $[x, p] = i$. Further, the Hamiltonian can be written as

$$(5.4) \qquad H = \frac{1}{2}\omega(a^\dagger a + aa^\dagger) = \omega(a^\dagger a + \frac{1}{2}) \equiv \omega(N + \frac{1}{2}).$$

Diagonalizing H is then equivalent to diagonalizing the Hermitian operator N, which is easily found to obey $[N, a^\dagger] = a^\dagger$ and $[N, a] = -a$. Together with $[a, a^\dagger] =$

1, these relations define a Lie algebra called the Heisenberg algebra, and by studying its representations we will determine the spectrum of N.

Let $|n>$ be a normalized eigenstate of N, so that $N|n>=n|n>$. We have

$$Na^\dagger|n> = (a^\dagger N + a^\dagger)|n> = (n+1)a^\dagger|n>, \quad (5.5)$$

$$Na|n> = (aN - a)|n> = (n-1)a|n>. \quad (5.6)$$

This says that when $a^\dagger$ (resp., a) acts on an eigenstate of N, it gives a new eigenstate with eigenvalue raised (resp., lowered) by one unit. In particular it would seem that by repeatedly applying a we could obtain eigenstates of N or H having negative eigenvalues. However, H is a sum of squares of Hermitian operators and as such cannot have negative eigenvalues. The only escape from this paradox is that repeated application of a must eventually terminate with a state obeying $a|\psi>=0$. Because a is in reality a first-order differential operator, this state is unique up to normalization.[7] By applying $a^\dagger$ we see that $N|\psi>=0$. We should therefore denote this state, called the ground state or vacuum state, as $|0>$ (not to be confused with 0, the zero vector in the Hilbert space). By repeated application of $a^\dagger$ to $|0>$ we obtain eigenstates of N with all nonnegative integers as eigenvalues. This is the complete eigenvalue spectrum of N, since if there were a state with nonintegral eigenvalue then repeated application of a would bypass $|0>$ and contradict the positivity of H.

If the state $|n>$ is normalized, we can compute the norm squared of $a^\dagger|n>$ as

$$<n|aa^\dagger|n> = <n|a^\dagger a + 1|n> = n+1. \quad (5.7)$$

It follows inductively that $|n>\equiv (n!)^{-1/2}(a^\dagger)^n|0>$ is a normalized eigenstate of N with eigenvalue n for each positive integer n. The action of the raising and lowering operators on these normalized states is

$$a^\dagger|n> = \sqrt{n+1}|n+1>, \quad a|n> = \sqrt{n}|n-1>. \quad (5.8)$$

The uniqueness of the ground state was shown above. This is a question about the irreducibility of this particular representation of the Heisenberg algebra in $L^2(\mathbf{R})$ which, as we saw, is not determined by the algebra alone but by the specific differential operator representing a. We can now show inductively that in fact each eigenspace of N is one-dimensional. Suppose n is the smallest eigenvalue for which there exists a normalized eigenstate $|n'>$ linearly independent of $|n>$. Then

$$n|n'> = a^\dagger a|n'> = \sqrt{n}a^\dagger|n-1> = n|n>, \quad (5.9)$$

a contradiction. This completes the determination of the spectrum of N as usually given in physics texts. However, since the spectrum of an unbounded operator may not be exhausted by its eigenvalues (equivalently, there may be more eigenfunctions in Ω^*), we should check that the states $|n>$ are already complete in $\mathcal{H}$. This is done in Glimm and Jaffe [10], for example.

---

[7] By solving the differential equation $a|\psi>=0$ we can obtain the corresponding wave function as $\psi(x) = C\exp(-m\omega x^2/2)$.

We thus have a ladder of simple eigenvectors of H, with eigenvalues $(n+\frac{1}{2})\omega$. A measurement of the energy of an oscillator in the state $|n>$ is certain to yield this result. Matrix elements $<m|A|n>$ of any operator $A(x,p)$ of physical interest are easily computed by reexpressing A in terms of a and $a^\dagger$ and using Eqs. (5.8). It is often stated that the oscillator can emit or absorb energy only in integer multiples of ω, by making a "quantum jump" between pairs of states $|n>$. To explain this, suppose that the oscillator interacts with some external system, such as an electric field, for some time interval $0 < t < T$. During this interval, the Hamiltonian of the oscillator is changed by the addition of a potential energy term describing the interaction. Before and after this interval the states $|n>$ are eigenstates of H, but during the interval they are not. Under time evolution, an initial state $|n>$ at $t = 0$ may evolve to a linear combination $\sum c_m |m>$ at $t = T$. If the interaction term in H is small, the coefficients c_m can be calculated by using perturbation theory to approximately diagonalize the new Hamiltonian. A measurement of the energy in the final state of the oscillator will yield the result $(m+\frac{1}{2})\omega$ with probability $|c_m|^2$ (and, according to the controversial "reduction of the state vector" postulate, the oscillator will be left in the corresponding state $|m>$, thus completing the jump). The difference between this energy and the initial energy, which is a multiple of ω, is absorbed from or emitted to the external system.

LECTURE 6
The Path Integral Formulation of Quantum Mechanics

The Hilbert space formalism for quantum mechanics corresponds to the Hamiltonian version of classical mechanics in that the Hamiltonian is the infinitesimal generator for the time evolution. We will now derive an expression for quantum mechanical time evolution in terms of the action of the various paths the system might follow classically. This Lagrangian formulation of quantum mechanics leads us to path integrals. Our treatment will be very standard; an instructive alternative approach can be found in [16].

We consider for simplicity a particle moving in one dimension, with Hamiltonian $H = p^2/2m + V(x)$. Suppose that at time t the particle is located at x, so that its state is the eigenstate $|x>$ of the operator x. We want to compute the probability amplitude that the particle will be found at position x' at a later time t', denoted $<x', t'|x, t>$. This can be computed as

(6.1) $$<x', t'|x, t> = <x'|e^{-iH(t'-t)}|x>.$$

In the Schrödinger picture the interpretation is that the initial state is $|x>$, we apply the time evolution operator to obtain the state $e^{-iH(t'-t)}|x>$ at time t', and we project this onto the eigenstate $|x'>$. In the Heisenberg picture, the state is initially and always the eigenstate $|x, t> = e^{iHt}|x>$ of the position operator at time t, $x(t) = e^{iHt}x(0)e^{-iHt}$, where $x(0)$ coincides with the standard position operator of the Schrödinger picture, and we take the component of this along the eigenstate $|x', t'> = e^{iHt'}|x'>$ of $x(t')$. In either case, we should not be surprised to encounter singular distributions in calculating this quantity in view of the singular nature of the eigenstates of position. From a mathematical point of view it should be regarded as the integral kernel of the time evolution operator. Namely, given any two states $|\phi>$ and $|\psi>$ we can insert two complete sets of position eigenstates to obtain

(6.2) $$<\phi|e^{-iH(t'-t)}|\psi> = \int dx\, dx' <\phi|x'><x'|e^{-iH(t'-t)}|x><x|\psi>$$
$$= \int dx\, dx'\, \phi^*(x') K(x, t; x', t') \psi(x).$$

The plan will be to compute $<x',t'|x,t>$ first for infinitesimal $\Delta t = t' - t$, and then to somehow integrate up to finite time intervals. We compute

$$\begin{aligned}
<x'|e^{-iH\Delta t}|x> &= <x'|e^{-i\Delta tV(x)}e^{-i\Delta tp^2/2m}[1+O(\Delta t^2)]|x> \\
&= e^{-i\Delta tV(x)}<x'|e^{-i\Delta tp^2/2m}|x> \\
&= e^{-i\Delta tV(x)}\int dpdp' <x'|p'><p'|e^{-i\Delta tp^2/2m}|p><p|x> \\
&= e^{-i\Delta tV(x)}\int \frac{dpdp'}{2\pi} e^{-i\Delta tp^2/2m}\delta(p'-p)e^{ip'x'}e^{-ipx} \\
&= e^{-i\Delta tV(x)}\int \frac{dp}{2\pi} e^{ip(x'-x)}e^{-i\Delta tp^2/2m},
\end{aligned}$$

(6.3)

where we have inserted complete sets of eigenstates of p in the third line. (By now the reader should distinguish effortlessly between the operators x and p and their eigenvalues x and p!) The resulting Gaussian integral is only convergent if we assume that Δt has a small negative imaginary part. We will discuss the meaning of this assumption extensively below, but for now we simply look up the integral and obtain

(6.4) $$<x',t'|x,t> = \left(\frac{m}{2\pi i\Delta t}\right)^{1/2} e^{im(x'-x)^2/2\Delta t}e^{-i\Delta tV(x)}.$$

This formula has a remarkable interpretation. Imagine that the particle travels at constant velocity from point x at time t to point x' at time t', so that its path in an xt plane is a straight line segment. The constant velocity must be $(x'-x)/\Delta t$, so that the kinetic energy is $m(x'-x)^2/2\Delta t^2$. When $<x',t'|x,t>$ is used as an integral kernel, the rapidly oscillating phase ensures that the dominant contributions for small Δt occur for x' near x. Then the potential energy is approximately $V(x)$ and the exponent in our formula is just the action for this straight-line path! That is,

(6.5)
$$<x',t'|x,t> = \left(\frac{m}{2\pi i\Delta t}\right)^{1/2} \exp i\int_t^{t'}[\frac{1}{2}m\dot{x}^2 - V(x)]dt = \left(\frac{m}{2\pi i\Delta t}\right)^{1/2} \exp iS,$$

for infinitesimal Δt.

To compute $<x',t'|x,t>$ for a finite time interval, we subdivide the interval by choosing intermediate times $t' > t_1 > t_2 > \cdots > t_n > t$ and insert complete sets of eigenstates of the Heisenberg operators $x(t_i)$ to obtain

(6.6)
$$<x',t'|x,t> = \int dx_1 \cdots dx_n <x',t'|x_1,t_1><x_1,t_1|x_2,t_2> \cdots <x_n,t_n|x,t>.$$

As $n \to \infty$, by our previous computation, each of these matrix elements is proportional to $\exp iS$, where S is the action for a straight-line path joining the appropriate points. Multiplying them together, we obtain $\exp iS$ with S the action for a zigzag path composed of straight line segments joining the intermediate points (x_i, t_i) (Figure 6.1). The integral over the x_i is an integral over the space of all

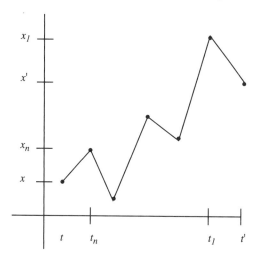

Figure 6.1. Piecewise linear path from x at t to x' at t'. The integration over the locations x_i of the corners approximates the Feynman path integral.

piecewise linear paths with corners at the intermediate times t_i. In the limit this should approach an integral over the space of all continuous paths from (x,t) to (x',t') with each path weighted according to its action, formally,

$$(6.7) \qquad <x',t'|x,t> = N \int [Dx] \exp i \int_t^{t'} [\tfrac{1}{2}m\dot{x}^2 - V(x)]dt,$$

with $[Dx]$ a formal measure and N a formal normalization constant. This path integral is the basic formula of this approach to quantum mechanics. Clearly our calculations have been heuristic, and much remains to be done to rigorously establish such a formula. One must prove that the limit $n \to \infty$ exists, and equals the matrix element we wanted to compute. This will involve a careful estimate of the errors made in our computation for "infinitesimal" Δt. Then one should prove that the limit of the measures on the spaces of piecewise linear paths exists and gives a measure on continuous paths. In fact, no one has ever succeeded in doing this, and the reason is closely related to our need to assume a small imaginary part for Δt.

The time evolution operator $\exp -iHt$ whose matrix elements we have been computing makes sense for certain complex, as well as real, values of t. Recall that the spectrum of H in general is bounded below but not above. Therefore $\exp -iHz$ is a very nice operator, analytic in z, for z in the lower half of the complex z plane. However, in the upper half plane it gives exponentially increasing weight to the higher eigenvalues of H and is too unbounded to possess an integral kernel $<x',t'|x,t>$. The real axis, containing the values of physical interest, is the boundary between these regions, and as such formal computations here can be dangerous. (By the spectral theorem, $\exp -iHz$ is strongly continuous for $\operatorname{Im} z \leq 0$ and we are interested in the continuous boundary values of an analytic function.) The small negative imaginary part assumed for Δt above simply amounts to computing the kernel by analytic continuation from slightly below the real axis. This turns out to be such a good idea that we will carry it to the extreme and compute the kernel by analytic continuation from the negative imaginary axis. We parametrize the imaginary axis by $z = -i\tau$ with τ real, and compute the kernel of $\exp -iHz = \exp -H\tau$. Retracing our derivation of the path integral, or just formally substituting $t = -i\tau$ in the result, gives the Feynman-Kac formula

$$<x',\tau'|x,\tau> \equiv <x'|e^{-H(\tau'-\tau)}|x> = N\int [Dx]\exp - \int_\tau^{\tau'} d\tau [\frac{1}{2}m\dot{x}^2 + V(x)]. \tag{6.8}$$

The integral is over all paths from (x,τ) to (x',τ'). This formula can indeed be rigorously proven by the means sketched previously. The important estimates are contained in the Trotter product formula,

$$e^{-(A+B)} = \lim_{n\to\infty} (e^{-A/n} e^{-B/n})^n \tag{6.9}$$

for suitable operators A and B. The resulting measure on the space of continuous paths in the case $V(x) = 0$ is (conditional) Wiener measure. Neither the constant N, the formal Lebesgue measure $[Dx]$, nor the exponential of the kinetic part of the action (because Wiener measure is concentrated on nondifferentiable paths) make sense separately, but only their product which is Wiener measure. The $\pi/2$ rotation of our attention from the real axis to the imaginary is called "Wick rotation to Euclidean space", named for the physicist G.C. Wick. We will see shortly how to view it as a literal rotation. Although at the moment we are not concerned with Lorentz invariance, in QFT we will be, and the term "Euclidean space" refers to the effect of the rotation in changing the indefinite Minkowski metric $t^2 - x^2 - y^2 - z^2$ to the definite Euclidean metric $-\tau^2 - x^2 - y^2 - z^2$. The quantity $\frac{1}{2}m\dot{x}^2 + V(x)$ is called the Euclidean Lagrangian (and sometimes coincides with the Hamiltonian). Our earlier "Minkowski" path integral formula is viewed as a mnemonic for the Feynman-Kac formula plus analytic continuation in $\Delta z = -i(\tau' - \tau)$ back to the real axis.

The path integral would not be the versatile tool that it is if it could only compute the kernel of the time evolution operator. It can also be used to compute matrix elements of the position operators in the Heisenberg picture. Consider the computation of

$$<x',t'|x(t_1)\cdots x(t_n)|x,t>, \tag{6.10}$$

where $t' > t_1 > \cdots > t_n > t$. Inserting complete sets of eigenstates of $x(t_i)$ at the intermediate times allows us to rewrite this as

$$\int dx_1 \cdots dx_n\, x_1 \cdots x_n <x',t'|x_1,t_1> \cdots <x_n,t_n|x,t>, \tag{6.11}$$

and the path integral representation for the remaining matrix elements gives

$$<x',t'|x(t_1)\cdots x(t_n)|x,t> = N\int [Dx]x(t_1)\cdots x(t_n)\exp iS, \tag{6.12}$$

where $x(t_i)$ in the integral denotes the location of the path $x(t)$ at time t_i. Note that the factors $x(t_i)$ on the right side of this equation are just numbers (actually functionals of the paths) and can be rearranged freely, while those on the left are noncommuting operators and must be arranged in decreasing order of their time arguments. That is, the path integral on the right automatically computes the matrix element of a product of $x(t)$ operators in a particular order. We define the time ordering symbol T so that $T[x(t_1)\cdots x(t_n)]$ is the product of the operators in decreasing time order, regardless of their order as written. Then we have the path integral formula,

$$(6.13) \qquad <x',t'|T[x(t_1)\cdots x(t_n)]|x,t> = N\int [Dx]\, x(t_1)\cdots x(t_n)\exp iS,$$

with no restriction on the order of the intermediate times t_i.

Even more cleverly, we can write a generating functional for all such matrix elements at once. We add a so-called source term $x(t)J(t)$ to the Lagrangian, where $J(t)$ is a given function with support in the interval $[t,t']$. The path integral with this new Lagrangian is a functional of $J(t)$, and we can take its functional derivatives. We obtain

$$<x',t'|T[x(t_1)\cdots x(t_n)]|x,t> =$$
$$(6.14) \qquad (-i)^n \frac{\delta^n}{\delta J(t_1)\cdots \delta J(t_n)} N\int [Dx]\exp i\int_t^{t'} dt\,[\tfrac{1}{2}m\dot{x}^2 - V(x) + xJ]\Big|_{J=0},$$

because each functional derivative on the right brings down a factor $x(t_i)$ into the integrand, reproducing our previous formula.

This may strike you as the ultimate in cleverness, but there is one more step to take. Even more useful than matrix elements between states like $|x,t>$ are matrix elements between the eigenstates of the Hamiltonian, especially matrix elements in the ground state $|0>$. The following trick allows us to compute these as well by means of path integrals. Let the source $J(t)$ now have support in a subinterval $[a,b]$ of $[t,t']$. Adding the source term to the Lagrangian will change the Hamiltonian as well, into, say, H^J. The corresponding unitary time evolution operator $U^J(t_1,t_2)$ between times t_1 and t_2 will not coincide with $\exp -iH(t_2-t_1)$ when the source is nonzero. Its matrix elements are

$$(6.15) \qquad \begin{aligned}<x',t'|x,t>^J &= <x'|U^J(t,t')|x> \\ &= <x'|e^{-iH(t'-b)}U^J(a,b)e^{-iH(a-t)}|x>.\end{aligned}$$

We now insert complete sets of eigenstates $|n>$ of H, with eigenvalues E_n (which are *not* generally eigenstates of H^J) to obtain

$$<x',t'|x,t>^J = \sum_{m,n}<x'|e^{-iH(t'-b)}|m><m|U^J(a,b)|n><n|e^{-iH(a-t)}|x>$$
$$(6.16) \qquad = \sum_{m,n}<x'|m><n|x>\, e^{-iE_m(t'-b)}e^{-iE_n(a-t)}<m|U^J(a,b)|n>.$$

Now we want to take the limits $t \to -\infty$, $t' \to +\infty$. This makes sense when we recall that $<x',t'|x,t>^J$ is the analytic continuation of the Euclidean matrix element $<x',\tau'|x,\tau>$, which has a representation as above with $e^{-iE_m\cdots}$ replaced by $e^{-E_m\cdots}$, so that we can let $\tau \to -\infty$, $\tau' \to +\infty$. Equivalently, we let $t \to -\infty$, $t' \to +\infty$ in the complex plane, along a line with a small negative slope rather than the real axis. We also assume, as is usual in QFT for just this reason, that a constant has been added to H to make the ground state energy equal to zero. Then in the limit only the ground state contributes to $<x',t'|x,t>^J$, which becomes

$$(6.17) \qquad <x',t'|x,t>^J = <x'|0><0|x> Z(J),$$

where, by definition, the vacuum functional or partition function $Z(J)$ is

$$(6.18) \qquad Z(J) = <0|U^J(a,b)|0> = <0|U^J(t,t')|0>.$$

We have the normalization condition $Z(0) = 1$, but in general $Z(J) \neq 1$ because $|0>$ is not an eigenstate of H^J when $J \neq 0$. Rewriting (6.17) as

$$(6.19) \qquad Z(J) = \frac{<x',t'|x,t>^J}{<x'|0><0|x>},$$

we obtain the path integral formula for $Z(J)$,

$$(6.20) \qquad Z(J) = N \int [Dx] \exp iS^J,$$

with N chosen so that $Z(0) = 1$. The integral is over all paths having fixed but arbitrary endpoints x', x at the times $\pm\infty$. Furthermore, repeating these arguments for the matrix elements (6.14) gives

$$(6.21) \qquad <0|T[x(t_1)\cdots x(t_n)]|0> = (-i)^n \left.\frac{\delta^n}{\delta J(t_1)\cdots \delta J(t_n)} Z(J)\right|_{J=0}.$$

$Z(J)$ is thus the generating functional for a whole series of ground state matrix elements which turn out to have direct physical meaning in QFT. They are called Green's functions, correlation functions, or n-point functions, and in a real sense they contain all the physical predictions of the theory.

Before we learn how to actually compute path integrals, let me add two comments about their importance in physics. First, the path integral formalism is the most elegant way to understand the connection between quantum and classical mechanics. With physical units restored, the path integral weight $\exp iS$ is actually $\exp iS/\hbar$. In the limit $\hbar \to 0$ it is reasonable to try to compute the integral by the method of stationary phase. The integral over paths should be dominated by the neighborhood of the path which extremizes S, which is precisely our Lagrangian formulation of classical mechanics. The asymptotic expansion of the path integral for small $\hbar$ is known as the semiclassical or WKB (Wentzel, Kramers, and Brillouin) approximation in physics. Second, the Euclidean path integral (6.8) has an interpretation in terms of statistical mechanics which leads to the modern convergence between the subjects of statistical mechanics and QFT. In a small nutshell, statistical mechanics is a probability theory in which a physical system at temperature T has a probability proportional to $\exp -E/kT$ of being found in a state having energy E, with k being Boltzmann's constant. The Euclidean path integrals for a particle can be reinterpreted as moments of this probability measure for some continuous one-dimensional system, e.g. $x(\tau)$ is the pressure or density in a one-dimensional gas at location τ and the Euclidean action is the energy of the gas. Note that τ has become a spatial coordinate while x is the value of a field at that location. The term "partition function" itself comes from statistical mechanics, where $Z(J)$ is related to the number of ways of partitioning the system's total energy among its constituent particles.

LECTURE 7
The Harmonic Oscillator via Path Integrals

We will now see how path integrals are actually computed, by calculating $Z(J)$ for the simple harmonic oscillator. Just as the oscillator is the only exactly solvable problem in physics, so this type of "Gaussian" path integral is the only one that can honestly be evaluated. As such our computation is fundamental for all applications of path integrals in QFT.

We want to compute

$$(7.1) \qquad Z(J) = N \int [Dx] \exp i \int dt\, (L + xJ), \qquad Z(0) = 1,$$

where

$$(7.2) \qquad L = -\frac{1}{2} mx \left(\frac{d^2}{dt^2} + \omega^2 \right) x.$$

This form of the Lagrangian differs from that of Eq. (1.8) by an integration by parts, like the forms in Eqs. (2.4, 2.5). It exposes the central role of the self-adjoint operator $\frac{d^2}{dt^2} + \omega^2$. To see how to proceed, compare the finite-dimensional integral over column vectors x in $\mathbf{R}^n$

$$(7.3) \qquad \int d^n x \exp(-x^T A x + J^T x),$$

with J a fixed vector and A a nonsingular symmetric matrix. (In fact, the finite-dimensional integrals over piecewise-linear paths which approximate the path integral have this form.) We complete the square

$$-x^T A x + J^T x = -(x - \tfrac{1}{2} A^{-1} J)^T A (x - \tfrac{1}{2} A^{-1} J) + \tfrac{1}{4} J^T A^{-1} J$$

and use the translation-invariance of Lebesgue measure to substitute $y = x - \frac{1}{2}A^{-1}J$, obtaining

(7.4) $$e^{J^T A^{-1} J/4} \int d^n y\, e^{-y^T A y}.$$

We can normalize the result to unity when $J = 0$ by simply dividing by the integral, so that the result is obtained from the translation-invariance of the measure without ever evaluating an integral.

Since the formal Lebesgue measure $[Dx]$ is also translation-invariant,[8] we can apply the same procedure to $Z(J)$ as soon as we know "the" inverse for the operator $A = \frac{d^2}{dt^2} + \omega^2$. That is, we need to solve the differential equation

(7.5) $$(\frac{d^2}{dt^2} + \omega^2)f(t) = g(t).$$

As we have previously mentioned, physicists like to express linear operators in terms of their integral kernels, so that the solution is written in terms of a Green's function $G(t, s)$ as

(7.6) $$f(t) = \int ds\, G(t, s) g(s),$$

and the statement that this inverts A becomes

(7.7) $$(\frac{d^2}{dt^2} + \omega^2)G(t, s) = \delta(t - s).$$

Not all solutions of this equation are of the form $G(t, s) = G(t - s)$, because of the freedom to add arbitrary solutions of the homogeneous equation, but I assert with the benefit of hindsight that the one we are seeking is. We compute it via its Fourier transform,

(7.8) $$\hat{G}(k) = \int dt\, G(t) e^{-ikt},$$

(7.9) $$G(t) = \int \frac{dk}{2\pi} \hat{G}(k) e^{ikt},$$

which satisfies

(7.10) $$(-k^2 + \omega^2)\hat{G}(k) = 1.$$

The solution to this equation in the sense of distributions is a bit tricky,

(7.11) $$\hat{G}(k) = P\frac{-1}{k^2 - \omega^2} + h(k)\delta(k^2 - \omega^2).$$

---

[8] That is, Wiener measure has the properties that would be expected if it were.

LECTURE 7. THE HARMONIC OSCILLATOR VIA PATH INTEGRALS

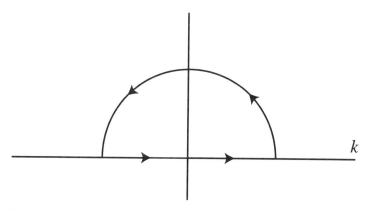

Figure 7.1. Integration contour in the k plane for computing $G(t)$ from its Fourier transform, $t > 0$.

The point is that the operator A has a kernel ("zero modes" to a physicist), so is not uniquely invertible until boundary conditions are specified for Eq. (7.5). $G(t)$ can be changed by addition of functions in the kernel, which are combinations of $e^{\pm i\omega t}$. Its Fourier transform is thus only determined up to distributions concentrated at $k = \pm \omega$. Yet another way to say this is to note that the inverse Fourier transform

$$(7.12) \qquad G(t) = \int \frac{dk}{2\pi} \frac{-1}{k^2 - \omega^2} e^{ikt}$$

is ambiguous because the integration contour (the real axis) passes through two poles at $k = \pm \omega$. A prescription for handling these poles amounts to a choice of boundary conditions and selects one of the possible Green's functions.

The appropriate boundary conditions in quantum mechanics are specified implicitly by the condition that the correlation functions obtained from $Z(J)$ should be related by analytic continuation to those given by a Euclidean path integral. The Euclidean path integral for the harmonic oscillator is calculated by the same technique of completing the square, but the relevant self-adjoint operator becomes

$$(7.13) \qquad -\frac{d^2}{d\tau^2} + \omega^2.$$

The corresponding Euclidean Green's function is, unambiguously,

$$(7.14) \qquad G_E(k) = \frac{1}{k^2 + \omega^2},$$

because this function has no poles on the real axis.[9] The Minkowski space Green's function we seek is the one related by analytic continuation to the Fourier transform of $G_E(k)$. To determine it, return to Eq. (7.12) and close the contour with a semicircle at infinity, in the upper half plane if $t > 0$ (Figure 7.1). This function

---

[9] One may wonder why the Euclidean Green's function is unambiguously determined when the operator (7.13) still has a kernel, spanned by the functions $\exp \pm \omega t$. The point is that these functions are not in L^2 or even in the set of distributions Ω^*.

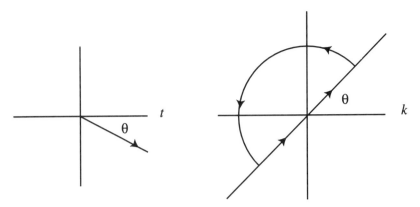

Figure 7.2. Wick rotation of the contour for computing $G(t)$, with t on the indicated line in the fourth quadrant.

$G(t)$ is analytic in some region of the complex t plane. For t in the fourth quadrant, say on the line indicated in Figure 7.2, the semicircle does contribute to the integral, but we can fix that by rotating the contour in the k plane, providing it does not cross any poles of the integrand. After a rotation by $\pi/2$ (the Wick rotation), t is on the negative imaginary axis and $G(t)$ becomes the Euclidean Green's function. Therefore, $G(t)$ is the analytic continuation of $G_E(\tau)$ provided only that the integration contour did not cross poles. This specifies the correct treatment of the poles on the real axis: the contour must pass below the one at $k = -\omega$ and above the one at $k = +\omega$, or equivalently the poles must be moved infinitesimally off the axis to $k = \pm\sqrt{\omega^2 - i\epsilon}$. The resulting Green's function,

$$(7.15) \qquad G(t) = \int \frac{dk}{2\pi} \frac{-1}{k^2 - \omega^2 + i\epsilon} e^{ikt},$$

often called the Feynman Green's function or Feynman propagator, is easily evaluated by residues as

$$(7.16) \qquad G(t) = \frac{i}{2\omega}[\theta(t)e^{-i\omega t} + \theta(-t)e^{i\omega t}],$$

where $\theta(t) = 0$ for $t < 0$, $\theta(t) = 1$ for $t > 0$ is the Heaviside step function rather than the Riemann theta function. Of course, we could easily have determined the correct $G(t)$ by explicitly evaluating $G_E(\tau)$ and analytically continuing, but the method used works in QFT also, and motivates the term "Wick rotation" as well. Comparing the explicit form of $G(t)$ with the time evolution operator e^{-iHt}, we can see why the Feynman propagator is often described as propagating positive energy states (ω) into the future ($t > 0$) and negative energy states ($-\omega$) into the past ($t < 0$). In QFT this is related to the behavior of antiparticles and can be used to give a physical derivation of the correct Green's function.

LECTURE 7. THE HARMONIC OSCILLATOR VIA PATH INTEGRALS

Armed with the correct Green's function, we easily complete the square and finish the computation of $Z(J)$:

(7.17)
$$Z(J) = \exp \frac{i}{2m} \int dt\, ds\, J(t) G(t-s) J(s).$$

By taking two functional derivatives we obtain the two-point correlation function

$$\begin{aligned}
<0|T[x(t)x(t')]|0> &= (-i)^2 \frac{\delta^2}{\delta J(t) \delta J(t')} Z(J)|_{J=0} \\
&= -\frac{i}{m} \frac{\delta}{\delta J(t')} [\int ds J(t) G(t-s)] Z(J)|_{J=0} \\
&= -\frac{i}{m} G(t-t') \\
\text{(7.18)} \quad &= \frac{1}{2m\omega} [\theta(t-t') e^{-i\omega(t-t')} + \theta(t'-t) e^{i\omega(t-t')}].
\end{aligned}$$

Note that the second functional derivative has to act on the $J(t)$ brought down by the first one in order to get a nonzero result upon setting $J = 0$. Up to a factor, the two-point function coincides with the Green's function. This is why correlation functions in general are called Green's functions.

There is a simple rule for computing the n-point correlation functions, called Wick's theorem, which is really a general fact about the moments of a Gaussian measure. When we compute the functional derivative

(7.19)
$$\frac{\delta^n}{\delta J(t_1) \cdots \delta J(t_n)} Z(J)|_{J=0},$$

the first derivative brings down a $J(t_1)$, among other things, from the exponential in $Z(J)$. Unless some other derivative acts on this $J(t_1)$, the result vanishes for $J = 0$. The nonvanishing terms arise from all possible ways of pairing the functional derivatives so that one acts on the $J(t)$ brought down by the other. The n-point function is a sum over all pairings of the product of factors $-\frac{i}{m} G(t_i - t_j)$ for each pairing. For example, the four-point function is

(7.20)
$$\begin{aligned}
<0|T[x(t_1)x(t_2)x(t_3)x(t_4)]|0> = &-\frac{1}{m^2} [G(t_1-t_2) G(t_3-t_4) \\
&+ G(t_1-t_3) G(t_2-t_4) + G(t_1-t_4) G(t_2-t_3)].
\end{aligned}$$

The n-point functions for n odd vanish since a complete pairing is not possible.

LECTURE 8
Quantum Field Theory: Free Fields

Quantum field theory, to a first approximation, is the relativistic quantum mechanics of systems having infinitely many degrees of freedom, such as fields. We want to construct a quantum mechanics of the electromagnetic field, or the Klein-Gordon field, just as we constructed the quantum mechanics of a particle moving in a potential. We will in fact discuss the Klein-Gordon equation so as to avoid the complications associated with nontrivial representations of the Lorentz group and with gauge invariance.

We recall the action

$$(8.1) \qquad S = \frac{1}{2}\int d^4x (\partial_\mu \phi \partial^\mu \phi - m^2 \phi^2),$$

which leads to the Klein-Gordon equation $(\partial_\mu \partial^\mu + m^2)\phi = 0$, and the corresponding Hamiltonian

$$(8.2) \qquad H = \frac{1}{2}\int d^3x (\pi^2 + |\nabla \phi|^2 + m^2 \phi^2),$$

where $\pi(x) = \partial^0 \phi(x)$. At a fixed time, the infinitely many coordinate and momentum variables $\phi(\mathbf{x}, t), \pi(\mathbf{x}, t)$ should be reinterpreted as Hermitian operators acting in a Hilbert space $\mathcal{H}$ and satisfying commutation relations modeled on their classical Poisson brackets,

$$(8.3) \qquad [\phi(\mathbf{x}, t), \pi(\mathbf{y}, t)] = i\delta^3(\mathbf{x} - \mathbf{y}),$$
$$(8.4) \qquad [\phi(\mathbf{x}, t), \phi(\mathbf{y}, t)] = [\pi(\mathbf{x}, t), \pi(\mathbf{y}, t)] = 0.$$

We are adopting the Heisenberg picture in which the operators are time-dependent and the canonical commutators hold only at equal times. Of course, we have the option of going to the Schrödinger picture, transferring the time dependence to the states. However, this obscures the Lorentz invariance of the theory since the Lorentz group mixes the time and space components of x, and therefore the

Heisenberg picture is generally more convenient in QFT. One can check that the Heisenberg equations of motion

$$\dot{\pi}(x) = -i[\pi(x), H], \quad \dot{\phi}(x) = -i[\phi(x), H] \tag{8.5}$$

are equivalent to the Klein-Gordon equation.

Unfortunately, the singular δ distribution in the canonical commutators shows that $\phi(x)$ and $\pi(x)$ cannot be simply operator-valued functions on $\mathbf{R}^4$, since then their commutator would also be such a function. We encountered this problem in the Hamiltonian formulation of classical field theory, and resolved it by choosing better coordinate and momentum functions on phase space. Similarly here we postulate that for each test function $f(\mathbf{x})$ on $\mathbf{R}^3$ there are time-dependent operators $\phi[f, t], \pi[f, t]$ formally given by

$$\phi[f, t] = \int d^3 x f(\mathbf{x}) \phi(\mathbf{x}, t), \quad \pi[f, t] = \int d^3 x f(\mathbf{x}) \pi(\mathbf{x}, t). \tag{8.6}$$

Then $\phi(x), \pi(x)$ should be viewed as operator-valued distributions, and the canonical commutators actually mean, for example,

$$[\phi[f, t], \pi[g, t]] = i \int d^3 x\, f(\mathbf{x}) g(\mathbf{x}). \tag{8.7}$$

It happens that this interpretation of the fields is correct in noninteracting, or free, field theories such as the Klein-Gordon theory, but that spatial smearing alone is not sufficient to produce well-defined operators in interacting field theories. In those cases we assume the existence of fields $\phi[f] = \int d^4 x f(x) \phi(x)$, and similarly for π, for each test function f on $\mathbf{R}^4$ rather than $\mathbf{R}^3$. In general one cannot make sense of equal time commutators in these theories.

We will now solve the Klein-Gordon, or free Hermitian scalar field, theory. This means that we will construct a Hilbert space $\mathcal{H}$ containing operator-valued distributions $\phi(x), \pi(x)$ satisfying the Klein-Gordon equation and the canonical commutators. Rather than producing these objects from thin air, I will give the heuristic derivation which is standard in the physics literature, and only then connect it with the precise mathematical definitions. The mathematician should try to appreciate how some familiar mathematical objects are described in the language and notation of physics.

If we view the Klein-Gordon equation as simply a PDE of a standard type for a real-valued function $\phi(x)$ on $\mathbf{R}^4$, the solution can be obtained easily by means of a Fourier transform:

$$\phi(x) = \int (dk)[a(\mathbf{k})e^{i(\mathbf{k}\cdot\mathbf{x} - E_\mathbf{k} t)} + a^\dagger(\mathbf{k})e^{-i(\mathbf{k}\cdot\mathbf{x} - E_\mathbf{k} t)}], \tag{8.8}$$

$$\pi(x) = \int (dk)(-iE_\mathbf{k})[a(\mathbf{k})e^{i(\mathbf{k}\cdot\mathbf{x} - E_\mathbf{k} t)} - a^\dagger(\mathbf{k})e^{-i(\mathbf{k}\cdot\mathbf{x} - E_\mathbf{k} t)}]. \tag{8.9}$$

Several comments are necessary concerning the notation here. The suggestively-named functions $a(\mathbf{k}), a^\dagger(\mathbf{k})$ on $\mathbf{R}^3$ are complex conjugates of each other, so that $\phi(x)$ is real. The Klein-Gordon equation is satisfied provided only that $E_\mathbf{k} =$

$\sqrt{|\mathbf{k}|^2 + m^2}$; we define a momentum 4-vector $k^\mu = (E_\mathbf{k}, \mathbf{k})$ which lies on the hyperboloid $k^2 = m^2$ in $\mathbf{R}^4$ (often referred to as the "mass shell"), so that $k \cdot x = k^\mu x_\mu = E_\mathbf{k} t - \mathbf{k} \cdot \mathbf{x}$. Finally, the Fourier integrals over $\mathbf{R}^3$ use a Lorentz-invariant measure for convenience,

$$(8.10) \qquad \int (dk) = \int \frac{d^3k}{(2\pi)^3 2E_\mathbf{k}} = \int \frac{d^4k}{(2\pi)^4} 2\pi \delta(k^2 - m^2) \theta(k^0),$$

where the last form makes the Lorentz invariance obvious.

In QFT we use the above solution to motivate a change of variables: at some arbitrary but fixed time $t = x^0$ we replace the variables $\phi(x), \pi(x)$ by the $a(\mathbf{k}), a^\dagger(\mathbf{k})$ introduced above. These are now adjoint operators for each $\mathbf{k}$ (actually distributions again), given explicitly by

$$(8.11) \qquad a^\dagger(\mathbf{k}) = \int d^3x\, e^{-ik \cdot x}[E_\mathbf{k} \phi(x) - i\pi(x)],$$

$$(8.12) \qquad a(\mathbf{k}) = \int d^3x\, e^{ik \cdot x}[E_\mathbf{k} \phi(x) + i\pi(x)].$$

Any resemblance to the change of variables we made in solving the simple harmonic oscillator is totally intentional. These new creation and annihilation operators are found to satisfy

$$(8.13) \qquad [a(\mathbf{k}), a^\dagger(\mathbf{p})] = (2\pi)^3 2E_\mathbf{k} \delta^3(\mathbf{k} - \mathbf{p}),$$
$$(8.14) \qquad [a(\mathbf{k}), a(\mathbf{p})] = [a^\dagger(\mathbf{k}), a^\dagger(\mathbf{p})] = 0.$$

In terms of these operators the Hamiltonian reads

$$(8.15) \qquad H = \frac{1}{2} \int (dk) E_\mathbf{k} [a^\dagger(\mathbf{k}) a(\mathbf{k}) + a(\mathbf{k}) a^\dagger(\mathbf{k})]$$

$$(8.16) \qquad = \int (dk) [E_\mathbf{k} a^\dagger(\mathbf{k}) a(\mathbf{k}) + 8\pi^3 E_\mathbf{k}^2 \delta^3(\mathbf{0})],$$

where the second line results from the formal application of the commutator (8.13).

In the meaningless expression $\int (dk) 8\pi^3 E_\mathbf{k}^2 \delta^3(\mathbf{0})$ we have encountered the first of the famous infinities of QFT, and the easiest to deal with. The Hamiltonian H should be viewed as describing uncountably many simple harmonic oscillators labeled by the vectors $\mathbf{k}$, and we expect that there will be a vacuum state $|0>$ satisfying $a(\mathbf{k})|0> = 0$ for all $\mathbf{k}$. The constant term involving $\delta^3(\mathbf{0})$ is infinite because it is the sum of the ground state energies $\omega/2$ of infinitely many oscillators [compare Eq. (5.4)]. We propose to subtract an infinite constant from H so that the energy of the vacuum state will be zero instead, so that we use simply

$$(8.17) \qquad H = \int (dk) E_\mathbf{k} a^\dagger(\mathbf{k}) a(\mathbf{k}).$$

Two common justifications for this procedure are the following.

(1) Even in classical mechanics $V(\mathbf{x})$ and hence H are arbitrary by an additive constant; only differences in energy are measurable and any state can be defined to have zero energy.

(2) The only function of a Hamiltonian is to generate the correct Heisenberg equations of motion. Since one can check that both forms (8.15) and (8.17) of H have the same commutators with $a(\mathbf{k})$ and $a^\dagger(\mathbf{k})$, they are equally acceptable.

In fact, our original expression (8.2) for H was only formal. It contained expressions such as $\pi(x)^2$ which, as products of distributions, are not generally meaningful. We are now defining what those products meant, with the interpretation of H in terms of oscillators as a guide, and by definition they do not include the infinite constant. A convenient way to formulate the definition is in terms of normal-ordered products. We say that a polynomial in creation and annihilation operators $P(a^\dagger, a)$ has been normal ordered, denoted $:P(a^\dagger, a):$, when each term has been reordered so that all $a^\dagger$'s stand to the left of all a's. This reordering is done by fiat, as if all operators commuted, and not by using the canonical commutation relations to interchange them. For example,

$$(8.18) \qquad \frac{1}{2} :a^\dagger(\mathbf{k})a(\mathbf{k}) + a(\mathbf{k})a^\dagger(\mathbf{k}): = a^\dagger(\mathbf{k})a(\mathbf{k}),$$

so that our modification of the Hamiltonian simply amounts to saying that all products in it should be interpreted as normal ordered. It is often the case that normal ordered products of distributions make sense when general products do not. It is a useful fact that the vacuum expectation value of a normal ordered product is always zero, which in our case means that the energy $<0|H|0>$ of the vacuum vanishes, as we wished.

The Hamiltonian H commutes with several other operators, which are therefore conserved quantities. As examples we have the number operator,

$$(8.19) \qquad N = \int (dk) a^\dagger(\mathbf{k}) a(\mathbf{k}),$$

and the momentum operator,

$$(8.20) \qquad \mathbf{P} = \int (dk) \mathbf{k} a^\dagger(\mathbf{k}) a(\mathbf{k}),$$

which, like H, annihilate the vacuum. We can combine H and $\mathbf{P}$ into a 4-momentum operator

$$(8.21) \qquad P^\mu = (H, \mathbf{P}) = \int (dk) k^\mu a^\dagger(\mathbf{k}) a(\mathbf{k}),$$

which is indeed the generator of the translation group in $\mathbf{R}^4$,

$$(8.22) \qquad [\phi(x), P^\mu] = i\partial^\mu \phi(x).$$

As in the harmonic oscillator, we expect that the eigenstates of H can be obtained by acting on the vacuum $|0>$ with finite products of the creation operators $a^\dagger(\mathbf{k})$

for various **k**. By using the commutation relations as we did then, we can verify that acting with $a^\dagger(\mathbf{k})$ on a simultaneous eigenstate of N and P^μ gives a new eigenstate with eigenvalues increased by unity for N and by k^μ for P^μ. Recall that $k^2 = m^2$, or $E_\mathbf{k}^2 = |\mathbf{k}|^2 + m^2$. This is nothing but the relation between the energy and momentum of a free particle of mass m in relativity. Although we did not have time to derive it earlier, we can make it plausible now. With physical units restored it reads $E_\mathbf{k} = \sqrt{|\mathbf{k}|^2 c^2 + m^2 c^4}$, and in the nonrelativistic limit $c \to \infty$ this becomes $E_\mathbf{k} = mc^2 + |\mathbf{k}|^2/2m$, showing the energy mc^2 corresponding to the (rest) mass as well as the Newtonian kinetic energy $|\mathbf{k}|^2/2m$.

We have been lead to the following remarkable conclusion. The eigenstates of H are of the form $a^\dagger(\mathbf{k}_1) \cdots a^\dagger(\mathbf{k}_n)|0>$. The total energy and momentum of this state are exactly those of a collection of n free particles of momenta $\mathbf{k}_i$, the operator N counts the particles, and the operators $a^\dagger(\mathbf{k}_i)$ and $a(\mathbf{k}_i)$ create and annihilate them. These particles are the quanta of the field, and this is the very deep connection between fields and particles in QFT: the stationary states of a free field theory are multiparticle states. The quantization of the electromagnetic field is technically harder but conceptually the same, and the resulting particles are photons. The Klein-Gordon equation describes spinless particles such as pi mesons. (The spin refers essentially to the highest weight of the representation of the Lorentz group by which the field transforms; it is zero for the trivial representation.)

Let us now translate the physicist's description of the Hilbert space into rigorous mathematics. The only thing wrong with the description above, really, is that the creation and annihilation "operators" are still operator-valued distributions and must be smeared in **k** to get actual operators. Physically, they create and annihilate free particles of definite momenta, and we recall that even in quantum mechanics the free particle states lie outside the Hilbert space $L^2(\mathbf{R}^3)$, in a rigged extension. So, rigorously: the Hilbert space $\mathcal{H}$ will be the space of all sequences of complex functions of the form

$$(8.23) \qquad F = \{F_0, F_1(\mathbf{k}), F_2(\mathbf{k}_1, \mathbf{k}_2), \ldots\},$$

where F_n is a symmetric function of n vectors from $\mathbf{R}^3$, such that the norm to be defined below is finite. The sequence $\{1, 0, 0, \ldots\}$ is the vacuum state previously denoted $|0>$. The sequence $\{0, 0, \ldots, F_n(\mathbf{k}_1, \ldots, \mathbf{k}_n), 0, \ldots\}$ is a smeared n-particle state which we would have previously written as

$$(8.24) \qquad \frac{1}{n!} \int (dk_1) \cdots (dk_n) F_n(\mathbf{k}_1, \ldots, \mathbf{k}_n) a^\dagger(\mathbf{k}_1) \cdots a^\dagger(\mathbf{k}_n)|0>,$$

where any antisymmetric part of F_n would not contribute to this equation because the $a^\dagger$s all commute. The scalar product is

$$(8.25) \qquad <F, G> = F_0^* G_0 + \sum_{n=1}^\infty \frac{1}{n!} \int (dk_1) \cdots (dk_n) F_n^*(\mathbf{k}_1, \cdots \mathbf{k}_n) G_n(\mathbf{k}_1, \cdots \mathbf{k}_n).$$

The nuclear subspace Ω of the rigged Hilbert space can be taken as the sequences with only finitely many nonzero entries, and each F_n decreasing very fast as any

$|\mathbf{k}_i| \to \infty$. For each test function $f(\mathbf{k})$ there is an operator

$$(8.26) \qquad a^\dagger[f] = \int (dk) f(\mathbf{k}) a^\dagger(\mathbf{k})$$

which acts on a sequence F by

$$(8.27) \qquad [a^\dagger[f]F]_n(\mathbf{k}_1,\dots,\mathbf{k}_n) = \sum_{j=1}^n f(\mathbf{k}_j) F_{n-1}(\mathbf{k}_1,\dots,\hat{\mathbf{k}}_j,\dots,\mathbf{k}_n),$$

where $\hat{\mathbf{k}}_j$ means that this argument is omitted. Similarly,

$$(8.28) \qquad [a[f]F]_n(\mathbf{k}_1,\dots,\mathbf{k}_n) = \int (dk) f(\mathbf{k}) F_{n+1}(\mathbf{k},\mathbf{k}_1,\dots,\mathbf{k}_n).$$

All these formulas follow from (8.24) and the canonical commutators (8.13).

Returning to the original Klein-Gordon fields $\phi(x)$, we construct for each test function $g(\mathbf{x})$ an operator $\phi[g,t] = \int d^3x\, g(\mathbf{x}) \phi(\mathbf{x},t)$ by using Eq. (8.8). We see that $\phi[g,t]$ is expressed in terms of $a[\hat g]$ and $a^\dagger[\hat g]$, where $\hat g$ is the Fourier transform of g. The resulting operators do satisfy the Klein-Gordon equation in the sense of distributions. The Hamiltonian of Eq. (8.17) exists as an operator in $\mathcal{H}$ and generates time evolution according to the Heisenberg equations.

The operator-valued distributions $\phi(x), a(\mathbf{k}), a^\dagger(\mathbf{k})$ themselves make sense as operators from Ω into Ω^* in the rigged Hilbert space. For example,

$$(8.29) \qquad a^\dagger(\mathbf{p})|0> = |1_\mathbf{p}> = \{0, (2\pi)^3 2E_\mathbf{p} \delta^3(\mathbf{k}-\mathbf{p}), 0, \dots\},$$

which is a sequence of distributions rather than functions. Normal-ordered products make sense in the same way, e.g.,

$$(8.30)\; a^\dagger(\mathbf{p})a(\mathbf{p})\{0, F_1(\mathbf{k}), 0, 0, \dots\} = \{0, (2\pi)^3 2E_\mathbf{p} F_1(\mathbf{p}) \delta^3(\mathbf{k}-\mathbf{p}), 0, 0, \dots\},$$

and of course $a^\dagger(\mathbf{p})a(\mathbf{p})|0> = 0$. However, other products do not exist in even this weak sense, e.g. formally $a(\mathbf{p})a^\dagger(\mathbf{p})|0> = (2\pi)^3 2E_\mathbf{p} \delta^3(\mathbf{0})|0>$, which was the difficulty that motivated us to introduce normal ordering in the first place.

A Hilbert space of this type, which arises in free field theory and rigorously consists of sequences of functions as above, is called a Fock space (after V.A. Fock). In the physics literature Fock spaces are defined by giving scarcely more than the commutators of the relevant creation and annihilation operators, and it is up to the reader to supply the sequence-space construction if desired. However, a little experience doing calculations in these spaces should convince mathematicians of the convenience of the physicist's way of describing them.

The path integral machinery for computing Green's functions applies in free field theories with barely a change in notation, since the relevant integrals are still

Gaussian. For the Klein-Gordon theory we define

$$Z(J) = N \int [D\phi] \exp iS, \tag{8.31}$$

$$S = \int d^4x [-\frac{1}{2}\phi(\partial_\mu \partial^\mu + m^2)\phi + J(x)\phi(x)], \tag{8.32}$$

where intuitively we integrate over all continuous functions $\phi(x)$ having specified but arbitrary limits at $t \to \pm\infty$. The calculation is again simply completing the square; all we need is the Green's function satisfying

$$(\partial_\mu \partial^\mu + m^2)G(x-y) = \delta^4(x-y), \tag{8.33}$$

which is

$$G(x-y) = \int \frac{d^4k}{(2\pi)^4} \frac{-1}{k^2 - m^2 + i\epsilon} e^{ik \cdot (x-y)}, \tag{8.34}$$

where the $i\epsilon$ prescription is fixed by the Euclidean Green's function $(k^2 + m^2)^{-1}$ and a Wick rotation of the k^0 axis. This Green's function can be evaluated in terms of Bessel functions, but its Fourier transform will be all we need. The partition function is

$$Z(J) = \exp \frac{i}{2} \int d^4x \, d^4y \, J(x)G(x-y)J(y), \tag{8.35}$$

and as usual the correlation functions are given by

$$<0|T[\phi(x_1)\cdots\phi(x_n)]|0> = (-i)^n \frac{\delta^n}{\delta J(x_1)\cdots\delta J(x_n)} Z(J)|_{J=0}. \tag{8.36}$$

Axioms

I will now present a set of axioms for QFT which reflect our (limited) experience with the Klein-Gordon field. Mathematicians find it reassuring to see that there actually are rules of the game which cannot be revised at will. In the 1960s, before renormalization was deeply understood, the mathematical foundation of QFT was felt to be insecure due to the cavalier treatment of formally infinite quantities. Its applicability beyond the subject of electrodynamics, for example to the strong interactions of particle physics, was also in doubt. The goal of axiomatic quantum field theory was to rigorously prove the folk theorems of the subject, and to learn which difficulties of the theory were inherent in its structure and which reflected unjustified approximations or calculational methods. Today it is somewhat disconnected from field theory as practiced by (most) physicists, but it is reassuring to see how much can be rigorously proven. Note that these axioms are intended to describe realistic quantum field theories in $\mathbf{R}^4$, so they refer to the Lorentz group and metric. They must be modified for topological and conformal field theories where spacetime may be compact and no metric is given.

Axiom 1. (*Quantum Mechanics*) States and observables of a quantum field are represented as usual by vectors and operators in a rigged Hilbert space $\mathcal{H}$, as per the axioms of quantum mechanics.

Axiom 2. (*Relativistic Invariance*) A continuous unitary representation of the (double cover of the) Poincaré group acts in the Hilbert space, an operator $U(\Lambda, a)$ representing the transformation $x' = \Lambda x + a$. We define the 4-momentum operator P^μ by $U(1, a) = \exp i P^\mu a_\mu$. Its eigenvalues k^μ lie in or on the forward light cone, $k^2 \geq 0$. There is a unique (up to a phase) Poincaré invariant vector $|0>$ called the vacuum.

Axiom 3. (*Fields*) For each test function $f(x)$ on $\mathbf{R}^4$ there is a set of operators $\phi_i[f] = \int d^4x f(x) \phi_i(x)$. All these operators and their adjoints are defined on a common domain Ω, which is dense in $\mathcal{H}$, closed under their action and that of $U(\Lambda, a)$, and is a linear set containing $|0>$. All the matrix elements $<\Psi|\phi_i[f]|\Psi'>$ are distributions in f. Under the Poincaré group the fields transform by

$$(8.37) \qquad U(\Lambda, a) \phi_i[f(x)] U^{-1}(\Lambda, a) = \sum_j M_{ij}(\Lambda^{-1}) \phi_j[f(\Lambda^{-1}(x-a))],$$

where M_{ij} is some matrix representation of $SL(2, \mathbf{C})$ and this equation holds when applied to any vector in Ω.

Axiom 4. (*Microscopic Causality*) If the supports of f and g are spacelike separated, i.e. $(x-y)^2 < 0$ for every x and y in the support of f and g respectively, then

$$(8.38) \qquad [\phi_i[f], \phi_j[g]]_\pm \equiv \phi_i[f]\phi_j[g] \pm \phi_j[g]\phi_i[f] = 0,$$

where the $-$ sign must occur if the operators $\phi_i[f]$ represent observables.

Axiom 5. (*Cyclicity of the Vacuum*) The set of vectors obtained by applying polynomials in the $\phi_i[f]$ and their adjoints to the vacuum is dense in Ω, hence in $\mathcal{H}$.

We did not verify microscopic causality for the scalar field, but it is easy to compute $[\phi(x), \phi(y)]$ from our solution and to verify that it is a Lorentz invariant function of $x - y$. Since spacelike separated points can be Lorentz transformed so that $x^0 = y^0$ (exercise), microscopic causality follows from the equal time commutators. This axiom has many important physical consequences. The uncertainty principle (exercise) places no restriction on the accuracy with which commuting operators can be measured. The fact that measurements at spacelike separated points do not interfere with each other reflects the fact that no signal can propagate between them (faster than light). This axiom also ensures that time-ordered correlation functions are Lorentz invariant, which is not obvious a priori. The instruction to arrange operators in order of their time arguments is not Lorentz invariant, because the time order of spacelike separated events is not invariant (exercise). But precisely in this case the operators commute, so their order is irrelevant. Finally, the spin-statistics theorem says that the $+$ sign occurs in axiom (4) for fields associated to double-valued representations of the Lorentz group. As a consequence, each of the four components of the quantized Dirac field obeys $\psi_i[f,t]^2 = 0$ (recall that in free field theories spatial smearing is enough). Therefore, in contrast to the scalar field, there is no vector in the Dirac Hilbert space describing two particles both in the same state f. This is the famous Pauli exclusion principle which is responsible for the shell structure of atoms and ultimately the stability of all matter! Particles obeying the exclusion principle are called fermions, while the others are bosons.

The cyclicity axiom (5) is a substitute for the properties of the creation and annihilation operators arising from canonical equal time commutation relations, which do not make sense when smearing over time as well as space is necessary. Since any vector can be approximated by fields acting on the vacuum, the diagonal vacuum matrix elements of operators determine all others. This partially accounts for the central importance of the n-point functions in field theory. The rest of the explanation will emerge in the next section.

The Dirac Field

As a second example of a free field theory satisfying these axioms, and to indicate the origin of the spin-statistics connection, I will briefly sketch the quantization of the Dirac equation. We recall the Lagrangian

$$(8.39) \qquad \mathcal{L} = \bar{\psi}_\beta (i(\gamma^\mu)_{\beta\alpha} \partial_\mu - m\delta_{\beta\alpha}) \psi_\alpha,$$

where for clarity I have introduced indices $\alpha, \beta = 1, 2, 3, 4$ labeling the components of ψ and the matrix entries of γ^μ. The momentum conjugate to ψ_α is by definition

$$(8.40) \qquad \pi_\alpha = \frac{\partial \mathcal{L}}{\partial \dot{\psi}_\alpha} = i\bar{\psi}_\beta (\gamma^0)_{\beta\alpha} = i\psi_\alpha^\dagger,$$

so the canonical commutation relations should be

$$(8.41) \qquad [\psi_\alpha(\mathbf{x},t), \psi_\beta^\dagger(\mathbf{y},t)]_\pm = \delta_{\alpha\beta} \delta^3(\mathbf{x}-\mathbf{y}),$$
$$(8.42) \qquad [\psi_\alpha(\mathbf{x},t), \psi_\beta(\mathbf{y},t)]_\pm = [\psi_\alpha^\dagger(\mathbf{x},t), \psi_\beta^\dagger(\mathbf{y},t)]_\pm = 0,$$

where I have left the choice of commutators or anticommutators open for now. The Hamiltonian is constructed as

$$(8.43) \qquad H = \int d^3x \, [\pi_\alpha \dot{\psi}_\alpha - \mathcal{L}] = \int d^3x \, \psi^\dagger(-i\gamma^0 \gamma^j \partial_j + m\gamma^0)\psi.$$

The general solution of the Dirac equation, obtained via Fourier transform, which will motivate our change of variables to creation and annihilation operators, is

$$(8.44) \qquad \psi(x) = \sum_{s=\pm 1/2} \int (dk) [b(\mathbf{k},s) u(\mathbf{k},s) e^{-ik\cdot x} + d^\dagger(\mathbf{k},s) v(\mathbf{k},s) e^{ik\cdot x}],$$
$$(8.45) \qquad \psi^\dagger(x) = \sum_{s=\pm 1/2} \int (dk) [b^\dagger(\mathbf{k},s) u^\dagger(\mathbf{k},s) e^{ik\cdot x} + d(\mathbf{k},s) v^\dagger(\mathbf{k},s) e^{-ik\cdot x}].$$

For the notation, review our discussion of the Klein-Gordon equation. Unlike the Klein-Gordon field $\phi(x)$ which was real and observable classically, hence assumed Hermitian in quantum mechanics, the Dirac field was complex even classically. Hence we do not take it to be Hermitian in QFT, which is why there are two distinct sets of creation and annihilation operators $b^\dagger, b$ and $d^\dagger, d$. We indeed have a solution to the Dirac equation provided that $k^2 = m^2$ and that the spinors u and v satisfy the Fourier-transformed Dirac equations

(8.46) $$(\gamma^\mu k_\mu - m)u(\mathbf{k}, s) = (\gamma^\mu k_\mu + m)v(\mathbf{k}, s) = 0.$$

That is, the four spinors $u(\mathbf{k}, \pm 1/2), v(\mathbf{k}, \pm 1/2)$ are the four eigenvectors of the matrix $\gamma^\mu k_\mu$, which has eigenvalues $\pm m$ as a consequence of

(8.47) $$(\gamma^\mu k_\mu)^2 = \gamma^\mu k_\mu \gamma^\nu k_\nu = \frac{1}{2}(\gamma^\mu \gamma^\nu + \gamma^\nu \gamma^\mu)k_\mu k_\nu = \eta^{\mu\nu} k_\mu k_\nu = k^2 = m^2.$$

They are conventionally normalized so that

(8.48) $$\bar{u}(\mathbf{k}, s)u(\mathbf{k}, s') = -\bar{v}(\mathbf{k}, s)v(\mathbf{k}, s') = 2m\delta_{ss'}.$$

The pair of spinors $u(\mathbf{k}, \pm 1/2)$ form a basis for a two-dimensional representation (highest weight $1/2$) of the $SO(3)$ subgroup of the Lorentz group which fixes the vector $\mathbf{k}$, and the same is true of the pair $v(\mathbf{k}, \pm 1/2)$. This is just the spatial rotation group in the rest frame of the particle, the inertial coordinate system in which $\mathbf{k} = 0$. This is what is meant by the statement that the Dirac equation describes particles of spin $1/2$.

In terms of the new creation and annihilation operators, the Hamiltonian becomes

(8.49) $$H = \sum_s \int (dk) E_\mathbf{k} [b^\dagger(\mathbf{k}, s)b(\mathbf{k}, s) - d(\mathbf{k}, s)d^\dagger(\mathbf{k}, s)].$$

The operators themselves satisfy the (anti)commutators

(8.50) $$[b(\mathbf{k}, s), b^\dagger(\mathbf{p}, s')]_\pm = [d(\mathbf{k}, s), d^\dagger(\mathbf{p}, s')]_\pm = (2\pi)^3 2E_\mathbf{k} \delta_{ss'} \delta^3(\mathbf{k} - \mathbf{p}),$$

with other (anti)commutators vanishing. We want to use these relations to redefine the Hamiltonian by changing the $dd^\dagger$ term into $d^\dagger d$ and dropping the resulting infinite constant. Because of the minus sign already present in H, we will get a Hamiltonian involving either the difference or the sum of the b and d number operators according to our choice of commutators or anticommutators for these operators respectively. The choice of commutators means that creating more d particles will lower the energy, so that the energy is unbounded below. In an interacting theory the vacuum would be unstable to the spontaneous creation of more and more such particles. Thus we are forced to quantize the Dirac field using anticommutators, which is the spin-statistics connection leading to the Pauli exclusion principle: the vanishing anticommutator of an operator with itself means that

(8.51) $$b^\dagger(\mathbf{k}, s)b^\dagger(\mathbf{k}, s) = d^\dagger(\mathbf{k}, s)d^\dagger(\mathbf{k}, s) = 0,$$

and it is impossible to create two particles of either b or d type with identical values of $\mathbf{k}, s$.

The Hamiltonian is now

(8.52) $$: H := \sum_s \int (dk) E_\mathbf{k}[b^\dagger(\mathbf{k}, s)b(\mathbf{k}, s) + d^\dagger(\mathbf{k}, s)d(\mathbf{k}, s)].$$

It commutes with the following conserved quantities: the particle number

(8.53) $$N = \sum_s \int (dk)[b^\dagger(\mathbf{k}, s)b(\mathbf{k}, s) + d^\dagger(\mathbf{k}, s)d(\mathbf{k}, s)],$$

the momentum

(8.54) $$\mathbf{P} = \sum_s \int (dk)\mathbf{k}[b^\dagger(\mathbf{k}, s)b(\mathbf{k}, s) + d^\dagger(\mathbf{k}, s)d(\mathbf{k}, s)],$$

the charge

(8.55) $$Q = e\sum_s \int (dk)[b^\dagger(\mathbf{k}, s)b(\mathbf{k}, s) - d^\dagger(\mathbf{k}, s)d(\mathbf{k}, s)],$$

which is in fact the charge associated with the conserved current $j^\mu = \bar\psi\gamma^\mu\psi$ we have seen before, but normalized to have eigenvalue e in the state $b^\dagger(\mathbf{k}, s)|0>$, and the z-component of spin,

(8.56) $$S_z = \sum_s \int (dk)s[b^\dagger(\mathbf{k}, s)b(\mathbf{k}, s) + d^\dagger(\mathbf{k}, s)d(\mathbf{k}, s)],$$

so called because this operator is part of the generator of rotations about the z axis in the $SO(3)$ subgroup mentioned above. Hence we obtain the particle interpretation of the Dirac field: $b^\dagger(\mathbf{k}, s)$ creates particles (electrons in the application to quantum electrodynamics) of mass m, momentum $\mathbf{k}$, spin component s, and electric charge e, while $d^\dagger(\mathbf{k}, s)$ creates their antiparticles (positrons), which have the same mass, momentum, and spin but opposite charge. We see that QFT predicts the existence of antiparticles as well as the spin-statistics connection. The mathematical description of the Dirac Fock space in terms of sequences of functions is left to the reader. The path integral formulation of the Dirac field, which involves fundamentally new ideas because of the anticommutation relations, is unfortunately beyond the scope of these lectures (but see those of Orlando Alvarez in this volume).

LECTURE 9
Interacting Fields, Feynman Diagrams, and Renormalization

Free field theories describe particles which do not interact with each other. The eigenstates of H, which change only by a phase factor under time evolution, contain particles whose momenta are constant for all time. Interacting particles are described by nonlinear field equations, whose Lagrangians contain higher powers of fields than quadratic. Continuing to avoid the mathematical complications of gauge invariance or nontrivial Lorentz representations, we take as our example the Klein-Gordon field with a quartic interaction,

$$(9.1) \qquad \mathcal{L} = \frac{1}{2}\partial_\mu \phi \partial^\mu \phi - \frac{1}{2}m^2 \phi^2 - \frac{1}{4!}\lambda \phi^4.$$

We choose a quartic rather than cubic interaction so that the Hamiltonian,

$$(9.2) \qquad H = \int d^3x [\frac{1}{2}\pi^2 + \frac{1}{2}|\nabla \phi|^2 + \frac{1}{2}m^2 \phi^2 + \frac{1}{4!}\lambda \phi^4],$$

is positive definite. To get a feeling for the effects of the quartic term, suppose we express it in terms of the $a, a^\dagger$ operators using the change of variables (8.8). It becomes a sum of terms having the structure

$$(9.3) \qquad \int (dk_1) \cdots (dk_4) a(\mathbf{k}_1) \cdots a^\dagger(\mathbf{k}_4) \delta^4(k_1 + \cdots - k_4).$$

That is, it can create or destroy any combination of four particles provided that the total energy and momentum do not change. By destroying two particles and then recreating them with different momenta it can describe collisions in which particles exchange momentum. It can also change the total particle number, allowing the description of processes in which new particles are created from the kinetic energy of others.

This so-called $\lambda \phi^4$ field theory can be rigorously constructed in two or three space-time dimensions, but not (thus far) in the physical case of four dimensions. What we know about it (assuming it exists) is a combination of general theorems of axiomatic field theory and perturbative series expansions in powers of λ around the free limit $\lambda = 0$. We will in fact develop a series expansion for the partition function which is formally given by the path integral

$$(9.4) \qquad Z(J) = \int [D\phi] \exp i \int d^4 x [\frac{1}{2} \partial_\mu \phi \partial^\mu \phi - \frac{1}{2} m^2 \phi^2 - \frac{1}{4!} \lambda \phi^4 + J\phi],$$

normalized to $Z(0) = 1$. (We know that the quadratic terms in this exponential define an honest Gaussian measure on a path space, but what is hard is to interpret the rest of the exponential as an integrable function with respect to this measure.) From experience with the corresponding one variable integral,

$$(9.5) \qquad \int dx \exp -(x^2 + \lambda x^4),$$

we should not expect such an expansion to converge, but ideally it would be an asymptotic expansion for the exact result which would give accurate numerical results for λ small. In quantum electrodynamics, where the expansion parameter is $e^2/\hbar c \approx 1/137$ (e is the charge of an electron), such perturbative calculations do agree with experimental measurements to as many as 12 decimal places.

Most experimental tests of QFT involve scattering experiments in which beams of known particles are created at accelerators and allowed to collide. One measures the probabilities that various types and numbers of particles emerge from the collision with various momenta. In our model field theory we ask for the probability amplitude to begin with m particles of 4-momenta k_i at $t = -\infty$ and to end with n particles of momenta p_i at $t = +\infty$, denoted $<\{p_i\}\text{out}|\{k_i\}\text{in}>$. This is viewed as an element of a matrix, the scattering or S-matrix, whose rows and columns are indexed by the stationary states of the free field theory. Intuitively we expect that any collection of particles will separate as $t \to \pm\infty$ into widely separated, hence noninteracting, particles (and perhaps bound clusters of particles, e.g. atoms). In field theory we express this by some statement such as $\phi(x) \sim Z^{1/2} \phi_{\text{out}}(x)$ for $t \to +\infty$, and $\phi(x) \sim Z^{1/2} \phi_{\text{in}}(x)$ for $t \to -\infty$, where ϕ is the interacting field, ϕ_{in} and ϕ_{out} are free fields obeying the Klein-Gordon equation, and Z is not the partition function but merely some normalization constant which can be shown to be the same for both limits. The in and out states appearing in the S-matrix element are created by the $a^\dagger$ operators in the Fourier expansions of $\phi_{\text{in}}, \phi_{\text{out}}$. The precise meaning of these operator limits is quite subtle, e.g. they certainly do not hold as strong operator equations, but only weakly as relations between certain of their matrix elements. Using them, one can derive the remarkable LSZ (Lehmann-Symanzik-Zimmermann) reduction formula expressing the S-matrix elements in terms of the correlation functions:

$$\begin{aligned}<\{p_i\}\text{out}|\{k_i\}\text{in}> &= (iZ^{-1/2})^{m+n} \int d^4 x_1 \cdots d^4 x_m d^4 y_1 \cdots d^4 y_n e^{i(p_j \cdot y_j - k_j \cdot x_j)} \\ (9.6) \qquad &\times (\Delta_{x_1} + m^2) \cdots (\Delta_{y_n} + m^2) <0|T[\phi(x_1) \cdots \phi(y_n)]|0>.\end{aligned}$$

In view of the fact that $\Delta + m^2$ becomes $-k^2 + m^2$ under Fourier transform, this says that the Fourier transform of the correlation function must have poles at

LECTURE 9. INTERACTING FIELDS

the mass shell values $p_i^2 = k_i^2 = m^2$, and the residues of these poles are the measurable scattering amplitudes! This accounts for the emphasis on the computation of correlation functions in field theory. As a check, one finds that in free field theory S conserves particle number and momenta.

Let us now derive the asymptotic expansion for $Z(J)$ and hence for the correlation functions. This is cleverly done by using the source $J(x)$ to replace the quartic term by four functional derivatives:

$$
\begin{aligned}
Z(J) &= \int [D\phi] \exp\left[\int d^4x \frac{-i\lambda}{4!} \phi^4\right] \exp i \int d^4x [\frac{1}{2}\partial_\mu\phi\partial^\mu\phi - \frac{1}{2}m^2\phi^2 + J\phi] \\
&= \int [D\phi] \exp\left[\int d^4x \frac{-i\lambda}{4!}\left(-i\frac{\delta}{\delta J(x)}\right)^4\right] \\
&\quad \times \exp i \int d^4x [\frac{1}{2}\partial_\mu\phi\partial^\mu\phi - \frac{1}{2}m^2\phi^2 + J\phi] \\
(9.7) \quad &= \exp\left[\int d^4x \frac{-i\lambda}{4!}\left(-i\frac{\delta}{\delta J(x)}\right)^4\right] Z_0(J).
\end{aligned}
$$

That is, $Z(J)$ is expressed as a series of successively higher functional derivatives of $Z_0(J)$, the partition function for the free theory evaluated in Eq. (8.35). The correlation functions follow by taking further derivatives,

$$
(9.8) \quad <0|T[\phi(x_1)\cdots\phi(x_n)]|0> = (-i)^n \frac{\delta^n}{\delta J(x_1)\cdots\delta J(x_n)} Z(J)|_{J=0}.
$$

They are expressed as power series in λ whose coefficients are various integrals of products of the Green's functions $G(x-y)$ appearing in $Z_0(J)$. Writing out these terms is a tedious combinatorial exercise which is greatly facilitated by Feynman diagrams, which are simply graphical representations of the various terms.

Recall from Wick's theorem that the nonvanishing terms in a repeated functional derivative such as (9.8) come from complete pairings of all the derivatives, the pairing of $\delta/\delta J(x)$ with $\delta/\delta J(y)$ leading to a factor $G(x-y)$ in the result. We represent each possible complete pairing by a graph in which there is a vertex labeled with each argument x of a functional derivative $\delta/\delta J(x)$. Vertices corresponding to paired derivatives are joined by a line. It follows that four lines will meet at each internal vertex [a vertex corresponding to an operator $\delta^4/\delta J(x)^4$ from the expansion of the exponential in Eq. (9.7)], while only one line emerges from an external vertex corresponding to an argument x_i in the correlation function being computed. The following "Feynman rules" specify the contribution to the correlation function (9.8) represented by a given graph. Each internal vertex in the graph represents a factor $-i\lambda$, so that the number of internal vertices in a graph is the order of the corresponding term in the power series expansion in λ. A line connecting points x and y represents a factor $-iG(x-y)$. The resulting expression is to be integrated over all internal vertex labels, leaving a function of the external points only. The correlation function to a given order in λ is the sum of the contributions of all graphs with that number of internal vertices or less.

For some graphs, there is an additional numerical "symmetry factor" in addition to the factors specified in these rules. It arises from an incomplete cancellation of

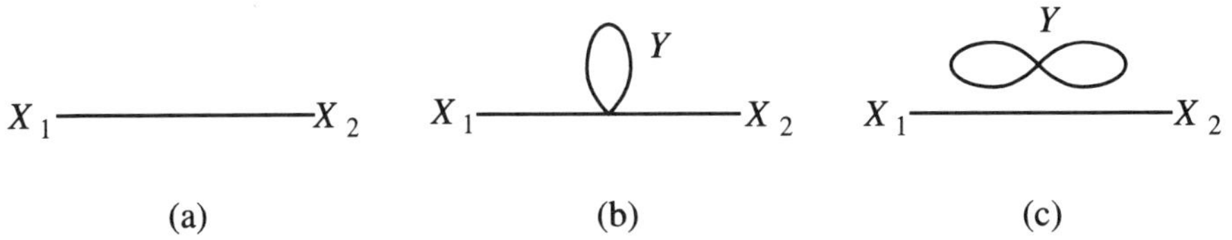

Figure 9.1. Feynman diagrams contributing to the two-point function to order λ.
(a) $-iG(x_1 - x_2)$.
(b) $\frac{\lambda}{2} \int d^4y G(x_1 - y) G(y - x_2) G(0)$.
(c) $\frac{\lambda}{8} G(x_1 - x_2) \int d^4y G^2(0)$.

two numerical factors already present in the series for $Z(J)$: the $1/4!$ in the $\lambda\phi^4$ interaction, and the $1/j!$ in the jth term of the expansion of the exponential in Eq. (9.7). The "normal" situation in a graph is for each internal vertex y to be connected to four distinct neighbor vertices $x_1, \ldots, x_4$. There are $4!$ distinct ways to choose which of the four derivatives $\delta^4/\delta J(y)^4$ are paired with the $\delta/\delta J(x_i)$, so by letting one graph denote all such choices we eliminate the factor $1/4!$. However, for some graphs this counting is not correct. Whenever an internal vertex is connected to fewer than four distinct neighbors, an extra numerical factor must be attached to the contribution of the graph. For example, in the graph of Figure 4b, there are 4×3 ways to connect x_1, x_2 to y, but then there is only one way to connect y to y. This graph has a factor $4 \times 3/4! = 1/2$. In general, this "symmetry factor" is $1/N$, where N is the order of the permutation group of the lines in the graph, leaving the vertices fixed. Similarly, the $1/j!$ at order j is supposed to be cancelled by the $j!$ ways of labelling the internal vertices with integration variables, so that by drawing only one of the labelings we can omit this factor. Again, there are exceptional situations in which different labelings do not give distinguishable graphs and a symmetry factor is left over, but we will not encounter these.

As an example, let us compute the two point function $<0|T[\phi(x_1)\phi(x_2)]|0>$ to second order in λ. Using Eqs. (9.7, 9.8), we have

$$<0|T[\phi(x_1)\phi(x_2)]|0> = (-i)^2 \frac{\delta^2}{\delta J(x_1)\delta J(x_2)} \left[1 - \frac{i\lambda}{4!} \int d^4x \frac{\delta^4}{\delta J(x)^4} \right.$$

(9.9)
$$\left. - \frac{\lambda^2}{4!4!} \int d^4x\, d^4y \frac{\delta^8}{\delta J(x)^4 \delta J(y)^4} + \cdots \right] Z_0(J)|_{J=0}.$$

The results of these derivatives to order λ are shown with the corresponding graphs in Figure 4. At zeroth order in λ we have the free field theory Green's function $-iG(x - y)$. At higher order we see at once that some diagrams are topologically connected while others have two (or more) connected components. Each component is a separate multiplicative factor in the corresponding expression; no factors $G(x - y)$ link points in different components. As I will now explain, we should omit the contribution of every disconnected diagram. This is because of the fact, which we have ignored until now, that $Z(J)$ was to be normalized so that $Z(0) = 1$. To accomplish this, all our formulas for correlation functions must be divided by $Z(0)$. In fact, we can explicitly exhibit a factor $Z(0)$ in the two-point function, because $Z(0)$ itself has a Feynman diagram expansion: it is a degenerate case of

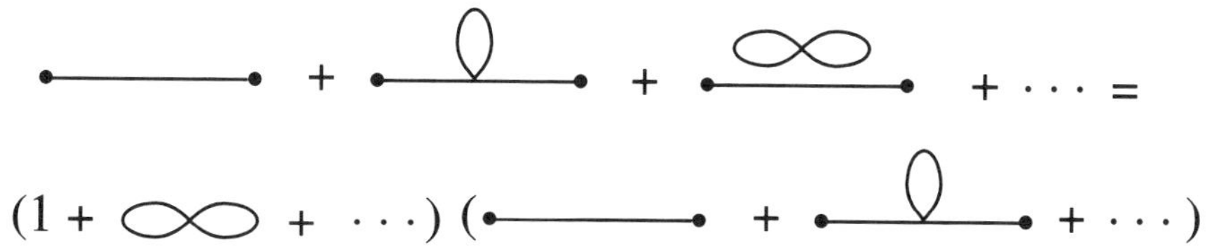

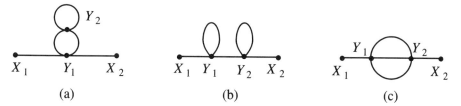

Figure 9.2. Factorization of the two-point function into the zero-point function $Z(0)$ and the series of connected diagrams.

Figure 9.3. Connected graphs for the two-point function at order λ^2.
(a) $\frac{i\lambda^2}{4} \int d^4y_1 d^4y_2 G(x_1 - y_1) G(y_1 - x_2) G^2(y_1 - y_2) G(0)$.
(b) $\frac{i\lambda^2}{4} \int d^4y_1 d^4y_2 G(x_1 - y_1) G(y_1 - y_2) G(y_2 - x_2) G^2(0)$.
(c) $\frac{i\lambda^2}{6} \int d^4y_1 d^4y_2 G(x_1 - y_1) G^3(y_1 - y_2) G(y_2 - x_2)$.

a "zero-point" correlation function and as such corresponds to diagrams having no external vertices at all. Pictorially, the two-point function we computed factors into two series as in Figure 5, where the first series is $Z(0)$ and the second contains only connected diagrams. A similar argument shows that ignoring the disconnected diagrams is the way to correctly normalize every correlation function. Unfortunately, the integral corresponding to the remaining connected diagram contains a factor $G(0)$ which is undefined.

The connected contributions to the two-point function at second order are shown in Figure 6. The second order contribution also involves divergent integrals because of the singularity of the Green's function $G(y_1 - y_2)$ as $y_1 \to y_2$. These are the rest of the notorious infinities of QFT — the serious ones this time. Obtaining sensible results from this series of divergent integrals is the goal of the program of renormalization. The mathematical origin of the divergences is the same as in free field theory: the presence of undefined formal products of distributions in the Lagrangian from which we computed the path integral. Renormalization can in fact be described in mathematical terms as the problem of defining these products of distributions, but we will take a more physical approach.

It will be helpful to study not the correlation functions themselves, but their Fourier transforms. Not only are these are the physically significant quantities according to the LSZ formula, but we only know $G(x - y)$ through its Fourier representation anyway. Mathematically it is standard to study products of distributions in terms of their symbols. Therefore we study the two point function in momentum space,

$$(9.10) \quad \hat{G}(p_1, p_2) \equiv \int d^4x_1 d^4x_2 <0|T[\phi(x_1)\phi(x_2)]|0> e^{-ip_1 \cdot x_1} e^{-ip_2 \cdot x_2}.$$

For example, computing this transform for the first order contribution of Figure 4b and inserting the integral representation of $G(x-y)$ gives

(9.11)
$$\int d^4x_1 \, d^4x_2 \, d^4x \, e^{-ip_1 \cdot x_1} e^{-ip_2 \cdot x_2}$$
$$\times \frac{\lambda}{2} \int \frac{d^4k_1 \, d^4k_2 \, d^4k}{(2\pi)^{12}} e^{ik_1 \cdot (x_1-x)} e^{ik_2 \cdot (x_2-x)} e^{ik \cdot 0}$$
$$\times \frac{-1}{(k_1^2 - m^2 + i\epsilon)(k_2^2 - m^2 + i\epsilon)(k^2 - m^2 + i\epsilon)}$$

Performing the integrals over x_1, x_2, x, this becomes

(9.12)
$$\frac{\lambda}{2} \int d^4k_1 \, d^4k_2 \, d^4k \, \delta^4(k_1 - p_1)\delta^4(k_2 - p_2)\delta^4(k_1 + k_2)$$
$$\times \frac{-1}{(k_1^2 - m^2 + i\epsilon)(k_2^2 - m^2 + i\epsilon)(k^2 - m^2 + i\epsilon)}.$$

Using the delta functions to do two more integrals, we obtain

(9.13)
$$\frac{\lambda}{2}\delta^4(p_1 + p_2)\frac{1}{(p_1^2 - m^2)(p_2^2 - m^2)} \int d^4k \, \frac{-1}{k^2 - m^2 + i\epsilon},$$

which we immediately recognize as quadratically divergent.

By applying the same transform to an arbitrary graph, we obtain a set of Feynman rules for directly writing down the series expansion of the Fourier transforms of the correlation functions. We draw the same diagrams as before, but instead of labeling the vertices we label each line with a 4-momentum variable k. The lines should be directed; with our Fourier transform conventions the momenta on the external lines (those containing an external vertex) should flow into the graph, but the directions on the internal lines are arbitrary. Each line represents a factor $i/(k^2 - m^2 + i\epsilon)$, and each vertex represents a factor $-i\lambda(2\pi)^4\delta^4(\sum k)$, where the delta function sets to zero the net momentum flowing into the vertex. Multiply by the symmetry factor $1/N$ and integrate the internal momenta with the measure $\prod d^4k/(2\pi)^4$. As in the above example, these rules always produce an overall delta function which expresses the conservation of the total momentum flowing into the graph, and there will always be explicit poles when an external momentum $p^2 = m^2$, as required by the LSZ formula. Often we omit these factors for brevity. As another example, Figure 7 shows a second order graph and its contribution.

To analyze and tame the divergences it is helpful to introduce a regularization, or cutoff, for the integrals. The simplest (though not computationally the most useful) way to do this is to fix a large number Λ and restrict each momentum integral $d^4k/(2\pi)^4$ to the region $k^2 < \Lambda^2$. Because this region is not compact when k^2 is defined with the Lorentz metric, we should perform the Wick rotation to Euclidean space on all the k^0 integrals first. This amounts to replacing every term k^2 in the Lorentz metric by $-k^2$ in the Euclidean metric. We then compute integrals over balls of radius Λ, and the results, which are finite but depend on Λ, are related by analytic continuation to the integrals of actual interest. The problem is to determine how singular the integrals are in the limit $\Lambda \to \infty$, and

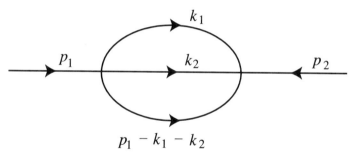

Figure 9.4. The second-order contribution of Figure 6c to the two-point function in the momentum space (Fourier transform) representation:
$(2\pi)^4 \delta^4(p_1 + p_2)(p_1^2 - m^2)^{-1}(p_2^2 - m^2)^{-1} \frac{-i\lambda^2}{6}$
$\times \int \frac{d^4 k_1 d^4 k_2}{(2\pi)^8}(k_1^2 - m^2 + i\epsilon)^{-1}(k_2^2 - m^2 + i\epsilon)^{-1}[(p_1 - k_1 - k_2)^2 - m^2 + i\epsilon]^{-1}.$

how to modify them to remove this divergence. Of course, the modification cannot be arbitrary, but should be physically motivated so that we know how it affects the physical interpretation of the finite results.

Let me remark that this regularization procedure is precisely what a mathematician might do to study singular products of distributions. For example, to study products of delta functions, we might replace them by delta sequences of smooth functions which become more localized at zero. One way to do this is to impose a cutoff $k^2 < \Lambda^2$ on the Fourier integral representation,

$$(9.14) \qquad \delta(x) = \int \frac{dk}{2\pi} e^{ikx}.$$

We could then study the singularity in the products of such smooth functions for $\Lambda \to \infty$.

We have estimated the "degree of divergence" of our integrals thus far by "naive power counting": each momentum integration d^4k is counted as Λ^4, each denominator $(k^2 - m^2)$ counts as Λ^{-2}, we multiply these factors to obtain Λ^D, and the integral is finite if $D < 0$ but divergent if $D \geq 0$. A basic result known as Weinberg's theorem [17] (due to Nobel Laureate — but not at the time — Steven Weinberg) justifies this procedure. It shows that the contribution to the integral from a region in which all integration variables are of order Λ does behave as this naive power of Λ (where Λ^0 is understood to mean $\ln \Lambda$). There may be contributions with different behavior from other regions, but they are associated with identifiable subgraphs of the Feynman diagram. This makes possible an inductive study of the divergences.

Let us return to the two-point function $\hat{G}(p_1, p_2)$ to order λ^2 which diverges as Λ^2 by our calculations and Weinberg's theorem. Observe that it is a Lorentz-invariant function of p_1, p_2 supported at $p_1 = -p_2$, therefore a function only of $p_1^2 = p_2^2 \equiv p^2$. By Taylor-expanding the integrands of the Feynman graphs in p, we can write

$$(9.15) \qquad \hat{G}(p^2) = A + Bp^2 + C(p^2)^2 + \cdots,$$

where $A, B, C, \ldots$ are certain integrals over internal momenta k_i obtained by differentiating the original integral and setting $p = 0$. The key point is that each derivative produces more momentum factors in the denominator, thereby lowering the degree of divergence according to Weinberg's theorem. Thus, A diverges as Λ^2, B diverges as $\ln \Lambda$, and higher coefficients are actually finite. The crucial result, which holds for all graphs and is well-known in distribution theory, is that the divergences can always be isolated in the form of a polynomial in the external momenta, in this case $A + Bp^2$. Subtracting this polynomial from $\hat{G}$ leaves a finite result, but we must understand the effect of such subtractions on the interpretation of the theory.

To do this, let us return to the Lagrangian of Eq. (9.1) and ask for the physical interpretation of the parameters m and λ. We know that when $\lambda = 0$ the particles are free and have mass m, which leads us to guess that in general m is their mass and λ is some measure of their interaction strength or charge. To confirm this, we should choose some measure of interaction strength, perhaps the average deflection angle of interacting beams of particles of some specified momentum, and compute it in terms of m and λ using Feynman diagrams. We might also deduce the particle mass in terms of m and λ by comparing deflection angles at various momenta. Given our experience, the result of the computation is likely to be divergent. So let us admit that we do not know the physical meaning of the parameters in the Lagrangian. To indicate this, let us rename them as the "bare" or "unrenormalized" mass m_0 and charge λ_0, in contrast to the "dressed", "physical", or "renormalized" mass m and charge λ which are defined and measured by actual experiments. We also rename the bare field ϕ_0, so that the bare Lagrangian is

$$\mathcal{L} = \frac{1}{2}\partial_\mu \phi_0 \partial^\mu \phi_0 - \frac{1}{2}m_0^2 \phi_0^2 - \frac{1}{4!}\lambda_0 \phi_0^4. \tag{9.16}$$

We define the renormalized field by $\phi = Z^{-1/2}\phi_0$, where Z is the constant in the LSZ formula (9.6), so that ϕ is the field which actually becomes free in the limits $t \to \pm\infty$. We assume that Z, m_0, and λ_0 can be related to the physical mass and charge by asymptotic series of the form

$$Z = 1 + a_1 \lambda + a_2 \lambda^2 + \cdots, \tag{9.17}$$
$$m_0 = m(1 + b_1 \lambda + b_2 \lambda^2 + \cdots), \tag{9.18}$$
$$\lambda_0 = \lambda(1 + c_1 \lambda + c_2 \lambda^2 + \cdots), \tag{9.19}$$

where the coefficients may depend on m, Λ, and on the fixed momentum in the experiment defining the physical charge. These series guarantee that in the free theory with $\lambda = 0$ we have $Z = 1$, $\lambda_0 = 0$, and m_0 equal to the physical mass.

Now, perturbation theory has given us an expression for $\hat{G}(p^2)$ in terms of m_0, λ_0, Λ through order λ_0^2. We demand that $Z^{-1}\hat{G}(p^2)$, the correlation function for the fields ϕ rather than ϕ_0, should be finite in the limit $\Lambda \to \infty$ when it is reexpressed in terms of the physical quantities m, λ using the series above to order λ^2. If it is possible to choose the coefficients a_i, b_i, c_i in these series so as to make all the n-point functions finite to all orders in λ in this way, our theory is called renormalizable. In fact, the $\lambda\phi^4$ theory is renormalizable, as is quantum electrodynamics.

One way to see how renormalization works is to substitute the series (9.17)–(9.19) into the bare Lagrangian (9.16), obtaining

$$\begin{aligned}
\mathcal{L} =\ & \frac{1}{2}\partial_\mu\phi\partial^\mu\phi - \frac{1}{2}m^2\phi^2 - \frac{1}{4!}\lambda\phi^4 \\
& + \frac{1}{2}(a_1\lambda + a_2\lambda^2 + \cdots)\partial_\mu\phi\partial^\mu\phi \\
& - \frac{1}{2}m[(a_1+b_1)\lambda + (a_2+b_2+a_1b_1)\lambda^2 + \cdots]\phi^2 \\
& - \frac{1}{4!}[(c_1+2a_1)\lambda + (c_2+2a_2+a_1^2+2a_1c_1)\lambda^2 + \cdots]\phi^4.
\end{aligned}$$

(9.20)

The substitution has produced many new interaction terms in the Lagrangian, called counterterms. When we expand $Z(J)$ in its perturbation series, the new terms will lead to new types of vertices in the Feynman diagrams. The counterterm proportional to $\partial_\mu\phi\partial^\mu\phi$ leads to a vertex where two lines meet representing a factor $p^2(a_1\lambda + \cdots)$ (p is the momentum flowing through the vertex; the p^2 arises from the Fourier transform of the two derivatives). The counterterm proportional to ϕ^2 is another vertex where two lines meet representing the factor $-i(a_1+b_1)m\lambda+\cdots$, and the ϕ^4 counterterm is a four-point vertex representing $-i(c_1+2a_1)\lambda+\cdots$. There will be new diagrams involving the two-point vertices making contributions of the form $A + Bp^2$ to $\hat{G}(p^2)$, and by choosing the coefficients correctly they will precisely cancel the divergences we found earlier. The four-point counterterm is needed to cancel divergences in the four-point function which we have not studied.

In order for this theory to be renormalizable it is clearly necessary that we do not continually encounter new divergences as we compute new Green's functions or as we proceed to higher orders in λ, because we only have three types of counterterms available to cancel them all. To see that this condition is satisfied, we will determine the degree of divergence of an arbitrary diagram having E external lines, V internal vertices, and I internal lines. There are I internal momentum integrals, and $V-1$ delta functions expressing momentum conservation at each vertex (the -1 is for the overall conservation of external momenta, which will not help us do any integrals), leaving $L = I - V + 1$ integrals to do (this is also the number of independent loops, or generators of the fundamental group, of the graph). There are I denominators containing momenta, so the graph behaves like Λ raised to the power $4L-2I = 4(I-V+1)-2I = 2I-4V+4$. Now we need a relation between V and I. Since four lines meet at each vertex, the total number of lines in the graph should be $4V$, but this counts the internal lines twice, so in fact $4V = E + 2I$. Substituting, the graph behaves like Λ^{4-E}. This crucial result shows that the divergences do not get worse as the graphs get larger. In fact, the two-point function is quadratically divergent at all orders, the four-point function is logarithmically divergent (that's what the third counterterm was for!) and the other Green's functions are "superficially convergent" (that is, if they diverge, it is only because they contain two- or four-point subgraphs). From here the proof of renormalizability is a (nontrivial!) combinatorial exercise of checking that the same counterterm that makes a given graph finite will also appear with the correct coefficient to cancel the divergences arising when this graph appears as a subgraph of a larger one.

Let me conclude this introduction to renormalization theory with several comments and hints concerning topics beyond the scope of these lectures.

(1) The key to the renormalizability of $\lambda\phi^4$ theory is our computation of the degree of divergence $4 - E$ for the E-point function. This depends on a delicate balance between the dimension of spacetime, which leads to numerator factors d^4k in Feynman integrals, and the power of ϕ in the interaction term of $\mathcal{L}$, which controls the number of lines meeting at each vertex and indirectly the number of momentum denominators in the integrals. One finds that $\lambda\phi^4$ theory is not renormalizable in more than four dimensions, and is superrenormalizable (diagrams actually become less divergent at higher orders in λ, and beyond a certain order no more counterterms are needed) in less than four dimensions. An easy way to understand this is to apply dimensional analysis to the Lagrangian of Eq. (9.1). In order for the action to be dimensionless, so that $\exp iS$ makes sense, we see from the kinetic term that ϕ must have dimensions of length^{-1}, or of momentum, in four dimensions, and then from the interaction term that λ is dimensionless. Since each Green's function has a fixed dimension, and λ is dimensionless, the momentum integrals at every order in λ have the same dimensions, and therefore the same degree of divergence. In contrast, in more than four dimensions, λ would have dimensions of momentum to some negative power, so that momentum integrals at higher orders must have more momenta in their numerators and hence higher degrees of divergence. In general, a dimensionless expansion parameter is the poor man's test for renormalizability.

(2) The only condition we have imposed on the counterterms a_i, b_i, c_i is that they contain terms with behavior Λ^2 and $\ln \Lambda$ which cancel the divergences of the Feynman integrals for $\Lambda \to \infty$. But we also want to compute the resulting finite Green's functions. Since a_i, b_i, c_i are still arbitrary by constants independent of Λ, the renormalized Green's functions are far from unique. How do we fix or understand this ambiguity? The answer is that we specify that m is the physical mass of the particles, and λ is their charge as measured in a specific experiment. When we compute the results of this experiment via Feynman diagrams and demand that they be equal to m and λ to each order of perturbation theory, this will fix the arbitrary constants in the counterterms to that order. If we choose another definition of the charge, as being the result of some other experiment, the constants will change. In general, there is a standard physical definition of the mass, while the meaning of the charge is more a matter of convention, but even this need not always be the case. For example, in quantum chromodynamics (QCD) we believe that quarks attract each other by such strong forces that they can never be isolated as free particles. In such a case, the mass of a free quark cannot be measured, and several indirect definitions of mass via various experiments are used. What we really have is a two-parameter family of quantum field theories for which we can choose coordinates m and λ in many convenient ways. The freedom to add constants to the counterterms is the invariance of this family under reparametrization, which is one form of what is called the renormalization group.

(3) Suppose we have fixed some experimental definitions of m and λ. In order to get a renormalized theory with these specific numerical values of m and λ, according to Eqs. (9.18,9.19) the bare parameters m_0 and λ_0 must flow in a specific manner with Λ as $\Lambda \to \infty$. (Of course, the asymptotic series may not have any validity for $\Lambda \to \infty$. We are really assuming that they are asymptotic expansions of some expressions which do make sense.) This one-parameter Abelian group action on the two-dimensional space of m_0 and λ_0 parametrized by Λ is another form of the

renormalization group. For a more profound discussion of renormalization in terms of this flow, see [**18**].

(4) After renormalization, every term in the perturbation series of every Green's function is finite. However, this says nothing about the convergence of the series itself. As mentioned already, in those two- and three-dimensional field theories which can be constructed rigorously, perturbation theory is only asymptotic to the exact Green's functions, and we hope/pretend this is the case in four dimensions. In some cases Borel summation of perturbation theory gives the exact Green's functions, while in other cases the series is not Borel summable. Nevertheless we know from experiment that the first few terms of the series in quantum electrodynamics give correct results to many decimal places. The mathematician reader should stop to absorb the fact that the renormalizability of electrodynamics is not just a theorem to a physicist: physics students must actually be trained to compute renormalized Green's functions correct down to the signs and factors of 2π!

(5) Recall that a Lagrangian with a continuous symmetry possesses a conserved current $j^\mu, \partial_\mu j^\mu = 0$, which is generally some quadratic expression in the fields. Green's functions involving this current, as given by Feynman integrals before regularization, will formally satisfy

$$(9.21) \qquad \partial_\mu <0|T[j^\mu(x)\phi(x_1)\cdots\phi(x_n)]|0> = 0.$$

However, the cutoff Feynman integrals may not satisfy this relation, and there may be no choice of counterterms for which the renormalized Green's functions satisfy it. [Because j^μ is quadratic in the fields, this Green's function is an $(n+2)$-point function with two arguments coinciding. Even after renormalization, Green's functions are normally singular when their arguments coincide. Further counterterms, amounting to a modification of the formal definition of j^μ, are required to make this Green's function finite. This is called the renormalization of composite operators.] In such a case the conservation law $\partial_\mu j^\mu = 0$ was only formal; the rigorously defined operator-valued distribution $j^\mu(x)$ is not conserved, and we say there is an anomaly. Such anomalies typically occur in theories involving Dirac fields and are deeply related to the Atiyah-Singer index theorem for the Dirac operator.

EXERCISES

Self-adjointness and time evolution in quantum mechanics

Let H be a complex separable Hilbert space. An operator A on H is a linear function from a linear subspace $D(A)$ to H; $A: D(A) \to H$ for short. The linear subspace $D(A)$ is called the domain of A. For the most part we are interested only in the case where $D(A)$ is dense in H. Key examples to keep in mind are differential operators on $L^2(\mathbf{R}^n)$, which cannot be defined on all of $L^2(\mathbf{R}^n)$ in any natural way.

Definition. Given two linear operators A and B on H, we say that $A \subset B$ if $D(A) \subset D(B)$ and $A = B|_{D(A)}$.

Definition. Given a densely defined operator A on H we define the adjoint operator $A^\dagger$ on H as follows: an element y in H is in $D(A^\dagger)$ iff there is a z in H such that $<Ax|y> = <x|z>$ for all x in $D(A)$. For such a y we define $A^\dagger y = z$. Remark: if A is a bounded operator on all of H then $D(A^\dagger) = H$ also.

Definition. A densely defined operator A on H is said to be symmetric if $A \subset A^\dagger$. This is equivalent to saying that $<Ax|y> = <x|Ay>$ for all x, y in $D(A)$.

Definition. A densely defined operator A on H is said to be self-adjoint or Hermitian if $A = A^\dagger$, i.e. $A \subset A^\dagger$ and $A^\dagger \subset A$.

1. Show that the following operators are symmetric but not self-adjoint.
   (a) Let $H = L^2([0,1])$ and $D(A) = C^\infty$ functions on $(0,1)$ with compact support, and $Af = f''$ for f in $D(A)$.
   (b) Let H and $D(A)$ be as above, but now take $Af = if'$.
   (c) Let $H = L^2(\Omega)$, where Ω is a bounded open subset of $\mathbf{R}^n$. Take $D(A) = C^\infty$ functions on Ω with compact support, and $Af(x) = \partial_i(a_{ij}(x)\partial_j f)(x)$, where the matrix $a_{ij}(x)$ is Hermitian with C^1 entries.

What is so special about self-adjoint operators? Part of the answer is contained in the following theorem.

Theorem (Stone and von Neumann). *Suppose that for each t in $\mathbf{R}$, $U(t)$ is a unitary operator on H, with $U(0) = 1$. Also assume that $U(t+s) = U(t)U(s)$*

for all s,t and that for each x in H we have $\lim_{t\to 0} U(t)x = x$. (Such a U is called a strongly continuous one-parameter group of unitary transformations, or a continuous symmetry for short.) Define a linear operator A on H as follows: $Ax = -i\lim_{t\to 0}[U(t)x - x]/t$, with $D(A)$ consisting of all x in H for which this limit exists and belongs to H. Then A is a densely defined self-adjoint operator called the infinitesimal generator of U. Conversely if A is a densely defined self-adjoint operator on H, there is a unique continuous symmetry for which A is the infinitesimal generator. (This $U(t)$ is denoted by $\exp itA$.) Furthermore, if $u(t) = e^{itA}u_0$ with u_0 in $D(A)$, then $u(t)$ is in $D(A)$ for all t and is the unique solution to the equation $\dot{u}(t) = iAu(t)$ with $u(0) = u_0$.

Therefore there is a 1-1 correspondence between continuous symmetries and self-adjoint operators A. We will see below that self-adjoint cannot be replaced by symmetric.

2. In each of the problems below a continuous symmetry $U(t)$ of H and a subspace V of H will be given. Let A denote the infinitesimal generator of $U(t)$. In each case show that $V \subset D(A)$ and find the action of A on functions in V.
   (a) $H = L^2(\mathbf{R}^3), U(t)f(x) = f(x + tv)$ for a fixed v in $\mathbf{R}^3$, and V is the C^1 functions with compact support on $\mathbf{R}^3$.
   (b) H and V are as in part (a). Set $U(t)f(x) = f(e^{-tB}x)$ where B is a skew symmetric 3×3 matrix. Also show that $U(t)$ is unitary for each t.
   (c) Let $H = L^2([0,2\pi])$ and $U(t)f(x) = U_\alpha(t)f(x) = \alpha^{n(x+t)}f(r(x+t))$, where $\alpha \in \mathbf{C}$ with $|\alpha| = 1$ and the functions n and r on $\mathbf{R}$ are defined as follows. For each s in $\mathbf{R}$ let $s = n(s)2\pi + r(s)$ where $n(s)$ is an integer and $0 \leq r(s) < 2\pi$. Let $V = V_\alpha = \{f \in C^1([0,2\pi]) : f(2\pi) = \alpha f(0)\}$. Also show that U_α is a continuous symmetry.

3. Let $H = L^2([0,2\pi])$ and $Af = -if'$ on $D(A) = C^1$ functions on $(0,2\pi)$ with compact support. (Note that by part 1b, A is symmetric but not self-adjoint.) For each $\alpha \in \mathbf{C}$ with $|\alpha| = 1$ let A_α denote the infinitesimal generator of U_α defined in problem 2c.
   (a) Show that $A \subset A_\alpha$. So A_α is a self-adjoint extension of A.
   (b) Let $u_0(x)$ be a positive C^∞ function with compact support in $(\pi/2, \pi)$ so that u_0 is in $D(A)$. Show that the unique solution to the partial differential equation $\partial_t u(t,x) = iAu(t,x) = \partial_x u(t,x)$ with $u(0,x) = u_0(x)$ is $u(t,x) = u_0(x+t)$. Notice for $t > 0$ that $u(t,\cdot)$ stays in $D(A)$ as long as $t < \pi/2$. But for t large enough the function $u(t,\cdot)$ is no longer zero at the left end point. This gives an indication of why A being symmetric is not good enough for producing continuous symmetries. For differential operators the choice of domain is essentially a choice of boundary conditions. A choice which makes the differential operator symmetric but not self-adjoint indicates that one has specified boundary conditions which are too restrictive. As a result there is no possible unitary time evolution which will maintain these boundary conditions. This is what happened in this problem. It is an interesting fact that the boundary conditions $f(2\pi) = \alpha f(0)$ with $|\alpha| = 1$ are the only possible boundary conditions which produce a self-adjoint extension of A.

Gaussian measure problems

A basic formula in the study of Gaussian measures is

$$\text{(9.22)} \qquad \int_{\mathbf{R}} e^{-ax^2/2} dx = \sqrt{2\pi/a}.$$

Definition. The measure μ on $\mathbf{R}^n$ given by

$$\text{(9.23)} \qquad d\mu(x) = Z^{-1} e^{-<Ax,x>/2} dx,$$

where A is a positive definite matrix and Z is a constant such that $\mu(\mathbf{R}^n) = 1$, is called the mean-zero Gaussian measure on $\mathbf{R}^n$ with covariance $G = A^{-1}$.

4. (a) Find the constant Z.
   (b) Show that $\int_{\mathbf{R}^n} e^{<p,x>} d\mu(x) = e^{<Gp,p>/2}$ for all p in $\mathbf{R}^n$.
   (c) Use analytic continuation to show that $\int_{\mathbf{R}^n} e^{i<p,x>} d\mu(x) = e^{-<Gp,p>/2}$.

   Definition. For any finite measure μ on $\mathbf{R}^n$, the function $\hat{\mu}(p) = \int_{\mathbf{R}^n} e^{i<p,x>} d\mu(x)$ is called the Fourier transform of μ.

   Fact: $\hat{\mu}$ uniquely determines the measure μ.
   (d) For v in $\mathbf{R}^n$ and $f : \mathbf{R}^n \to \mathbf{R}$ "nice", set $\partial_v f(x) = <v, \nabla f(x)>$. Show that

$$\text{(9.24)} \qquad \int_{\mathbf{R}^n} \partial_v f(x) d\mu(x) = \int_{\mathbf{R}^n} <Av, x> d\mu(x)$$

   for nice f. Use this to show that $\partial_v^\dagger = -\partial_v + <Av, \cdot>$.
   (e) Show that $\int_{\mathbf{R}^n} <v,x><w,x> d\mu(x) = <Gv,w>$ for all v,w in $\mathbf{R}^n$.
   (f) (*Wick's theorem.*) Show more generally that

$$\text{(9.25)} \qquad \int_{\mathbf{R}^n} <v_1,x> \cdots <v_{2n},x> d\mu(x) = \sum <Gv_{i_1}, v_{j_1}> \cdots <Gv_{i_n}, v_{j_n}>,$$

   where the sum is over all pairings $(i_1, j_1), \ldots, (i_n, j_n)$ of the integers $\{1, 2, \ldots, 2n\}$. Hint: integrate by parts or use part (b).
   (g) Suppose that F and H are bounded measurable functions on $\mathbf{R}^n$ of the form $F(x) = f(\{<v_i, x>\})$ and $H(x) = h(\{<w_j, x>\})$, where the v_i and w_j are a finite collection of vectors in $\mathbf{R}^n$ such that $<Gv_i, w_j> = 0$ for all i, j. Show that

$$\text{(9.26)} \qquad \int_{\mathbf{R}^n} F(x) H(x) d\mu(x) = \int_{\mathbf{R}^n} F(x) d\mu(x) \int_{\mathbf{R}^n} H(x) d\mu(x).$$

   You may use the fact that F may be approximated in $L^2(d\mu)$ by finite linear combinations of functions of the form $\exp i \sum \alpha_k <v_k, x>$ where the α_k are real numbers.

5. (*Gaussian measures and the heat equation.*) Keep the same notation as in problem 4. Let L be the second order differential operator $L = (1/2)G_{ij}\partial_i\partial_j$. Let e^{tL} be the solution operator to the partial differential equation $u_t = Lu$ (using $L^2(dx)$ functions here). That is, if $u(t,x) = (e^{tL}f)(x)$, where f is a C^2 function on $\mathbf{R}^n$ such that all its derivatives up to second order are in $L^2(dx)$, then $u(t,x)$ is the unique function satisfying $u_t = Lu$, $u(0,x) = f(x)$, and $u(t,\cdot)$ is in $L^2(dx)$ for each time t. Facts: e^{tL} extends to a bounded operator on $L^2(dx)$ for $t \geq 0$, $e^{tL}(L^2(dx)) \subset C^\infty(\mathbf{R}^n)$ for $t > 0$, and e^{tL} for $t < 0$ is not defined on all of $L^2(dx)$ and is unbounded. Show that

$$(9.27) \qquad e^{tL}f(x) = \int_{\mathbf{R}^n} f(x + t^{1/2}y) d\mu(y) = (p_t * f)(x),$$

where

$$(9.28) \qquad p_t(x) = Z_t^{-1} e^{-\langle Ax, x\rangle/2t},$$

with Z_t chosen so that $\int_{\mathbf{R}^n} p_t(x) dx = 1$. In particular we conclude that $\int_{\mathbf{R}^n} f(x) d\mu(x) = e^L f(0)$. This formula is useful for computing Gaussian integrals in cases where f is a nice function like a polynomial or an exponential. In these cases $e^{tL}f$ can be computed by using the power series expansion for e^{tL}.

6. (*Normal ordering.*) Normal ordering is a standard operation in QFT. There is a function version of normal ordering and an operator version. In this problem we explore the function version. Fact: loosely speaking, if $t > 0$ and f is a function for which $e^{tL}f$ makes sense, then $e^{tL}f(x)$ is real analytic in x.

Definition. Given a function f on $\mathbf{R}^n$, $:f:(x) = e^{-L}f(x)$ is called the normal ordered version of f.

From the fact above, for this to make sense f must at least be analytic. More precisely f must be of the form $f = e^L g$ for some function g.

(a) Show that $:f:$ is defined if f is a polynomial in x, or if $f(x) = e^{\langle p,x\rangle}$. Also find $:e^{\langle p,x\rangle}:$.
(b) Show that $:\langle v_1, x\rangle \cdots \langle v_m, x\rangle: = \partial_{\alpha_1} \cdots \partial_{\alpha_m} :e^{\alpha_i \langle v_i, x\rangle}:$ at $\alpha = 0$ in $\mathbf{R}^m$.
(c) Show that $\partial_v :f: = :\partial_v f:$.
(d) Let M_f denote the multiplication operator $M_f g = fg$. Show that $[e^{tL}, M_{\langle v, \cdot\rangle}] = te^{tL}\partial_{Gv}$ by showing that both sides satisfy the same ordinary differential equation with the same initial condition.
(e) Use parts (c) and (d) to show that $:\langle v, \cdot\rangle f: = \langle v, \cdot\rangle :f: - \partial_{Gv} :f:$. A nicer way to write this formula is $:\langle v, \cdot\rangle f: = \partial^\dagger_{Gv} :f:$. As a corollary of this formula we see that $:\prod_i \langle v_i, \cdot\rangle: = (\prod_i \partial^\dagger_{Gv})1$.
(f) Let Q_n be the orthogonal projection onto the subspace of $L^2(d\mu)$ orthogonal to the polynomials in x of degree n or less. Show that for $(n+1)$-homogeneous polynomials of the form $f = \prod_{i=1}^{n+1} \langle v_i, \cdot\rangle$ we have $:f: = f - Q_n f$. This is sometimes how normal ordering is defined. Hint: use the last formula of part (e).

(g) Find a graphical technique for computing integrals of the form $\int_{\mathbf{R}^n} :f_1: \cdots :f_n: d\mu$ where each of the f_i are homogeneous polynomials.

(h) Use part (g) to show that

$$(9.29) \quad \int_{\mathbf{R}^n} :\prod_{i=1}^{k} <v_i,\cdot>::\prod_{i=1}^{l} <w_i,\cdot>: d\mu = \delta_{kl} \sum \prod_{i=1}^{k} <Gv_i, w_{\sigma(i)}>,$$

where the sum is over all permutations σ of $\{1, 2, \ldots, k\}$.

7. In this problem we turn to the operator version of normal ordering.

Notation. Set $a(v) = \partial_{Gv}$ and $a^\dagger(v) = \partial^\dagger_{Gv}$ for each v in $\mathbf{R}^n$. The $a(v)$ are called annihilation operators and the $a^\dagger(v)$ are creation operators. Also set $\phi(v) = a(v) + a^\dagger(v) = M_{<v,\cdot>}$, the so-called field operators.

(a) Show that $[a(v), a^\dagger(w)] = <Gv, w> 1$.

Definition. Let O be an operator expressed as a linear combination of products of annihilation and creation operators. The normal ordered version of O, denoted $:O:$, is the operator formed by reordering each term of O so that all creation operators stand to the left of all annihilation operators.

Example. If $O = a(v)a^\dagger(w)$ then $:O: = a^\dagger(w)a(v)$.

Remark. The notation is misleading. The normal ordering operation is not a function on operators but on the formal expression of an operator in terms of creation and annihilation operators. For example by part (a) we can write O of the above example as $O = a^\dagger(w)a(v) + <Gv, w> 1$ which is already in normal ordered form. So unless $<Gv, w> = 0$ we obtain distinct results by normal ordering different expressions of the same operator.

(b) Suppose that A and B are bounded operators on a Hilbert space such that the operator $C = [A, B]$ commutes with both A and B. Show that $e^{A+B} = e^A e^B e^{-[A,B]/2}$. Hint: let $\psi(t) = e^{-tA} e^{t(A+B)}$ and show that $\dot{\psi}(t) = e^{-tA} B e^{tA} \psi(t)$. Then show that $e^{-tA} B e^{tA} = B - tC$ by differentiation, and solve for ψ.

(c) Suppose that $f(x)$ is an analytic "nice" function on $\mathbf{R}^n$ and that $f(x) = \sum f_\alpha \prod <v_i, x>^{\alpha_i}$, where $v_1, \ldots, v_n$ is a basis for $\mathbf{R}^n$, f_α are complex numbers, and the sum is over all multi-indices $\alpha = (\alpha_1, \ldots, \alpha_n)$ with each α_i a nonnegative integer. Notice that the multiplication operator M_f may be written as $M_f = \sum f_\alpha \prod \phi(v_i)^{\alpha_i}$. Show that $:M_f: = M_{:f:}$ where $:f:$ was defined in the previous problem. Hint: first take $f(x) = e^{<v,x>}$ so that $M_f = e^{\phi(v)}$. Show that $:e^{\phi(v)}: = e^{-<Gv,v>/2} e^{\phi(v)}$ using part (b). Disregard the fact that $a(v)$ and $a^\dagger(v)$ are unbounded operators.

8. (*Relations between $L^2(d\mu)$ and "Fock space" on $\mathbf{R}^n$*.) For notational simplicity we now take $A = I$, so that $G = I$ also. Let $S(\mathbf{R}^n)$ denote the complexified symmetric tensor algebra on $\mathbf{R}^n$. For $v_1, \ldots, v_k$ and $w_1, \ldots, w_l$ in $\mathbf{R}^n$ set $<v_1 \cdots v_k, w_1 \cdots w_l> = \delta_{kl} \sum \prod <v_i, w_{\sigma(i)}>$, where the sum is over all permutations of $\{1, 2, \cdots, k\}$. Extend this form to $S(\mathbf{R}^n)$ by sesquilinearity to get an

inner product. The symmetric Fock space on $\mathbf{R}^n$ is the completion of $S(\mathbf{R}^n)$ with respect to this inner product, and is denoted $F(\mathbf{R}^n)$.
  (a) Let $U: F(\mathbf{R}^n) \to L^2(d\mu)$ be given by $U(v_1 \cdots v_k) =: \prod <v_i, \cdot>$ extended by linearity and continuity to $F(\mathbf{R}^n)$. Show that U is a unitary isomorphism of Hilbert spaces.
  (b) Define for v in $\mathbf{R}^n$, $e^v = \sum_{k=0}^{\infty} v^k/k!$ in $F(\mathbf{R}^n)$. Show that $U(e^v) =: e^{<v,\cdot>}$. This formula could also be used to define U.
  (c) Show that $U^{-1}a^{\dagger}(v_0)U(v_1 \cdots v_k) = v_0 v_1 \cdots v_k$, and that

$$(9.30) \quad U^{-1}a(v_0)U(w_0 w_1 \cdots w_k) = \sum_{i=0}^{k} <v_0, w_i> w_0 \cdots \hat{w}_i \cdots w_k,$$

where the caret means omission of the vector w_i. Compare Equations (8.27, 8.28).

9. (*Gaussian measure on infinite dimensional Hilbert space.*) Let H be the Hilbert space of square summable infinite real sequences, $x = (x_1, x_2, \dots)$. Let $<x, y> = \sum_i x_i y_i$ be the inner product on H. Also let V denote the vector space of all real sequences. We say that x in V has finite support if $x_i = 0$ for all sufficiently large i.

Theorem. *There exists a unique measure μ on V such that $\hat{\mu}(p) \equiv \int_V e^{i<p,x>} d\mu(x) = e^{-<p,p>/2}$ for all sequences p of finite support. Informally this is the measure $d\mu(x) = Z^{-1} e^{-<x,x>/2} \prod_{i=1}^{\infty} dx_i$.*

  (a) Suppose that $F(x) = f(x_1, \dots, x_n)$ is a function on V depending on only a finite number of variables. Show that $\int_V F(x) d\mu(x) = \int_{\mathbf{R}^n} f(y) d\mu_n(y)$, where μ_n is the Gaussian measure on $\mathbf{R}^n$ with covariance 1.
  (b) Let $\alpha \in V$ be a sequence of positive terms. Let $H_\alpha = \{x \in V: \sum \alpha_i^2 x_i^2 < \infty\}$. Prove that $\mu(H_\alpha) = 1$ if $\sum \alpha_i^2 < \infty$ and $\mu(H_\alpha) = 0$ otherwise. Outline:
    (i) Notice that $1_{H_\alpha}(x) = \lim_{\epsilon \to 0} \exp{-\epsilon^2 \sum_{i=1}^{\infty} \alpha_i^2 x_i^2}$, so that

$$(9.31) \quad \mu(H_\alpha) = \lim_{\epsilon \to 0} \int_V \exp(-\epsilon^2 \sum_{i=1}^{\infty} \alpha_i^2 x_i^2) d\mu(x)$$

  by the dominated convergence theorem.
    (ii) Again by the dominated convergence theorem,

$$(9.32) \quad \int_V \exp(-\epsilon^2 \sum_{i=1}^{\infty} \alpha_i^2 x_i^2) d\mu(x) = \lim_{n \to \infty} \int_V \exp(-\epsilon^2 \sum_{i=1}^{n} \alpha_i^2 x_i^2) d\mu(x).$$

    (iii) Compute this last integral explicitly.

Remark. From part (b) above we see that $\mu(H) = 0$! This can also be seen from the strong law of large numbers, which implies that $\lim_{n \to \infty} \sum_{i=1}^{n} x_i^2/n = 1$ for almost every x in V with respect to μ.

(c) For v in H let $v^n = (v_1, v_2, \ldots, v_n, 0, 0, \ldots)$. Show that $\lim_{n \to \infty} <v^n, \cdot>$ exists in $L^2(d\mu)$. We abuse notation and denote the limit as $<v, \cdot>$ which is an $L^2(d\mu)$ almost everywhere defined function on V.

(d) Show that for v, w in H, $\int_V <v, x><w, x> d\mu(x) = <v, w>$ and $\int_V e^{i<v,x>} d\mu(x) = e^{-<v,v>/2}$.

Remark. Because of the above result, even though $\mu(H) = 0$ the function $<v, \cdot>$ is still a well-defined element of $L^2(d\mu)$ as an almost everywhere defined function on V. Furthermore the usual Gaussian integral formulas still hold. Because of this most of what was done in Exercises $4 - -8$ above is valid for any Hilbert space H provided $<v, \cdot>$ is interpreted properly.

10. In this problem we study Wiener measure which is described informally by

$$(9.33) \qquad d\mu(x) = Z^{-1} \exp\left[-\frac{1}{2} \int_0^1 x'(s)^2 ds\right] \prod_{0 \leq s \leq 1} dx(s),$$

where the measure μ is on continuous functions on $[0, 1]$ such that $x(0) = 0$, and Z is a normalization constant so that μ has total weight unity.

(a) Set $<f, g> = \int_0^1 f(s)g(s) ds$ and $|f|^2 = <f, f>$ for f and g functions on $[0, 1]$. Do an informal computation by completing the square to show that

$$(9.34) \qquad \int e^{<f,x'>} d\mu(x) = e^{-|f|^2/2}.$$

(b) Let h be a continuous function and set $f(s) = -\int_s^1 h(u) du$, so that $f' = h$ and $f(1) = 0$. Use this f in part (a) and do an integration by parts to conclude that

$$(9.35) \qquad \int e^{<h,x>} d\mu(x) = e^{<Gh,h>/2},$$

where $Gh(s) \equiv \int_0^1 \min(s, u) h(u) du$.

Remark. The function $\min(s, u)$ is the Green's function for the differential operator $A = -d^2/ds^2$ with a Dirichlet boundary condition at 0 and a Neumann boundary condition at 1. The functions $u_n(s) = C_n \sin(n\pi + \pi/2)s$ for $n \geq 0$ form an orthonormal basis of eigenfunctions of A with eigenvalues $\lambda_n = (n\pi + \pi/2)^2$. From this information it follows from the previous problem that $\mu(L^2(ds)) = 1$ but $\mu(L_1^2(ds)) = 0$, where $L_1^2(ds)$ denotes the functions on $[0, 1]$ with a derivative in L^2. N. Wiener proved the following stronger result.

Theorem. *There is a unique probability measure μ on $V = \{x \in C([0, 1]): x(0) = 0\}$ such that*

$$(9.36) \qquad \int e^{<h,x>} d\mu(x) = e^{-<Gh,h>/2}$$

for all continuous functions h on $[0,1]$.

(The continuous functions $x(s)$ are often called Brownian or Wiener paths.) Since this is the prototypical Euclidean path integral, we will show more directly that the differentiable functions have measure zero. This will also serve to introduce you to stochastic calculus.

(c) Set $h(s) = \sum_{i=1}^{n} h_i \delta(s - s_i)$ with $s_i \in [0,1]$ and h_i real. Plug this h into Eq. (9.36) above to conclude that $[x(s_1), \ldots, x(s_n)]$ is distributed as the Gaussian measure on $\mathbf{R}^n$ with covariance $G_{ij} = \min(s_i, s_j)$.

(d) Let $P = \{0 = s_0 < s_1 < \cdots < s_n = s\}$ be a partition of the interval $[0, s]$. Define the function $V_P(x) = \sum_{i=1}^{n} [x(s_i) - x(s_{i-1})]^2$ for x in V. Show that $V_P(x) \to s$ in $L^2(d\mu)$ as the mesh of P goes to zero. This says that the quadratic variation of a typical path up to s is s. Show that this cannot happen if the path is C^1.

Remark. Part (d) can be written suggestively as $dx(s)^2 = ds$, whereas in ordinary calculus one has $dx(s)^2 = 0$. As a result the calculus formula $df(x(s)) = f'(x(s))dx(s)$ must now be replaced by $df(x(s)) = f'(x(s))dx(s) + (1/2)f''(x(s))ds$ for Brownian paths. This is called Ito's formula and is one of the basic ingredients of stochastic calculus. Of course the formula must be interpreted properly.

Miscellaneous physical problems

11. (*Special Relativity.*)
    (a) Show that the linear transformation $x' = \Lambda x$, with

$$(9.37) \qquad \Lambda = \begin{pmatrix} \gamma & -\gamma\beta & 0 & 0 \\ -\gamma\beta & \gamma & 0 & 0 \\ 0 & 0 & 1 & 0 \\ 0 & 0 & 0 & 1 \end{pmatrix},$$

with $|\beta| < 1$, $\gamma = (1 - \beta^2)^{-1/2}$ is a proper Lorentz transformation, and that its inverse is obtained by replacing β by $-\beta$. Show that the spatial (x', y', z') coordinate axes are parallel to the (x, y, z) axes, and that the origin of the primed coordinate system is observed in the unprimed system to move along the positive x axis at velocity β. This Lorentz transformation is called a boost along the positive x axis at velocity β.

(b) (*Time dilation.*) In the primed coordinate system, two events are observed to occur at the same location but separated by the time interval T. Show that in the unprimed system they are separated by the time interval γT.

(c) (*Relativity of simultaneity.*) In the primed coordinate system, two events are observed to occur simultaneously at points on the x' axis a distance L apart. Show that in the unprimed system these events are separated by the time interval $\gamma \beta L$. Which one occurs first? Conversely, suppose that in the unprimed system two events occur at points on the x axis, separated by a distance L and a time interval T with $T^2 - L^2 < 0$. For such "spacelike separated" events, show that β can be chosen to make the events simultaneous in the primed frame.

(d) (*Length contraction.*) A ruler is at rest in the primed system, extending along the x' axis from the origin to $x' = L$. To measure its length, an observer in the unprimed system will note the location of its endpoints at some fixed time (her own t coordinate!) and measure the distance between them. Show that she will obtain L/γ.

12. (*The Lorentz group and $SL(2,\mathbf{C})$.*)
As in Eq. (3.1), we associate to each vector x^μ in $\mathbf{R}^4$ the 2×2 Hermitian matrix

$$(9.38) \qquad X = \begin{pmatrix} x^0 + x^3 & x^1 - ix^2 \\ x^1 + ix^2 & x^0 - x^3 \end{pmatrix}.$$

Now define a map from $SL(2,\mathbf{C})$ to the proper Lorentz group by sending $A \in SL(2,\mathbf{C})$ to the Lorentz transformation $X' = AXA^\dagger$. Show that this is in fact a surjective homomorphism with kernel $\{\pm 1\}$.

13. (*Heisenberg's uncertainty principle.*)
Let $|\psi>$ be any vector in $L^2(\mathbf{R})$, and x and p the usual quantum-mechanical operators satisfying $[x,p] = i$. Then the Hermitian operators $\tilde{x} \equiv x - <\psi|x|\psi>$ and $\tilde{p} \equiv p - <\psi|p|\psi>$ measure the deviations of x and p from their mean values in the state $|\psi>$. Note that $[\tilde{x},\tilde{p}] = i$. Now define $(\Delta x)^2 \equiv <\psi|\tilde{x}^2|\psi>$ and $(\Delta p)^2 \equiv <\psi|\tilde{p}^2|\psi>$, which measure the mean square deviations of x and p from their average values. Use Schwarz' inequality to prove that

$$(9.39) \qquad (\Delta x)^2 (\Delta p)^2 \geq |<\psi|\tilde{x}\tilde{p}|\psi>|^2.$$

By writing $\tilde{x}\tilde{p}$ as the sum of Hermitian and skew Hermitian parts, deduce the uncertainty principle,

$$(9.40) \qquad (\Delta x)^2 (\Delta p)^2 \geq 1/4,$$

or, with physical units restored,

$$(9.41) \qquad (\Delta x)^2 (\Delta p)^2 \geq \hbar^2/4.$$

Show that the ground state of the simple harmonic oscillator is actually a minimum uncertainty state, with equality in this formula.

14. (*Harmonic oscillator Green's function.*) Use our solution of the simple harmonic oscillator in terms of creation and annihilation operators to compute the Green's function

$$(9.42) \qquad G(t-t') = im <0|T[x(t)x(t')]|0>,$$

thus verifying that analytic continuation from the Euclidean space path integral does select the correct Green's function. Also show directly from the Heisenberg equation of motion for $x(t)$ that the Green's function must satisfy

$$(9.43) \qquad (\frac{d^2}{dt^2} + \omega^2)G(t-t') = \delta(t-t').$$

15. (*Angular momentum and SO(3).*) The Lagrangian $L = \frac{1}{2}m|\dot{\mathbf{x}}|^2 - V(|\mathbf{x}|)$ describes a particle moving in $\mathbf{R}^3$ in a spherically symmetric, or central, potential — one which depends only on the particle's distance from the origin. It is invariant under the usual action of the rotation group $SO(3)$ by 3×3 orthogonal matrices multiplying $\mathbf{x}$.

    (a) Show that the conserved quantities associated by Noether's theorem to the generators of rotations about the x, y, and z axes are just the components of the angular momentum vector $\mathbf{L} = \mathbf{x} \times \mathbf{p}$:

$$(9.44) \qquad L_1 = L_x = yp_z - zp_y,$$
$$(9.45) \qquad L_2 = L_y = zp_x - xp_z,$$
$$(9.46) \qquad L_3 = L_z = xp_y - yp_x,$$

    where $\mathbf{p} = m\dot{\mathbf{x}}$.

    (b) In quantum mechanics, where $[x_j, p_k] = i\delta_{jk}$, verify that $[L_j, x_k] = i\epsilon_{jkl}x_l$, $[L_j, p_k] = i\epsilon_{jkl}p_l$, and $[L_j, L_k] = i\epsilon_{jkl}L_l$, where ϵ_{jkl} is ± 1 according as jkl is an even or odd permutation of 123, and is zero if j, k, l are not all distinct. This type of commutation relation with L_j is characteristic of any triple of operators which transform like the components of a vector under rotations. Note that $v_k \to v_k + (i\theta)i\epsilon_{jkl}v_l$ is precisely the change in the kth component of a vector under an infinitesimal rotation by angle θ about the x_j axis.

    (c) Verify that $\mathbf{L}$ is indeed conserved in quantum mechanics by checking that $[H, L_j] = 0$, where the Hamiltonian is $H = |\mathbf{p}|^2/2m + V(|\mathbf{x}|)$. You will find it useful that $[f(\mathbf{x}), p_j] = i\partial_j f(\mathbf{x})$.

    (d) $SO(3)$ also acts on $L^2(\mathbf{R}^3)$ by $Af = f \circ A^{-1}$, where $A \in SO(3), f \in L^2(\mathbf{R}^3)$. Show that the L_j are the generators of this representation by showing that, e.g., $1 - i\theta L_z$ acts as an infinitesimal rotation by angle θ about the z axis.

16. (*The hydrogen atom.*)

    Warning: this problem involves a great deal of tedious calculation! You don't have to do it all!

    The hydrogen atom consists of an electron of mass m and charge e moving in the central electrostatic field of a proton, which we assume to be fixed at the origin. The Hamiltonian for the electron is

$$(9.47) \qquad H = \frac{|\mathbf{p}|^2}{2m} - \frac{e^2}{|\mathbf{x}|}.$$

    H acting in $L^2(\mathbf{R}^3)$ has a discrete spectrum of negative eigenvalues giving the atomic energy levels which we will compute. (There is also a continuous spectrum of positive eigenvalues which describe a free electron coming in from infinity with kinetic energy equal to the eigenvalue, scattering off the proton's electric field, and moving out to infinity again. These will not concern us.)

(a) H has a large symmetry algebra. In addition to the conserved angular momentum, $[H, \mathbf{L}] = 0$, there is also the *Runge-Lenz vector* $\mathbf{R}$, the Hermitian operator

$$\mathbf{R} = \frac{1}{2m}(\mathbf{p} \times \mathbf{L} - \mathbf{L} \times \mathbf{p}) - \frac{e^2}{|\mathbf{x}|}\mathbf{x}, \tag{9.48}$$

for example,

$$R_z = \frac{1}{m}(zp_x^2 + zp_y^2 - xp_xp_z - yp_yp_z + ip_z) - e^2 z(x^2 + y^2 + z^2)^{-1/2}. \tag{9.49}$$

Show that $[H, \mathbf{R}] = 0$. In classical mechanics, the electron would follow an elliptical orbit with the origin at one focus, and this vector points from the origin to the nearer vertex of the ellipse. Its length is proportional to the eccentricity of the orbit.

(b) Show further that

$$\mathbf{R} \cdot \mathbf{L} = \mathbf{L} \cdot \mathbf{R} = 0, \tag{9.50}$$

$$[L_j, R_k] = i\epsilon_{jkl}R_l, \tag{9.51}$$

$$|\mathbf{R}|^2 = e^4 + \frac{2}{m}H(|\mathbf{L}|^2 + 1), \tag{9.52}$$

$$[R_j, R_k] = -\frac{2i}{m}H\epsilon_{jkl}L_l. \tag{9.53}$$

(c) Let $D \subset L^2(\mathbf{R}^3)$ be the span of the eigenvectors of H having negative eigenvalues. We define a Hermitian operator $\mathbf{K}$ with domain D by

$$\mathbf{K} = (-\frac{m}{2H})^{1/2}\mathbf{R}. \tag{9.54}$$

Further defining $\mathbf{M} = (\mathbf{L} + \mathbf{K})/2$, $\mathbf{N} = (\mathbf{L} - \mathbf{K})/2$, show that

$$[M_j, M_k] = i\epsilon_{jkl}M_l, \tag{9.55}$$

$$[N_j, N_k] = i\epsilon_{jkl}N_l, \tag{9.56}$$

$$[M_j, N_k] = 0, \tag{9.57}$$

$$(9.58) \qquad |\mathbf{M}|^2 = |\mathbf{N}|^2 = (|\mathbf{L}|^2 + |\mathbf{K}|^2)/4,$$

$$(9.59) \qquad H = -\frac{me^4}{2(4|\mathbf{M}|^2 + 1)}.$$

(d) According to part (c), the operators $\mathbf{M}$ and $\mathbf{N}$ acting in D generate a (unitary) representation of the Lie algebra $su(2) \times su(2)$. Suppose we decompose D as the direct sum of irreducible representations of this algebra. The Casimir operators $|\mathbf{M}|^2$ and $|\mathbf{N}|^2$ commute with the generators and are therefore constant on each irreducible representation by Schur's lemma. Therefore, again by (c), H is also constant on each irrep, so these are the eigenspaces of H. Only those irreps having $|\mathbf{M}|^2 = |\mathbf{N}|^2$ occur, and the value of each of these Casimir operators is $j(j+1)$ in the $2j+1$-dimensional irrep of $su(2)$ having highest weight $j = 0, \frac{1}{2}, 1, \frac{3}{2}, \ldots$. Defining the *principal quantum number* $n = 2j + 1 = 1, 2, \ldots$, we deduce that the irrep with highest weights (j, j) is an n^2-dimensional eigenspace of H with eigenvalue $-me^4/2n^2$. In fact, each possible value of n occurs in the decomposition of D, exactly once. To prove this one must construct operators like the $a^\dagger$ and a of the harmonic oscillator which raise and lower the eigenvalue of H. This is even more tedious than what you've already been through. Including these new operators enlarges the $su(2) \times su(2)$ symmetry of the hydrogen atom to $so(4, 2)$. See, if you like, [19]. (With physical units restored, the energies of the stationary states of the hydrogen atom are $-me^4/2\hbar^2 n^2 = -13.6/n^2$ electron volts. The ground state energy of -13.6 eV can be measured directly as the negative of the energy required to ionize the atom, and of course the energy differences between states can be measured with extreme precision from the frequencies of photons emitted during a transition between the states. This precision is more than sufficient to measure the corrections to the energies you've derived, arising from such effects as the motion of the proton, the spin of the electron, and the non-instantaneous transmission of the electric force between them, but that's another story.)

17. (*Feynman diagrams.*) Draw the Feynman diagrams contributing to the four point function or vertex function $< 0|T[\phi(x_1)\phi(x_2)\phi(x_3)\phi(x_4)]|0 >$ in the $\lambda \phi^4$ scalar field theory up to order λ^2. Write down the corresponding terms of the perturbation series as well as their Fourier transforms. How divergent are the integrals?

BIBLIOGRAPHY

1. V. Arnold, *Mathematical Methods of Classical Mechanics*, second edition, Springer-Verlag 1989.
2. H. Goldstein, *Classical Mechanics*, second edition, Addison-Wesley 1980.
3. A.Z. Capri, *Nonrelativistic Quantum Mechanics*, Benjamin-Cummings 1985.
4. R. Shankar, *Principles of Quantum Mechanics*, Plenum Press 1980.
5. R.P. Feynman and A.R. Hibbs, *Quantum Mechanics and Path Integrals*, McGraw-Hill 1965.
6. P. Ramond, *Field Theory: A Modern Primer*, second edition, Addison-Wesley 1989.
7. L.H. Ryder, *Quantum Field Theory*, Cambridge University Press 1985.
8. R.J. Rivers, *Path Integral Methods in Quantum Field Theory*, Cambridge University Press 1987.
9. J. Collins, *Renormalization*, Cambridge University Press 1984.
10. J. Glimm and R. Jaffe, *Quantum Physics*, second edition, Springer-Verlag 1987.
11. N.N. Bogolubov, A.A. Logunov, and R.T. Todorov, *Introduction to Axiomatic Quantum Field Theory*, Reading: Benjamin 1975.
12. I.M. Singer, "Some remarks on the Gribov ambiguity," Commun. Math. Phys. **60**, 7–12 (1978).
13. M. Henneaux and C. Teitelboim, *Quantization of Gauge Systems*, Princeton University Press 1992.
14. D. Nemeschansky, C. Preitschopf, and M. Weinstein, "A BRST primer," Ann. Phys. **183:2**, 226–268 (1988).
15. B. Kostant and S. Sternberg, "Symplectic reduction, BRS cohomology, and infinite-dimensional Clifford algebras," Ann. Phys. **176:1**, 49–113 (1987).
16. E. Farhi and S. Gutmann, "The functional integral constructed directly from the Hamiltonian", MIT preprint CTP#1943, 1991, submitted to Nucl. Phys. B.
17. W.E. Caswell and A.D. Kennedy, "Asymptotic behavior of Feynman integrals: Convergent integrals", Phys. Rev. **D28**, 3073–3089 (1983).
18. J. Polchinski, "Renormalization and effective Lagrangians", Nucl. Phys. **B231**, 269–295 (1984).
19. B.G. Wybourne, *Classical Groups for Physicists*, Wiley, New York 1974.

Lectures on Quantum Mechanics and the Index Theorem

Orlando Alvarez

Lectures on Quantum Mechanics and the Index Theorem

Orlando Alvarez

Abstract. These lectures are intended for an audience of mathematicians with a background in undergraduate quantum mechanics. They were given at the Regional Geometry Institute in Park City, Utah during the summer of 1991. They are a follow up to the introductory quantum mechanics lectures given by J. Rabin. These lectures cover a variety of topics which appear in current applications of quantum field theory to mathematics. The path integral derivation of the Atiyah-Singer index formula for the Dirac operator is used to illustrate the ideas.

Introduction

The past fifteen years have seen a growing interaction between mathematics and theoretical physics. The methods we physicists use seem mysterious and *ad-hoc* to many mathematicians. The purpose of these lectures is to apply what you learned in the introductory quantum mechanics course by Jeff Rabin to a geometrically interesting question. I wanted to do a non-trivial path integral example and also teach you a bit about supersymmetry. The Atiyah-Singer formula for the index of an elliptic operator is the prime candidate that achieves both goals and is also of significant mathematical importance in itself. Many computations are done in detail to familiarize you with the "practical" everyday computations done in physics. Rigor is not the order of the day and I urge you to try to fill in the details.

By now, much of the material covered in these lectures is standard. For this reason, I will not give references in the text but I do provide brief references to the original sources.

---

[1]This work was supported in part by the National Science Foundation under grant PHY90-21139, and by the Director, Office of Energy Research, Office of High Energy and Nuclear Physics, Division of High Energy Physics of the U.S. Department of Energy under Contract DE-AC03-76SF00098. Department of Physics, University of California at Berkeley, Berkeley, CA 94720 and Theoretical Physics Group, Lawrence Berkeley Laboratory, University of California, Berkeley, CA 94720.

Current address: Department of Physics, University of Miami, P.O. Box 248046, Coral Gables, FL 33124.

I would like to thank Herb Clemens, Dan Freed, Karen Uhlenbeck and the RGI staff for their efforts in organizing the first Park City RGI summer school. It was a very worthwhile experience and truly enjoyable. I would also like to thank Rob Harrington who kindly proofread the notes and assisted me at Park City. My sincerest thanks to Luanne Neumann for taking my mostly computational hand written lectures notes and TEXing a first draft. Finally, I would like to thank my friends Dan Freed, Is Singer and Paul Windey for teaching me much of what I know about index theory.

This work was supported in part by the National Science Foundation under grant PHY90-21139, and by the Director, Office of Energy Research, Office of High Energy and Nuclear Physics, Division of High Energy Physics of the U.S. Department of Energy under Contract DE-AC03-76SF00098.

References

The Atiyah-Singer index theorem is thoroughly discussed in

- M. Atiyah and I. Singer. Index of elliptic operators, I. *Ann. Math.*, **87**:484, 1968.
- M. Atiyah and I. Singer. Index of elliptic operators, III. *Ann. Math.*, **87**:546, 1968.

The quantum mechanical path integral approach to the index theorem is discussed in

- L. Alvarez-Gaumé. Supersymmetry and the Atiyah-Singer index theorem. *Comm. Math. Phys.*, **90**:161, 1983.
- D. Friedan and P. Windey. Supersymmetric derivation of the Atiyah-Singer index theorem and the chiral anomaly. *Nucl. Phys.*, **B235**:395, 1984.
- E. Witten. unpublished, 1983, see M. Atiyah's exposition in the *Astérisque* volume in honor of H. Weyl.

LECTURE 1
Canonical Quantization of Bosonic Systems

Lagrangian and Hamiltonian mechanics

The prototypical bosonic system is the motion of a point particle moving on a d-dimensional manifold M. The trajectory of the particle is described by a curve $q(t)$ on M. The dynamics of the system is specified by a function of the position and the velocity[2], $L(q(t), \dot q(t))$, called the *Lagrangian*. Nowadays we say that L is a function on the tangent bundle TM of M. The *classical action* $I[q] = \int dt\, L(q(t), \dot q(t))$ is the integral of the Lagrangian along the trajectory[3]. The *Action Principle* states that the equations of motion are the Euler-Lagrange equations

$$(1.1) \qquad \frac{\partial L}{\partial q^j} - \frac{d}{dt}\left(\frac{\partial L}{\partial \dot q^j}\right) = 0$$

which determine the extrema of the classical action. In the above $(q^1, q^2, \ldots, q^d)$ are local coordinates on M. The *canonically conjugate momenta* p is defined by

$$(1.2) \qquad p_j = \frac{\partial L}{\partial \dot q^j}.$$

The *Hamiltonian* is a function on the cotangent bundle T^*M of M and it is defined to be the Legendre transform of the Lagrangian:

$$(1.3) \qquad H(q, p) = p_j \dot q^j - L(q, \dot q).$$

We follow the Einstein summation convention where one implicitly sums over repeated indices.

---

[2] Time derivatives are often denoted by an overdot.
[3] There is an historical convention in physics of using square brackets rather than parentheses to specify the arguments of a function on an infinite dimensional space.

Example 1.1. The dynamics of the the motion of a particle on $\mathbb{R}^d$ is specified by a function on M called the potential $V(q)$. The associated dynamical quantities are

$$L(q,\dot{q}) = \frac{1}{2}\dot{q}^2 - V(q), \tag{1.4}$$

$$p_j = \dot{q}^j, \tag{1.5}$$

$$H(q,p) = \frac{1}{2}p^2 + V(q). \tag{1.6}$$

The Hamiltonian version of the action principle states that the extrema of

$$I[q,p] = \int dt\, \left(p_j \dot{q}^j - H(q,p)\right) \tag{1.7}$$

are Hamilton's equations of motion

$$\dot{q} = \frac{\partial H}{\partial p}, \tag{1.8}$$

$$\dot{p} = -\frac{\partial H}{\partial q}. \tag{1.9}$$

The Hamiltonian action principle leads to an elementary observation about the associated symplectic structure. The Poincaré form $\theta = p_j dq^j$ may be determined by examining the first term of equation (1.7) and observing that

$$\int dt\, p\dot{q} = \int_\gamma \theta \tag{1.10}$$

where γ is the trajectory. The associated symplectic form ω is simply given by

$$\omega = d\theta = dp_j \wedge dq^j. \tag{1.11}$$

If an action is already in Hamiltonian form then we can immediately identify the symplectic structure. The symplectic manifold associated with the motion of a particle on a manifold M is the cotangent bundle T^*M.

Given a symplectic manifold with local coordinates $(z^1, z^2, \cdots, z^{2n})$ and Poincaré form $\theta = A_\alpha(z) dz^\alpha$, the symplectic form is

$$\omega = \frac{1}{2}\omega_{\alpha\beta}(z) dz^\alpha \wedge dz^\beta, \tag{1.12}$$

where

$$\omega_{\alpha\beta}(z) = \partial_\alpha A_\beta - \partial_\beta A_\alpha.$$

Define the inverse matrix $\omega^{\alpha\beta}$ by

$$\omega^{\alpha\beta}\omega_{\beta\gamma} = \delta^\alpha{}_\gamma. \tag{1.13}$$

LECTURE 1. CANONICAL QUANTIZATION OF BOSONIC SYSTEMS

The Poisson bracket is defined by

(1.14) $$\{f, g\}_{PB} = \omega^{\alpha\beta}(z) \frac{\partial f}{\partial z^\alpha} \frac{\partial g}{\partial z^\beta} .$$

Example 1.2. A particle moves on the real line. The associated symplectic manifold $T^*\mathbb{R}$ has coordinates $(z^1, z^2) = (q, p)$.

$$\begin{aligned}
\theta &= p\,dq , \\
\omega &= dp \wedge dq , \\
\omega_{12} &= -1 , \\
\omega_{21} &= +1 , \\
\omega^{12} &= +1 , \\
\omega^{21} &= -1 , \\
\{f, g\}_{PB} &= \omega^{12} \frac{\partial f}{\partial q} \frac{\partial g}{\partial p} + \omega^{21} \frac{\partial f}{\partial p} \frac{\partial g}{\partial q} \\
&= \frac{\partial f}{\partial q} \frac{\partial g}{\partial p} - \frac{\partial f}{\partial p} \frac{\partial g}{\partial q} , \\
\{q, p\}_{PB} &= 1 .
\end{aligned}$$

Naive quantization

In its simplest form, *quantization* is the process of implementing the Poisson bracket algebra on a Hilbert space $\mathcal{H}$. To be more precise, q and p are to be implemented as operators[4] $\hat{q}$ and $\hat{p}$ on $\mathcal{H}$ such that $[\hat{q}^j, \hat{p}_k] = i\delta^j{}_k$. For example, if $M = \mathbb{R}$ then $\mathcal{H} = L^2(\mathbb{R})$ and

(1.15) $$\hat{q} = \text{multiplication by } q ,$$

(1.16) $$\hat{p} = -i\frac{\partial}{\partial q} .$$

Noether's theorem

Let $L(q, \dot{q})$ be the Lagrangian which specifies a motion on M. A Lie group G with Lie algebra $\mathfrak{g}$ acts on M. The action is given by

$$q \xrightarrow{g} q + \epsilon \langle F(q), X \rangle + 0(\epsilon^2) ,$$

where $g = \exp \epsilon X$. The angular brackets denote the pairing of an element of the Lie algebra $\mathfrak{g}$ with an element of the dual space $\mathfrak{g}^*$. Since $\langle F(q), X \rangle$ is the vector field on M induced by $X \in \mathfrak{g}$, it follows that $F(q)$ is a section of $TM \otimes \mathfrak{g}^*$. If the Lagrangian L is invariant under the action of G the following theorem holds.

Theorem 1.3 (Noether's Theorem I). *Let* $Q = \sum_j p_j F^j(q)$, *where* $F(q) = \{F^1(q), \ldots, F^d(q)\}$ *in a local coordinate system. The following are consequences of the G-invariance of the Lagrangian*

---

[4] Given a classical observable B, the associated quantum mechanical operator will often be denoted by $\hat{B}$.

One SHO	Several SHO's
$\hat{H} = \frac{1}{2}\omega(\hat{a}^\dagger \hat{a} + \hat{a}\hat{a}^\dagger)$,	$\hat{H} = \frac{1}{2}\omega \sum_j (\hat{a}_j^\dagger \hat{a}_j + \hat{a}_j \hat{a}_j^\dagger)$,
$\hat{N} = \hat{a}^\dagger \hat{a}$	$\hat{N}_i = a_i^\dagger a_i$,
$[\hat{a}, \hat{a}] = 0$,	$[\hat{a}_i, \hat{a}_j] = 0$,
$[\hat{a}^\dagger, \hat{a}^\dagger] = 0$,	$[\hat{a}_i^\dagger, \hat{a}_j^\dagger] = 0$,
$[\hat{a}, \hat{a}^\dagger] = 1$	$[\hat{a}_i, \hat{a}_j^\dagger] = \delta_{ij}$,
$[\hat{N}, \hat{a}] = -\hat{a}$,	$[\hat{N}, \hat{a}_j] = -\hat{a}_j$,
$[\hat{N}, \hat{a}^\dagger] = +\hat{a}^\dagger$,	$[\hat{N}, \hat{a}_j^\dagger] = +\hat{a}_j^\dagger$.

Table 1.1. Facts and conventions for simple harmonic oscillators.

1. Q is a constant of the motion, i.e., $\dfrac{dQ}{dt} = 0$.
2. Q generates transformations on M, i.e., $F(q)^j = \{q^j, Q\}_{PB}$.
3. Let $Q_X = \langle Q, X \rangle$ for $X \in \mathfrak{g}$ then $\{Q_X, Q_Y\}_{PB} = Q_{[X,Y]}$.

Exercise 1.4. Do the following exercises.
  1. Let ξ_X be the vector field on TM induced by $X \in \mathfrak{g}$. Show that $[\xi_X, \xi_Y] = \xi_{[X,Y]}$.
  2. Show that
  $$\left\langle F^i(q), X \right\rangle \left\langle \frac{\partial F^j}{\partial q^i}, Y \right\rangle - \left\langle \frac{\partial F^j}{\partial q^i}, X \right\rangle \left\langle F^i(q), Y \right\rangle = \left\langle F^j(q), [X,Y] \right\rangle .$$
  3. Prove Noether's Theorem I.

Exercise 1.5 (Angular Momentum). Let $x \in \mathbb{R}^d$ where $x = (x_1, \ldots, x_d)$, and consider the Lagrangian $L = \frac{1}{2}\dot{x}^2 = \frac{1}{2}\dot{x}_j \dot{x}_j$ which is invariant under $SO(d)$ transformations on x. Under an infinitesimal $SO(d)$ rotation $R_{ij} = \delta_{ij} + \epsilon \omega_{ij} + O(\epsilon^2)$, where $\omega_{ij} + \omega_{ji} = 0$, show that the associated Noether charge is $Q_{ij} = x_i p_j - x_j p_i$. Explicitly write down these Noether charges for $d = 3$ and verify that they are the components of the angular momentum vector. What are the Noether charges associated with translations?

Theorem 1.6 (Noether's Theorem II). *Assume that L is not quite invariant under $g = e^{\epsilon X}$ but rather $L \to L + \epsilon \dfrac{d}{dt}\langle \mathcal{S}, X \rangle + O(\epsilon^2)$. The quantity $Q = \sum p_j F^j(q) - \mathcal{S}$ is a constant of the motion and is also the generator of the symmetry.*

Exercise 1.7. Prove Noether's Theorem II.

Exercise 1.8 (The Hamiltonian and time translations). Consider a Lagrangian $L(q, \dot{q})$ which does not depend explicitly on time. Show that under the action of time translations $t \to t + \epsilon$ the conserved charge is the Hamiltonian $Q = H = p\dot{q} - L$.

Review of the simple harmonic oscillator

It is useful to collect some facts[5] about the simple harmonic oscillator (SHO) in Table 1.1. The Hamiltonian for the simple harmonic oscillator is

---

[5] See J. Rabin's lectures or any introductory quantum mechanics book.

LECTURE 1. CANONICAL QUANTIZATION OF BOSONIC SYSTEMS

$$\hat{H} = \frac{1}{2}\left(\hat{p}^2 + \omega^2 \hat{x}^2\right).$$

The raising (creation) operator $\hat{a}^\dagger$ and the lowering (annihilation) operator $\hat{a}$ are defined by

$$\hat{a} = \frac{\omega\hat{x} + i\hat{p}}{\sqrt{2}},$$

$$\hat{a}^\dagger = \frac{\omega\hat{x} - i\hat{p}}{\sqrt{2}}.$$

The Hilbert space of a simple harmonic oscillator is the simplest bosonic Fock space. Since the Hamiltonian is positive semidefinite there exists a state $|0\rangle$ which is annihilated by $\hat{a}$. An orthonormal basis for the Fock space is given by the states $|n\rangle = (\hat{a}^\dagger)^n |0\rangle / \sqrt{n!}$.

state	N	E
$|0\rangle$	0	$\left(0 + \frac{1}{2}\right)\omega$
$|1\rangle$	1	$\left(1 + \frac{1}{2}\right)\omega$
$|2\rangle$	2	$\left(2 + \frac{1}{2}\right)\omega$
$\vdots$	$\vdots$	$\vdots$

Such a Fock space has a particle interpretation due to the additive nature of the energy E when one adds an "excitation" by augmenting N. One interprets the ground state[6] $|0\rangle$ as a state with no particles and energy $\frac{1}{2}\omega$. The state $|1\rangle$ is a one particle (quanta) state with energy ω relative to the ground state, the state $|2\rangle$ is a two particle state with energy $\omega + \omega = 2\omega$ relative to the ground state, *etc.*

---

[6]The ground state of a system is also often referred to as the *vacuum*.

LECTURE 2
Canonical Quantization of Fermionic Systems

The basics

In the early days of quantum mechanics it was realized that there were two categories of particles: *bosons* with "integral spin" ($s = 0, 1, 2, 3, \dots$); and *fermions* with "half-integral spin" ($s = 1/2, 3/2, 5/2, \dots$). Fermions satisfy the Pauli exclusion principle which states that there can be at most one fermion in a given quantum mechanical state.

In classical physics, bosons are described by classical field theory. The most famous example being the Maxwell equations for the electromagnetic field. The fundamentals for quantizing classical fields were formulated after the discovery of quantum mechanics by imitating the quantization of particle motion. Surprisingly, the formulation of the quantum equations of motion for spin 1/2 particles due to Pauli was also discovered very early. It took a long time before the classical description for spin 1/2 objects was formulated and even longer before it became universally accepted.

Consider an electron fixed at a point in three space in the presence of a magnetic field **B**. This is a two state system where the z-component of angular momentum can either be $+1/2$ or $-1/2$. Because it is a two state system the Hilbert space is given by $\mathcal{H} = \mathbb{C}^2$. The Hamiltonian is a hermitian operator on $\mathbb{C}^2$. These are easily characterized by the Pauli matrices which represent the Lie algebra of $SU(2)$, the cover of $SO(3)$, on $\mathbb{C}^2$:

$$(2.1) \qquad \sigma_x = \begin{pmatrix} 0 & 1 \\ 1 & 0 \end{pmatrix}, \qquad \sigma_y = \begin{pmatrix} 0 & -i \\ i & 0 \end{pmatrix}, \qquad \sigma_z = \begin{pmatrix} 1 & 0 \\ 0 & -1 \end{pmatrix}.$$

$\{\sigma_x, \sigma_y, \sigma_z\}$ and the identity matrix constitute a basis for hermitian 2×2 matrices. Pauli postulated that the Hamiltonian for the system is given by

$$(2.2) \qquad H = -\mathbf{B} \cdot \sigma = -(B_x \sigma_x + B_y \sigma_y + B_z \sigma_z).$$

In the above, I have set a whole bunch of physically important parameters[7] equal to one. To be more precise, the Hamiltonian is actually given by $H = -\mathbf{B}\cdot\mu$ where the magnetic moment of the electron is given by

$$\mu = g\,\frac{e\hbar}{2mc}\,\frac{\sigma}{2}\,.$$

The "fudge factor" g is called the gyromagnetic ratio of the electron and it is approximately equal to 2. The Pauli equation is simply

$$i\,\frac{\partial \psi}{\partial t} = -(\mathbf{B}\cdot\sigma)\,\psi \quad \text{where} \quad \psi = \begin{pmatrix} \psi_+ \\ \psi_- \end{pmatrix}\,.$$

One of the reasons why we believe in quantum field theory is that the theoretical calculation of the gyromagnetic ratio of the electron agrees with experiment to twelve significant digits. Quantum electrodynamics leads to a perturbative expansion for g in the fine structure constant[8] $\alpha = e^2/(4\pi\hbar c) \approx 1/137$. The theoretical computation of g gives

$$(g-2)/2 = a_1\frac{\alpha}{\pi} + a_2\left(\frac{\alpha}{\pi}\right)^2 + a_3\left(\frac{\alpha}{\pi}\right)^3 + a_4\left(\frac{\alpha}{\pi}\right)^4 + \cdots\,.$$

The zeroth order value of 2 was calculated by Dirac around 1928 and was one of the early triumphs of quantum theory. The order α correction was calculated by Schwinger in 1948 with the result $a_1 = 1/2$ and was one of the early successes of renormalization theory. Subsequent computations have been herculean efforts. The theoretical computation of order α^4 involves evaluating 729 Feynman diagrams. Some of the diagrams are 10 dimensional parametric integrals with the integrand consisting of more than 20,000 rational functions constructed from the integration variables (the size of the FORTRAN source code for such an integrand exceeds 350 KB).

The gyromagnetic ratios for the electron e^- and its anti-particle, the positron e^+, are experimentally determined to be

$$\begin{aligned}(g(e^-)-2)/2 &= 1\,159\,652\,188.4(4.3)\times 10^{-12}\,,\\ (g(e^+)-2)/2 &= 1\,159\,652\,187.9(4.3)\times 10^{-12}\,.\end{aligned}$$

The experimental error is in parentheses and represents the uncertainty "in the last digits". The PCT theorem of quantum field theory requires the gyromagnetic ratios to be the same for particles and anti-particles. The theoretical determination is given by

$$(g(\text{theory})-2)/2 = 1\,159\,652\,140.7(5.3)(4.1)(27.1)\times 10^{-12}\,.$$

The first two errors come from theoretical uncertainties in a_3 and a_4, the third error comes from the uncertainty in the value of α. The best experimental value

---

[7] We use standard high energy physics units where $\hbar = c = 1$.
[8] In this article we use Heaviside-Lorentz units which are at times called rationalized CGS units.

LECTURE 2. CANONICAL QUANTIZATION OF FERMIONIC SYSTEMS   283

One fermion	Several fermions
$\hat{H} = \frac{1}{2}\omega(\hat{a}^\dagger \hat{a} - \hat{a}\hat{a}^\dagger)$,	$\hat{H} = \frac{1}{2}\sum_j \omega_j \left(\hat{a}_j^\dagger \hat{a}_j - \hat{a}_j \hat{a}_j^\dagger\right)$,
$\hat{N} = \hat{a}^\dagger \hat{a}$,	$\hat{N} = \sum_j \hat{a}_j^\dagger \hat{a}_j$,
$\{\hat{a},\hat{a}\} = 0$,	$\{\hat{a}_i, \hat{a}_j\} = 0$,
$\{\hat{a}^\dagger,\hat{a}^\dagger\} = 0$,	$\{\hat{a}_i^\dagger, \hat{a}_j^\dagger\} = 0$,
$\{\hat{a},\hat{a}^\dagger\} = 1$,	$\{\hat{a}_i, \hat{a}_j^\dagger\} = \delta_{ij}$,
$[\hat{N},\hat{a}] = -\hat{a}$,	$[\hat{N},\hat{a}_j] = -\hat{a}_j$,
$[\hat{N},\hat{a}^\dagger] = +\hat{a}^\dagger$,	$[\hat{N},\hat{a}_j^\dagger] = +\hat{a}_j^\dagger$.

Table 2.1. Facts and conventions for fermionic harmonic oscillators.

of α comes from the quantized Hall effect and is approximately given by $\alpha^{-1} = 137.035\,997\,9(32)$. Theory and experiment agree within 1.7 standard deviations. This is why we believe in quantum field theory!

Exercise 2.1 (Properties of the Pauli matrices). Show the following:
1. $\sigma_a \sigma_b = \delta_{ab} + i\epsilon_{abc}\sigma_c$.
2. $[\sigma_a, \sigma_b] = 2i\epsilon_{abc}\sigma_c$.
3. $\exp(i\theta \hat{n} \cdot \vec{\sigma}) = \cos\theta + i\hat{n} \cdot \vec{\sigma} \sin\theta$, where $\hat{n} \cdot \hat{n} = 1$.
4. Verify $\exp(i\theta \hat{n} \cdot \vec{\sigma}) \in SU(2)$.

Exercise 2.2 (The precessing electron). An electron is in a uniform magnetic field **B**. Solve the Pauli equation for arbitrary initial conditions. Assume that initially the electron is in the state

$$\psi(0) = \frac{1}{\sqrt{2}} \begin{pmatrix} 1 \\ 1 \end{pmatrix},$$

and that $\mathbf{B} = B_0 z$. What is $\psi(t)$? If A is any operator, define the expectation value[9] of A at time t by $\langle A \rangle(t) = (\psi(t), A\psi(t))$. Compute $\langle \sigma_x \rangle(t)$ and $\langle \sigma_y \rangle(t)$ and show that the electron precesses. If you put in all the factors you will see that the precession frequency depends on g. This is a key ingredient in the experimental determination of g.

The fermionic oscillator

We now turn to the question of what is the classical system whose canonical quantization is the Pauli equation. Early in the history of quantum mechanics, Jordan and Wigner discovered the canonical anticommutation relations which describe fermionic systems. We first discuss a system of a single fermion. The physically relevant question is on how one constructs a Hilbert space where there are two states, "unoccupied" and "occupied". One considers operators $\hat{a}$ and $\hat{a}^\dagger$ which *anticommute*. In Table 2.1, the left column describes the algebra for a single fermionic oscillator algebra. The right column describes several non-interacting fermionic oscillators. The operator $\hat{a}$ decreases fermion number by one and the operator $\hat{a}^\dagger$ increases fermion number by one. The Hilbert space generated by the creation and

---
[9] See J. Rabin's lectures.

annihilation operators is known as fermionic Fock space and it is simply $\mathbb{C}^2$ in this simple example:

(2.3)
state	N	E
$|0\rangle$	0	$-\frac{1}{2}\omega$
$|1\rangle$	1	$+\frac{1}{2}\omega$

In the case of n oscillators, the fermionic Fock space is of dimension 2^n and is isomorphic to an exterior algebra with n generators.

Quantization of fermionic systems

It is important to ask the question whether there is a classical system whose canonical quantization is the one fermionic SHO system described above. It took many years for an unusual quantization method involving Grassmann variables to be accepted as both a useful and necessary tool in theoretical physics. Before tackling the fermionic problem we return to the quantization of the bosonic SHO in the "first order" or Hamiltonian action formulation. Consider a classical system with variables a and $\bar{a}$, governed by the classical action

$$(2.4) \qquad I = \int dt(i\bar{a}\dot{a} - H),$$

where $H(a,\bar{a}) = \frac{1}{2}\omega(\bar{a}a + a\bar{a})$. The reality of the action requires $\bar{a}$ to be the complex conjugate of a. The Poincaré form is $\theta = i\bar{a}\,da$ and the symplectic form is $\omega = id\bar{a} \wedge da$. The Poisson bracket relations for this system are

$$\{a, i\bar{a}\}_{PB} = 1,$$
$$\{a, a\}_{PB} = 0,$$
$$\{\bar{a}, \bar{a}\}_{PB} = 0.$$

From the previous discussion of canonical quantization one would like to implement the above Poisson bracket algebra in a Hilbert space where one has operators $\hat{a}$ and associated adjoint $\hat{a}^\dagger$ satisfying the commutator algebra $[\hat{a}, i\hat{a}^\dagger] = i$. This is just our old bosonic Fock space associated with the SHO.

To quantize the fermionic SHO we introduce classical Grassmann variables satisfying the following relations:

$$a\bar{a} + \bar{a}a = 0,$$
$$aa + aa = 0,$$
$$\bar{a}\bar{a} + \bar{a}\bar{a} = 0.$$

The action for the fermionic oscillator mimics equation (2.4) for the bosonic oscillator

$$I = \int_{t_1}^{t_2} dt\, i\bar{a}\dot{a} - \frac{1}{2}\omega(\bar{a}a - a\bar{a}), \tag{2.5}$$

$$= \int dt\, i\bar{a}\dot{a} - \omega\bar{a}a. \tag{2.6}$$

The classical equations of motion are taken to the the classical Euler-Lagrange equations associated with the above action with the proviso that one carefully handles the anticommuting nature of the variables. Below is a sample computation to emphasize the importance of the ordering of the variables:

$$\begin{aligned}
\delta I &= \int_{t_1}^{t_2} dt\, i(\delta\bar{a})\dot{a} + i\bar{a}\delta\dot{a} - \omega(\delta\bar{a})a - \omega\bar{a}\delta a \\
&= \int_{t_1}^{t_2} dt\, \delta\bar{a}(i\dot{a} - \omega a) + i\frac{d}{dt}(\bar{a}\delta a) - i\dot{\bar{a}}\delta a - \omega\bar{a}\delta a \\
&= i\bar{a}\delta a\Big|_{t_1}^{t_2} + \int dt\, \delta\bar{a}(i\dot{a} - \omega a) + \int dt(-i\dot{\bar{a}} - \omega\bar{a})\delta a.
\end{aligned}$$

If one assumes δa is fixed at the endpoints then the associated Euler-Lagrange equations are

$$\begin{aligned}
i\dot{a} - \omega a &= 0, \\
-i\dot{\bar{a}} - \omega\bar{a} &= 0.
\end{aligned}$$

Note that the first equation is the complex conjugate of the second equation.

There exists a canonical formulation for Grassmann systems that requires developing a knowledge of Grassmann differential forms. The following exercise provides the necessary background.

Exercise 2.3. (Differential calculus for Grassmann variables)
1. As a warm-up we study some elementary properties of the differential algebra of ordinary polynomials. Let $x_1, \ldots, x_n$ be n commuting variables. Consider the polynomial algebra $P[x_1, \ldots, x_n]$ over a nice field such as $\mathbb{R}$ or $\mathbb{C}$. Note that the polynomial algebra is infinite dimensional. On this algebra we want to define the exterior differential algebra Λ^* with polynomial coefficients. To do this we introduce new objects $dx_1, dx_2, \ldots, dx_n$ and the following multiplication rules:
   C0) $x_i x_j = x_j x_i$,
   C1) $dx_i dx_j = -dx_j dx_i$,
   C2) $x_i(dx_j) = (dx_j)x_i$,
   C3) associativity and distributivity.
   Define an operator d by

$$d = \sum_{j=1}^{n} dx_j \frac{\partial}{\partial x^j},$$

with $\frac{\partial}{\partial x_i}(dx_\ell) = 0$. I have deliberately not introduced wedges for the exterior product since the commutativity rules are explicit. Verify that the above defines a consistent exterior algebra with d satisfying

(a) $d^2 = 0$,

(b) If α has "degree r" and β has "degree s" then

$$\alpha\beta = (-1)^{rs}\beta\alpha,$$
$$d(\alpha\beta) = (d\alpha)\beta + (-1)^r \alpha(d\beta).$$

Note that the differential algebra is a finite dimensional algebra with polynomial coefficients.

2. We now proceed to develop the same ideas for Grassmann variables. Let $\theta_1, ..., \theta_n$ be n anti-commuting variables:

G0) $\theta_i \theta_j = -\theta_j \theta_i$.

Consider the polynomial algebra $P[\theta_1, ..., \theta_n]$ over a nice field such as $\mathbb{R}$ or $\mathbb{C}$.

(a) What is its dimension?

(b) Is there a natural grading on $P[\theta_1, ..., \theta_n]$?

We now wish to define a differential algebra with polynomial coefficients. First we introduce objects $d\theta_1, ..., d\theta_n$ which satisfy the following commutation rules

G1) $d\theta_i d\theta_j = d\theta_j d\theta_i$,

G2) $\theta_i(d\theta_j) = (d\theta_j)\theta_i$,

G3) associativity and distributivity.

3. Verify that this defines a consistent infinite dimensional algebra with polynomial coefficients. We will now turn this into a differential algebra. Firstly we define $\frac{\partial}{\partial \theta_j}$ by

A) $\frac{\partial}{\partial \theta_j} \theta_k = \delta_{jk}$,

B) $\frac{\partial}{\partial \theta_\ell}(\theta_j \theta_k) = \delta_{\ell j}\theta_k - \theta_j \delta_{\ell k}$,

C) linearity, associativity, etc.

Note that $\frac{\partial}{\partial \theta_j}$ is an anti-derivation on $P[\theta_1, ..., \theta_n]$. Define the differential d by

$$d = \sum_{j=1}^n d\theta_j \frac{\partial}{\partial \theta_j}$$

where $\frac{\partial}{\partial \theta_j}(d\theta_k) = 0$.

Do the following exercises:

(a) What is the Leibniz rule satisfied by d?

(b) Show $d^2 = 0$.

The Hamiltonian dynamics of a Grassmann system is constructed by mimicking the symplectic geometry of ordinary manifolds and using the calculus developed in the previous exercise. The Poincaré form and symplectic form may be read off action (2.5):

$$\theta = i\bar{a}\,da\,,\tag{2.7}$$
$$\omega = d\theta = id\bar{a} \wedge da$$
$$= ida \wedge d\bar{a}\,.\tag{2.8}$$

This immediately leads to the Poisson bracket relations

$$\{a,a\}_{PB} = 0\,,$$
$$\{\bar{a},\bar{a}\}_{PB} = 0\,,$$
$$\{a,i\bar{a}\}_{PB} = 1\,.$$

The *principle* for the quantization of fermionics systems is that Grassmannian classical systems are quantized via anticommutation relations rather than commutation relations. Applying this principle to the above means that we have to look for a Hilbert space with operators $\hat{a}$ and $\hat{a}^\dagger$ satisfying

$$\{\hat{a},\hat{a}\} = 0\,,\tag{2.9}$$
$$\{\hat{a}^\dagger,\hat{a}^\dagger\} = 0\,,\tag{2.10}$$
$$\{\hat{a},i\hat{a}^\dagger\} = i\,.\tag{2.11}$$

Clifford algebra fermions

A variant of the above quantization procedures leads to Clifford algebras. This happens when the variables are not complex such as a and $\bar{a}$ in the case previously discussed. Let ψ_1, ψ_2 and ψ_3 be classical real (hermitian) Grassmann variables satisfying

$$\{\psi_i,\psi_j\} = 0,\quad \psi_i^\dagger = \psi_i\,,\quad (\psi_i\psi_j)^\dagger = \psi_j\psi_i\,.$$

Consider the classical Lagrangian given by

$$\mathcal{L} = \frac{1}{2}i\psi_j\dot{\psi}_j - \frac{i}{2}\epsilon_{ijk}B_i\psi_j\psi_k\,.$$

Note that $\mathcal{L}$ is hermitian and that the Hamiltonian is given by

$$H = \frac{i}{2}\epsilon_{ijk}B_i\psi_j\psi_k\,.$$

One can easily verify that the associated Poincaré form and symplectic forms are given by

$$\theta = \frac{1}{2}i\psi_j d\psi_j\,,$$
$$\omega = d\theta = \frac{1}{2}id\psi_j \wedge d\psi_j\,.$$

This leads to the Poisson brackets $\{\psi_j, i\psi_k\}_{PB} = i\delta_{jk}$. The quantization principle tells us that we have to look for hermitian operators $\hat{\psi}$ on a Hilbert space satisfying the algebra

$$\{\hat{\psi}_j, \hat{\psi}_k\} = \delta_{jk} .$$

The above is the Clifford algebra on $\mathbb{R}^3$. Our first task is to find the associated Hilbert space. Choose $\hat{\psi}_j = \sigma_j/\sqrt{2}$. By a classical theorem this is the unique choice up to equivalence for an irreducible finite dimensional representation of the Clifford algebra. We immediately see that the Hilbert space is $\mathcal{H} = \mathbb{C}^2$. The Hamiltonian

$$H = \frac{i}{2} B_j \epsilon_{jkl} \hat{\psi}_k \hat{\psi}_l$$

may easily be rewritten as

$$H = -\frac{1}{2} \mathbf{B} \cdot \boldsymbol{\sigma} ,$$

which is just up to normalization the Pauli Hamiltonian discussed in equation (2.2).

The Clifford algebra for $\mathbb{R}^{2n}$ acts on a Hilbert space of dimension 2^n. The Clifford algebra for $\mathbb{R}^{2n+1}$ acts on a Hilbert space of dimension 2^n.

Exercise 2.4 (Representation theory of Clifford algebras on $\mathbb{R}^{2n}$). In this exercise we exploit properties of fermionic Fock spaces to construct the irreducible representations of a Clifford algebra on $\mathbb{R}^{2n}$. Given the algebra $\{\gamma_\mu, \gamma_\nu\} = 2\delta_{\mu\nu}$ where $\mu, \nu = 1, 2, ..., 2n$, how do we construct a representation of the algebra in terms of hermitian matrices? We will solve the problem by using fermion creation and annihilation operators.

Define operators $a_1, a_2, ..., a_n$ and $a_1^\dagger, a_2^\dagger, ..., a_n^\dagger$ by

$$\begin{aligned}
a_1 &= \tfrac{1}{2}(\gamma_1 + i\gamma_2), & a_1^\dagger &= \tfrac{1}{2}(\gamma_1 - i\gamma_2), \\
a_2 &= \tfrac{1}{2}(\gamma_3 + i\gamma_4), & a_2^\dagger &= \tfrac{1}{2}(\gamma_3 - i\gamma_4), \\
&\vdots & &\vdots \\
a_n &= \tfrac{1}{2}(\gamma_{2n-1} + i\gamma_{2n}), & a_n^\dagger &= \tfrac{1}{2}(\gamma_{2n-1} - i\gamma_{2n}).
\end{aligned}$$

1. Calculate $\{a_i, a_j\}$, $\{a_i, a_j^\dagger\}$ and $\{a_i^\dagger, a_j^\dagger\}$.
2. What is the Hilbert space $\mathcal{H}$ and what is its dimensionality?
3. Consider the number operator $N = \sum_{j=1}^n a_j^\dagger a_j$ and define the "chirality operator"
   $\Gamma = \exp(i\pi N)$, show that
   (a) $\Gamma^2 = 1$
   (b) $\Gamma \gamma_\mu + \gamma_\mu \Gamma = 0$
   (c) Let $P_\pm = \tfrac{1}{2}(1 \pm \Gamma)$, show P_+ and P_- are projection operators and that $\mathcal{H} = \mathcal{H}_+ \oplus \mathcal{H}_-$ in an obvious notation. The vectors in $\mathcal{H}_+$ are called positive chirality spinors and the vectors in $\mathcal{H}_-$ are called negative chirality spinors.

LECTURE 2. CANONICAL QUANTIZATION OF FERMIONIC SYSTEMS

(d) Show $\Gamma = \alpha\gamma_1\gamma_2...\gamma_{2n}$ and determine α. Hint: $\exp(i\pi a_j^\dagger a_j) = (1-2a_j^\dagger a_j)$.
Remark: In physics, one is most interested in $2n = 4$ and one often sees Γ referred to as γ_5. The notation is often abused and γ_5 is often used for what we called Γ independent of the dimensionality.

4. Let $\sigma_{\mu\nu} = -\frac{1}{4}(\gamma_\mu\gamma_\nu - \gamma_\nu\gamma_\mu)$. Verify that the $\{\sigma_{\mu\nu}\}$ are a basis for a representation[10] of the Lie algebra of $SO(2n)$. Show that the subspaces $\mathcal{H}_\pm$ are invariant under $SO(2n)$. Can you show that they are actually irreducible?

5. We will now discuss what spinors have to do with "spin $\frac{1}{2}$". The essence of this question may be understood by studying the $2n = 2$ case[11]. Consider the rotation

$$R(\theta) = \begin{pmatrix} \cos\theta & \sin\theta \\ -\sin\theta & \cos\theta \end{pmatrix}$$

acting on a vector $V = (V_1, V_2)$. Let $V^\pm = V_1 \pm iV_2$ and show that $R(\theta) : V^\pm \mapsto e^{\pm i\theta}V^\pm$. Look at the space of spinors and consider the states $|0\rangle$, and $|1\rangle = a^\dagger |0\rangle$. Work out the following transformation law

$$R(\theta) : \begin{cases} |0\rangle & \mapsto ? \\ |1\rangle & \mapsto ? \end{cases}$$

Do you now understand why one says that spinors have spin $\frac{1}{2}$?

6. In the previous problem we saw both spin one and spin $\frac{1}{2}$ examples. This is an important physical concept because it occurs in nature. The photon is described by a gauge field which is basically a one-form which may be identified with a vector because we have a metric. Electrons as we just saw are spin $\frac{1}{2}$ objects. There is also a very important spin 2 object in nature, the *graviton*, which mediates the gravitational interaction. In Einstein's theory the graviton corresponds to the metric tensor which is a symmetric tensor. Consider the symmetric traceless tensor

$$h = \begin{pmatrix} a & b \\ b & -a \end{pmatrix}$$

and work out the transformation law for $a \pm ib$ under $R(\theta)$.

Exercise 2.5 (Representation theory of Clifford algebras on $\mathbb{R}^{2n+1}$). Use the previous exercise to construct the representation theory for Clifford algebras on $\mathbb{R}^{2n+1}$.

Exercise 2.6 (Spinors, differential forms, and Kähler manifolds). We are going to do some elementary linear algebra which leads to a deep result in Kähler manifold theory. What we are going to show is that the space of spinors is almost the same as $\Lambda^{(0,\bullet)}$. This is surprising since spinors have "half integer" spin while differential forms have integer spin. These facts are closely related to the natural embedding of $U(n)$ in $SO(2n)$.

---

[10] The spinor representation you are discovering is actually only a projective representation of $SO(2n)$. It is a true representation of the universal cover $Spin(2n)$.

[11] Convince yourself of this fact!

1. View $\mathbb{R}^{2n}$ as $\mathbb{C}^n$ with the following coordinate identifications $z_1 = x_1 + ix_2$, $z_2 = x_3 + ix_4, \ldots, z_n = x_{2n-1} + ix_{2n}$. Consider $\Lambda^{(p,q)}(\mathbb{C}^n)$, the differential forms which are homogeneous of degree p in dz and of degree q in $d\bar{z}$. Let $\epsilon(d\bar{z}^j)$ be exterior multiplication by $d\bar{z}^j$ and let $\iota(d\bar{z}^j) = \epsilon(d\bar{z}^j)^\dagger$ where the adjoint is defined by the natural metric on $\mathbb{C}^n$. The metric allows us to identify the tangent space with the cotangent space. Thus one can easily show that the adjoint of $\epsilon(d\bar{z}^j)$ is interior multiplication by $\partial/\partial \bar{z}^j$. Note that the "interior multiplication" $\iota(d\bar{z}^j)$ basically instructs one to erase a $d\bar{z}^j$ and put in the appropriate sign. Compute $\{\iota(d\bar{z}^j), \epsilon(d\bar{z}^k)\}$. Note that $\iota(d\bar{z}^j)1 = 0$ and the exterior algebra $\Lambda^{(0,\bullet)}(\mathbb{C}^n) = \bigoplus_{q=0}^n \Lambda^{(0,q)}(\mathbb{C}^n)$ is generated by the action of ϵ on 1.

2. The construction of $\Lambda^{(0,\bullet)}$ described above is identical with the construction of spinors S on $\mathbb{C}^n$ if one makes the following identifications

$$\begin{array}{ccc} |0\rangle & \leftrightarrow & 1 \\ a_j^\dagger & \leftrightarrow & \epsilon(d\bar{z}^j) \\ a_k & \leftrightarrow & \iota(d\bar{z}^k) \end{array}$$

Note that the algebra of $a, a^\dagger$ is isomorphic to the algebra of ι, ϵ. Thus there is a natural identification of S with $\Lambda^{(0,\bullet)}$.

3. The identification we made is natural as vector spaces but it is not quite natural if we consider $U(n) \subset SO(2n)$. Note that $\dim \mathfrak{u}(n) = n^2$ and it is generated by n^2 anti-hermitian operators. Define n^2 operators by $X_{j\ell} = a_j^\dagger a_\ell$ and show that

$$[X_{ij}, X_{k\ell}] = \delta_{jk} X_{i\ell} - \delta_{\ell i} X_{kj}.$$

Remember that $\mathfrak{gl}(n)$ is the complexification of $\mathfrak{u}(n)$. Let E_{ij} be the $n \times n$ matrix which is zero everywhere except 1 in the ij-entry. Show that

$$[E_{ij}, E_{k\ell}] = \delta_{jk} E_{i\ell} - \delta_{\ell i} E_{kj}.$$

Thus the skew combinations $i(X_{ij} + X_{ij}^\dagger)$ and $(X_{ij} - X_{ij}^\dagger)$ form a basis for the Lie algebra of $U(n)$. Let's try to understand the action of the torus of $U(n)$ on $d\bar{z}^j$. The toral element

$$t_j = \begin{pmatrix} 1 & & & & & & \\ & \ddots & & & & & \\ & & 1 & & & & \\ & & & e^{i\theta_j} & & & \\ & & & & 1 & & \\ & & & & & \ddots & \\ & & & & & & 1 \end{pmatrix}$$

acts on $d\bar{z}^k$ by $t_j \cdot d\bar{z}^k = e^{-i\theta_j \delta_{jk}} d\bar{z}^j$. Show that (no summation convention)

$$e^{-i\theta_j a_j^\dagger a_j} a_k^\dagger e^{i\theta_j a_j^\dagger a_j} = e^{-i\theta_j \delta_{jk}} a_k^\dagger.$$

LECTURE 2. CANONICAL QUANTIZATION OF FERMIONIC SYSTEMS

Thus the operators that represent the maximal torus of $U(n)$ are $e^{i\theta_j a_j^\dagger a_j}$. Let's look at the maximal torus of $SO(2n)$. We might as well restrict ourselves to rotations in the 12–plane. Consider the vector

$$\begin{pmatrix} \cos\theta & \sin\theta \\ -\sin\theta & \cos\theta \end{pmatrix} \begin{pmatrix} V_1 \\ V_2 \end{pmatrix}.$$

From a previous exercise we know that

$$V_1 \pm iV_2 \to e^{\pm i\theta}(V_1 \pm iV_2).$$

Consider the operator $e^{\theta\sigma_{12}}$ on the representation space. Note that $\sigma_{12} = i(a_1^\dagger a_1 - \frac{1}{2})$ and thus $e^{\theta\sigma_{12}} = e^{i\theta(a_1^\dagger a_1 - 1/2)}$. Thus each toral generator has a shift of $1/2$ when comparing our realizations of $U(n)$ and $SO(2n)$. You can check that such a shift does not occur if you compare, say σ_{13} with $i(a_3^\dagger a_1 + a_1^\dagger a_3)$ and $(a_3^\dagger a_1 - a_1^\dagger a_3)$. The diagonal element

$$U = \begin{pmatrix} e^{i\theta_1} & & \\ & \ddots & \\ & & e^{i\theta_n} \end{pmatrix}$$

has determinant $\det U = \exp(i\sum \theta_j)$. In principle we have computed the transition functions required for the construction of spinors on a complex manifold. If the manifold is Kähler, *i.e.*, the Riemannian connection is compatible with the natural hermitian connection, and admits a spin structure then we have actually shown that $S \otimes K^{1/2} \approx \Lambda^{(0,\bullet)}$ where $K^{1/2}$ is the square root of the canonical bundle. The transition functions for K are "$\det U$".

LECTURE 3
Supersymmetry

Getting spin 1/2 particles to move

Our successful quantization of the Pauli Hamiltonian is unsatisfactory because we know that electrons move. How do we modify the above procedure to describe a moving spin 1/2 particle. The answer is quite simple. Just add the standard bosonic Lagrangian for a free particle to our Grassmann quantization. Let x_μ be coordinates for $\mathbb{R}^d$. Consider the Lagrangian

$$（3.1） \quad \mathcal{L} = \underbrace{\frac{1}{2}\dot{x}_\mu \dot{x}_\mu}_{\text{translation}} + \underbrace{\frac{i}{2}\psi_\mu \dot{\psi}_\mu}_{\text{spin}} .$$

The Poisson bracket algebra for this system is $\{x_\mu, x_\nu\}_{PB} = 0$, $\{p_\mu, p_\nu\}_{PB} = 0$, $\{x_\mu, p_\nu\}_{PB} = \delta_{\mu\nu}$ and $\{\psi_\mu, \psi_\nu\}_{PB} = \delta_{\mu\nu}$. The associated Hilbert space is $\mathcal{H} = L^2(\mathbb{R}^d) \otimes \mathbb{C}^{2^N}$, where $N = \lfloor d/2 \rfloor$. One easily verifies that under quantization, the Hamiltonian is given by $H = \frac{1}{2}\hat{p}^2 = \frac{1}{2}\Delta$ where Δ is the Laplacian[12] on $\mathbb{R}^d$.

We have to verify that the above truly describes a spin 1/2 system. The way do to this is to use Noether's theorem to compute the generators for rotations. Under an infinitesimal rotation matrix $\delta_{\mu\nu} + \omega_{\mu\nu} + O(\omega^2)$ where ω is a skew real matrix, one has

$$(3.2) \quad \delta x_\mu = \omega_{\mu\nu} x_\nu ,$$
$$(3.3) \quad \delta \psi_\mu = \omega_{\mu\nu} \psi_\nu .$$

The associated Noether charge is

$$J_{\mu\nu} = x_\mu p_\nu - x_\nu p_\mu - \frac{i}{2}(\psi_\mu \psi_\nu - \psi_\nu \psi_\mu) .$$

---

[12] I always choose the sign of generalized Laplacians such that the eigenvalues are asymptotically positive. For example, in the simplest case $\Delta = -\partial_x^2$.

For example, if we go to $d = 3$ we have $J_{12} = xp_y - yp_x - \frac{i}{2}(\psi_1\psi_2 - \psi_2\psi_1)$ which under quantization becomes $\hat{J}_{12} = \hat{x}\hat{p}_y - \hat{y}\hat{p}_x - i\hat{\psi}_1\hat{\psi}_2$. Using our previous results we conclude that

$$\tag{3.4} J_{12} = \underbrace{\hat{x}\hat{p}_y - \hat{y}\hat{p}_x}_{L_{12}} + \underbrace{\frac{1}{2}\sigma_z}_{S_{12}} \,.$$

This is the famous decomposition of "total" angular momentum J_{12} into "orbital angular momentum" L_{12} and into "spin angular momentum" S_{12}. To better understand the above we give a standard physics discussion on the relation between orbital angular momentum and spin angular momentum and subsequently relate it to well known mathematical constructs. We restrict ourselves to $\mathbb{R}^3$ to parallel the discussion in quantum mechanics books. The generalization to arbitrary dimension is straightforward.

The typical Hilbert space in quantum mechanics is a function space such as $L^2(\mathbb{R}^3)$, therefore the discussion that follows has not only a quantum mechanical interpretation, but also an interpretation in classical field theory where the classical fields belong to some type of function space.

Let us begin with the simplest case where we have an ordinary function space. The rotation group $SO(3)$ acts on $\mathbb{R}^3$ and therefore induces an action on $C^\infty(\mathbb{R}^3)$. Let $R(\theta, \hat{n})$ be a rotation by angle θ around the $\hat{n}$ axis, $\hat{n} \cdot \hat{n} = 1$. The infinitesimal action of a rotation on $\mathbb{R}^3$ is given by $R(\delta\theta, \hat{n})\vec{x} = \vec{x} + \delta\theta\, \hat{n} \times \vec{x} + O\left((\delta\theta)^2\right)$. To determine the action on a function, a physicist usually makes the following statement, "The value of the transformed function f' at the transformed point is the value of the function f at the original point." If the transformed point is given by $\vec{x}' = R\vec{x}$ then $f'(\vec{x}') = f(\vec{x})$. Simple algebra immediately tells us that $f'(\vec{x}) = f(R^{-1}\vec{x})$. The infinitesimal version of this statement is simply (up to corrections of $O((\delta\theta)^2)$)

$$\begin{aligned}\delta f(\vec{x}) &= f'(\vec{x}) - f(\vec{x}) \\ &= f(\vec{x} - \delta\theta\hat{n} \times \vec{x}) - f(\vec{x}) \\ &= -\delta\theta(\hat{n} \times \vec{x}) \cdot \vec{\nabla} f + O(\delta\theta^2) \\ &= -\delta\theta \epsilon_{ijk} n^i x^j \partial_k f \\ &= -\delta\theta \hat{n} \cdot (\vec{x} \times \vec{\nabla})f\,.\end{aligned}$$

The generator of rotations $\vec{L}$ is defined by

$$\tag{3.5} \vec{L} = -i\vec{x} \times \vec{\nabla} = \vec{x} \times \vec{p}\,,$$

where in the last equality we have used the quantum mechanical definition of momentum given in equation (1.16). It is easy to verify that the above satisfies the angular momentum algebra $[L_j, L_k] = i\epsilon_{jkl}L_l$ which is, of course, isomorphic to the Lie algebra of $SO(3)$. The operator $\vec{L}$ is called the "orbital" angular momentum operator because it depends explicitly on the location $\vec{x}$ of the particle, see (3.5).

The situation becomes much more interesting for a vector. Slightly modifying our previous discussion, a physicist would say, "The value of the transformed vector

$\vec{V}'$ at the transformed point is the rotation of the vector $\vec{V}$ at the original point: $\vec{V}'(\vec{x}') = RV(\vec{x})$." This means that the relationship between the transformed vector and the original vector is simply given by

$$\vec{V}'(\vec{x}) = R\vec{V}(R^{-1}\vec{x}) \,.$$

Notice that the rotation R enters in two places. At this stage we might as well generalize the above. Let ρ be a representation of $SO(3)$ (or of its cover $SU(2)$). Let $\rho(R(\theta,\hat{n})) = \exp(-i\theta\hat{n}\cdot\vec{S})$. One can easily verify that $[S_j, S_k] = i\epsilon_{jkl}S_l$. Assume a field Φ transforms as

$$\Phi'(\vec{x}) = \rho(R)\Phi(R^{-1}\vec{x}) \,.$$

The infinitesimal transformation law for the field is simply given to order $(\delta\theta)^2$ by

$$\begin{aligned}
\delta\Phi &= (1 - i\delta\theta\,\hat{n}\cdot\vec{S})(1 - \delta\theta\,\hat{n}\cdot(\vec{x}\times\vec{\nabla}))\Phi - \Phi \\
&= (1 - i\delta\theta\,\hat{n}\cdot\vec{S})(1 - i\delta\theta\,\hat{n}\cdot\vec{L})\Phi - \Phi \\
&= \delta\theta\,\hat{n}\cdot\left(\vec{L} + \vec{S}\right)\Phi \\
&= \delta\theta\,\hat{n}\cdot\vec{J}\Phi \,.
\end{aligned}$$

Operator $\vec{J}$ is called the "total angular momentum operator". Note that $\vec{L}$ depends on position and is called the "orbital" angular momentum operator. The terminology originates with the orbits of electrons in an atom. The operator $\vec{S}$ is independent of position and is called the "intrinsic" or "spin" angular momentum operator.

Returning to our Noether theorem result, we see that equation (3.4) is the total angular momentum operator for a particle with *spin* 1/2. Therefore Lagrangian (3.1) describes a freely moving spin one-half particle[13].

The simplest supersymmetry

The variation of the Lagrangian under arbitrary deformations of the field is given by

$$\delta L = \dot{x}_\mu\delta\dot{x}_\mu + \frac{i}{2}(\delta\psi_\mu)\dot{\psi}_\mu + \frac{i}{2}\psi_\mu\delta\dot{\psi}_\mu \,.$$

We wish to show that a very special type of deformation leaves the action invariant. Let ϵ be a hermitian anticommuting parameter, $\epsilon^\dagger = \epsilon$, and consider the following variation of the fields

(3.6) $$\delta x_\mu = i\epsilon\psi_\mu \,,$$
(3.7) $$\delta\psi_\mu = -\epsilon\dot{x}_\mu \,.$$

---

[13] Lagrangian (3.1) is a special case where both the orbital angular momentum and the spin angular momentum are independently constants of the motion. The reason is that the Lagrangian is invariant under independent rotations to the x and ψ variables. In the general interacting case, the Lagrangian is only invariant under simultaneous rotations of x and ψ.

To evaluate the variation of the Lagrangian under the above it is important to remember that x is an even element of the Grassmann algebra, and that ψ and ϵ are odd elements.

$$\begin{aligned}
\delta L &= i\dot{x}_\mu \epsilon \dot{\psi}_\mu - \frac{i}{2}\epsilon \dot{x}_\mu \dot{\psi}_\mu - \frac{i}{2}\psi_\mu \epsilon \dot{x}_\mu \\
&= i\dot{x}_\mu \epsilon \dot{\psi}_\mu - \frac{i}{2}\epsilon \dot{x}_\mu \dot{\psi}_\mu - \frac{i}{2}\frac{d}{dt}(\psi_\mu \epsilon \dot{x}_\mu) + \frac{i}{2}\dot{\psi}_\mu \epsilon \dot{x}_\mu \\
&= -\frac{i}{2}\frac{d}{dt}(\psi_\mu \epsilon \dot{x}_\mu) .
\end{aligned}$$

The conserved charge may be computed using Noether's Theorem II

$$\epsilon Q = i\epsilon p_\mu \psi_\mu = i\epsilon \psi_\mu p_\mu = i\epsilon \psi_\mu \dot{x}_\mu .$$

Q is called the supercharge and it is the generator of supersymmetry.

We now proceed to canonically quantize the supersymmetric system in dimension $d = 2n$ for simplicity. The basic canonical (anti)commutation relations are $[\hat{x}_\mu, \hat{p}_\nu] = i\delta_{\mu\nu}$ and $\{\hat{\psi}_\mu, \hat{\psi}_\nu\} = \delta_{\mu\nu}$. If we define $\hat{\psi}_\mu = \gamma_\mu/\sqrt{2}$ then the anticommutation relations may be rewritten in the conventional physics form $\{\gamma_\mu, \gamma_\nu\} = 2\delta_{\mu\nu}$ used to specify the Dirac γ-matrices. The full algebra of operators is implemented in the Hilbert space $\mathcal{H} = L^2(\mathbb{R}^{2n}) \otimes \mathbb{C}^N$ where $N = 2^n$. The first factor reflects the translation properties and the second factor the spin properties. Note that the Noether charge quantized operator $\hat{Q}$ may be written as

$$\begin{aligned}
\hat{Q} &= i\hat{\psi}_\mu \hat{p}_\mu \\
&= i\frac{\gamma_\mu}{\sqrt{2}}\left(-i\frac{\partial}{\partial x^\mu}\right) \\
&= \frac{1}{\sqrt{2}}\gamma^\mu \frac{\partial}{\partial x^\mu} .
\end{aligned}$$

The last line is the Dirac operator on $\mathbb{R}^{2n}$. In the physics literature, the Dirac operator is denoted by $\slashed{\partial}$ and the conventional normalization is

$$\slashed{\partial} = \gamma^\mu \frac{\partial}{\partial x^\mu} .$$

Thus the Noether charge of the quantized supersymmetric particle moving in $\mathbb{R}^{2n}$ is simply the Dirac operator on $\mathbb{R}^{2n}$.

It is interesting and useful to compute the supersymmetry transformation of the Noether charge:

$$\begin{aligned}
\delta Q &= i(\delta \psi_\mu)\dot{x}^\mu + i\psi_\mu \delta \dot{x}^\mu \\
&= i(-\epsilon \dot{x}^\mu)\dot{x}^\mu + i\psi_\mu(i\epsilon\dot{\psi}_\mu) \\
&= -i\epsilon \dot{x}^\mu \dot{x}^\mu + \epsilon \psi_\mu \dot{\psi}_\mu \\
&= -2i\epsilon \left[\frac{1}{2}\dot{x}^\mu \dot{x}^\mu + \frac{i}{2}\psi_\mu \dot{\psi}_\mu\right] .
\end{aligned}$$

LECTURE 3. SUPERSYMMETRY

We have discovered the remarkable fact that the supersymmetry transformation of Q is the supersymmetric Lagrangian L. With some additional information about the supersymmetry algebra one can exploit this fact usefully.

We have to understand the algebra of supersymmetry transformations. In other words, what are the "commutation relations" of the supersymmetry algebra. Observe that the "δ" operation in equations (3.6) and (3.7) is even with respect to the Grassmann algebra due to the presence of the anticommuting parameter ϵ. To determine the structure of the algebra we first perform a supersymmetry transformation by parameter ϵ_1, follow it up by a supersymmetry transformation by parameter ϵ_2, and subsequently compute the commutator.

$$x^\mu \xrightarrow{\epsilon_1} x^\mu + i\epsilon_1 \psi_\mu$$
$$\xrightarrow{\epsilon_2} x^\mu + i\epsilon_2 \psi^\mu + i\epsilon_1(\psi_\mu - i\epsilon_2 \dot{x}_\mu)$$
$$= x^\mu + i(\epsilon_1 + \epsilon_2)\psi^\mu - i\epsilon_1 \epsilon_2 \dot{x}^\mu ,$$
$$\psi^\mu \xrightarrow{\epsilon_1} \psi^\mu - \epsilon_1 \dot{x}^\mu$$
$$\xrightarrow{\epsilon_2} \psi^\mu - \epsilon_2 \dot{x}^\mu - \epsilon_1(\dot{x}^\mu + i\epsilon_2 \dot{x}^\mu)$$
$$= \psi^\mu - (\epsilon_1 + \epsilon_2)\dot{x}^\mu - i\epsilon_1 \epsilon_2 \dot{\psi}^\mu .$$

Therefore using an obvious notation we conclude that the commutator is given by

$$(3.8) \qquad (\delta_{\epsilon_2}\delta_{\epsilon_1} - \delta_{\epsilon_1}\delta_{\epsilon_2}) = -2i\epsilon_1\epsilon_2 \frac{\partial}{\partial t} .$$

According to Noether's theorem the above commutation relation is implemented in phase space via the Poisson bracket. The equation above states that the commutator of two supersymmetry transformations is the generator of translations it follows that the Poisson bracket of two supersymmetry transformations should be the Hamiltonian, see Exercise (1.8). We have to test whether

$$\{\epsilon_2 Q, \epsilon_1 Q\}_{PB} \stackrel{?}{=} -2i\epsilon_1\epsilon_2 H .$$

This is a straightforward exercise as seen below

$$\{\epsilon_2 i p_\mu \psi_\mu, \epsilon_1 i p_\nu \psi_\nu\}_{PB} = (+i)^2 p_\mu p_\nu \epsilon_2 \epsilon_1 \{\psi_\mu, \psi_\nu\}_{PB}$$
$$= -p_\mu p_\nu \epsilon_2 \epsilon_1 (-i\delta_{\mu\nu})$$
$$= i\epsilon_2 \epsilon_1 p^2$$
$$= -2i\epsilon_1 \epsilon_2 H .$$

Under quantization we see that

$$\begin{aligned}
\{\hat{Q},\hat{Q}\} &= 2\hat{Q}^2 \\
&= 2(i\hat{p}_\mu \hat{\psi}_\mu)(i\hat{p}_\nu \hat{\psi}_\nu) \\
&= -2\hat{p}_\mu \hat{p}_\nu \hat{\psi}_\mu \hat{\psi}_\nu \\
&\ -2\hat{p}_\mu \hat{p}_\nu \cdot \frac{1}{2}\left(\hat{\psi}_\mu \hat{\psi}_\nu + \hat{\psi}_\nu \hat{\psi}_\mu\right) \\
&= -p_\mu p_\nu \delta_{\mu\nu} = -p^2 = -2\hat{H}\ .
\end{aligned}$$

We conclude that the supersymmetry algebra satisfies $\{\hat{Q},\hat{Q}\} = -2\hat{H}$ or $\hat{Q}^2 = -\hat{H}$. Note that our $\hat{Q}$ is anti-hermitian.

The supersymmetry algebra satisfied by L and Q may be written in the form

$$\delta Q = -2i\epsilon L\ , \tag{3.9}$$

$$\delta L = \epsilon \frac{1}{2} \frac{dQ}{dt}\ . \tag{3.10}$$

The above should be compared to equations (3.6) and (3.7). The transformation laws are very similar except that the role of "bosons" and "fermions" have been reversed. The supersymmetry algebra representation given by (3.6) and (3.7) is called a bosonic multiplet, and the one given by (3.9) and (3.10) is called a fermionic multiplet. A most important observation is that the last element of the fermionic multiplet transforms as a total derivative and therefore is a candidate for a Lagrangian whose action is invariant under supersymmetry.

The Dirac equation on a manifold

The model of Section (3) is easily generalized to a Riemannian manifold M with metric in local coordinates given by $ds^2 = g_{\mu\nu}(x)dx^\mu \otimes dx^\nu$. The inner product of vectors X and Y will be often written as $\langle X, Y \rangle$. The key observation needed for the generalization is that supersymmetry transformations (3.6) and (3.7) are valid on an arbitrary manifold. Namely, the supersymmetry transformation rule is automatically covariant. In fact, ψ transforms like a section of the tangent bundle. More precisely, at time t, $\psi(t)$ may be taken to be a vector in $TM_{x(t)}$ with the *proviso* that $\psi(t)$ is an odd element of a Grassmann algebra.

Exercise 3.1 (Covariance of supersymmetry transformation laws). Assume we change local coordinates on M from x to x' given by $x^\mu = f^\mu(x')$. Postulate that ψ transforms under the change of coordinates as a vector field:

$$\psi^\mu = \frac{\partial x^\mu}{\partial x'^\nu} \psi'^\nu\ .$$

Verify that the supersymmetry transformation laws (3.6) and (3.7) are covariant. Note that the covariance of the supersymmetry transformation laws is only true due to the anticommuting nature of ψ.

The key to constructing a supersymmetric Lagrangian on M is to exploit equation (3.9). It is easy to generalize the flat space Q to a curved space Q since in flat space Q is simply the inner product of the velocity with the fermion ψ. The previous exercise demonstrates that ψ may be taken to have the transformation law of a vector, therefore,

$$Q = i \langle \dot{x}, \psi \rangle = i g_{\mu\nu}(x) \dot{x}^\mu \psi^\nu$$

is a well defined invariant object. To construct the Lagrangian we use (3.9):

$$\begin{aligned} L &= \frac{1}{2} g_{\mu\nu}(x) \dot{x}^\mu \dot{x}^\nu \\ &\quad + \frac{i}{2} g_{\mu\nu}(x) \psi^\mu \left(\frac{d}{dt} \psi^\nu + \dot{x}^\lambda \Gamma^\nu_{\lambda\mu}(x) \psi^\mu \right) \\ &= \frac{1}{2} \langle \dot{x}, \dot{x} \rangle + \frac{i}{2} \left\langle \psi, \frac{D\psi}{Dt} \right\rangle . \end{aligned}$$ (3.11)

The Riemannian connection associated with the metric g is specified by the Christoffel symbols

$$\Gamma^\nu_{\lambda\mu} = \frac{1}{2} g^{\nu\rho} \left(\partial_\lambda g_{\rho\mu} + \partial_\mu g_{\lambda\rho} - \partial_\rho g_{\lambda\nu} \right) .$$

The object $D\psi/Dt$ is simply the covariant derivative along the curve $x(t)$. One can easily verify that Lagrangian (3.11) changes by a total derivative under a supersymmetry calculation. The associated Noether charge is up to a proportionality factor the Dirac operator $\slashed{D}$ on M after quantization. All of this is most easily seen by working in Riemann normal coordinates where many results reduce to flat space computations.

Exercise 3.2. Construct the Lagrangian above.

The connection one forms $\Gamma^\mu{}_\nu = dx^\lambda \, \Gamma^\mu_{\lambda\nu}$ can be used to define the Riemann curvature two form $\mathcal{R}^\mu{}_\nu$ via the equation

$$\mathcal{R}^\mu{}_\nu = d\Gamma^\mu{}_\nu + \Gamma^\mu{}_\sigma \wedge \Gamma^\sigma{}_\nu .$$

The components of the Riemann curvature two form are defined by

$$\mathcal{R}^\mu{}_\nu = \frac{1}{2} R^\mu{}_{\nu\rho\sigma} \, dx^\rho \wedge dx^\sigma .$$

Exercise 3.3. Derive the Euler-Lagrange equations associated to Lagrangian (3.11).

Exercise 3.4. Use Noether's Theorem to work out an explicit expression for the Dirac operator on M in local coordinates.

LECTURE 4
The Index of Operators

The basics

Let M be a Riemannian manifold and let $E_\pm$ be vector bundles with metrics over M. Given an elliptic differential operator $\mathcal{D}: C^\infty(E_+) \longrightarrow C^\infty(E_-)$, the index of $\mathcal{D}$, written $\operatorname{Index}\mathcal{D}$, is defined by the equation

$$\operatorname{Index}\mathcal{D} = \dim \ker \mathcal{D} - \dim \operatorname{coker} \mathcal{D} \tag{4.1}$$
$$= \dim \ker \mathcal{D} - \dim \ker \mathcal{D}^\dagger. \tag{4.2}$$

Theorem 4.1. *The index of $\mathcal{D}$ is a homotopy invariant with respect to "reasonable perturbations" of the operator $\mathcal{D}$.*

The proof is relatively simple. First one observes that $\ker \mathcal{D} = \ker \mathcal{D}^\dagger \mathcal{D}$ and $\ker \mathcal{D}^\dagger = \ker \mathcal{D}\mathcal{D}^\dagger$. Pick a basis $\{\varphi_n\}$ of orthonormal eigensections of $\mathcal{D}^\dagger \mathcal{D}$ with $(\mathcal{D}^\dagger \mathcal{D})\varphi_n = \lambda_n \varphi_n$. If $\lambda_n > 0$ then define $\psi_n = \mathcal{D}\varphi_n/\sqrt{\lambda_n}$. It is easy to verify that $(\mathcal{D}\mathcal{D}^\dagger)\psi_n = \lambda_n \psi_n$ and that $(\psi_n, \psi_m) = \delta_{nm}$. We conclude that there is a natural isomorphism between $(\ker \mathcal{D})^\perp$ and $(\ker \mathcal{D}^\dagger)^\perp$. The pairing fails only on the kernels. If one perturbs the operator $\mathcal{D}$ then sections can only leave or enter the respective kernels pair wise because of the necessitated pairing in the orthogonal complements to the respective kernels.

Theorem 4.2 (Heat kernel formula for the index). *The index may be written as*

$$\operatorname{Index}\mathcal{D} = \operatorname{Tr} e^{-\beta \mathcal{D}^\dagger \mathcal{D}} - \operatorname{Tr} e^{-\beta \mathcal{D}\mathcal{D}^\dagger}, \tag{4.3}$$

where $\beta > 0$. The above is independent of β.

Exercise 4.3. Prove the previous theorem.

Supersymmetry and the index

The heat kernel formula for the index is the beginnings of using supersymmetry to access the index theorem. The setup involves some cosmetic changes to the discussion of the previous section. Let $E = E_+ \oplus E_-$ and let

$$iQ = \begin{pmatrix} 0 & \mathcal{D}^\dagger \\ \mathcal{D} & 0 \end{pmatrix}$$

act on E. Define operators H and Γ by the equations

$$H = (iQ)^2 = \begin{pmatrix} \mathcal{D}^\dagger \mathcal{D} & 0 \\ 0 & \mathcal{D}\mathcal{D}^\dagger \end{pmatrix}, \quad \Gamma = \begin{pmatrix} 1 & 0 \\ 0 & -1 \end{pmatrix}.$$

One immediately obtains the result that

(4.4) $$\operatorname{Index} \mathcal{D} = \operatorname{Tr} \Gamma e^{-\beta H}.$$

We can apply this general discussion to spinors. Let $S_\pm(M)$ be the chiral spin bundles on M. Choose E to be the total spin bundle: $E = S(M) = S_+(M) \oplus S_-(M)$ and let Q be the Dirac operator on M. Choose Γ to the the "chirality matrix" γ_5. Then the index of the chiral Dirac operator acting on sections of $S_+(M)$ is the number of positive chirality "zero modes" minus the number of negative chirality "zero modes". A "zero mode" is an abusive abbreviation for a solution in the kernel of the operator. Physicists also abbreviate things by speaking about the index of the Dirac operator, meaning the index of the chiral Dirac operator. One often sees the notation $\operatorname{Index} Q$. The strategy we will pursue for obtaining the Atiyah-Singer formula is to write down a path integral representation for $\operatorname{Tr} \Gamma e^{-\beta H}$.

LECTURE 5
Path Integrals

Path integral for the simple harmonic oscillator

The simple harmonic oscillator is defined by the Minkowski space Lagrangian

(5.1) $$L = \frac{1}{2}\dot{q}^2 - \frac{1}{2}\omega^2 q^2$$

with energy eigenvalues $E_n = \left(n + \frac{1}{2}\right)\omega$. The partition function $Z = \operatorname{Tr} e^{-\beta H}$ is easily computed

$$\operatorname{Tr} e^{-\beta H} = \sum_{n=0}^{\infty} e^{-\beta(n+\frac{1}{2})\omega}$$

$$= \frac{1}{2\sinh\left(\frac{1}{2}\beta\omega\right)}.$$

We will now perform the same calculation using path integral methods. The partition function may be written in the form

$$\operatorname{Tr} e^{-\beta H} = \int dx \, \langle x | e^{-\beta H} | x \rangle$$

$$= \int_{PP} [\mathcal{D}q] \, \exp\left[-\int_0^\beta dt \, \frac{1}{2}\left(\dot{q}^2 + \omega^2 q^2\right)\right]$$

$$= \left[\det_{PBC}\left(-\frac{d^2}{dt^2} + \omega^2\right)\right]^{-1/2}.$$

where PP stands for "periodic paths" and PBC stands for "periodic boundary conditions". There are several pertinent remarks I will make about the above.

In an infinite dimensional space, such as the space of paths, gaussian integrals are defined by analogy with finite dimensional gaussian integrals. Consider $\mathbb{R}^n$ with the standard inner product and let

$$[\mathcal{D}x] = \frac{d^n x}{(2\pi)^{n/2}}.$$

If A is a real symmetric positive definite matrix then

$$(\det A)^{-1/2} = \int [\mathcal{D}x] \exp\left(-\frac{1}{2}\sum A_{ij}x^i x^j\right).$$

If we can define the determinant for certain classes of operators in infinite dimensional spaces then we can define gaussian integrals.

Exercise 5.1. Compute

$$\int [\mathcal{D}x] \exp\left(-\frac{1}{2}\sum A_{ij}x^i x^j + \sum J_i x^i\right).$$

Remember that the Hamiltonian in this problem is $H = \frac{1}{2}\Delta + \frac{1}{2}\omega^2 q^2$. The heat kernel or diffusion kernel is given by

$$\langle x | e^{-\frac{1}{2}\beta\Delta} | y \rangle = \frac{1}{(2\pi\beta)^{1/2}} e^{-(x-y)^2/2\beta}.$$

Physically this is the probability density that a particle with initial certainty of being at y diffuses in time β to point x. The kernel $\langle x | e^{-\beta H} | y \rangle$ is the probability of diffusing from y to x in time β in the presence of a potential $\frac{1}{2}\omega^2 q^2$ which tries to keep the particle near the origin. If the time β is small, then a particle diffusing from x and back to x cannot wander very far away from x and therefore it should be possible to treat the potential term as constant. This argument indicates that one should have an asymptotic expansion of the form

$$\langle x | e^{-\beta H} | x \rangle \sim \frac{1}{\sqrt{2\pi\beta}}\left(1 - \frac{1}{2}\beta\omega^2 x^2 + O(\beta^2)\right).$$

The justification for the above is that in a diffusion process, a random walk, the root mean square separation grows like $\sqrt{\beta}$ and therefore a Taylor series expansion of the heat equation will be in powers of $\sqrt{\beta}$. Additionally, there is a symmetry in the problem $q \to -q$ and thus the expansion parameter is actually β.

To evaluate the determinant with PBC we first write down an expansion for $q(t)$ in terms of a complete set of orthonormal eigenfunctions

$$q(t) = \frac{1}{\sqrt{\beta}} \sum_{n=-\infty}^{\infty} q_n e^{(2\pi in/\beta)t}.$$

The reality of q imposes the constraint $\bar{q}_n = q_{-n}$. Eigenvalues of $-\frac{d^2}{dt^2} + \omega^2$ are simply

$$\left(\frac{2\pi n}{\beta}\right)^2 + \omega^2.$$

We are now in a position to expand the determinant. We will do a whole bunch of formal manipulations which I urge you to attempt to justify:

$$\det\left(-\frac{d^2}{dt^2}+\omega^2\right) = \prod_n \left(\left(\frac{2\pi n}{\beta}\right)^2 + \omega^2\right)$$

$$= \omega^2 \left(\prod_{n=1}^{\infty}\left[\left(\frac{2\pi n}{\beta}\right)^2 + \omega^2\right]\right)^2$$

$$= \left(\omega \prod_{n=1}^{\infty}\left(\frac{2\pi n}{\beta}\right)^2 \prod_{n=1}^{\infty}\left(1+\left(\frac{\beta\omega}{2\pi n}\right)^2\right)\right)^2$$

$$= \underbrace{\left[\prod_{n=1}^{\infty}\left(\frac{2\pi n}{\beta}\right)^2\right]^2}_{\det'\left(-\frac{d^2}{dt^2}\right)} \left(\omega \prod_{n=1}^{\infty}\left(1+\left(\frac{\beta\omega}{2\pi n}\right)^2\right)\right)^2 .$$

The object $\det' A$ will always stand for the determinant of the operator A with the "zero modes" excluded. Our formal manipulations seem to indicate that

$$\det\left(-\frac{d^2}{dt^2}+\omega^2\right) = \det'\left(-\frac{d^2}{dt^2}\right) \times \left[\omega \prod_{n=1}^{\infty}\left(1+\left(\frac{\beta\omega}{2\pi n}\right)^2\right)\right]^2 .$$

Our formal manipulations are quite fruitful. Note that the second term above is absolutely convergent and that the $\det'$ term is formally divergent. Our naive conclusion is that $\det(-d^2/dt^2 + \omega^2)$ is divergent only because the $\det'$ term is divergent. If we can get a handle on the $\det'$ term them maybe we can figure out how to make sense of $\det(-d^2/dt^2 + \omega^2)$.

We will compute $\det'(-d^2/dt^2)$ in two different ways. The first way exploits the equivalence between the Hamiltonian formulation and the path integral formulation. The second way is an introduction to zeta function regularization. Observe that

$$\text{Tr}\, e^{-\beta \hat{p}^2/2} = \int_{PP} [\mathcal{D}q]\, e^{-\int_0^\beta dt\, \dot{q}^2/2} .$$

The left hand side of the above has a simple representation,

$$\int dx\, \langle x | e^{-\beta \Delta/2} | x \rangle = \left(\int dx\right) \frac{1}{\sqrt{2\pi\beta}} .$$

Note that $\int dx$ is formally infinite. We now work on the right hand side path integral by observing that the computation involves the introduction of an expansion in terms of a complete set of orthonormal eigenfunctions for the laplacian:

$$q(t) = \frac{1}{\sqrt{\beta}} \sum_{n=-\infty}^{\infty} q_n e^{2\pi i n t/\beta} . \tag{5.2}$$

Introducing this expansion into the path integral allows one to do all the integrals except for the orthonormal "zero mode". This leads to the expression

$$\int \frac{dq_0}{\sqrt{2\pi}} \left[\det{}'\left(-\frac{d^2}{dt^2}\right)\right]^{-1/2} .$$

The integral over the orthonormal "zero mode" may be related to the ordinary coordinate q by observing that in (5.2) a change in q_0 induces a change in q given by $dq = dq_0/\sqrt{\beta}$. Inserting this into the path integral expression above we find

$$\left(\int dx\right) \sqrt{\frac{\beta}{2\pi}} \left[\det{}'\left(-\frac{d^2}{dt^2}\right)\right]^{-1/2} .$$

Comparing our two evaluations we conclude

$$\left[\det{}'\left(-\frac{d^2}{dt^2}\right)\right]^{-1/2} = \frac{1}{\beta} .$$

We now discuss a different derivation based on zeta function regularization. Let $\mathcal{D}$ be a positive semi-definite elliptic operator of degree r with eigenvalues $\{\lambda_n\}$. Naively one concludes that

$$\log \det{}' \mathcal{D} = \operatorname{Tr}' \log \mathcal{D} = \sum{}' \log \lambda_n , \tag{5.3}$$

where the prime denotes the omission of the "zero modes". The zeta function of the operator $\mathcal{D}$ is defined by

$$\zeta_\mathcal{D}(s) = \sum{}' \frac{1}{\lambda_n^s} .$$

The zeta function is convergent if $r \cdot \operatorname{Re} s > d$. Therefore $\zeta_\mathcal{D}(s)$ is an analytic function of s for $\operatorname{Re} s$ sufficiently positive. Note that formally

$$\left.\frac{d\zeta_\mathcal{D}(s)}{ds}\right|_{s=0} = -\sum{}' \log \lambda_n .$$

We define

$$\log \det{}' \mathcal{D} = -\left.\frac{d\zeta_\mathcal{D}(s)}{ds}\right|_{s=0} \tag{5.4}$$

where $\zeta_\mathcal{D}(s)$ is defined by analytic continuation.

LECTURE 5. PATH INTEGRALS

Let $\Delta = -d^2/dt^2$ with periodic boundary conditions, then

$$\zeta_\Delta(s) = \sum_{n \neq 0} \left(\frac{2\pi n}{\beta}\right)^{-2s}$$

$$= 2\left(\frac{\beta}{2\pi}\right)^{2s} \zeta(2s),$$

where $\zeta(s) = \sum_{n=1}^{\infty} n^{-s}$ is the ordinary Riemann zeta function. Computing the derivative we find

$$\zeta'_\Delta(0) = 2 \cdot 2 \cdot \underbrace{\zeta(0)}_{-\frac{1}{2}} \cdot \log \frac{\beta}{2\pi} + 2 \cdot 2 \cdot \underbrace{\zeta'(0)}_{-\frac{1}{2}\log 2\pi}$$

$$= -2\log \beta .$$

This is in exact agreement for our previous computation of $\det'(-d^2/dt^2)$. Thus we have a finite expression for the determinant with the highly suggestive result

(5.5) $$\det\left(-\frac{d^2}{dt^2} + \omega^2\right) = \left[(\beta\omega) \prod_{n=1}^{\infty}\left(1 + \left(\frac{\beta\omega}{2\pi n}\right)^2\right)\right]^2 .$$

Thus we have reached the following conclusion. Using the energy eigenvalues we see that

$$\mathrm{Tr}\, e^{-\beta H} = \frac{1}{2\sinh\left(\frac{1}{2}\beta\omega\right)} .$$

From the path integral we have computed that

$$\mathrm{Tr}\, e^{-\beta H} = \left[\beta\omega \prod_{n=1}^{\infty}\left(1 + \left(\frac{\beta\omega}{2\pi n}\right)^2\right)\right]^{-1} .$$

Comparing both expressions and using the equivalence of the Hilbert space and path integral formulations of quantum mechanics leads to a physicist's proof of Euler's product representation for the sine function

$$\sin x = x \prod_{n=1}^{\infty}\left(1 - \left(\frac{x}{\pi n}\right)^2\right) .$$

The simple harmonic oscillator problem may be also be solved by using the Wick rotated version of the first order formulation previously discussed

$$\mathrm{Tr}\, e^{-\beta H} = \int [\mathcal{D}a\mathcal{D}\bar{a}] \exp\left[-\int_0^\beta dt\, (\bar{a}\dot{a} + \omega\bar{a}a)\right]$$

$$= \left[\det\left(\frac{d}{dt} + \omega\right)\right]^{-1} .$$

Exercise 5.2. Compute $\det(\frac{d}{dt} + \omega)$ with periodic boundary conditions. Use the fact that the partition function for a simple harmonic oscillator SHO is real and positive to argue that

$$\det\left(\frac{d}{dt} + \omega\right) = \det\left(-\frac{d}{dt} + \omega\right).$$

Does this lead to an unambiguous determination for the sign of the determinant?

Exercise 5.3. Compute $\det(\frac{d}{dt} + \omega)$ with anti-periodic boundary conditions.

We will now attempt to construct a path integral formulation for fermions. Look back at (2.3) and compute the partition function for a fermionic harmonic oscillator

$$\operatorname{Tr} e^{-\beta H} = e^{-\beta(-\frac{1}{2}\omega)} + e^{-\beta(+\frac{1}{2}\omega)}$$

(5.6)
$$= 2\cosh\frac{\beta\omega}{2}.$$

It is also convenient to introduce the operator[14]

$$(-1)^N = \begin{pmatrix} 1 & 0 \\ 0 & -1 \end{pmatrix}$$

and compute the "twisted" trace

$$\operatorname{Tr}(-1)^F e^{-\beta H} = e^{-\beta(-\frac{1}{2}\omega)} + (-1)e^{-\beta(\frac{1}{2}\omega)}$$

(5.7)
$$= 2\sinh\frac{\beta\omega}{2}.$$

We now make the following observations

$$\operatorname{Tr}(-1)^F e^{-\beta H} = \det_{PBC}\left(\frac{d}{dt} + \omega\right),$$

$$\operatorname{Tr} e^{-\beta H} = \det_{APBC}\left(\frac{d}{dt} + \omega\right).$$

where APBC stands for anti-periodic boundary conditions. This strongly suggest a definition for gaussian integrals for anticommuting variables. Thus if a, $\bar{a}$ are anticommuting variables it appears as if

$$\int [\mathcal{D}a\mathcal{D}\bar{a}]\exp\left[-\int_0^\beta dt\,(\bar{a}\dot{a} + \omega\bar{a}a)\right] = \det\left(\frac{d}{dt} + \omega\right),$$

where the determinant is taken with appropriate boundary conditions. Note that putting periodic boundary conditions on the fermionic path integral is equivalent

---

[14] The operator $(-1)^N$ is often called $(-1)^F$ since F is often used for the fermion number operator.

to an insertion of the operator $(-1)^N$ in the trace and anti-periodic boundary conditions simply give the trace. This is different than in the bosonic case. We now develop a theory of fermionic integration which will justify the observations above.

Theory of fermionic integration

In its simplest guise integration is a linear functional on a function space. If we have a Grassmann algebra generated by one variable θ we define

$$\int d\theta \, 1 = 0 ,$$

$$\int d\theta \, \theta = 1 .$$

Extend the above rules by linearity in an obvious way and this gives us a theory of integration on a single anticommuting variable. Note that the rule of integration is to pick out the top order term. Likewise, if we have n Grassmann variables $\{\theta_1, \theta_2, \ldots, \theta_n\}$ we define

$$\int d\theta_1 \cdots d\theta_n \, 1 = 0 ,$$

$$\int d\theta_1 \cdots d\theta_n \, \theta_j = 0 ,$$

$$\vdots$$

$$\int d\theta_1 \cdots d\theta_n \, \theta_1 \theta_2 \cdots \theta_n = 1 .$$

Again this picks out the top order form.

Exercise 5.4. (Jacobian for fermionic integration)
1. Let θ be a single anticommuting variable. Let $\theta' = \alpha\theta$, α a scalar. Let $d\theta' = J d\theta$ where J is the Jacobian. Show that $J = 1/\alpha$. This is the opposite of bosonic variables.
2. Let $\{\theta_1, \ldots, \theta_n\}$ be anticommuting and let $\theta'_j = \sum_i \theta_i L^i{}_j$. Show

$$d\theta'_1 \cdots d\theta'_n = (\det L)^{-1} d\theta_1 d\theta_2 \ldots d\theta_n.$$

These rules are sufficient to define gaussian Grassmann integrals. Let $\{\theta_1, \ldots, \theta_n\}$ again be Grassmann variables and let A be a skew $n \times n$ matrix. Assume n is even, $n = 2m$. The gaussian integral may be rewritten in the form

$$\int d\theta_1 \cdots d\theta_n \, \exp\left(\frac{1}{2}\sum_{ij} A_{ij}\theta_i\theta_j\right) = \int d\theta_1 \cdots d\theta_n \, \frac{1}{m!}\left(\frac{1}{2}\sum A_{ij}\theta_i\theta_j\right)^m$$

$$= \text{Pf}(A) ,$$

where Pf(A) is the Pfaffian of the matrix A. To understand the Pfaffian a bit better one observes that the matrix A may be block diagonalized

$$A = \begin{pmatrix} 0 & a_1 & & & \\ -a_1 & 0 & & & \\ & & 0 & a_2 & \\ & & -a_2 & 0 & \\ & & & & \ddots \end{pmatrix}$$

via a special orthogonal transformation. Since the diagonalization transformation has determinant one this will not affect the fermionic path integral and it follows that Pf(A) = $a_1 a_2 \ldots a_m$, and consequently (Pf(A))^2 = det A.

Exercise 5.5. (Gaussian integration with fermionic variables)
1. If θ is a single anticommuting variable and β is an anticommuting constant, then why should one impose that the measure be translationally invariant, i.e., if $\theta \to \theta + \beta$ then $d\theta \to d\theta$.
2. If $\{\theta_1, \ldots, \theta_n\}$ are anticommuting, A_{ij} is a skew matrix, and if $\{\eta_1, \ldots, \eta_n\}$ are also anticommuting, then compute

$$\int d\theta_1 \ldots d\theta_n \, \exp\left(\frac{1}{2}\sum_{ij} A_{ij}\theta_i\theta_j + \sum_k \eta_k \theta_k\right).$$

Exercise 5.6. Let $\{b_1, \ldots, b_n, c_1, \ldots, c_n\}$ be anticommuting variables and let A be an $n \times n$ matrix. Compute

$$\int db_1 dc_1 \ldots db_n dc_n \, \exp\left(\sum b_i A_{ij} c_j\right).$$

LECTURE 6
The Atiyah-Singer formula

Atiyah-Singer formula for the Dirac operator

Let us now look in detail at the path integral formulation of our supersymmetric system. I remind you that $\hat{H} = (i\hat{Q})^2$ and $\hat{Q}$ was the Dirac operator on the target spin manifold M with $\dim M = 2n = d$. The index $\mathcal{I}$ for the Dirac operator is given by

$$\mathcal{I} = \operatorname{Tr} \Gamma e^{-\beta H} = \operatorname{Tr}(-1)^F e^{-\beta H}$$
$$= \int_{PBC} [\mathcal{D}x][\mathcal{D}\psi]\, e^{-\int_0^\beta dt\, L}$$

where the Euclidean space Lagrangian is given by

$$L = \frac{1}{2} g_{\mu\nu} \dot{x}^\mu \dot{x}^\nu + \frac{1}{2} g_{\mu\nu}(x) \psi^\mu \frac{D\psi^\nu}{Dt}.$$

Remember that the periodic boundary conditions correspond to an insertion of $(-1)^F$ in the trace operation. Another important observation is that the periodic boundary conditions respect the supersymmetry transformations (3.6) and (3.7). Therefore the path integral being computed respects supersymmetry. After Wick rotation to Euclidean space, the supersymmetry transformations may be written in the form $\delta x^\mu = i\epsilon \psi^\mu$ and $\delta \psi^\mu = -i\epsilon \dot{x}^\mu$. It is necessary that supersymmetry be respected because we need the states to be paired to successfully compute the index. Since $\mathcal{I}$ is independent of β we will do the calculation in the limit $\beta \downarrow 0$. This will allow us to use the method of steepest descent. The following identity

$$\int_0^\beta dt \left(\frac{1}{2} g_{\mu\nu}(x) \dot{x}^\mu \dot{x}^\nu + \frac{1}{2} g_{\mu\nu}(x) \psi^\mu \frac{D\psi^\mu}{Dt} \right)$$
$$= \int_0^1 ds \left(\frac{1}{\beta} \frac{1}{2} g_{\mu\nu}(x) \frac{dx^\mu}{ds} \frac{dx^\nu}{ds} + \frac{1}{2} g_{\mu\nu}(x) \psi^\mu \frac{D\psi^\nu}{Ds} \right)$$

tells us that as $\beta \downarrow 0$ any classical solution which has $\dot{x} \neq 0$ will be exponentially suppressed because the Boltzmann factor in the path integral is of the form $e^{-\int L}$. Therefore, for small β, the important critical points will be the constant maps.

Our configuration space for the bosonic fields is $L(M)$, the space of loops on M. To implement the steepest descent method we need to determine the critical set $\mathcal{M}$. The classical equations of motion are

$$-g_{\lambda\mu}(x)\frac{D\dot{x}^\mu}{Dt} + \frac{1}{2}R_{\mu\nu\lambda\rho}\psi^\mu\psi^\nu\dot{x}^\rho = 0,$$
$$\frac{D\psi^\mu}{Dt} = \frac{d\psi^\mu}{dt} + \dot{x}^\lambda\Gamma^\mu_{\lambda\nu}\psi^\nu = 0.$$

These are the equations for a spinning particle in a gravitational field. Note that $x = \mathrm{constant}$, and $\psi = \mathrm{constant}$ are critical points. Thus the bosonic critical set $\mathcal{M}$ contains the constant maps, i.e., $M \subset \mathcal{M}$. The equation for x is essentially the geodesic equation. If we set $\psi = 0$ by fiat then we get the geodesic equation. The important lesson is that a spinning particle, $\psi \neq 0$, does not move along a geodesic. If $\pi_1(M) \neq 0$ then in general one will have non-contractible geodesics. These will not be considered in our computation because their contribution to the path integral will be exponentially suppressed as $\exp(-c/\beta)$ where the constant c is related to the minimum length geodesic in the appropriate sector.

Exercise 6.1. Consider the Euclidean space SUSY Lagrangian on M

$$L_E = \frac{1}{2}\langle \dot{x}, \dot{x}\rangle + \frac{1}{2}\left\langle \psi, \frac{D\psi}{Dt}\right\rangle$$

and work out the classical equations of motion.

Exercise 6.2 (Steepest descent $\hbar$ expansion). Consider the one dimensional integral

$$Z = \int_{-\infty}^{\infty} \frac{dx}{\sqrt{2\pi\hbar}}\, e^{-f(x)/\hbar}$$

where the function f looks like Figure 6.1.

We are interested in the asymptotic expansion for Z as $\hbar \to 0$. The standard approximation is called the method of steepest descent. Define $x = x_0 + \sqrt{\hbar}y$ and expand $f(x)$ in a Taylor series around x_0, where $f'(x_0) = 0$,

$$f(x) = f(x_0) + \sqrt{\hbar}\, yf'(x_0) + \frac{1}{2}\hbar y^2 f''(x_0)$$
$$+ \frac{1}{3!}\hbar^{3/2}\, y^3 f'''(x_0) + \frac{1}{4!}\hbar^2\, y^4 f^{(4)}(x_0) + \cdots$$

LECTURE 6. THE ATIYAH-SINGER FORMULA

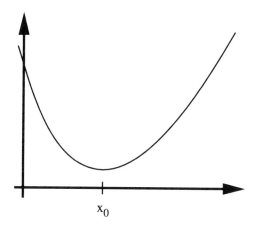

Figure 6.1. Shape of the function f in the steepest descent evaluation of the integral. The minimum occurs at x_0.

Inserting the Taylor series into the integrand one obtains

$$Z = e^{-f(x_0)/\hbar} \int_{-\infty}^{\infty} \frac{dy}{\sqrt{2\pi}} \exp\left[-\frac{1}{2}f''(x_0)y^2 \right.$$
$$\left. + \left(\hbar^{1/2}\frac{1}{3!}f'''(x_0)y^3 + \hbar\frac{1}{4!}f^{(v)}(x_0)y^4 + ...\right)\right].$$

Define moments by

$$\langle y^n \rangle = \frac{\int \frac{dy}{\sqrt{2\pi}} e^{-1/2 f''(x_0)y^2} y^n}{\int \frac{dy}{\sqrt{2\pi}} e^{-1/2 f''(x_0)y^2}}.$$

We may write the following expansion for Z:

$$Z = \frac{e^{-1/\hbar f(x_0)}}{\sqrt{f''(x_0)}} \left\langle \exp\left[-\hbar^{1/2}\frac{1}{3!}f^{(3)}(x_0)y^3 - \hbar\frac{1}{4!}f^{(4)}(x_0)y^4 + ...\right]\right\rangle.$$

Note that the terms in angular brackets are of the form $1 + O(\hbar)$ because $\langle y^3 \rangle = 0$, therefore, the quadratic expansion around the critical point x_0 already give us something of $O(\hbar^0)$. This is very important in our index calculation because the role of $\hbar$ is played by the time β. Thus a quadratic approximation to the path integral about the critical point should give us something which is of $O(\beta^0)$. Since the corrections are $O(\beta)$, we know they must vanish by general principles, *i.e.*, the index is independent of β. To compute the index via path integrals all we require is to evaluate the contributions of the quadratic fluctuations around the critical point.

We now proceed to compute the contribution to the path integral of the quadratic fluctuations[15] around the critical points, i.e., the classical solutions to the Euler-Lagrange equations. As mentioned before the dominant contribution to the path integral comes from the constant modes, x_0 and ψ_0. Choose Riemann normal coordinates at x_0; this means that local coordinates are chosen such that the form of a generic metric is as close as possible to the Euclidean metric. In practice this means one can choose $g_{\mu\nu}(x_0) = \delta_{\mu\nu}$ and $\partial_\lambda g_{\mu\nu}(x_0) = 0$. Let $\xi(t)$ and $\eta(t)$ be the fluctuations about the constant solution:

$$x^\mu = x_0^\mu + \xi^\mu,$$
$$\psi^\mu = \psi_0^\mu + \eta^\mu.$$

Note that $dx^\mu = d\xi^\mu$ and $d\psi^\mu = d\eta^\mu$ which will be important in discussing the measure in the path integral. The action may be written to quadratic order as

(6.1) $$I_2 = \int_0^\beta dt \left(\frac{1}{2} \frac{d\xi^\mu}{dt} \frac{d\xi^\mu}{dt} + \frac{1}{2} \eta^\mu \frac{d\eta^\mu}{dt} + \frac{1}{2} \tilde{\mathcal{R}}_{\mu\nu}(x_0) \xi^\mu \frac{d\xi^\nu}{dt} \right)$$

where $\tilde{\mathcal{R}}_{\mu\nu}(x_0) = \frac{1}{2} R_{\mu\nu\rho\sigma}(x_0) \psi_0^\rho \psi_0^\sigma$.

Exercise 6.3. Derive the above expression for the Lagrangian describing the the fluctuations to quadratic order. In the above, the Riemann tensor identity $R_{\mu\nu\rho\sigma} = R_{\rho\sigma\mu\nu}$ was used.

To compute the index all we will need is the quadratic order approximation for the steepest descent method as indicated in Exercise (6.2). The index is simply given by

$$\mathcal{I} = \int [\mathcal{D}\xi][\mathcal{D}\eta] \, e^{-I_2}.$$

To compute the above we introduce the Fourier expansions

$$\xi^\mu = \frac{1}{\sqrt{\beta}} \sum_{n=-\infty}^{\infty} \xi_n^\mu e^{2\pi i n t/\beta},$$

$$\eta^\mu = \frac{1}{\sqrt{\beta}} \sum_{n=-\infty}^{\infty} \eta_n^\mu e^{2\pi i n t/\beta}.$$

Note that action (6.1) involves the quadratic forms

$$-\delta_{\mu\nu} \frac{d^2}{dt^2} + \tilde{\mathcal{R}}_{\mu\nu} \frac{d}{dt} \quad \text{on} \quad \xi,$$
$$\delta_{\mu\nu} \frac{d}{dt} \quad \text{on} \quad \eta.$$

---

[15] The expansion around the critical point set corresponds to locally decomposing the manifold into the critical point set and the normal bundle to the critical point set.

Note that the $n = 0$ modes have eigenvalue zero with respect to the above, thus, the path integral becomes

$$\mathcal{I} = \int \prod_{\mu=1}^{d} \frac{d\xi_0^\mu}{\sqrt{2\pi}} d\eta_0^\mu \left[\det{}'\left(\delta_{\mu\nu}\frac{d}{dt}\right)\right]^{+1/2} \left[\det{}'\left(-\delta_{\mu\nu}\frac{d^2}{dt^2} + \widetilde{\mathcal{R}}_{\mu\nu}(x_0)\frac{d}{dt}\right)\right]^{-1/2}$$

$$= \int \prod_{\mu=1}^{d} \frac{d\xi_0^\mu}{\sqrt{2\pi}} d\eta_0^\mu \left[\det{}'_{PBC}\left(-\delta_{\mu\nu}\frac{d}{dt} + \widetilde{\mathcal{R}}_{\mu\nu}(x_0)\right)\right]^{-1/2} .$$

How do we evaluate $\det'$ above? Note that $\mathcal{R}_{\mu\nu}(x_0)$ is an even element of the Grassmann algebra, therefore, we will momentarily treat it as an ordinary number in order to evaluate the determinant. The final answer will be written in terms of invariant objects. Treat $\widetilde{\mathcal{R}}_{\mu\nu}$ as an antisymmetric matrix, block diagonalize it to obtain

$$(6.2) \qquad \widetilde{\mathcal{R}} = \begin{pmatrix} 0 & y_1 & & & \\ -y_1 & 0 & & & \\ & & \ddots & & \\ & & & 0 & y_n \\ & & & -y_n & 0 \end{pmatrix},$$

and let us concentrate on a 2×2 block. Notice that the operator $-\delta_{\mu\nu}\frac{d}{dt} + \mathcal{R}_{\mu\nu}(x_0)$ is skew and real, therefore, the eigenvalues come in complex conjugate pairs and there is a natural definition for the determinant. By studying the spectrum one easily sees that

$$\det{}'\begin{pmatrix} -\frac{d}{dt} & +y_1 \\ -y_1 & -\frac{d}{dt} \end{pmatrix} = \det{}'\left(-\frac{d^2}{dt^2} + y_1^2\right)$$

$$= \frac{\det\left(-\frac{d^2}{dt^2} + y_1^2\right)}{y_1^2}$$

$$= \left[\frac{2\sinh \beta y_1/2}{y_1}\right]^2 ,$$

where in the last line we used our previous computation of the determinant in the simple harmonic oscillator problem, see (5.5). Thus we conclude that the index may be written as

$$(6.3) \qquad \mathcal{I} = \int \prod_{\mu=1}^{d} \frac{d\xi_0^\mu}{\sqrt{2\pi}} d\eta_0^\mu \prod_{j=1}^{d/2} \frac{y_j/2}{\sinh \beta y_j/2} .$$

The product may be written in the form

$$\frac{1}{\beta^{d/2}} \det \left(\frac{\beta\widetilde{\mathcal{R}}/2}{\sinh \frac{1}{2}\beta\widetilde{\mathcal{R}}} \right)^{1/2}.$$

Since $\widetilde{\mathcal{R}}^p = 0$ for $p > d/2$, it follows that the above has a finite Taylor series in $\widetilde{\mathcal{R}}$, therefore, there is no ambiguity as to its meaning.

We just computed the contributions due to the quadratic fluctuations near a given critical point. We have to remember that if there is more than one critical point then the steepest descent approximation instructs us that we should sum over all critical points. For example, the critical point set $\mathcal{M}$ for x contains the manifold M itself. All critical points not in M have an exponentially small contribution to the path integral thus we neglect them. We have to sum over all points in M, i.e., integrate over M. Formula (6.3) involves an integration over the "zero mode" ξ_0. This is fortuitous because the "zero mode" corresponds to translations in the tangent space at x_0 and therefore is essentially the variable of integration. More precisely, our expansion for x was of the form

$$x = x_0^\mu + \frac{1}{\sqrt{\beta}} \xi_0^\mu + \cdots$$

This tells us that changing x_0 is essentially the same as changing ξ_0, i.e., we can generate the same dx if we chose $dx_0^\mu = d\xi_0^\mu/\sqrt{\beta}$. It is also useful to make the same transformation for the anticommuting variables. Remembering that the Jacobian is reciprocal to bosonic variables we obtain $d\psi_0^\mu = \sqrt{\beta}\, d\eta_0^\mu$. Inserting this into expression (6.3) we obtain

$$\mathcal{I} = \int \left(\prod_{\mu=1}^{d} \frac{dx_0^\mu}{\sqrt{2\pi}} d\psi_0^\mu \right) \frac{1}{\beta^{d/2}} \det \left(\frac{\beta\widetilde{\mathcal{R}}}{\sinh \beta\widetilde{\mathcal{R}}/2} \right)^{1/2}.$$

To put this in a more conventional form make the change of variables $\psi_0^\mu = \chi_0^\mu/\sqrt{2\pi\beta}$ with the measure changing as $d\psi_0^\mu = \sqrt{2\pi\beta}d\chi_0^\mu$. Observe that

$$\beta\widetilde{\mathcal{R}}_{\mu\nu} = \frac{1}{2\pi} \cdot \frac{1}{2} \mathcal{R}_{\mu\nu\rho\sigma}(x_0) \chi_0^\rho \chi_0^\sigma.$$

The index may be written as

$$\mathcal{I} = \int \prod_{\mu=1}^{d} dx_0^\mu \, d\chi_0^\mu \, \det \left(\frac{\frac{1}{2} \cdot \frac{1}{2\pi} \cdot \frac{1}{2} \mathcal{R}_{\mu\nu\rho\sigma} \chi_0^\rho \chi_0^\sigma}{\sinh \frac{1}{2} \cdot \frac{1}{2\pi} \cdot \frac{1}{2} \mathcal{R}_{\mu\nu\rho\sigma}(x_0) \chi_0^\rho \chi_0^\sigma} \right)^{1/2}.$$

This is the Atiyah-Singer formula for the index of the Dirac operator. We can turn it into notation more familiar to mathematicians by observing that the "instruction" $\int \prod dx d\chi$ acting on an integrand is a physicist's notation for integration of differential forms. The $d\chi$ tells us that we only get a contribution from the term which is homogeneous of degree d in χ. Since the χ's are anticommuting the expression is automatically fully antisymmetrized. The dx term tells us to do an ordinary integration. This is exactly what it means to integrate a differential form. Let us now rewrite the above in a more familiar notation. Introduce the notation of differential forms by defining the curvature two form

$$\mathcal{R}_{\mu\nu} = \frac{1}{2} R_{\mu\nu\rho\sigma}(x) dx^\rho \wedge dx^\sigma \,.$$

Our expression for the Atiyah-Singer formula may be written in the form

(6.4) $$\mathcal{I} = \int_M \det \left(\frac{\frac{1}{2} \cdot \frac{1}{2\pi} \mathcal{R}}{\sinh \frac{1}{2} \cdot \frac{1}{2\pi} \mathcal{R}} \right)^{1/2} \,.$$

The above is the traditional form for the Atiyah-Singer formula for the index of the Dirac operator. A short hand notation is often used. Use the formal block diagonalization idea discussed in conjunction with equation (6.2)

(6.5) $$\mathcal{R} = \begin{pmatrix} 0 & x_1 & & & \\ -x_1 & 0 & & & \\ & & \ddots & & \\ & & & 0 & x_n \\ & & & -x_n & 0 \end{pmatrix}$$

and introduce the $\hat{A}$-genus of the manifold via the formal expansion

(6.6) $$\hat{A}(M) = \prod_{j=1}^{d/2} \frac{x_j/2}{\sinh x_j/2} \,.$$

Theorem 6.4 (Atiyah-Singer Index Formula). *The Atiyah-Singer formula for the index of the Dirac operator is*

(6.7) $$\mathcal{I} = \int_M \hat{A}(M) \,.$$

LECTURE 7
Global Topological Issues

Characteristic classes

Let E be a vector bundle over a manifold M. For our purposes a characteristic class is a device which assigns a cohomology class in M to the vector bundle E. The assignment essentially proceeds along the following line. Let ω be a connection on E with curvature Ω. The curvature satisfies the Bianchi identity $0 = D\Omega = d\Omega + [\omega, \Omega]$ where D is the covariant differential. A consequence of the above is that the differential form $\text{Tr}\,\Omega^p$ is a closed form and the ring generated by such forms may be associated to elements of $H^*(M, \mathbb{R})$. Just as important is the following key observation: if ω and ω' are two connections and if $\eta(\omega)$ and $\eta(\omega')$ are the respectively associated closed forms made out of the curvatures then $\eta(\omega) - \eta(\omega') = d\tau$ for some form τ. Therefore, the cohomology class of η is independent of the choice of connection on the vector bundle and is a topological invariant. By being clever, one can actually assign to every vector bundle integral cohomology classes. The most famous of the integer characteristic classes is the Euler class which on two manifolds can be represented as a curvature integral via the Gauss-Bonnet Theorem. Two other famous integer characteristic classes are the Pontrjagin classes and the Chern classes.

The *total Pontrjagin class* $p(E) \in H^*(M, \mathbb{Z})$ of a real $O(k)$ bundle E with curvature Ω is defined by

$$
\begin{aligned}
p(E) &= \det\left(I - \frac{1}{2\pi}\Omega\right) \\
&= 1 + p_1(E) + p_2(E) + \cdots .
\end{aligned}
\tag{7.1}
$$

The form $p_k(E)$ is a $4k$-form. The Pontrjagin class of a manifold is the Pontrjagin class of its tangent bundle, $p(M) = p(TM)$.

The $\hat{A}$-genus may be expanded in terms of Pontrjagin classes of the manifold with the result

$$
\hat{A}(M) = 1 - \frac{1}{24}p_1(M) + \frac{1}{5760}(7p_1(M)^2 - 4p_2(M)) + \cdots .
\tag{7.2}
$$

Since the Pontrjagin classes are integer classes we see that the index of the Dirac operator on a four manifold must be a multiple of 24. Note that if $\dim M \neq 4k$, where $k \in \mathbb{Z}_+$, then the index of the Dirac operator vanishes.

If E is a complex vector bundle of rank r over M then the *total Chern class* in $H^*(M, \mathbb{Z})$ of the vector bundle is defined by

$$c(E) = \det\left(I + \frac{i}{2\pi}\Omega\right)$$
$$= 1 + c_1(E) + c_2(E) + \cdots,$$

where $c_k(E)$ is a $2k$-form. Applying our formal diagonalization procedure to Ω leads to the formal expression

$$\frac{i}{2\pi}\Omega = \begin{pmatrix} x_1 & & & \\ & x_2 & & \\ & & \ddots & \\ & & & x_r \end{pmatrix}$$

with the following formal expansion for the total Chern class of the form $c(E) = \prod(1 + x_i)$.

Another useful characteristic class is the Chern character defined by

$$ch(E) = \operatorname{Tr} \exp\left(\frac{i}{2\pi}\Omega\right) = \sum_{j=1}^{r} e^{x_j}.$$

The Chern character may be related to the Chern classes in a straightforward way:

$$ch(E) = \operatorname{rank} E + c_1(E) + \frac{1}{2}(c_1^2(E) - 2c_2(E)) + \cdots.$$

If E and F are vector bundles then the characteristic classes we have defined satisfy the following relations

$$\begin{aligned} c(E \oplus F) &= c(E) \wedge c(F), \\ ch(E \oplus F) &= ch(E) + ch(F), \\ ch(E \otimes F) &= ch(E) \wedge ch(F). \end{aligned}$$

Exercise 7.1 (The four sphere). Consider S^4 with the standard metric and compute the Pontrjagin class and the index of the Dirac operator.

Exercise 7.2 (Projective space). Look up the Fubini-Study metric in a differential geometry book or a complex manifolds book and compute the Chern class for the tangent bundle (see paragraph below) for $\mathbb{CP}^1$ and $\mathbb{CP}^2$. Show that $\int_{\mathbb{CP}^2} p_1 = 3$. Can you compute the Chern class for an arbitrary $\mathbb{CP}^n$?

Let us apply the above to the standard example of $\mathbb{CP}^n$, complex projective n-space. Let $T^{(1,0)}(\mathbb{CP}^n)$ be the $(1,0)$ part of the complexified tangent bundle then $c(T^{(1,0)}(\mathbb{CP}^n)) = (1+x)^{n+1}$. It follows from this that the Pontrjagin class of the tangent bundle is given by $p(T(\mathbb{CP}^n)) = (1+x^2)^{n+1} = 1 + (n+1)x^2 + \cdots$. We immediately see that $c_1 = (n+1)x$, $c_2 = n(n+1)x^2/2$ and $p_1 = c_1^2 - c_2 = (n+1)x^2$.

As an exercise, we will try to compute the index of the Dirac operator on $\mathbb{CP}^2$. The Atiyah-Singer formula states that

$$\mathcal{I} = \int_{\mathbb{CP}^2} \hat{A} = -\frac{1}{24} \cdot \int p_1 = -\frac{1}{24} \cdot 3 = -\frac{1}{8}.$$

This is a startling result. The index is supposed to be an integer yet we have found a fraction. The implication is that there is something wrong with our path integral calculation. Actually, there is nothing wrong with our path integral calculation. The reason for this paradox is that one can show that it is impossible to define spinors globally on $\mathbb{CP}^2$. This is usually stated by saying that $\mathbb{CP}^2$ does not have a spin structure. Morally we had no right to attempt the calculation above and the fractional value given by the index formula is a warning. A remarkable fact as we will see below is that the path integral is smart enough to tell us when a spin structure exists! The normal bundle quadratic deformation analysis we just did cannot be the full answer since it only dealt with local questions.

Global properties of the path integral

We go back to the full Lagrangian (3.11) and study some global properties of the path integral. For simplicity we assume that M is connected and simply connected. We can do the fermionic integral exactly obtaining

$$(7.3) \qquad \mathcal{I} = \int_{L(M)} [\mathcal{D}x] \, e^{-\int_0^\beta dt \frac{1}{2}\langle \dot{x}, \dot{x} \rangle} \, \mathrm{Pf}\left(\frac{D}{Dt}\right)(x(t)).$$

This is an averaging of the Pfaffian with respect to Wiener measure. The question is whether $\mathrm{Pf}(D/Dt)$ may be consistently defined over $L(M)$. The real skew operator D/Dt has imaginary eigenvalues which come in complex conjugate pairs $\pm i\lambda_n$, where λ_n is real. First we define $\mathrm{Pf}(D/Dt)$ at a *generic* fiducial loop γ_0. At a generic loop, zero is not an eigenvalue of D/Dt. Take the positive imaginary eigenvalues at γ_0 and simply define $\mathrm{Pf}(D/Dt)(\gamma_0) = \prod_{\lambda_n > 0} \lambda_n(\gamma_0)$. We emphasize that the eigenvalues depend on the curve γ_0. Consider a homotopy of loops[16] γ_s, $s \in [0,1]$, beginning at the loop γ_0 which returns to $\gamma_0 = \gamma_1$ at $s = 1$. The eigenvalues vary as the parameter s goes from 0 to 1, see Figure 7.1. This flow of the eigenvalues is known as *spectral flow*. If an eigenvalue crosses zero then we will have that $\mathrm{Pf}(D/Dt)(\gamma_1) = -\mathrm{Pf}(D/Dt)(\gamma_0)$. This means that the Pfaffian of D/Dt is not well defined on $L(M)$. This sign ambiguity implies that our path integral is ill defined. If $\pi_1(L(M)) = 0$ then this sign ambiguity cannot happen. Elementary homotopy theory tells us that $\pi_1(L(M)) = \pi_2(M)$ because of our assumptions about M. Since the Pfaffian in question is ambiguous up to a sign we see that the obstruction is in $H^2(M, \mathbb{Z}/2\mathbb{Z})$. The obstruction is called the second Stiefel-Whitney class.

The Stiefel-Whitney classes are $\mathbb{Z}/2\mathbb{Z}$ cohomology classes. If E is a real vector bundle of rank k over a connected manifold M, $\dim M = d = 2n$, then it is

---

[16] This homotopy of loops on $L(M)$ is equivalent to a map from a two-torus to M.

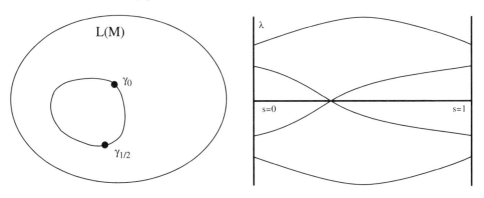

Figure 7.1. The homotopy γ_s of loops and the flow of eigenvalues.

possible to define Stiefel-Whitney classes $w_i \in H^i(M, \mathbb{Z}/2\mathbb{Z})$ for $i = 1, \ldots, d-1$. Additionally, the highest class $w_d \in H^d(M, \mathbb{Z}/2\mathbb{Z})$ is the reduction modulo 2 of an integer class. The total Stiefel-Whitney class is defined to be

$$w(E) = 1 + w_1(E) + \cdots + w_d(E).$$

For a complex vector bundle, the total Stiefel-Whitney class is the reduction modulo 2 of the total Chern class. For $\mathbb{CP}^n$, the total Stiefel-Whitney class is given by

$$\begin{aligned} w(\mathbb{CP}^n) &= (1+x)^{n+1} \mod 2 \\ &= \left(1 + (n+1)x + \frac{n(n+1)}{2}x^2 + \cdots\right) \mod 2. \end{aligned}$$

The integer class x in $\mathbb{CP}^n$ has degree two and therefore $w_2(\mathbb{CP}^2) = 3x \mod 2 = x \mod 2 \neq 0$. One cannot define spinors on $\mathbb{CP}^2$. This is in agreement with our previous computation of a "fractional index" for the Dirac operator on $\mathbb{CP}^2$. Note that there is no trouble defining spinors on $\mathbb{CP}^1$ because $w_2(\mathbb{CP}^1) = 2x \mod 2 = 0$.

Exercise 7.3. Clean up all the loopholes in the above discussion.

Lectures on Axiomatic Topological Quantum Field Theory

Frank Quinn

Lectures on Axiomatic Topological Quantum Field Theory

Frank Quinn

Abstract. Basic ideas and simple examples of topological quantum field theories are described. The examples include Chern-Simons theories with finite gauge group from the functional integral point of view (with "integrals" which are finite sums). Another example illustrates constructions using category theory, similar to the constructions on 3-manifolds using representations of "quantum groups". The main novelty beyond giving formal definitions and at least sketching proofs is the systematic investigation of natural ring structures and tensor products in modular theories. This gives derivations of "fusion rule algebras" and modular functors (rational conformal field theories) from more basic structures. There are several algebraic classifications, and for example most TQFT on 1 + 1-dimensional manifolds are seen to be non-modular. Sample calculations are emphasized and exercises are included.

Topological quantum field theories (TQFT) are a recent development in the interface between physics and mathematics. The mathematical interest in them comes from the hope that they will disclose new phenomena, or at least offer efficient organization of previously studied invariants like the Jones invariants of links, or the Donaldson invariants of 4-manifolds. The physical interest comes from their value as examples in which extensive calculations are possible. They also shed light on mathematical structures involved in more realistic theories. It seems fair to say that from both points of view this is still an exploration—in a state of flux with the best applications (one hopes) still to come.

These notes are intended to provide an introduction through simple examples and abstract structure. We do not carry out the more sophisticated constructions, but indicate how these are related to the simple examples. The tools are topology and algebra, rather than the analysis used in the physically inspired examples. And in line with their origin in lectures[2] directed to graduate students some of the examples are presented in perhaps too much detail, exercises are given, there are lots

---

[1] Virginia Polytechnic Institute and State University Blacksburg VA 24061-0123
E-mail address: quinn@math.vt.edu

[2] From lectures given at the Regional Geometry Institute in Park City, and the IMA summer school in Evanston, summer 1991

© 1995 American Mathematical Society

of drawings, and many proofs are only sketched. The references were not complete when these notes were written, and will be long out of date when published.

Outline

The first section presents a rough idea of the characteristic properties of a TQFT. We do eventually get a formal definition (in §6), but since it is fairly elaborate, and since simple examples can be discussed without it, it is postponed as long as possible. Briefly a TQFT is defined on "spaces" usually denoted Y, and "spacetimes" usually denoted X. It associates to a space a module $Z(Y)$ over a coefficient ring, often the complex numbers. A spacetime interpolating between two spaces Y_1, Y_2 induces a homomorphism $Z_X \colon Z(Y_1) \to Z(Y_2)$. The most characteristic property of a TQFT is the behavior with respect to disjoint union:

$$Z(Y_1 \sqcup Y_2) = Z(Y_1) \otimes_R Z(Y_2).$$

The tensor product forshadows the appearance of a lot of abstract algebra.

In section 2 we explain why topologists might be interested in these theories. The most interesting possibilities have to do with unstable phenomena in smooth 4-manifolds (eg. the smooth Poincaré conjecture) and finite 2-complexes (eg. the Andrews-Curtis conjecture). Developments to date mostly concern invariants for 3-manifolds and links, though this has not yet substantial consequences. Eventually the main importance will probably be their use as a take-off point for nontopological versions in physics and geometry.

The third and fourth sections give the first examples. The simplest is based on the Euler characteristic, and is defined for all finite CW complexes. Next (in §4) is the "finite total homotopy" example suggested by Kontsevich [**K**]. This is the analog for finite gauge group of the functional integral used by Witten [**Wi1**] to predict the "Chern-Simons" TQFT on 3-manifolds.

In section 5 the first moderately involved example is described: the TQFT of §4 is elaborated by "twisting" by a characteristic number defined using a cohomology class in a classifying space. This corresponds to a nontrivial Lagrangian in the functional integral, and has been considered in [**DW**], [**FQ**]. The resulting TQFT is quite a bit more complicated structurally, and we consider it from two points of view. First the "answer" is given. This is a reasonably explicit description in terms of choices, and to illustrate what such descriptions involve it is implemented for oriented surfaces and 2-dimensional cohomology classes in classifying spaces of finite groups. This answer is too dependent on choices, and too arbitrary looking, to be satisfactory as a definition (to a mathematician, anyway). The second point of view is natural but abstract, involving inverse limits of functors. The "answer" is then obtained from this by computing inverse limits. A brief introduction to inverse limits is included since these are a basic tool in many of the constructions. At the end of the section we discuss how the analytic constructions are analogous to these models: geometric quantization being subsituted for inverse limits, etc.

In section 6 a formal definition of TQFT is given. This is abstract, and very general about the "spaces" and "spacetimes" on which the field theory is defined.

This generality is driven by examples. In the geometric theories defined on manifolds we want to allow for extra structure, singularities, etc. without having to tinker with the basic definition. The simpler examples are defined more generally, often on all finite CW complexes, and to understand them it is important to consider them in their natural setting. Also, there are TQFT-like structures which are not even defined on spaces: "field theories" on groups or algebras for example.

Section 7 describes the elementary structure of a TQFT. For example each state module $Z(Y)$ has a nondegenerate bilinear pairing on it, and in consequence is a finitely generated projective module. A TQFT as defined here is shown to be determined by its "vacuum vectors." This relates the definition to the original one suggested by Atiyah, which is seen to be slightly overdetermined. Ambialgebras make their first appearance, and are shown to characterize TQFT in low dimensions: TQFT on $1+1$-manifolds and $0+1$-complexes both correspond to commutative ambialgebras.

The most interesting part of the development starts in §8, where we begin consideration of modular TQFT. In modular theories the boundary objects may have "corners" and unions $Y_1 \cup_W Y_2$ identified along corners are defined. A corner object W has an associated boundary object $W \times I$ and $Z(W \times I)$ receives a natural ring structure. The characteristic disjoint union property extends in the following way: if $W \subset \partial Y$ then $Z(Y)$ has a natural module structure over the ring $Z(W \times I)$, and then

$$Z(Y_1 \cup_W Y_2) = Z(Y_1) \otimes_{Z(W \times I)} Z(Y_1).$$

From the functional integral point of view this property results from locality of the action. Section 8 begins with a rough description of the basic structure of a modular theory and an indication of how the ring and module structures are derived from it. The focus of the section is on examples. Modular extensions of the examples of §§3–5 are described, as well as a nearly trivial example based on automorphism groups which forshadows the categorical constructions of §10.

Section 9 begins with a more formal definition of modular theories. The basic structure (following from the axioms) is then described, beginning with special ambialgebra structures on $Z(W \times I)$ for corner objects W. "Special" means the trace ambialgebra has a unit. On $1+1$-manifolds the structure on $Z(\text{pt} \times I)$ gives a classification: modular TQFT correspond exactly to special ambialgebras. The associated nonmodular TQFT is determined by the trace ambialgebra. It follows that most $1+1$-manifold TQFT are not modular, since most commutative ambialgebras cannot be realized as trace ambialgebras.

The next topic is nondegenerate forms. The nondegenerate pairings on $Z(Y)$ over $R = Z(\phi)$ are extended to pairings over the ring $Z(\partial Y \times I)$. This permits construction of a TQFT on the boundary category using ranks, which is a generalization of the Verlinde "fusion rule algebra."

Morita equivalence is discussed next. This comes from a phenomenon in algebra where different rings have isomorphic categories of modules. It is used to change the corner ambialgebras without changing the associated nonmodular TQFT. This, in turn, is used to relate modular TQFT to the modular functors of Segal and others:

roughly a modular functor is the state module functor of a modular TQFT on 2+1-manifolds which has commutative corner ambialgebras. Any (complex coefficient) modular TQFT is shown to be Morita equivalent to one of these.

The final topic of the section concerns the Verlinde basis for $Z(W \times S^1)$, when W is a corner object. Here the basis appears as the primitive idempotents in a commutative semisimple ambialgebra. This leads to the definition of "labeled link invariants" which appear prominently in some other treatments. Another application of the basis is a matrix description of the induced symmetries of $Z(S^1 \times S^1)$ in a modular 2+1-manifold TQFT. A consequence is the Verlinde theorem, stated in the physics literature as "the S-matrix diagonalizes the fusion rules."

The final section, 10, concerns the construction of modular TQFT using category theory. This works in low dimensions: 1+1-complexes and 2+1-manifolds. In the manifold case it was pioneered by Reshetikhin and Turaev with categories derived from the representation category of a Hopf algebra, particularly the "quantum group" $SU_q(2)$ when q is a root of unity. A more axiomatic construction basically in terms of representations of mapping class groups was initiated by Moore and Seiberg, and developed by Walker, Crane, Kohno, and many others. Here we work out explicit calculations in a TQFT defined on 1+1-dimensional CW complexes using a category derived from the mod 5 representations of the algebraic group $SL(2)$. This example shows clearly the kind of manipulations involved both in other 1+1- complex TQFT and in the analogous constructions on 2+1-manifolds.

Algebraic facts are described in appendex A. This begins with a review of (algebraic) tensor products and nondegenerate pairings. We then discuss traces and ranks. A detailed discussion is given of ambialgebras. This is an algebraic structure which is similar to, but simpler than, a Hopf algebra, and which appears frequently in TQFT. Theories on surfaces, for example, are equivalent to ambialgebras. Ambialgebras are new here, although equivalent structures have been used previously, and the structure theory is closely related to the representation theory of groups. Semisimple ambialgebras over the complex numbers are studied in detail, since they occur as corner ambialgebras in modular TQFT.

LECTURE 1
A Rough Idea

We explain the basic ideas behind the words "topological quantum field theory." This will be formalized in section 5.

On a general heuristic level a "field theory" associates a set of "fields" to a "space," referred to here as a boundary. One tends to think of the boundaries as manifolds of some fixed dimension d, especially $d = 2$ or 3. But as noted above it will be helpful to be much more general about this. These boundaries occur as boundaries of "spacetimes" which we think of as manifolds of dimension $d+1$. When the boundary of a spacetime is divided into two parts, the "incoming" boundary and the "outgoing" boundary, then the field theory gives a way to begin with fields on the incoming boundary and propagate them across the spacetime to give fields on the outgoing boundary.

Figure 1.1. The standard picture

If the state module of fields on Y is denoted $Z(Y)$, and X is a spacetime with incoming boundary Y_1 and outgoing boundary Y_2, then X determines a function

$$Z_X : Z(Y_1) \to Z(Y_2).$$

We note the convention of denoting the functions with a subscript: $Z(Y)$ is a set, Z_X is a function.

The functions associated to spacetimes should have a composition property: if we have spacetimes

$$Y_1 \xrightarrow{X_1} Y_2 \xrightarrow{X_2} Y_3$$

then glueing X_1 and X_2 along Y_2 gives a spacetime going from Y_1 to Y_3. The associated function should be the composition:

$$Z_{X_1 \cup_{Y_2} X_2} = Z_{X_2} Z_{X_1}. \tag{1.2}$$

Eventually we define a "bordism category" with objects the boundaries Y and morphisms $X\colon Y_1 \to Y_2$ the spacetimes with incoming boundary Y_1 and outgoing Y_2. A spacetime with boundary divided in this way is called a "bordism" in conformity with topological usage. A TQFT will be a functor defined on this category, so the composition property (1.2) is just the usual composition property of a functor.

We next explain "topological." The characteristic property of topological field theories is that a product spacetime $I \times Y$ gives the identity function

$$Z_{I \times Y} = \mathrm{id} \colon Z(Y) \to Z(Y).$$

Thus if Z_X is nontrivial it must be because X itself is topologically nontrivial. This contrasts with the physically interesting theories which depend on a Riemannian metric and other data on X, and which can give highly nontrivial Z_X on cylinders $I \times Y$.

Finally "quantum." A characteristic property of quantum field theories is that they take disjoint unions of spaces to tensor products of state modules: $Z(Y_1 \sqcup Y_2) \simeq Z(Y_1) \otimes Z(Y_2)$. Note that for tensor products to make sense the sets $Z(Y)$ must have an algebraic structure (be a module over a ring). In contrast, "classical" theories tend to take unions to cartesian products, which requires no special structure.

LECTURE 2
Topological Opportunities for TQFT

Atiyah [**A1, A2**] has suggested several settings in which TQFT might be valuable. We recall some of these, and add a few.

The most important possiblity concerns invariants of smooth 4-manifolds. Here the spaces would be 3-manifolds (perhaps with restrictions, or extra structure) and the spacetimes would be 4-manifolds. Invariants of closed 4-manifolds would be obtained by thinking of the manifold as a spacetime from the empty space to itself. $Z(\phi)$ is the coefficient ring, the complex numbers in the geometric examples, so $Z_X \colon Z(\phi) \to Z(\phi)$ can be regarded as a complex number. If the manifold has boundary then by declaring it to be outgoing we get an element $Z_X(1) \in Z(\partial X)$. Or thinking of it as incoming gives a homomorphism (functional) $Z_X \colon Z(\partial X) \to Z(\phi)$. (We are ignoring orientation considerations here).

S. Donaldson has defined invariants using (classical) solutions to the self-dual Yang-Mills equations. These have revealed amazing aspects of smooth 4-manifolds unique to that dimension. They also are qualitatively different from previous smooth invariants and an order of magnitude harder to work with. Most of the calculations to date involve algebraic geometry, so are limited to algebraic surfaces.

Atiyah suggested there might be a TQFT-like formulation of the Donaldson invariants. This might greatly help in understanding and evaluating them, especially using the composition property: decompositions of the 4-manifold into pieces would be reflected in decompositions of the invariants. Some of this has been implemented by A. Floer and others who defined appropriate "fields" (at least for rational homology 3-spheres) and Donaldson, who associated homomorphisms to 4-manifolds with these as boundaries. Witten has also described a Lagrangian which formally predicts a quantum formulation using functional integral heuristics. Unfortunately this has not helped with actual technical constructions, so far.

We point out another heuristic reason to think quantum-type theories might have an advantage here. The strange behavior of smooth 4-manifolds vanishes after connected sums with $S^2 \times S^2$. Suppose X_1 and X_2 are 4-manifolds which are predicted to be diffeomorphic by the high-dimensional theory, which gives a satisfactory solution to analogous problems in all dimensions above 4. Then for some k the connected sums with k copies of $S^2 \times S^2$ are in fact diffeomorphic. This is not a deep result, but it "explains" why classical theories have been unsuccessful with the problem. Suppose $\mathcal{I}(X)$ is some sort of invariant of 4-manifolds which takes values in a group G. Typically this is multiplicative with respect to connected sum, or the deviation is not too complicated, so imagine $\mathcal{I}(X\#S^2\times S^2) = \mathcal{I}(X)\mathcal{I}(S^2\times S^2)$. Denote $\mathcal{I}(S^2\times S^2)$ by $s\in G$, then iteration gives $\mathcal{I}(X\#kS^2\times S^2) = \mathcal{I}(X)s^k$. If X_1 and X_2 become diffeomorphic after k-fold stabilization then we get

$$\mathcal{I}(X_1)s^k = \mathcal{I}(X_2)s^k.$$

But this equation takes place in a group. Multiplying by the inverse s^{-k} gives $\mathcal{I}(X_1) = \mathcal{I}(X_2)$, and the invariant fails to detect the difference between the two.

Quantum theories are likely to have the same multiplicative property because connected sum is close to disjoint union, which is multiplicative. The new advantage is that the product takes place in a ring rather than a group, so not all elements are invertable. In particular theories with $Z_{S^2\times S^2} = 0$ seem promising.

We mention a problem about which even less is known, but which is simpler to state and serves as a useful model for 4-manifolds. This is the Andrews-Curtis conjecture about deformations of 2-complexes. A 2-complex is obtained by beginning with some points, joining these by arcs, and then attaching 2-dimensional disks using maps $S^1 \to X$ (finitely many in each case). A "2-deformation" is obtained by a sequence of moves of two types. The first changes the attaching maps of the 2-cells by homotopy. The second is either an expansion, which attaches D^1 or D^2 by a point or arc respectively, or a collapse which removes a D^k attached in this way. These moves do not change the homotopy equivalence class of the 2-complex. The conjecture asks: suppose there is a deformation from X_1 to X_2 in which expansions and collapses of arbitrary dimension are allowed (ie. a simple homotopy equivalence). Is there then a 2-deformation?

As with smooth structure questions about 4-manifolds, the answer is known to be "yes" for the analogous question in any other dimension ([**Wa**]). The expectation is that the answer will be no in dimension 2, and there are lots of proposals for conterexamples. But no invariants have been found to distinguish between them.

This problem has an instability very like smooth 4-manifolds: if there is an arbitrary deformation from X_1 to X_2 then for some k there is a 2-deformation from $X_1 \vee kS^2$ to $X_2 \vee kS^2$ (here $\vee kS^2$ denotes the 1-point union with k copies of the 2-sphere). So a 2-complex acts much like half of a 4-manifold. The same heuristic argument given above "explains" why this question has resisted solution with classical-type invariants, and why quantum invariants might be better.

The study of deformation classes of 2-complexes thus seems a good model problem for diffeomorphism classes of 4-manifolds, as long as we are willing to be flexible about "boundaries" and "spacetimes." Here the boundaries are 1-complexes and the spacetimes are (deformation classes of) 2-complexes. For different reasons 2-complexes also give a good model for the construction of TQFT on 3-manifolds.

The constructions which have been most successful involve 3-manifolds. For these theories the boundaries are 2-manifolds, and the spacetimes are 3-manifolds with boundary. (A little more data is usually required). Witten [**Wi1**] predicted the existence of such theories via the functional integral heuristic, and detailed constructions from a rather different approach have been supplied by Reshetikhin and Turaev [**RT**] and Walker [**Wk**], with contributions by many others, eg. Crane [**C**], Moore and Seiberg [**MS1**], Kirby and Melvin [**KM**], Freed and Gompf [**FG**],

Kohno [**Ko**]. These theories are intriguing, and have given interesting insights into previously known invariants (particularly extensions of the Jones polynomial), but have yet to yield dramatic new results.

Questions about 3-manifolds do not have an instability like those noted above for 4-manifolds and 2-complexes, so there may be less unique opportunity here for TQFT. Also, to a very large extent 3-manifolds are determined by their fundamental groups, so in principle this provides a complete algebraic invariant. However fundamental groups are hard to study directly. For example they come in terms of a presentation, and there is no algorithm to determine from a presentation if the group is even nontrivial. So in a sense invariants of 3-manifolds are invariants for a special class of groups, and their value comes in skimming off useful information, or in being more accessible to computation. The advantages of TQFT may thus be quantitative rather than qualitative, and it is hard to anticipate when such differences might turn out to be crucial.

In fact major topological opportunities for TQFT seem limited, nearly by process of elimination: classical approaches work very well in other dimensions and for other types of problems. Probably the most exciting mathematical opportunities for quantum theories are in settings with more geometric structure, e.g. symplectic manifolds and their kin. However these can not be "topological" in the strong sense used here. The topological theories are likely to appear as starting points, limiting cases, and sources of intuition.

LECTURE 3
The Euler Theory

In the first two examples the boundaries are finite CW complexes, so we will associate a space of fields $Z(Y)$ to any such Y. The spacetimes are finite CW complex pairs (X, Y). The associated bordisms have Y divided into (disjoint) components Y_1 and Y_2, and this is denoted by $(X; Y_1, Y_2)$ or $X : Y_1 \to Y_2$. The first subset is the "incoming" boundary and the second the "outgoing." To these we associate homomorphisms $Z_{(X;Y_1,Y_2)}$, or just Z_X if the subsets are clear from the context. These homomorphisms will depend only on the homotopy equivalence class of X relative to Y. Note these "boundaries" need not have any special topological properties like the boundary of a manifold, but are just subcomplexes.

3.1. Euler characteristics

The first example is derived from the Euler characteristic. If X is a finite CW comples this is an integer defined by

$$\chi(X) = \sum_{i=0}^{\infty} (-1)^i \operatorname{rank} H_i(X).$$

The finiteness hypothesis on X implies this is a finite sum of integers. It can also be obtained directly from the cells of X: suppose n_i is the number of cells in X of dimension i, Then $\chi(X) = \sum_{i=0}^{\infty}(-1)^i n_i$. Extend this to pairs by defining $\chi(X, Y)$ to be the alternating sum of ranks of *relative* homology $H_i(X, Y)$. Again this can be expressed in terms of cells: the alternating sum of numbers of cells in X but not in Y.

Euler characteristics are additive. In particular if $X_1 \supset Y_1 \sqcup Y_2$ and $X_2 \supset Y_2$ and we form the union $X_1 \cup_{Y_2} X_2$, then

(3.2) $$\chi(X_1 \cup_{Y_2} X_2, Y_1) = \chi(X_1, Y_1) + \chi(X_2, Y_2).$$

This can be verified using the long exact sequence in homology, and excision, or by just counting cells. To make this look multiplicative, so it will fit in the TQFT context, we exponentiate. Let u be a unit in a commutative ring, and consider $u^{\chi(X)}$.

The Euler theory based on the unit $u \in R$ is defined by: the coefficient ring is R and

(3.3)
$$Z^{\chi,u}(Y) = R, \text{ all finite CW } Y$$
$$Z_X^{\chi,u}: Z^{\chi,u}(Y_1) \to Z^{\chi,u}(Y_2) \text{ is multiplication by } u^{\chi(X,Y_1)}.$$

The multiplicative property of state modules is the standard isomorphism $R \otimes_R R \simeq R$ given by $r \otimes s \mapsto rs$. The composition property (1.2) results from exponentiating the addition property (3.2). There is also a product property for spacetimes: $Z_{X_1 \sqcup X_2} = Z_{X_1} \otimes Z_{X_2}$.

Exercise 3.4.
1. Let Y be a finite complex, and consider the product $[0,1] \times Y$ as a spacetime with boundary the ends $Y_0 \sqcup Y_1 = \{0\} \times Y \sqcup \{1\} \times Y$. Let the entire boundary be incoming, so the outgoing boundary is empty. The induced homomorphism $Z_{(I \times Y; Y_0 \sqcup Y_1, \phi)}$ induces a bilinear form $\lambda: Z(Y) \otimes Z(Y) \to R$; evaluate this.
2. Let T denote the graph with three edges with a single end in common. Consider this as a spacetime with boundary the three outer endpoints. Multiply by Y, and denote the endpoint copies of Y by Y_1, Y_2, Y_3. Let two be incoming and one be outgoing, then $Z_{(T \times Y; Y_1 \sqcup Y_2)}$ gives $m: Z(Y) \otimes Z(Y) \to Z(Y)$ which is the product for a ring structure on $Z(Y)$. Conversely if we let two copies of Y be outgoing and one incoming we get a coproduct $\Delta: Z(Y) \to Z(Y) \otimes Z(Y)$. Evaluate these. Find a unit and counit, and verify these form an ambialgebra structure (see A.3, and note they do not form a Hopf algebra).

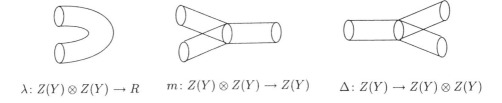

$\lambda: Z(Y) \otimes Z(Y) \to R$ $m: Z(Y) \otimes Z(Y) \to Z(Y)$ $\Delta: Z(Y) \to Z(Y) \otimes Z(Y)$

Figure 3.5 The bilinear form, product and coproduct on $Z(Y)$.

3. A somewhat similar construction can be done in the context of manifolds. Let P denote an $(d+1)$-disk with two open balls removed. The result is a manifold with boundary three copies of S^d. Consider two as incoming and one outgoing to get a product structure on $Z(S^d)$. Describe this, and the corresponding coproduct.

In this construction the ring R is arbitrary, but for physically oriented people the favorite choice is the complex numbers $R = \mathbf{C}$. The universal example is the Laurant polynomial ring $\mathbf{Z}[t, t^{-1}]$ with unit $u = t$. The example based on $u \in R$ is obtained from the universal one by the ring map $\mathbf{Z}[t, t^{-1}] \to R$ defined by evaluating polynomials at u.

The skew Euler TQFT

In the definition (3.3) the relative Euler characteristic is used, rel the incoming boundary. The analogous construction using the outgoing boundary also gives a TQFT. It is not really new since it is isomorphic to the original theory, but the isomorphism is nontrivial. We introduce it here because analogs will come up later.

Define, for $X\colon Y_1 \to Y_2$,

(3.5)
$$Z^{s\chi,u}(Y) = R, \text{ all finite CW } Y$$
$$Z_X^{s\chi,u}\colon Z^{\chi,u}(Y_1) \to Z^{\chi,u}(Y_2) \text{ is multiplication by } u^{\chi(X,Y_2)}.$$

Exercise 3.6. For a finite CW complex Y define an isomorphism $\tau_Y\colon Z^{s\chi,u}(Y) \to Z^{\chi,u}(Y)$ by multiplication by $u^{\chi(Y)}$. Show that this induces an isomorphism of TQFT, ie. show that for every bordism X the diagram commutes:

$$\begin{array}{ccc} Z^{s\chi,u}(Y_1) & \xrightarrow{Z_X^{s\chi,u}} & Z^{s\chi,u}(U_2) \\ {\scriptstyle \tau_{Y_1}}\downarrow & & \downarrow{\scriptstyle \tau_{Y_2}} \\ Z^{\chi,u}(Y_1) & \xrightarrow{Z_X^{\chi,u}} & Z^{\chi,u}(Y_2). \end{array}$$

LECTURE 4
Finite Total Homotopy TQFT

This example expands on one mentioned by Kontsevich [**K**], studied in special cases by Dijkgraaf and Witten [**DW**] and G. Segal [**S**] and worked out in detail in [**Q**] and [**FQ**]. Fix a space B. The state module of Y is the rational vector space with basis the set of homotopy classes of maps from Y to B:

(4.1) $$Z^B(Y) = \mathbf{Q}[Y, B].$$

Taking disjoint unions of spaces gives cartesian product of homotopy classes of maps, and taking the rational vector space generated by this gives a tensor product:

$$\mathbf{Q}[Y_1 \sqcup Y_2, B] = \mathbf{Q}([Y_1, B] \times [Y_2, B]) = \mathbf{Q}[Y_1, B] \otimes_{\mathbf{Q}} \mathbf{Q}[Y_2, B].$$

If f_i are maps on Y_i, then this identification is given explicitly by $f_1 \sqcup f_2 \mapsto f_1 \otimes f_2$. This is the multiplicative property required of state modules with coefficient ring $\mathbf{Q}$.

We will see later that the state modules in a TQFT must be finitely generated, so it will be necessary to restrict B so $[Y, B]$ is a finite set. Technically we will need the finiteness in defining the morphisms. For this we use the following:

Definition 4.2. A space B is said to have *finite total homotopy* if it has finitely many components, and if b is a basepoint then each $\pi_i(B, b)$ is finite, and only finitely many of these are nonzero.

The example to have in mind is the classifying space B_G for a finite group G. This is connected, has $\pi_1 = G$, and all other homotopy groups trivial.

Lemma. *If B has finite total homotopy and Y is a finite CW complex then the set of homotopy classes $[Y, B]$ is finite.*

The proof is by induction on the number of cells in Y, using the long exact homotopy sequence of a fibration in the induction step.

To proceed with the definition, suppose $(X; Y_1, Y_2)$ is a finite CW triad so we must define $Z_X : Z(Y_1) \to Z(Y_2)$. Let $[f_1] \in [Y_1, B]$ be one of the generators of

$Z(Y_1)$, then

(4.3) $$Z_X([f_1]) = \sum_{[f_2]} \mu_{X,f_1,f_2}[f_2]$$

where the sum is over all $[f_2] \in [Y_2, B]$. So far this is just the statement that when bases are given then homomorphisms are represented by matrices. The mathematics comes in the definition of the coefficients μ. Before giving this we recall the conditions they must satisfy. First μ should depend only on the homotopy type of X, Y_i and f_i. Second it should be multiplicative with respect to disjoint unions, and finally it must satisfy the composition condition (1.2), which translates into product relations among the matrix entries for different X, etc. These conditions essentially determine μ. Thus the definition should be regarded as the result of solving these relations rather than as a lucky guess which happens to work. Having said this we define

(4.4) $$\mu_{X,f_1,f_2} = \#^\pi \mathrm{Map}_{f_1}(X, B)_{[f_2]}$$

and explain the notation. First "Map" denotes the space of maps from X to B whose restriction to Y_1 is equal to f_1, and whose restriction to Y_2 is homotopic to f_2. Note the unsymmetrical treatment of the two boundaries. The other ingredient is the *homotopy order*, $\#^\pi M$ defined for spaces with finite total homotopy. If M is empty define this to be 0. If b is a basepoint for M define it to be the alternating product of the orders of the homotopy groups:

(4.5) $$\#^\pi(M, b) = \Pi_{i=1}^\infty (\#\pi_i(M,b))^{(-1)^i} = (\#\pi_1)^{-1}(\#\pi_2)(\#\pi_3)^{-1}\cdots.$$

Note that the finite total homotopy assumption implies that the terms are finite and all but finitely many are 1, so this is a well-defined rational number. (This is why we took the coefficient ring to be $\mathbf{Q}$.) Finally note that this only depends on the component of M containing the basepoint. If M is not connected choose basepoints $b_1, \ldots, b_n$, one in each component, and define $\#^\pi M$ to be the sum over components $\sum_i \#^\pi(M, b_i)$.

This invariant applies to $M = \mathrm{Map}_{f_1}(X, B)_{[f_2]}$ because a slight generalization of the lemma shows this has finite total homotopy if B has finite total homotopy and $(X; Y_1, Y_2)$ is a finite CW triad. Finally we note that the mapping space uses a representative f_1 for the homotopy class, but that the homotopy order does not depend on this choice.

There is a slight reformulation of the definition of Z_X which we will find useful. The expression in (4.3) is a sum over $[f_2]$ and the coefficients are further sums over components of the mapping spaces. Combining these gives

(4.6) $$Z_X^B([f_1]) = \sum_{[F] \in [X, f_1; B]} \#^\pi(\mathrm{Map}_{f_1}(X, B), F)[F|Y_2].$$

Here the sum is over homotopy classes rel Y_1 of $F: X \to B$ which are f_1 on Y_1. As in (4.5) the coefficient $\#^\pi(\text{Map}_{f_1}(X,B), F)$ denotes the homotopy order of the component containing F.

We briefly outline the proof of the composition property. First, if $M \to N \to P$ is a fibration of spaces with finite total homotopy and P connected, then the homotopy order multiplies: $\#^\pi N = (\#^\pi M)(\#^\pi P)$. This follows from the long exact homotopy sequence of a fibration. The other ingredient is that if $Y_1 \subset X_1$ then the restriction map $\text{Map}_{f_1}(X_1 \sqcup_{Y_2} X_2, B) \to \text{Map}_{f_1}(X_1, B)$ is a fibration, and the fiber over a map $F: X_1 \to B$ is the space $\text{Map}_{F|Y_2}(X_2, B)$. Combining this with the product formula for $\#^\pi$ shows the matrix entries for $Z_{X_1 \cup X_2}$ and $Z_{X_2} Z_{X_1}$ are equal. See [Q] for details.

We give sample calculations. Consider S^1 as a bordism from a single point to the empty set, and suppose B is connected with basepoint b. Then $Z(\text{pt}) = \mathbf{Q}[b]$. The homomorphism

$$Z(\text{pt}) \xrightarrow{Z_{(S^1; \text{pt}, \phi)}} Z(\phi)$$

is multiplication by the homotopy order $\#^\pi \text{Map}_b(S^1, B) = \#^\pi \Omega B$, where ΩB denotes the loop space. There is a fibration

$$\Omega B \to \mathcal{P} B \to B$$

where $\mathcal{P} B$ is the space of paths beginning at b, and is contractible. Since it is contractible and B is connected the multiplicative property for fibrations shows $\#^\pi \Omega B = (\#^\pi B)^{-1}$. So $Z_{(S^1; \text{pt}, \phi)}$ is multiplication by $(\#^\pi B)^{-1}$.

For another example consider the circle as a bordism from the empty set to itself. This induces multiplication by $\#^\pi \text{Map}(S^1, B)$. There is a fibration

$$\Omega B \to \text{Map}(S^1, B) \to B$$

so the homotopy order is the product of those of the base and fiber. But above we found a fibration with the same base and fiber, and contractible total space. Thus $\text{Map}(S^1, B)$ must have the same homotopy order as a contractible space, and $Z_{(S^1; \phi, \phi)}$ is the identity.

Exercise 4.7. (Refer to 3.4 for notation, and use (4.6)).
1. Describe the bilinear form on $Z(Y)$ induced by $I \times Y$ regarded as a bordism from two copies of Y to ϕ.
2. Describe the product and coproduct structures on $Z(Y)$ induced by $T \times Y$.
3. Suppose B is connected. Then on the subcategory of *graphs*, ie. boundaries finite sets of points and spacetimes 1-dimensional complexes, the TQFT Z^B has the same values as an Euler characteristic theory $Z^{\chi, u}$, for some unit $u \in \mathbf{Q}$. Find u. By tensoring together Z^B and the Euler TQFT with the inverse unit u^{-1} we get a TQFT which is constant on graphs. Describe this TQFT explicitly by modifying the coefficients (4.4) and (4.6).
4. If $B = K(A, n)$, an Eilenberg-MacLane space for a finite abelian group A, then

$$\pi_i \text{Map}_{f_1}(X, B) = H^{n-i}(X, Y_1; A)$$

(if the space is nonempty). Using this determine $Z(S^n)$ and the ambialgebra structure on this induced by the twice-punctured disk as in 3.4(3).

5. Suppose B is connected and has finite total homotopy. Define n to be the product of the orders of all homotopy groups. Show that the orders of the homotopy groups inverted in (4.5) all divide powers of n. Conclude that instead of $\mathbf{Q}$ we can use a ring S as coefficients if n is a unit in S (eg. $\mathbf{Z}[\frac{1}{n}]$ or $\mathbf{Z}/p$ when $(n,p) = 1$).

The skew theory

As for the Euler theory in (3.5), there is a skew version of this definition which interchanges the roles of the incoming and outgoing boundary. Again the result is isomorphic to the original. The definition (4.4) is easier to compute in examples, but the skew version fits better with inverse limits in §5.

For a finite complex Y and bordism $X \colon Y_1 \to Y_2$ define

(4.8)
$$Z^{s,B}(Y) = \mathbf{Q}[Y, B]$$
$$Z_X^{s,B}([f_1]) = \sum_{[f_2] \in [Y_2, B]} \#^\pi \mathrm{Map}_{f_2}(X, B)_{[f_1]} [f_2]$$

where the sum is over representatives of homotopy classes $[Y_2, B]$

Exercise 4.9.
1. Define $\tau_Y \colon Z^{s,B}(Y) \to Z^B(Y)$ (both are $\mathbf{Q}[Y, B]$) by

$$\tau_Y([f]) = \#^\pi \mathrm{Map}\,(Y, B)_{[f]}[f].$$

Show that this gives an isomorphism of TQFT in the sense described in (3.6). Note that the commutativity of the diagram implies that (4.8) does in fact satisfy the definition of a TQFT. (Hint: consider fibrations of $\mathrm{Map}\,(X, B)_{[f_1 \sqcup f_2]}$ over $\mathrm{Map}\,(Y, B)_{[f_1]}$ and $\mathrm{Map}\,(Y, B)_{[f_2]}$.)
2. Compute the ring structure on $Z^{s,B}(Y)$ induced by $T \times Y$, and compare with 4.7(2). Note that although the rings are isomorphic the skew version is in a more convenient form.

Witten's integral for finite gauge group

Here we think of the finite sum definition of Z_X as an "integral" over a finite measure space. This gives a finite analog of the functional integral used by Witten in [**Wi1**] to predict 3-manifold invariants, but with trivial Lagrangian. We modify it to get examples with nontrivial Lagrangian in the next section. See [**FQ**] or [**Q**] for more detail.

Witten's Chern-Simons integral [**Wi1**] is (ignoring "Wilson lines")

(4.10)
$$Z_X = \int_{\mathcal{A}/\mathcal{G}} D\mathcal{A} \exp(i\mathcal{L})$$

where $\mathcal{L} = \int_X (\cdots)$, and $(\cdots)$ is a 3-form associated to a connection. The integration in (4.10) is to take place over the set $\mathcal{A}/\mathcal{G}$ of all principal $SU(n)$ bundles with connections, modulo the equivalence relation induced by bundle isomorphism (gauge equivalence), and $D\mathcal{A}$ indicates a measure on this set.

If we replace the Lie group $SU(n)$ by a finite group G, then the following changes occur. First every bundle has a unique connection, so $\mathcal{A}/\mathcal{G}$ becomes just the set of isomorphism classes of bundles, or equivalently the homotopy classes $[X, B_G]$. Second since this set is finite, the integral becomes a finite sum, and a measure is determined simply by a function giving the volume of each point. Finally the Lie algebra of G is trivial, so the "3-form" $(\cdots)$ associated to the connection is trivial, and $\mathcal{L} = 0$. Putting these together (4.10) becomes a finite sum

$$\sum_{f \in [X, B_G]} D\mathcal{A}([f]).$$

This agrees with the homomorphism Z_X in (4.6) when the boundary is empty, if we identify the measure with the homotopy order: $D\mathcal{A}([f]) = \#^\pi(\operatorname{Map}(X, B_G), f)$. (When $B = B_G$ this is also seen below to be the inverse of the number of bundle automorphisms of the bundle associated to f: $(\#\operatorname{Aut}(f))^{-1}$). This identification is reasonable since this "measure" is uniquely determined by the formal properties of a TQFT (at least if defined on CW complexes, not just manifolds). It cannot actually be proved to be right because $D\mathcal{A}$ is not defined in general: the functional integral is a huristic device.

4.11. Calculations for finite groups

Here we describe maps into B_G in terms of homomorphisms into G, so that examples can be worked out. The first object is to describe $\pi_0 \operatorname{Map}_f(X, B_G)$.

First recall that if X is connected and x_0 is a basepoint, then homotopy classes of maps $f : X \to B_G$ which preserve basepoints correspond bijectively to homomorphisms $f_* : \pi_1(X, x_0) \to G$. In the terminology above this is $\pi_0 \operatorname{Map}_b(X, B_G)$, where b is a basepoint in B_G. Changing the basepoint changes the homomorphism by conjugation by an element of G, so the free homotopy classes are

$$[X, B_G] = \hom(\pi_1 X, G)/G,$$

where G acts by conjugation. If $Y \subset X$ is connected and f is specified on Y then choose the basepoint of X inside Y. It therefore stays fixed, so maps F give well-defined homomorphisms $F_* : \pi_1(X, x) \to G$. The condition that F restrict to f on Y corresponds to a commutativity condition for the homomorphisms: homotopy classes of such F correspond to homomorphisms so that the composition $\pi_1(Y, x) \to \pi_1(X, x) \to G$ is f_*.

Now we describe what happens if Y has two components Y_0, Y_1. The general case is similar, but will not be needed for the examples. Choose basepoints $y_i \in Y_i$ and assume that the fixed map f_i takes these to the basepoint in B_G. Choose y_0 as the basepoint of X. Then as above a map $F : X \to B_G$ corresponds to a homomorphism of the fundamental group. Join the y_i by an arc in X, then the image loop in B determines an element of G. These determine a bijection

(4.12) $$\pi_0 \mathrm{Map}_{\cup f_i}(X, B_G) \to \{(\tau, g)\}.$$

Here $\tau \colon \pi_1(X, x) \to G$ is a homomorphism, g is an element in G, and these satisfy a compatibility condition corresponding to the requirement that $F|Y_i = f_i$: the composition

$$\pi_1(Y_i, y_i) \to \pi_1(X, x) \xrightarrow{\tau} G$$

is f_{0*} if $i = 0$, and is $g(f_1)_* g^{-1}$ if $i = 1$. (I would like to thank I. Bobcheva for explaining this to me.)

Exercise 4.13.
1. Use the above to show that if $F \colon X \to B_G$ and $Y \subset X$ are given then

$$\pi_1(\mathrm{Map}_{F|Y}(X, B_G), F) = \begin{cases} \{1\} \text{ if } Y \neq \phi \\ \text{centralizer of } F_*(\pi_1 X) \subset G \text{ if } Y = \phi \end{cases}$$

$$\pi_i(\mathrm{Map}_{F|Y}(X, B_G), F) = 0, \text{ if } i > 1$$

(Hint: begin with $\pi_i(\mathrm{Map}_{F|Y}(X, B_G), F) = \pi_0 \mathrm{Map}_F(X \times S^i \cup Y \times D^{i+1}; B_G)$, where the subscript F indicates that the restriction to $X \times \{\mathrm{pt}\} \subset X \times S^i$ is required to be F.

2. If $f \colon Y \to B_G$ is given, use the above to describe $\#^\pi \mathrm{Map}_f(X, B_G)$.
3. $Z^B(S^1) = \mathbf{Q}[G/\mathrm{conj}]$, the module generated by conjugacy classes in G. Show that the product and unit, coproduct and counit for $Z^B(S^1)$ induced by the spacetime $D^2 - 2B^2$ (see 3.4) is

$$m([f], [g]) = \sum_{h \in G} [fhgh^{-1}]$$

$$e = \frac{1}{\#G}$$

$$\Delta([f]) = \sum_{h \in G} [h] \otimes [h^{-1} f]$$

$$\epsilon([f]) = \delta_{[f],[1]} = \begin{cases} 1 \text{ if } [f] = [1] \\ 0 \text{ otherwise} \end{cases}$$

According to the structure theory this is a commutative ambialgebra, see A.3 and determines the restriction of the TQFT to surfaces, see 7.16. Those familiar with representation theory should relate this to the character ring of G, [**BtD**].

4. Repeat this for $Z(S^1 \times Y)$ using $(D^2 - 2B^2) \times T$.

LECTURE 5
Twisted Finite Homotopy TQFT

Although the literal finite analog of Witten's functional integral (4.8) has Lagrangian $\mathcal{L} = 0$, a better model is obtained by "twisting" it by adding a nontrivial $\mathcal{L}$. The one in (4.7) has the form $\int_X (\cdots)$ where $(\cdots)$ is a 3-form on the 3-manifold X. Think of this as a singular cochain, so integration becomes evaluation of the cochain on the fundamental class of X. With this in mind we:
1. modify the spacetimes and boundaries to have fundamental classes (we still do not require them to be manifolds);
2. fix a cohomology class $[\alpha] \in H^*(B; A)$, where A is an abelian group (written multiplicatively),
3. so each "bundle" $f \colon X \to B$ gives an element $f^*[\alpha]([X]) \in A$ which we think of as $\exp(i\mathcal{L})$, and
4. integrate this over all bundles.

The most interesting part is finding an appropriate state module, and implementing the final step to define homomorphisms between these.

A remark on item (3) above: if A is the nonzero complex numbers $A = \mathbf{C}_\times \subset \mathbf{C}$ then $f^*[\alpha]([X])$ is literally of the form $e^{i\mathcal{L}}$. This however is too special: it does not display enough of the possible complications, and nontrivial examples are too remote. Instead we use an arbitrary group A written multiplicatively and formally "exponentiate" by forming the rational group ring $\mathbf{Q}[A]$. Thus the coefficient ring is $\mathbf{Q}[A]$ rather than $\mathbf{C}$. If $A = \mathbf{C}_\times$, $\mathbf{R}_\times$, or S^1 then there is a ring homomorphism $\mathbf{Q}[A] \to \mathbf{C}$ and a $\mathbf{C}$-coefficient theory is obtained by change of rings. This is similar to the use of the universal Euler theory with coefficient ring $\mathbf{Z}[t, t^{-1}]$ rather than $\mathbf{C}$, see the end of §3. It is also similar to Drinfeld's treatment of quantum groups in [**D1**] where the deformation parameter is a variable in a power series ring rather than a complex number.

Since the section is a long one we outline the contents. First oriented complexes, characteristic numbers, and holonomy are defined. Then a description of the TQFT in terms of choices, the "answer," is given in 5.5 and 5.6. This is too dependent on the choices and too unmotivated to be satisfactory as a definition, but we use it to work out some examples. Beginning in 5.24, a more invariant and natural mathematical description is given. The dependence on the choice of cocycle is described in 5.35. Finally, in 5.36, the differences between this model development and the sophisticated analytic examples are discussed.

5.1. Oriented complexes

Fix an integer d, then we define boundaries and spacetimes with appropriate "orientation" homology classes.

The boundaries are $(Y, [Y])$ where Y is a finite CW complex, $[Y] \in H_d(Y)$, and $H_{d+1}(Y) = 0$. The spacetimes are $(X, Y, [X])$, where $[X] \in H_{d+1}(X, Y)$, and $H_{d+1}(Y) = 0$. The boundary of a spacetime is the object $(Y, \partial[X])$. Notice that oriented manifolds are examples.

A bordism $X \colon Y_1 \to Y_2$ is, as above, a spacetime $(X, Y, [X])$ with a homotopy equivalence $Y \simeq Y_1 \sqcup Y_2$. However there is a sign change in the orientation of the incoming boundary: the equivalence takes $\partial[X]$ to $-[Y_1] + [Y_2]$. This is so that when we compose bordisms by gluing the outgoing boundary of one to the incoming boundary of the other then the orientation classes fit together to give an orientation for the union. The vanishing hypothesis $H_{d+1}(Y) = 0$ ensures that the orientation of the union is well-defined.

5.2. Characteristic numbers

Suppose $\alpha \colon C_{d+1}(B) \to A$ is a cochain representing a cohomology class in $[\alpha] \in H^{d+1}(B; A)$. As above, if $X \to B$ is a map from an oriented complex with empty boundary then we can get an element of A by pulling back and evaluating on the orientation: $f^*[\alpha]([X])$. On the chain level this is described by: let $x \in C_{d+1}(X)$ be a representative chain for $[X]$, then $f^*[\alpha]([X])$ is the image of x under the composition

$$C_{d+1}(X) \xrightarrow{f_*} C_{d+1}(B) \xrightarrow{\alpha} A.$$

When B is a classifying space, so a map classifies a bundle on X, such elements are called "characteristic numbers." In this empty-boundary case the element depends only on the homotopy class of f and the cohomology class.

Now suppose $(X, Y, [X])$ is an oriented complex with Y nonempty. The orientation $[X]$ is then a relative class, so the evaluation $f^*[\alpha]([X])$ is no longer well-defined on the homology level. However we can get an element by making choices and using the chain level description given above. Fix a particular map f, a chain $x \in C_{d+1}(X)$ representing $[X]$, and a cocycle α representing the cohomology class. Then $f^*\alpha(x)$ is defined. The next lemma describes how this depends on the choices.

Lemma 5.3. *Suppose $f' \colon X \to B$ is a map homotopic to f rel Y, and x' is a chain representing $[X]$ with $\partial x' = \partial x$. Then $(f')^*\alpha(x') = f^*\alpha(x)$.*

Proof. We show the independence in x; the independence up to homotopy is basically similar.

If $\partial x' = \partial x$ then $x - x'$ defines a cycle in $C_{d+1}(X)$, so a homology class in $H_{d+1}(X)$. Consider the fragment of the long exact sequence of the pair (X, Y)

$$\to H_{d+1}(Y) \to H_{d+1}(X) \to H_{d+1}(X, Y) \to$$

The image of the homology class $[x - x']$ in $H_{d+1}(X, Y)$ is 0 because both chains represent the class $[X] \in H_{d+1}(X, Y)$. Therefore by exactness $[x - x']$ is the image of a class in $H_{d+1}(Y)$. But this has been assumed to vanish, so $[x - x'] = 0$. Thus there is a chain $z \in C_{d+2}(X)$ with $\partial z = x - x'$. Since $f^*\alpha$ is a cochain it evaluates to 0 on a boundary:

$$0 = f^*\alpha(\partial z) = f^*\alpha(x - x') = f^*\alpha(x) - f^*\alpha(x')$$

so the two elements are equal. □

5.4. Holonomy groups. Suppose $[\alpha] \in H^{d+1}(B; A)$, $(Y, [Y])$ is a d-dimensional oriented complex, and $f\colon Y \to B$ is a map. The "holonomy" subgroup $\mathrm{Hmy}(f) \subset A$ is defined to be the collection of all possible $F^*[\alpha]([S^1 \times Y])$ where F is an extension of f to $S^1 \times Y$:

$$\mathrm{Hmy}(f) = \{F^*[\alpha]([S^1 \times Y]), \text{ where } F\colon S^1 \times Y \to B, F|Y = f\}.$$

We use the abbreviation "Hmy" for this because "Hol" is used as an abbreviation for "holomorphic." Note that since $S^1 \times Y$ has no boundary these elements depend only on the homotopy class of f, and the homology and cohomology classes. Also, if B has finite total homotopy then there are only finitely many homotopy classes, so the holonomy group is finite.

The answer

Fix a space with finite total homotopy B, and a cocycle α representing a class in $H^{d+1}(B; A)$.

If $(Y, [Y])$ is a boundary object, choose representatives $\{f_i\}$ for the homotopy classes $[Y, B]$ and a representative cycle $y \in C_d(Y)$ for $[Y]$. These choices determine an isomorphism

$$(5.5) \qquad Z^\alpha(Y) \simeq \sum_{[f_i]\in[Y,B]} \mathbf{Q}[A/\mathrm{Hmy}(f_i)][f_i].$$

This is the sum over representatives f_i, of the $\mathbf{Q}[A]$ modules isomorphic to $\mathbf{Q}[A/\mathrm{Hmy}(f_i)]$ with generator the homotopy class $[f_i]$. Basis elements in the sum are written as $[b][f_i]$, where $[b]$ denotes the image of $b \in A$ in the quotient. We caution that although both sides of (5.5) are independent of the choices of representatives f_i, the isomorphism does depend on them. This dependence prevents (5.5) from being used as the definition.

Next suppose $X\colon Y_1 \to Y_2$ is a bordism, we have representatives $f_{1,i}$ and $f_{2,j}$ chosen as above, and $y_i \in C_d(Y_i)$ are representative cycles for the orientations. Then the isomorphisms in (5.5) take the induced homomorphism to

$$(5.6) \qquad Z_X([b]_1[f_{1,i}]) = \sum_{\substack{a\in\mathrm{Hmy}(f_{1,i}) \\ [F]\in[X,f_{1,i};B]}} \frac{\#^\pi(\mathrm{Map}_{f_{1,i}}(X,B), F)}{\#\mathrm{Hmy}(f_{1,i})} [abF^*\alpha(x)]_2[F|Y_2]$$

We begin the explanation with the index F. These are representatives of rel-Y_1 homotopy classes of maps $X \to B$ which are $f_{1,i}$ on Y_1, and we choose representatives whose restrictions to Y_2 are one of the fixed representatives $f_{2,j}$. Thus each term in the sum is of the form $r[c]_2[f_{2,j}]$, with r rational, $c \in A$, and $[-]_2$ indicating the image in $A/\operatorname{Hmy}(f_{2,j})$. This does give an element in the description (5.5) of $Z(Y_2)$.

The rational coefficient involves the same homotopy order used in the untwisted version (4.6), and the inverse of the order of a holonomy group, which comes from an averaging process.

The first part of the A term is ab. b comes from the input, $[b]_1[f_{1,i}]$. Since it represents an element of the quotient by the holonomy group, b is only well-defined up to multiplication by elements of $\operatorname{Hmy}(f_{1,i})$. But it appears as a product ab and is summed over all $a \in \operatorname{Hmy}(f_{1,i})$, so the result is well-defined in A.

The characteristic number $F^*\alpha(x)$ is defined, as in (5.3), using a chain representing $[X]$ with $\partial x = -y_1 + y_2$. This is not quite well-defined in A. Lemma 5.3 shows it depends only on the homotopy class rel $Y_1 \cup Y_2$. It is specified to be $f_{1,i} \cup f_{2,j}$ on $Y_1 \cup Y_2$, but represents a rel-Y_1 homotopy class. Different choices might give different homotopy classes rel the whole boundary. However they differ essentially by a homotopy on Y_2 of $f_{2,j}$ to itself. The difference this makes in the characteristic number is therefore of the form $R^*\alpha[S^1 \times Y_2]$, where $R\colon S^1 \times Y_2 \to B$ restricts to $f_{2,j}$ on Y_2. This is an element in the holonomy group of $f_{2,j}$. Since we are taking the image in the quotient by this group (as indicated by the brackets $[-]_2$) the image is well-defined. □

We observe that if α is trivial then this reduces to the previous definition (4.6) tensored with the identity on $\mathbf{Q}[A]$. The holonomy groups are all trivial so these terms drop out, as does the characteristic number.

Exercise.
1. Determine the dependence of the description of the state module in (5.5) on the choices. Set $X = Y \times I$, let the incoming boundary be choices y, $\{f_i\}$ on $Y \times \{0\}$, let the outgoing boundary be choices y', $\{f_i'\}$ on $Y \times \{1\}$, and determine Z_X using (5.6).
2. Show this is at least well-defined in the weak sense that if the same choice are used on each end then Z_X is the identity.

5.7. Calculations for finite groups

Here we work out examples for $B = B_G$ and $d = 1$. For this we need ways to specify representatives of homotopy classes, explicit descriptions of 2-cocycles, and ways to evaluate pullback cocycles on chains. Even in this extremely simple case this is a laborious program.

We use a specific model for B_G, called the (unnormalized) bar construction. This is a CW complex obtained as the geometric realization of a simplicial set (or more accurately a Δ-set), but we describe it directly without using the simplicial machinery.

There is a single 0-simplex, and for $n > 0$ the n-simplices are indexed by elements of the n-fold product $\times^n G$. Think of these as labels for the edges of a geometric simplex in the following way. The vertices of the standard n-simplex Δ^n are ordered, and will be denoted $0, \ldots, n$. Take $(g_1, \ldots, g_n) \in \times^n G$, and let g_i label

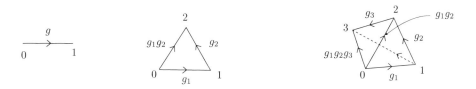

Figure 5.8 Labeled simplices.

the edge from $i-1$ to i. More generally if $i < j$ then the edge from i to j is labeled by the product $g_{i+1}g_{i+2}\cdots g_j$. See (5.8) for examples.

Define boundary operations $\partial_i\colon \times^n G \to \times^{n-1}G$ by restricting to the face opposite the ith vertex. Thus for a 2-simplex (g_1, g_2)

$$\partial_0(g_1, g_2) = g_2$$
$$\partial_1(g_1, g_2) = g_1 g_2$$
$$\partial_2(g_1, g_2) = g_1$$

Together these boundary operations describe attaching maps for the n-cells $\Delta^n \times (\times^n G)$ in the union of cells of dimension$< n$. Explicitly we proceed by induction, starting with a single point as 0-skeleton. If the n-skeleton is constructed then get the $(n+1)$-skeleton by attaching $\Delta^{n+1} \times (\times^{n+1}G)$: if $t \in \Delta^{n+1}$ lies in the ith face, identified as a copy of Δ^n, then identify $\{t\} \times (g_1, \ldots, g_{n+1})$ with $\{t\} \times \partial_i(g_1, \ldots, g_{n+1})$. It must be checked that this is well-defined for points which lie in two faces.

5.9. Representatives for homotopy classes. Now represent S^1 as an interval Δ^1 with the ends identified. Then each element $g \in G$ determines a map $f_g\colon S^1 \to B_G$ which takes Δ^1 to the 1-cell $\Delta^1 \times (g)$. Homotopy classes $[S^1, B_G]$ correspond to conjugacy classes in G, so to get representative maps choose representatives g_i for the conjugacy classes and form the associated maps f_{g_i}. Note that to actually carry this out requires a way to enumerate elements and conjugacy classes in G. This is often possible, but there is no algorithm for getting an enumeration (at least not from a general presentation of G).

More generally if Y is a 1-complex choose a collection of edges whose complement is contractible, say $I_1, \ldots, I_k$. Identify each with Δ^1, and construct maps by mapping each special edge to a 1-cell. This associates a map to a sequence of elements $g_1, \ldots, g_k$, and these represent the based homotopy classes of maps to B_G. Freely homotopic maps differ by conjugation (the whole sequence is conjugated by a single element). Therefore representatives for these are obtained by choosing representatives for the conjugacy classes of sequences.

Cocycles

The next objective is to describe 2-cocycles in B_G. We use the *cellular* chains: for a CW complex B the group $C_n^c(B)$ is the free abelian group generated by the n-cells of B. When it is constructed from simplices as above the boundary homorphism on an n-simples σ is given by

$$(5.10) \qquad \partial \sigma = \sum_{i=0}^{n} (-1)^i \partial_i \sigma$$

where $\partial_i \sigma$ is identified with an $(i-1)$-simplex and therefore a basis element in $C_{n-1}^c(B)$.

Since the 2-cells of B_G are indexed by $G \times G$, a (cellular) 2-cochain is a homomorphism $\mathbf{Z}[G \times G] \to A$, or equivalently a function

$$\alpha\colon G \times G \to A.$$

This is a cocycle if the composition with $\partial\colon C_3^c \to C_2^c$ is trivial. Using the expression (5.10) for the boundary this gives the *cocycle relation* (recall that we are writing A multiplicatively).

$$(5.11) \qquad \alpha(g_1, g_2 g_3)\alpha(g_2, g_3) = \alpha(g_1, g_2)\alpha(g_1 g_2, g_3).$$

Rather than manipulating the cocycle relation directly we will use a connection between group extensions and group cohomology. A 2-dimensional cohomology class $[\alpha] \in H^2(B_G, A)$ is equivalent to an isomorphism class of central extensions of G by A

$$(5.12) \qquad 1 \to A \to \hat{G} \xrightarrow{\rho} G \to 1.$$

This is a short exact sequence so that elements of A commute with everything in $\hat{G}$. Two such are isomorphic if there is an isomorphism of exact sequences which is the identity on A and G.

One way to see the connection with cohomology is that a cohomology class is represented by a map to an Eilenberg-MacLane space $B_G \to K(A, 2)$. Extend this as a fibration (up to homotopy type) $F \to B_G \to K(A, 2)$, then the sequence (5.12) comes from the long exact homotopy sequence of the fibration.

There is a more explicit connection. Choose a section of the projection $\hat{G} \to G$, and denote it by $g \mapsto \hat{g}$. Choose $\hat{1} = 1$. Denote the product in $\hat{G}$ with a dot: eg. $\hat{f} \cdot \hat{g}$. Then there is a bijection of sets $A \times G \to \hat{G}$ defined by $(a, g) \mapsto a \cdot \hat{g}$. The group structure on $\hat{G}$ can be expressed in terms of these coordinates: define a function $\alpha\colon G \times G \to A$ by

$$(5.13) \qquad \widehat{g_1 g_2} = \alpha(g_1, g_2) \cdot \hat{g}_1 \cdot \hat{g}_2.$$

Then in $A \times G$ the product is given by

$$(5.14) \qquad (a_1, g_1) \cdot (a_2, g_2) = (a_1 a_2 \alpha(g_1, g_2)^{-1}, g_1 g_2).$$

The usual treatment, eg. MacLane [**McL1, p. 111**], has α rather than α^{-1} in (5.14), and the α is on the other side in (5.13). The convention here works better in expressions for Z_X.

Conversely given a function α the expression (5.14) defines a product structure on $A \times G$. The requirement that this product be associative is equivalent to the cocycle condition (5.11), and in this case the resulting group fits canonically into an extension (5.12).

Now relate simplices in B_G to the product in $\hat{G}$. A 2-simplex (f, g) has edges f, g, fg. The value of α on this simplex is the correction term required to change the product of lifts $\hat{f} \cdot \hat{g}$ to $\widehat{fg}$. Or equivalently to convert $\hat{f}$ to $(\widehat{fg})\hat{g}^{-1}$, etc.

We give an example of an extension. Let C_n denote the cyclic group of order n, written multiplicatively, and consider $G = C_n \times C_n$. This has a presentation $C_n \times C_n = <a, b \mid 1 = a^n = b^n = [a,b]>$, where $[a,b] = aba^{-1}b^{-1}$ denotes the commutator. Define

(5.15) $\qquad \hat{G} = <a, b, t \mid 1 = a^n = b^n = [a,t] = [b,t],$ and $[a,b] = t>$

Then adding the relation $t = 1$ gives a homomorphism $\hat{G} \to G$. The kernel A of this is central, and is a cyclic group of order n generated by t. To see t has order n note $ab = tba$, so $b = a^n b = t^n b a^n = t^n b$.

Exercise 5.16.
1. Verify that associativity of the product (5.14) is equivalent to the cocycle condition (5.11).
2. Use the relation (5.13) to find a cocycle classifying the extension

$$\mathbf{Z} \xrightarrow{n} \mathbf{Z} \to \mathbf{Z}/n.$$

3. Show that every element of the group $\hat{G}$ of (5.15) can be written uniquely in the form $t^i a^j b^k$ with $0 \le i, j, k < n$.
4. Define a section $G \to \hat{G}$ by $\widehat{a^j b^k} = t^0 a^j b^k$, and use this and 5.13 to show that the function

$$\alpha(a^i b^j, a^k b^\ell) = t^{-jk}$$

is a cocycle representing the class in $H^2(C_n \times C_n; C_n)$ which classifies this extension.

We are ready to start using all this. Let S^1 have the usual orientation.

Lemma 5.17. *Suppose $g \in G$, and let $g\colon S^1 \to B_G$ denote the associated map. Then the holonomy group is*

$$\mathrm{Hmy}\,(g) = \{[\hat{h}, \hat{g}]_{\hat{G}} \mid [h, g]_G = 1\},$$

where $[h, g]_G$ denotes the commutator $hgh^{-1}g^{-1}$ in G. (5.5) becomes

$$Z^\alpha(S^1) \simeq \sum_{[g] \in G/\mathrm{conj}} \mathbf{Q}[A/\mathrm{Hmy}\,(g)][g] = \mathbf{Q}[\hat{G}/\mathrm{conj}].$$

Note that if the commutator $[h,g]_G$ is trivial then $[\hat{h},\hat{g}]$ lies in A. Also, since the extension is central this does not depend on the choices of lifts $\hat{h}, \hat{g}$. As an example note that if $\hat{G}$ is abelian Hmy$(g) = 1$, and correspondingly $\hat{G}/\text{conj} = \hat{G}$. This expression for S^1 can easily be extended to general 1-complexes.

Proof. The holonomy is defined in (5.4) to be the collection of all $F^*\alpha([S^1 \times S^1])$ for maps F which restrict to g on the second coordinate. Such extensions correspond to extensions of $\pi_1(S^1) \to G$ to $\pi_1(S^1 \times S^1) \to G$. Since $\pi_1(S^1 \times S^1)$ is free abelian on two generators such extensions correspond exactly to elements h which commute with g, ie. the centralizer of g. Cut $S^1 \times S^1$ open into a square and subdivide into two triangles, as in 5.18.

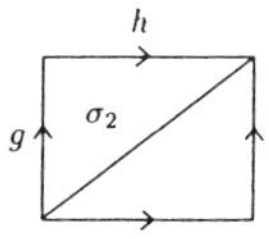

Figure 5.18 A simplicial structure and map on $S^1 \times S^1$.

Since the orientation of σ_2 is reversed the orientation of $S^1 \times S^1$ is represented by $\sigma_1 - \sigma_2$. Using the standard maps in the (based) homotopy classes gives a representative for F which takes σ_1 to the simplex (h,g) in B_G, and σ_2 to (g,h). Applying F_* to the orientation gives $(h,g) - (g,h)$ in $C_2(B_G)$. Applying α gives $\alpha(h,g)\alpha(g,h)^{-1}$. This gives the correction factor to convert the lift $\widehat{ghg^{-1}h^{-1}}$ to $\hat{g}\cdot\hat{h}\cdot\hat{g}^{-1}\cdot\hat{h}^{-1} = [\hat{g},\hat{h}]_{\hat{G}}$. We therefore get the commutators described in the lemma.

For the description of $Z^\alpha(S^1)$, note the quotient $A/\text{Hmy}(g)$ is $A/\{[\hat{h},\hat{g}]_{\hat{G}} \mid [h,g]_G = 1\}$, which is exactly the inverse image of the conjugacy class $[g]$ by the induced function $\hat{G}/\text{conj} \to G/\text{conj}$. This gives a natural bijection of $\cup_{[g]} A/\text{Hmy}(g) = \hat{G}/\text{conj}$. □

Exercise 5.19.
1. Show that the holonomy group of the identity element is always trivial.
2. Let G be $C_n \times C_n$ and $\hat{G}$ the extension of (5.15). Show that the holonomy of $a^i b^j$ is the subgroup of $A \simeq C_n$ generated by $t^{\gcd(i,j,n)}$, where "gcd" indicates the greatest common divisor of the nonzero elements. Use this to determine the order of Hmy$(a^i b^j)$.
3. In particular show that if n is prime then Hmy$(g) = A$ if $g \neq 1$, and Hmy$(1) = \{1\}$. Show this gives a bijection $\hat{G}/\text{conj} = (G-1) \cup A$.
4. Show (for general G, $\hat{G}$) that the sesquilinear form $Z(S^1) \otimes Z(S^1) \to \mathbf{Q}[A]$ is given by
$$\lambda(f_i, f_j) = \delta_{i,j} \sum_{\{h \mid [h,f_i]=1\}} [\hat{h},\hat{g}]_{\hat{G}}.$$

We now undertake the description of the multiplication
$$m = Z_{D^2 - 2B^2} : Z^\alpha(S^1) \otimes_{\mathbf{Q}[A]} Z^\alpha(S^1) \to Z^\alpha(S^1)$$

induced by the twice-punctured disk as in (3.4) and (4.13). According to the structure results of §7 this is part of a commutative ambialgebra structure which characterizes Z^α on surfaces.

Choose as cycle representing $[S^1]$ the generator y of the cellular 1-chains of the standard CW structure with one 1-cell. Choose representatives $\{g_i\}$ for the conjugacy classes of G, which correspond to homotopy classes $[S^1, B_G]$. These choices induce the isomorphisms of (5.5) and (5.17):

$$Z^\alpha(S^1) \simeq \sum_{[g_i] \in G/\mathrm{conj}} \mathbf{Q}[A/\mathrm{Hmy}\,(g_i)][g_i] = \mathbf{Q}[\hat{G}/\mathrm{conj}].$$

Generators in the middle description are denoted $[b][g_i]$, where $b \in A$ and $[b]$ indicates the image in the quotient A/Hmy.

Fix conjugacy classes $[g_1], [g_2]$ in G, and consider maps $D^2 - 2B^2 \to B_G$ which agree with the representatives of these classes on two of the boundary circles. Homotopy classes rel $2S^1$ are classified by elements $h \in G$ representing the homotopy class of the restriction to an arc joining the circles, see (5.20).

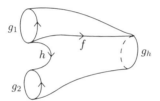

Figure 5.20 Description of $D^2 - 2B^2 \to B_G$.

The homotopy order of the components of this mapping space $\#^\pi(\mathrm{Map}_{g_1 \cup g_2}(D^2 - 2B^2, B_G), h)$ is 1, see (4.13). The third circle is taken to the conjugacy class $[g_1 h g_2 h^{-1}]$. Denote by g_h the chosen representative of this class, and choose $f \in G$ which conjugates $g_1 h g_2 h^{-1}$ to g_h. This determines a homotopy of the third circle to the map g_h.

Cut $D^2 - 2B^2$ open along the labeled arcs to obtain a 7-gon with labeled edges. Subdivide this into 5 triangles as shown in (5.21).

This describes a specific map F into B_G. The orientation cycle of $D^2 - 2B^2$ is represented by the chain $x = -\sigma_1 + \sigma_2 - \sigma_3 - \sigma_4 + \sigma_5$, and F takes this to $(h, g_2) + (hg_2h^{-1}, h) - (g_1, hg_2h^{-1}) - (fg_h f^{-1}, f) + (f, g_h)$. Apply α to this to get $F^*\alpha(x)$. The $F^*\alpha(x)[F|Y_2]$ part of (5.6) is then

$$\left[\alpha(h, g_2)^{-1}\alpha(hg_2h^{-1}, h)\alpha(g_1, hg_2h^{-1})^{-1}\alpha(fg_h f^{-1}, f)^{-1}\alpha(f, g_h)\right][g_h].$$

We could subsitute this into (5.6) and declare it finished, but we tidy it up by using the product in $\hat{G}$. The values of α on the simplices in the 7-gon are the correction factors needed to convert the value of the section on the product $\hat{g}_h = (f^{-1}g_1 hg_2 h^{-1}f)\hat{\;}$ to the product of sections $\hat{f}^{-1} \cdot \hat{g}_1 \cdot \hat{h} \cdot \hat{g}_2 \cdot \hat{h}^{-1} \cdot \hat{f}$. Putting this in

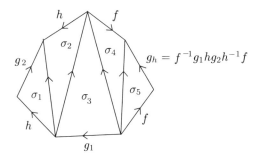

Figure 5.21 Simplicial structure and map to B_G.

for $F^*\alpha(x)[F|Y_2]$ and thinking of it in $\mathbf{Q}[\hat{G}/\text{conj}]$ so we can drop the conjugation by $\hat{f}$, we get

$$(5.22) \quad m([\hat{g}_1],[\hat{g}_2]) = \sum_{h \in G} \frac{1}{\#\text{Hmy}(g_1 \cup g_2)} \sum_{a \in \text{Hmy}(g_1 \cup g_2)} [a \cdot \hat{g}_1 \cdot \hat{h} \cdot \hat{g}_2 \cdot \hat{h}^{-1}].$$

If $\alpha = 1$, so the extension is trivial, then this becomes the product structure described in 4.13(3), tensored with $\mathbf{Q}[A]$.

We take this a little further and write it more completely in terms of $\hat{G}$, but there seems to be little benefit in this. Interpret $a \in \text{Hmy}(g_i)$ as the correction factor required to go from $\hat{g}_i$ to $\hat{f} \cdot \hat{g}_i \cdot \hat{f}^{-1}$ for some $f \in G$ which commutes with g_i. Break the "a" in (5.22) into a product of factors for g_1 and g_2 then both get replaced by such a conjugation. After conjugation of the whole product these can be absorbed into the h term. Replacing $\hat{g}_i$ by $x_i = b \cdot \hat{g}_i$ we get

$$m([x_1],[x_2]) = \sum_{h \in G} \frac{1}{\#\{\tilde{h}\}} \sum_{\tilde{h}} [x_1 \cdot \tilde{h} \cdot x_2 \cdot \tilde{h}^{-1}],$$

where the sum is over $\tilde{h} \in \hat{G}$ of the form $y_1 \hat{h} y_2$ for some y_i so that the image $\rho(y_i)$ commutes with g_i. This is not a very natural expression, but recall that the "answers" (5.5) and (5.6) are not very natural, and we cannot avoid at some point seeing the effects of the dependence on choices.

Exercise 5.23. Let G be the product of cyclic groups $C_n \times C_n$ and $\hat{G}$ the extension by C_n described in (5.16). Note G is abelian so $G = G/\text{conj}$.

1. Show that the multiplication in $Z^\alpha(S^1)$ considered as a quotient of $\mathbf{Q}[\hat{G}]$ is

$$m(a^{i_1}b^{j_1}, a^{i_2}b^{j_2}) = \sum_{0 \leq r,s < n} \frac{\ell}{n} \sum_{k=1}^{n/\ell} t^{j_1 i_2 + s i_2 - r j_2 + k\ell} a^{i_1+i_2} b^{j_1+j_2}$$

where ℓ denotes the least common denominator of the nonzero terms in (n, i_1, i_2, j_1, j_2).

2. In particular describe this when n is prime.
3. Describe in similar terms the rest of the ambialgebra structure: the coproduct

$$\Delta \colon Z^\alpha(S^1) \to Z^\alpha(S^1) \otimes_{\mathbf{Q}[A]} Z^\alpha(S^1),$$

the unit, and the counit.

5.24. The construction

Here we describe the source of the formulas (5.5) and (5.6) for Z^α in a natural but somewhat abstract setting.

The plan is to define a "precursor" TQFT by adding enough data to make characteristic numbers well-defined. We then "integrate out" the dependence on the extra data. The "integration" proceeds in two steps. First we define a category from all the choices of data for a fixed Y. The state module is obtained from the TQFT on this category, in this case by an inverse or direct limit. The second step associates homomorphisms Z_X to bordisms $X \colon Y_1 \to Y_2$. The morphisms in the precursor theory and the data associated to the inverse limit give homomorphisms, and Z_X is obtained by taking a weighted sum (integral) of these.

The precursor TQFT

We begin with the definition of boundaries and spacetimes with enough data added to make characteristic numbers well-defined. Define

Boundary objects. These are (Y, y, f) where $y \in C_d(Y)$ is a representative chain for the orientation $[Y]$, and $f \colon Y \to B$,

Spacetimes. $(X, Y, [X], y, f, [F])$ so that $(X, Y, [X])$ is a relatively oriented complex as in (5.1), $y \in C_d(Y)$ is a chain representing $\partial[X]$, $f \colon Y \to B$ is a map, and $[F]$ is a rel-Y homotopy class of maps $X \to B$ which restrict to f on Y.

If $X \colon Y_1 \to Y_2$ is a bordism, then define

(5.25)
$$\hat{Z}^\alpha(Y, y, f) = \mathbf{Q}[A]$$
$$\hat{Z}^\alpha_X = \text{multiplication by } F^*\alpha(x)$$

where $x \in C_{d+1}(X)$ represents $[X]$ and has $\partial x = -y_1 + y_2$. According to Lemma 5.3 $F^*\alpha(x)$ is well-defined and independent of the choice of x, or F in the relative homotopy class.

It is easily seen that (5.25) defines a TQFT on these boundaries and spacetimes. The tensor property comes from the natural isomorphism $\mathbf{Q} \otimes_{\mathbf{Q}} \mathbf{Q} = \mathbf{Q}$ as in the Euler theory in §3.

Exercise 5.26. Verify the composition property (1.2) for this TQFT. For this suppose (Y, y, f) is the outgoing boundary of X_1 and incoming in X_2, and:
1. show if x_1, x_2 represent the orientations of X_1 and X_2, then $x_1 + x_2$ represents the orientation of $X_1 \cup_{Y_2} X_2$.

2. show $[F_1 \cup F_2]$ is a well-defined rel-boundary homotopy class, and

$$(F_1 \cup F_2)^*\alpha(x_1 + x_2) = (F_1^*\alpha(x_1))(F_2^*\alpha(x_2)).$$

We now begin the construction of Z^α from $\hat{Z}^\alpha$, beginning with state modules. Fix a boundary object $(Y, [Y])$. Define a category with objects all possible choices of extra data on Y, and morphisms the choices of spacetime data on $Y \times I$. Specifically $\mathcal{D}(Y)$ has objects (y, f) where y is a chain representing $[Y]$ and $f: Y \to B$. A morphism $[F]: (y_1, f_1) \to (y_2, f_2)$ is a rel-boundary homotopy class of maps to B which agree with f_1 on $Y \times \{0\}$, and with f_2 on $Y \times \{1\}$.

The precursor TQFT restricts to give a functor on this category to the category of modules over $\mathbf{Q}$ (rational vector spaces). We want to extract a single $\mathbf{Q}$-module from this functor, and for this we use inverse or direct limits. We describe limits first in terms of an abstract universal property, since this will be used to construct the homomorphisms Z_X. A more concrete description is then given which leads to the "answer" (5.5) for the state modules.

Definition 5.27. [McL2] Suppose $\hat{Z}: \mathcal{D} \to \text{Mod}(\mathbf{Q})$ is a functor from a (small) category to modules over $\mathbf{Q}$. An *inverse limit* for this functor is
1. a $\mathbf{Q}$-module $\varprojlim_\mathcal{D} \hat{Z}$
2. for every object $y \in \mathcal{D}$ a homomorphism $q_y: \varprojlim_\mathcal{D} \hat{Z} \to \hat{Z}(y)$ such that for any morphism $f: y_1 \to y_2$ in the category the diagram commutes:

$$\begin{array}{ccc} \varprojlim_\mathcal{D} \hat{Z} & \xrightarrow{q_{y_1}} & \hat{Z}(y_1) \\ \downarrow \text{id} & & \downarrow \hat{Z}_f \\ \varprojlim_\mathcal{D} \hat{Z} & \xrightarrow{q_{y_2}} & \hat{Z}(y_2) \end{array}$$

3. the module and homomorphisms are universal with respect to this property: if L and $r_y: L \to \hat{Z}(y)$ also satisfy (2) then there is a unique $s: L \to \varprojlim_\mathcal{D} \hat{Z}$ so that the r_y factor: $r_y = q_y s$.

Since inverse limits exist and are unique (up to natural isomorphism) we can define

(5.28) $$Z^\alpha(Y) = \varprojlim_{\mathcal{D}(Y)} \hat{Z}^\alpha$$

The uniqueness up to natural isomorphism follows directly from the definition. The existence comes from the following explicit model. A point in the inverse limit is a "section" of $\hat{Z}$: a function s of objects in $\mathcal{D}$ such that $s(y) \in \hat{Z}(y)$. These are required to be invariant under $\mathcal{D}$ morphisms in the sense that if $f: y_1 \to y_2$ is a morphism then $s(y_2) = \hat{Z}_f(s(y_1))$.

There is an analogy for this which will be used later. Think of the objects of $\mathcal{D}$ as points in a space, and the vector spaces $\hat{Z}(y)$ as the fibers of a bundle over this space. Think of the morphisms between the fibers as being parallel translation associated to a connection on the bundle. Then the inverse limit is the space of sections of the bundle which are horizontal with respect to the connection.

In the category of interest, $\mathcal{D}(Y)$, all the morphisms are isomorphisms. In this case the model can be made more explicit. Define the *holonomy group* at an object y by

$$\mathrm{Hmy}\,(y) = \{\hat{Z}_f \mid f\colon y \to y \text{ is in } \mathcal{D}\}.$$

This group acts on $\hat{Z}(y)$. Note that if s is an element of the inverse limit then the element $s(y) \in \hat{Z}(y)$ is fixed by the action of this group. Therefore evaluation gives a homomorphism to the fixed set

$$\varprojlim_{\mathcal{D}} \hat{Z} \to \mathrm{fix}\,(\hat{Z}(y), \mathrm{Hmy}\,(y))$$

Next define $\pi_0 \mathcal{D}$ to be the equivalence classes of objects, where two are equivalent if there is a morphism between them. Choose a representative object y_i for each component $i \in \pi_0(\mathcal{D})$, then evaluating at each of these defines a homomorphism to the product:

$$(5.29) \qquad \varprojlim_{\mathcal{D}} \hat{Z} \to \overline{\bigoplus}_{i \in \pi_0 \mathcal{D}} \mathrm{fix}\,(\hat{Z}(y_i), \mathrm{Hmy}\,(y_i))$$

Here $\overline{\oplus}$ denotes the direct product, with elements arbitrary sequences of elements in the factors. The direct sum $\oplus$ consists of sequences with all but finitely many entries equal to 0.

Lemma. *If all the morphisms in $\mathcal{D}$ are isomorphisms then the homomorphism (5.29) is an isomorphism.*

This is proved by constructing an inverse. Suppose an element in the product is given, so $s_i \in \mathrm{fix}\,(\hat{Z}(y_i), \mathrm{Hmy}\,(y_i))$ for all i. Define a section $y \mapsto s(y)$ as follows: let i be the component of $\mathcal{D}$ containing y, then there is a morphism $f\colon y_i \to y$. Define $s(y) = \hat{Z}_f(s_i)$. The fact that s_i is fixed by the holonomy implies that $s(y)$ is well-defined, and satisfies the coherence condition for sections in $\varprojlim_{\mathcal{D}} \hat{Z}$. $\square$

One final modification is required to get the "answer" described in (5.5). The homomorphisms $\hat{Z}_X$ act on $\mathbf{Q}[A]$ by multiplication by an element of A, so the holonomy is the subgroup $\mathrm{Hmy}\,(y) \subset A$ described in (5.4). By composing the inclusion of the fixed set into $\mathbf{Q}[A]$ and projecting to the quotient we get a natural map

$$(5.30) \qquad \mathrm{fix}\,(\hat{Z}(y), \mathrm{Hmy}\,(y)) \to \mathbf{Q}[A/\mathrm{Hmy}\,(y)\,]$$

Lemma. *This is an isomorphism, with inverse given by*

$$[a] \mapsto \frac{1}{\#\mathrm{Hmy}\,(y)} \sum_{b \in \mathrm{Hmy}\,(y)} ba.$$

Note we are using the fact that the holonomy groups are finite. Also, this inverse uses a choice a representing an equivalence class $[a]$ in $A/\mathrm{Hmy}\,(y)$. The choice is well-defined up to elements in $\mathrm{Hmy}\,(y)$, so the indicated sum is well-defined.

Finally combining (5.29) and (5.30) with the definition of $Z^\alpha(Y)$ gives the answer described in (5.5).

Exercise 5.31. The *direct* limit $\varinjlim_{\mathcal{D}} \hat{Z}$ also associates a module to a functor, and is characterized by the universal property obtained by reversing all the morphisms in (5.27). In particular there are homomorphisms $p_y\colon \hat{Z}(y) \to \varinjlim_{\mathcal{D}} \hat{Z}$ for every object y.

1. If $\hat{Z}\colon \mathcal{D} \to \mathrm{Mod}\,(\mathbf{Q})$ is a functor, consider the quotient $(\oplus_{y\in \mathcal{D}}\hat{Z}(y))/C$ where C is the submodule generated by elements of the form $c - \hat{Z}_f(c)$ for c in some $\hat{Z}(y_1)$ and $f\colon Y_1 \to Y_2$ a morphism in $\mathcal{D}$. Show the natural homomorphism $p'_y\colon \hat{Z}(y) \to (\oplus_y \hat{Z}(y))/C$ satisfies the universal property so it provides an explicit model for the direct limit (and proves that it exists).
2. Suppose all morphisms in $\mathcal{D}$ are isomorphisms, and choose an object y_i in each component $i \in \pi_0 \mathcal{D}$ as in (5.28). Show this gives an isomorphism

$$\oplus_{i\in \pi_0 \mathcal{D}}(\hat{Z}(y_i)/\mathrm{Hmy}\,(y_i)) \xrightarrow{\simeq} \varinjlim_{\mathcal{D}} \hat{Z}.$$

3. Show that if $\pi_0 \mathcal{D}$ is finite and all holonomy groups $\mathrm{Hmy}\,(y)$ are finite then the natural map from inverse to direct limit

$$\varprojlim_{\mathcal{D}} \hat{Z} \to \varinjlim_{\mathcal{D}} \hat{Z}$$

is an isomorphism. (Hint: see (5.30)).

According to this if we use direct limits to construct Z^α we get the same result. Indeed, these exercises show that the answer (5.5) is really given in direct limit form. However when the two limits do not agree the inverse limit seems to be the more fundamental.

5.32. Induced homomorphisms

Next we define homomorphisms $Z_X\colon Z(Y_1) \to Z(Y_2)$ associated to a bordism of oriented complexes $X\colon Y_1 \to Y_2$. Homomorphisms *into* an inverse limit are obtained from the universal property: we construct for every set of data $(y_2, f_2) \in \mathcal{D}(Y_2)$ a homomorphism $Z_X\colon Z(Y_1) \to \hat{Z}(Y_2, y_2, f_2)$ satisfying the coherence condition in (5.27). Then the universal property implies these factor as $q_{y_2, f_2} Z_X$ for a unique Z_X.

Suppose (X, F) is a bordism in the precursor TQFT from (Y_1, y_1, f_1) to (Y_2, y_2, f_2), obtained by adding data (F) to the bordism X. A homomorphism $Z_X\colon Z(Y_1) \to \hat{Z}(Y_2, y_2, f_2)$ is obtained by composing the projection of the inverse limit $q_{y_1, f_1}\colon Z(Y_1) \to \hat{Z}(Y_1, y_1, f_1)$ with the induced homomorphism $\hat{Z}_{(X,F)}$. There may be many of these, corresponding to different F. We get the homomorphism we want by taking a weighted sum (integral) over all equivalence classes of such F. Thus Z_X is specified by:

$$(5.33) \qquad q_{y_2,f_2} Z_X = \sum_{[y_1,f_1,F]} \#^\pi(\mathrm{Map}_{f_1}(X,B)F)\, \hat{Z}_{(X,F)} q_{y_1,f_1}.$$

The sum is over equivalence classes of (y_1, f_1, F), where $F\colon X \to B$ restricts to f_i on Y_i, and y_1 represents the orientation $[Y_1]$. The equivalence relation is homotopy of F rel Y_2.

We show the terms in the sum on the right side of (5.33) are well-defined on equivalence classes. If $(F, y_1, f_1) \sim (F', y_1', f_1')$, so there is a homotopy of F to F' rel Y_2, then restriction to Y_1 gives a homotopy $f_1 \sim f_1'$ which we denote by h. If we compose (X, F') with $(Y \times I, h)$ then the homotopy can be regarded as a homotopy rel $Y_1 \cup Y_2$ from F to this composition. Thus in the precursor theory $\hat{Z}_{(X,F)} = \hat{Z}_{(X,F')} \hat{Z}_{(Y \times I, h)}$. On the other hand h is a morphism $h\colon (y_1, f_1) \to (y_1', f_1')$ in the category $\mathcal{D}(Y_1)$, so the coherence property of the inverse limit asserts that $q_{(y_1', f_1')} = \hat{Z}_h q_{(y_1, f_1)}$. Putting these together we get

$$\hat{Z}_{(X,F)} q_{(y_1,f_1)} = \hat{Z}_{(X,F')} \hat{Z}_{(Y \times I, h)} q_{(y_1,f_1)} = \hat{Z}_{(X,F')} q_{(y_1', f_1')}$$

so this composition depends only on the equivalence class.

We assert that (5.33) does define homomorphisms satisfying the axioms of a TQFT on oriented complexes. For details we refer to [Q], but outline what is involved. First, coherence with respect to Y_2 must be verified to be sure the formula does induce a homomorphism to the inverse limit. Then the composition and product properties must be verified. These come from the corresponding properties of the precursor theory and the proof used in the untwisted version.

Notice that the Y_2 end is held fixed in the characteristic number part of (5.33), while in the homotopy order coefficient the Y_1 end is held fixed. The homotopy order convention is a matter of convenience: reversing ends gives a skew version as in (4.8), which is an isomorphic theory but harder to calculate with. The characteristic number convention is a result of using the inverse limit. If the direct limit is used then the universal morphisms are reversed and the incoming end gets fixed. Exercise (5.31) implies this gives the same theory.

We expand on the use of the direct limit, since it is this form which is most convenient for calculation and is used in the answer (5.6). As before we consider the functor $\hat{Z}$ on the category $\mathcal{D}(Y)$. The natural morphisms for the direct limit are $p_{y,f}\colon \hat{Z}(Y, y, f) \to \varinjlim_\mathcal{D} \hat{Z}$. A homomorphism *from* the direct limit is determined by its compositions with these. Thus Z_X is specified by

$$(5.34) \qquad Z_X\, p_{y_1,f_1} = \sum_{[y_2,f_2,F]} \#^\pi(\mathrm{Map}_{f_1}(X,B)F)\, p_{y_2,f_2} \hat{Z}_{(X,F)},$$

where the sum is over rel-Y_1 homotopy classes of maps F, and $\hat{Z}(X, F)$ is the characteristic number $F^*\alpha(x)$.

Now the answer (5.6) can be assembled. This begins in

$$\sum_{f_1, i} \mathbf{Q}[A/\mathrm{Hmy}\,(f_{1,i})][f_{1,i}],$$

which is the expression given for $Z(Y_1)$. According to (5.31) this is identified with the direct limit $\varinjlim_{(Y_1)} \hat{Z}$. The sum of precursor groups $\sum_i \hat{Z}(Y_1, y, f_{1,i}) = \sum_i \mathbf{Q}[A][f_{1,i}]$ maps to this by the natural maps $p_{f_{1,i}}$. Averaging over the holonomy gives a 1-sided inverse,

$$[a] \mapsto \frac{1}{\#\mathrm{Hmy}\,(f_{1,i})} \sum_{b \in \mathrm{Hmy}\,(f_{1,i})} ab : \mathbf{Q}[A/\mathrm{Hmy}\,(f_{1,i})] \to \mathbf{Q}[A] = \hat{Z}(Y_1, y, f_{1,i}).$$

Now compose with the expression (5.34) for the composition $Z_X\, p_{y_1, f_1}$, to get (5.6).

5.35 Dependence on the cocycle

The TQFT Z^α is defined for a cocycle $\alpha \in C^{d+1}(B; A)$. A different cocycle representing the same cohomology class gives an isomorphic theory as follows. Another representative differs from α by the coboundary of a d-cochain $\delta\beta$. This gives an isomorphism of precursor theories $\hat{Z}^\alpha(Y, y, f) \to \hat{Z}^{\alpha+\delta\beta}(Y, y, f)$ by multiplication by $f^*(\beta)(y)$. (Recall the state modules $\hat{Z}$ are all $\mathbf{Q}[A]$.) This induces an isomorphism of the limit modules, or alternately on the explicit description (5.5).

We can now use the holonomy machinery developed above to analyse the dependence. Define a category with objects $(d+1)$-cocycles representing the cohomology class, and morphisms d-cochains. For each Y the construction above gives a functor from this category to $\mathbf{Q}$-modules. We get something independent of choice by taking direct or inverse limits, and can "calculate" this in terms of holonomy using (5.29) or (5.31). The part genuinely independent of the choice of cocycle is the fixed set, or quotient, of the holonomy.

This category has a single component, so the limit is the quotient or fixed set of the holonomy on a single copy of $\sum_{\{f_i\}} \mathbf{Q}[A/\mathrm{Hmy}\,(f_i)]$. The holonomy group is a sum of terms which act on the components in this decomposition. The term in the holonomy corresponding to the summand for f_i is

$$\{f_i^*(\beta)(y) \mid \beta \in C^d(B, A) \text{ and } \delta\beta = 0\} \subset A/\mathrm{Hmy}\,(f_i).$$

Since y is closed this depends only on cohomology classes, and this can be described as the subgroup of $A/\mathrm{Hmy}\,(f_i)$ obtained by evaluating $H^d(B; A)$ on the image of the orientation $f_*([Y])$. In the examples this is often nontrivial f, so the construction does essentially depend on α.

5.36. The Chern-Simons TQFT

Here we roughly sketch the construction of the Chern-Simons TQFT on oriented surfaces and 3-manifolds, for compact Lie groups G. We indicate the modifications necessary for non-discrete G, and indicate where the development is currently bogged down.

When $G = SU(n)$ the Lagrangian used by Witten [**W1**] is the Chern-Simons 3-form exponentiated and integrated over the manifold:

$$\exp(i\mathcal{L}) = \int_X \exp\left(\frac{ik}{4\pi} \text{Tr}(A \wedge dA - \frac{2}{3} A \wedge A \wedge A)\right).$$

This is defined for a connection A on a G-bundle. To put this in our context let B_G be a classifying space for G so that the universal bundle has a universal connection. This means that a map $f: Y \to B_G$ pulls back the universal bundle to give a bundle with a connection over Y, and up to connection-preserving isomorphism any bundle and connection can be obtained this way. The usual Grassmanian model for $SU(n)$ does this, and there are easy abstract arguments that such a space always exists. Then the expression inside the integral defines an S^1-valued 3-cochain α on B_G. If $f: X \to B_G$ classifies the bundle and connection, and X is a 3-manifold with orientation chain x, then the integral itself is the characteristic number $f^*\alpha(x)$.

For general G there are "classical Chern-Simons" cochains $\alpha_k \in C^3(B_G; S^1)$ corresponding to classes $k \in H^4(B_G; \mathbf{Z})$. These are described for finite G in [**DW**] and for general compact G in [**F**]. If G is simply connected then $H^4(B_G; \mathbf{Z})$ is infinite cyclic. In this case k may be regarded as an integer, and is usually called the "level." Many approaches to this subject use representations of Lie algebras. These can be thought of as using the corresponding simply-connected group, so have integral levels. When G is finite there is a natural isomorphism $H^{d+2}(B_G; \mathbf{Z}) \simeq H^{d+1}(B_G; S^1)$ coming from the long exact sequence induced by exact sequence of coefficient groups $1 \to \mathbf{Z} \to \mathbf{R} \to S^1 \to 1$ and the fact that the real cohomology is trivial. Thus in the model the cochains are closed and the "level" corresponds to the associated cohomology class. A major difference in the general case is that the cochains α_k are not closed.

In the model the first step is to define a precursor theory $\hat{Z}$ on a category with enough extra data to make characteristic numbers well-defined. Here, similarly, we define the "preclassical" TQFT on a category of objects with a map to B_G and orientation data. The orientation data on spacetimes must be more rigid than in the model because the cochain is not closed, and Lemma 5.3 does not apply. On this category the state modules are constant equal to $\mathbf{C}$, and the induced homomorphisms are multiplication by the characteristic number.

The next step is to define a "classical" Chern-Simons TQFT. The objective is to get a category in which there is a topology on the extra data on surfaces Y, and the possible extra data on spacetimes $Y \times I$ correspond to paths in this space. For example a map $Y \times I \to B_G$ is a path in the space of maps $Y \to B_G$, so the category in which the extra data is a map to B_G has this property. The extra orientation data in the preclassical theory does not. The category should also be close enough to the preclassical one that characteristic numbers are still essentially well-defined.

The "classical" version is then defined by inverse limits removing the dependence on the difference between the two categories. The result is a TQFT whose state modules are still 1-dimensional complex vector spaces, but no longer canonically identified with **C**. There are several choices for this intermediate category. One includes a map to B_G as part of the data. This is convenient technically, but depends on the specific version of the classifying space B_G used. The more common version identifies maps to B_G whose pullback bundles (with connections) are gauge equivalent. The extra data on an object is then a gauge equivalence class of G-bundle with connection. This does not depend on a particular choice of B_G, but is harder to work with because the topology and the cocycle are less accessible.

In most treatments (for restricted classes of G) the preclassical step is skipped, and the classical theory on objects with gauge equivalence classes of bundles is constructed directly. The general case is described in Freed [**F**].

By analogy with the model the next step should be to construct state modules $Z(Y)$ for surfaces by taking an inverse limit of the classical theory over the category of extra data on Y. This does not quite work, so we recast it as a bundle problem to describe what does happen. In passing to the classical theory we have obtained a topology on the space $D(Y)$ of extra data on Y. Associated to each point $\alpha \in D(Y)$ we have the state module $\hat{Z}(Y, \alpha)$ which is a 1-dimensional complex vector space. These fit together to define a complex line bundle over $D(Y)$ (in general this is technically difficult and involves Deligne cohomology, see [**F**]). Paths in $D(Y)$ correspond to extra data on $Y \times I$, so the classical theory associates homomorphisms to paths. The composition property implies these are the parallel transports associated to a connection on the line bundle. The inverse limit of the functor therefore can be described as the space of sections of the bundle which are horizontal with respect to the connection (see the comments after 5.27). Since the cocycle is not closed the connection is not flat, and the only horizontal section is trivial. Nontrivial results are obtained by restricting to subsets with symplectic structures and using geometric quantization instead of horizontal sections.

The extra data on Y in the classical theory is essentially the gauge equivalence classes of bundles with connections. (A gauge equivalence is a connection-preserving bundle isomorphism.) Consider the subset of equivalence classes of flat connections. Analytic and physical reasons for doing this are discussed in [**Wi1**,§3]. The basic idea is that this subset is a symplectic reduction of a space of bundles with connections by the action of the group of gauge isomorphisms. It is a finite dimensional stratified set (manifold with singularities). The line bundle with connection is "prequantum" in the sense that the curvature of the connection is a symplectic form at least over the nonsingular part. The Chern-Simons state modules are then supposed to be defined by geometric quantization of this symplectic manifold and line bundle. There are two versions of quantization: holomorphic and real. The real version is closer to the constructions used here, but has not been fully implemented (see [**JW**]). The idea is that one chooses a suitable lagrangian foliation, and the state module consists of (discontinuous) sections of the line bundle which are horizontal when restricted to each leaf of the foliation. In terms of inverse limits this is the inverse limit over the subcategory obtained by restricting the morphisms to paths which lie in a leaf. When comparing with 5.29 note π_0 of this category is the set of connected components of leaves.

The second approach to quantization involves complex structures and holomorphic sections. Choose a complex structure on the surface Y, and denote it by J. Then (when $G = SU(n)$) the space of flat connections corresponds to stable holomorphic $G_{\mathbf{C}}$ bundles on (Y, J), where $G_{\mathbf{C}}$ denotes the complexification [**AB**]. In particular the space of connections itself has a complex structure. The line bundle over this constructed in the classical theory is holomorphic. Define the space $\tilde{Z}(Y, J)$ to be the holomorphic sections of this line bundle. This construction is described in more detail in [**ADPE**], [**H**], and [**Wi3**], and an equivalent version is given in [**S**].

The output of this construction depends on a choice of complex structure. As usual one wants to remove this dependence by taking an inverse limit, and again this is expressed in terms of sections of bundles. The module $\tilde{Z}(Y, J)$ as a function of the complex structure J can be regarded as fibers of a vector bundle over the space of such structures. Morphisms in the category again correspond to parallel translation along curves, so can be specified infinitesimally by giving a connection on the bundle. The appropriate connection has been described from several points of view, [**H**], [**ADPW**], [**S**], and has been shown to be projectively flat (the holonomy is given by scalar multiplication). The space of structures is connected, so the inverse limit is the subspace of $\tilde{Z}(Y, J)$ fixed by the holonomy.

One final modification is required. The inverse limit just described is trivial, since a nontrivial scalar multiplication leaves only 0 fixed. Thus for the boundaries and spacetimes originally considered (oriented surfaces and 3-manifolds), this approach gives the trivial TQFT. To repair this a little structure is added to avoid the scalar holonomy. These enhanced objects are described as "extended" surfaces, etc. by Walker [**Wk**], abstracting some of the properties of null-bordisms. The spacetime objects are described in terms of structures on tangent bundles by Witten [**W4**] and Atiyah [**A3**]. A similar but slightly more refined structure is called "rigging" by Segal [**S**].

The final result of all this is the state module $Z(Y^+)$, where the "+" indicates the extra structure just described. These objects (modules, without the homomorphisms Z_X needed for a full TQFT) have been explored in the theoretical physics literature as "rational conformal field theories." The "rational" refers to finite dimensionality. "Conformal" signifies that they are defined on surfaces with complex structure, but depend only on the conformal equivalence class of the structure. Roughly speaking these arise from attempts to construct $1 + 1$ dimensional field theories. These constructions give spaces of fields on the 1-manifolds, but rather than a single homomorphism Z_Y associated to a surface, one gets a vector space of such homomorphisms. This vector space then serves as the state module of a $2 + 1$ dimensional theory. Conformal field theories are described in the lectures [**MS**] and the references given there.

We have sketched above how state modules $Z(Y)$ are defined for surfaces using Chern-Simons type invariants in a Lie group G. The situation is less clear for the homomorphisms Z_X induced by a 3-manifold. The definition (5.33) and the explicit form (5.6) involve sums over homotopy classes. For general G the straightforward analogs are integrals over gauge equivalence classes of connections as indicated in (4.10). This has been hard to make sense of directly. Instead the Z_X are reconstructed from properties of the state modules, eg. by Walker [**Wk**]. The idea is that a Morse function can be used to decompose a bordism into a composition of simpler ones which involve only a few kinds of "moves" on surfaces. Therefore by describing what happens under these moves the entire TQFT can be determined. This will be discussed in greater detail in §9.

LECTURE 6
Axioms

In this section we describe axioms for topological quantum field theories. A system of axioms guides constructions, clarifies what is common and what is unique about particular examples, and makes classification theorems possible. After some years of experiment we have an elaboration of the proposal of Atiyah [**A1**] which does this, but it is rather complicated. The problem comes in the description of domains on which the theories are defined. Traditional theories like homology theories tend to all be defined for the same objects, eg. finite CW complexes. In contrast each class of TQFT tends to have its own domain of definition, and their behavior strongly reflects the structure of the domain.

We begin with the definition of "domain categories." Bordism categories are manufactured from these, and TQFT are defined to be functors on bordism categories. The axioms for a domain category mainly formalize the properties of examples seen in earlier sections: finite complexes (§§3, 4), oriented finite complexes (5.1), oriented finite complexes together with a map to a classifying space (5.24), and oriented manifolds. There is one important feature which has not yet played much of a role, namely the involution on categories of oriented objects defined by reversing orientations. This is incorporated below as the involution "bar."

6.1. Definition of domain category

A *domain category* is a pair of categories, "spacetimes" $\mathcal{M}$ and "boundaries" $\partial \mathcal{M}$, together with certain functors and operations.

1. (unions and involutions) Each of $\mathcal{M}$ and $\partial \mathcal{M}$ have operations denoted $\sqcup$, commutative and associative with unit ϕ. Each has an involution, denoted by "bar (X)" or by $\overline{X}$, which commutes with $\sqcup$ and leaves the unit invariant;
2. (boundary and cylinders) there are functors $\partial \colon \mathcal{M} \to \partial \mathcal{M}$ and $\times I \colon \partial \mathcal{M} \to \mathcal{M}$ which commute with unions and the involutions, and preserve units. There is a natural morphism $i \colon \partial(Y \times I) \to \overline{Y} \sqcup Y$;
3. (glueing) Suppose $X \in \mathcal{M}$ and $f \colon \partial X \to (Y \sqcup \overline{Y}) \sqcup Y'$ is a morphism in $\partial \mathcal{M}$. Then there is a functorially associated object $\sqcup_Y X$ together with a natural morphism $\sqcup_Y f \colon \partial(\sqcup_Y X) \to Y'$. Further, glueing
   (i) commutes with unions: $(\sqcup_{Y_1} X_1) \sqcup (\sqcup_{Y_2} X_2) = \sqcup_{(Y_1 \sqcup Y_2)}(X_1 \sqcup X_2)$;
   (ii) commutes with involutions: $\overline{\sqcup_Y X} = \sqcup_{\overline{Y}} \overline{X}$;

(iii) commutes with other glueings: if $Y = Y_1 \sqcup Y_2$ then $\sqcup_{(Y_1 \sqcup Y_2)} X = \sqcup_{Y_1} (\sqcup_{Y_2} X)$;
4. (boundary collars) if $X \in \mathcal{M}$ and $Y \in \partial \mathcal{M}$ then
   (i) there is a morphism $X \sqcup_{\partial X} \partial X \times I \to X$ (notation explained below) whose boundary is the natural morphism id $\sqcup_{\partial X} i$ of 3.; and
   (ii) there is a morphism $(Y \times I) \sqcup_Y (Y \times I) \to Y \times I$ whose boundary commutes with the natural morphisms of boundaries to $\overline{Y} \sqcup Y$.

These are the axioms for a domain category, to the first approximation. In fact the "disjoint union" operations are not literally associative and commutative, but only up to natural isomorphism. This means the technically complete version of the definition takes place in the (rather elaborate) context of symmetric monoidal categories. Fortunately for our purposes it seems to be harmless to read "up to natural isomorphism" as "equal" and not be concerned with this refinement.

We are using the notation $\sqcup_Y X$ for the object obtained by identifying two copies of Y in the boundary of X. But we caution that this depends on the morphism $f \colon \partial X \to Y \sqcup \overline{Y} \sqcup Y'$, so when there might be confusion the notation $\sqcup_Y (X, f)$ is used. We also use the usual notation $X' \sqcup_Y X''$ when $X = X' \sqcup X''$ and the boundary of each piece contains one of the copies of Y.

Finally we remark on the collar axiom (4). In the manifold category this is the theorem that the boundary of a manifold has a neighborhood isomorphic to $I \times \partial X$. In the CW category it is the "homotopy extension theorem" which asserts that the projection $X \cup_Y (I \times Y) \to X$ is a homotopy equivalence rel Y. It is this axiom which makes the theories "topological," and fails in geometric settings. For example adding a collar to the boundary of a Riemannian manifold changes the volume, so cannot yield an isometric manifold.

Examples

Three types of examples, with variations, are sketched. The first is the paradigm example of manifolds. Next are categories of topological spaces, and finally purely algebraic examples involving modules over a ring. The most important examples for us are 3-manifolds $SDiff^{2+1}$, 2-complexes Cx^{1+1}, and variations on these.

6.2. Manifolds

$SDiff^{d+1}$ denotes the domain category whose spacetimes are compact smooth oriented $(d+1)$-manifolds with boundary, and the boundaries are the closed oriented d-manifolds. The prefix "S" indicates that orientations are part of the the data, as in $SO(n)$. Variations on this include:
1. Piecewise linear or topological manifolds. There is no essential difference between these and smooth manifolds up to dimension 3. However they diverge dramatically in dimension 4 and some of this divergence is detected by the TQFT-like Donaldson theory.
2. Compact smooth oriented manifolds together with a trivialization of twice the tangent bundle, as a spin bundle, or a "rigging," or an "extension." These are structures used to get nontrivial 2+1-dimensional Chern-Simons theories, see (5.36). Morphisms in the boundary category are required to

preserve the structure exactly (not just up to homotopy) so that glueing is well defined.
3. Compact spaces which are manifolds except for specified singularities (huristically, "black holes").
4. Compact oriented manifolds with orientation-preserving actions of a fixed compact Lie group.

In all of these the disjoint union, glueing, collars, etc. are the obvious variations on the ones in the basic manifold category.

6.3. Topological spaces

1. FHty denotes the category with spacetimes the finite CW complex pairs, and boundaries ∂FHty the finite CW complexes. This is the domain on which the examples discussed in sections 3 and 4 are defined.
2. The category $F\text{Cx}^{d+1}$ has objects which are compact CW complex pairs (X, Y) (a CW structure structure is part of the data), with X $(d+1)$-dimensional and Y d-dimensional. Morphisms are maps homotopic to "deformations" through $(d+1, d)$-complex pairs. Deformations through 2-complexes is defined in discussion of the Andrews-Curtis conjecture in section 2, and other dimensions are similar. (Dimension 2 is the most interesting one, and in particular the Andrews-Curtis conjecture lives in the morphisms of spacetimes in $F\text{Cx}^{1+1}$.)
3. SHty^{d+1} denotes the "oriented complexes" defined in (5.1). Specifically these are finite CW pairs $(X, Y, [X])$ such that $H_{d+1}(Y; \mathbf{Z}) = 0$, together with a $[X] \in H_{d+1}(X, Y; \mathbf{Z})$. Morphisms are homotopy equivalences of pairs which preserve the homology class. The boundary category has objects single spaces Y such that $H_{d+1}(Y; \mathbf{Z}) = 0$, together with a class in $H_d(Y; \mathbf{Z})$. The bar involution reverses the sign of the homology class, just as in the manifold case.
4. Further structure can be added, for example the precursor theory in (5.24) is defined on a domain category of oriented complexes with maps to a fixed space B. Objects of the boundary category have a map to B as part of the data, and morphisms are required to commute with these. This makes glueings well-defined. Spacetime objects come with rel-boundary homotopy classes of maps to B. One point in using homotopy classes rather than specific maps is so the collaring axiom will hold.

Exercise 6.4. Suppose $(X, Y \sqcup \overline{Y} \sqcup Y', [X])$ is an object of SHty^{d+1}. Locate an appropriate orientation class for the glueing $(X/Y, Y')$, and show that it is well-defined.

6.5. Algebraic examples

There are a number of purely algebraic examples in which the objects are algebras or modules over a ring. Some of these appear during the construction or analysis of more topological examples. In any case they illustrate the generality of the definition. Here we describe a domain category of algebras; trace groups and ranks will be used to define a TQFT on an analogous category of ambialgebras in the section on modular theories.

Alg(S) is the category of algebras over a commutative ring with involution S. The boundary category has objects finitely generated algebras with antiinvolution over S. The "union" operation is tensor product: $R_1 \sqcup R_2 = R_1 \otimes_S R_2$, and the ring itself is the unit (the "empty set"). The involution takes an algebra to the opposite algebra.

The spacetimes are pairs (R, P) where R is an algebra and P is a finitely generated projective R-module. Glueing along an algebra R in the boundary is defined by tensoring over R. More explicitly suppose an object has boundary $R \sqcup \overline{R} \sqcup R'$. Then the object is of the form $(R \otimes R^{\mathrm{op}} \otimes R', P)$. The R^{op} factor defines a right R-module structure on R, and $\sqcup_R P$ is defined by dividing out the difference between the left and right R structures. This construction on a bimodule is denoted $\otimes_R P$ in the appendex.

Exercise 6.6. Find a $\times I$ functor $\partial \mathrm{Alg}(S) \to \mathrm{Alg}(S)$ which satisfies the collaring axioms.

6.7. A nonexample

The main condition which makes these axioms "topological," as opposed to geometric, is the collaring axiom. To illustrate this we consider Riemannian manifolds, which is the setting for many of the physically more interesting theories.

Restrict to Riemannian manifolds whose metrics are products near the boundary. The space-time category consists of compact oriented $(d+1)$-dimensional Riemannian manifolds, with morphisms the isometries. The boundary category is the closed oriented d-dimensional manifolds, again with isometries as morphisms. These have disjoint unions, a boundary functor, and satisfy the glueing hypothesis. We could define a functor $\times I$ by taking the product metric with some fixed metric on I. This satisfies all the axioms except collaring.

There is still some collaring-like structure in this category, obtained by considering the forgetful functor to the smooth domain category $S\mathrm{Diff}^{d+1}$. Inverse images of collars $Y \times I$ in the image category give "collar categories" of objects in the geometric category. Using the differentiable collaring theorem we get an action of the category over $\partial X \times I$ on the objects over X. A structure of this sort has been used to good effect by Segal [1], and something like this occurs in the Chern-Simons theories for compact groups discussed in 5.36: the "precursor" theory is defined on a geometric category which does not satisfy the collaring axiom.

6.8. Bordism categories

A bordism between d-manifolds Y_1 and Y_2 is a $(d+1)$-manifold with boundary (isomorphic to) $\overline{Y}_1 \sqcup Y_2$. Here we use the analog of this in a general domain category to construct another category.

The definition

Suppose $\mathcal{M}$ is a domain category. Then Bord$(\mathcal{M})$ is the category with the same objects as $\partial \mathcal{M}$, and morphisms $Y_1 \to Y_2$ equivalence classes of data (X, f, Y_1, Y_2), with $X \in \mathcal{M}$ and $f\colon \partial X \to \overline{Y}_1 \sqcup Y_2$. The equivalence relation is generated by morphisms $G\colon X \to X'$ in $\mathcal{M}$ so that

$$\begin{array}{ccc} \partial X & \xrightarrow{\partial G} & \partial X' \\ {\scriptstyle f}\downarrow & & \downarrow{\scriptstyle f'} \\ \overline{Y}_1 \sqcup Y_2 & \xrightarrow{=} & \overline{Y}_1 \sqcup Y_2 \end{array}$$

commutes.

As before we say Y_1 is the *incoming* boundary of the bordism, and Y_2 is *outgoing*. Note that the incoming boundary has "reversed orientation."

Composition of bordism morphisms is defined by glueing. Suppose $(X_1, f_1)\colon Y_1 \to Y_2$ and $(X_2, f_2)\colon Y_2 \to Y_3$. Then the glueing $(X_1, f_1) \sqcup_{Y_2} (X_2, f_2)$ is defined, and comes with a natural morphism $\sqcup_{Y_2}(f_1 \sqcup f_2)\colon \partial(X_1 \sqcup_{Y_2} X_2) \to \overline{Y}_1 \sqcup Y_3$. Define the composition by $(X_2, f_2) \circ (X_1, f_1) = (X_1 \sqcup_{Y_2} X_2, \sqcup_{Y_2}(f_1 \sqcup f_2))$. This is well-defined on equivalence classes by naturality of glueing.

Associativity of the composition comes from the associativity of glueing. The equivalence class of the product $(Y \times I, i)$ is the identity morphism $Y \to Y$. This follows from the existence of collaring morphisms.

6.9. Operations

The bordism category inherits operations from the domain category. First, the union $\sqcup$ induces a commutative and associative operation on $\mathrm{Bord}(\mathcal{M})$, also denoted $\sqcup$. The bar involution also carries over: it is the same as the operation in $\partial \mathcal{M}$ on objects, and on morphisms $\overline{(X, f)} = (\overline{X}, \overline{f})$. It is an involution, and commutes with $\sqcup$.

6.10. Mapping cylinders

There are special bordisms constructed from morphisms in the boundary category, which we think of as mapping cylinders. If $f\colon Y_1 \to Y_2$ then define $\mathrm{cyl}(f) = (Y_1 \times I, (\mathrm{id} \sqcup f)i, Y_1, Y_2)$. This is $Y_1 \times I$ made into a bordism from Y_1 to Y_2 via the boundary morphism

$$\partial(Y_1 \times I) \xrightarrow{i} \overline{Y}_1 \sqcup Y_1 \xrightarrow{\mathrm{id} \sqcup f} \overline{Y}_1 \sqcup Y_2.$$

This is functorial (the cylinder of a composition is the composition of the cylinders) and commutes with the $\sqcup$ operations and involutions. Cylinders are isomorphisms in the bordism category, with inverse obtained by interchanging the ends and applying bar.

Finally we come to the central object of the lectures:

6.11. Definition of TQFT

Suppose $\mathcal{M}$ is a domain category and R a commutative ring. Then a *topological quantum field theory* on $\mathcal{M}$ with coefficient R is a multiplicative functor $Z\colon \mathrm{Bord}(\mathcal{M}) \to \mathrm{Mod}(R)$

"Multiplicative" means there is a natural isomorphism $Z(Y_1 \sqcup Y_2) \simeq Z(Y_1) \otimes_R Z(Y_2)$, and similarly for morphisms, and Z preserves the unit in the sense that $Z(\phi) = R$ and the homomorphism induced by the empty bordism is the identity $R \to R$. Usually we simply write this isomorphism as an equality, because this accurately suggests the naturality and avoids a lot of notation. It also fits with our convention of treating the $\sqcup$ operations as exactly associative and commutative. However we caution that in examples this is one of the first things to break down: there are many "submultiplicative" theories with a natural homomorphism $Z(Y_1) \otimes Z(Y_2) \to Z(Y_1 \sqcup Y_2)$ which is not always an isomorphism. To discuss these, and for complete proofs, the naturality conditions must be spelled out. For this see [**Q**].

A TQFT is *hermitian* if there is an involution "bar" on the coefficient ring R and a multiplicative natural tranformation of order two, $\theta: Z\mathrm{bar} \to \mathrm{bar}\, Z$. Such a transformation is called a "hermitian structure" for Z.

Here "bar" indicates both the involution in the domain category and the involution on the category of R-modules induced by the involution on R, see A1.3. A natural transformation consists of homomorphisms $\theta_Y : Z(\overline{Y}) \to \overline{Z(Y)}$ for which the diagram

(6.12)
$$\begin{array}{ccc} Z(\overline{Y}_1) & \xrightarrow{\theta_{Y_1}} & \overline{Z(Y_1)} \\ \downarrow Z_{\overline{X}} & & \downarrow \overline{Z_X} \\ Z(\overline{Y}_2) & \xrightarrow{\theta_{Y_2}} & \overline{Z(Y_2)} \end{array}$$

commutes for any bordism $X: Y_1 \to Y_2$. This is multiplicative if $\theta_{Y \sqcup Y'} = \theta_Y \otimes \theta_{Y'}$, and it preserves the unit in the sense that $\theta_\phi: R \to R$ is the involution on the ring, considered as an isomorphism $R \to \overline{R}$. Finally "order two" means $\overline{\theta_{\overline{Y}}}\theta_Y = \mathrm{id}$. Note this implies that θ_Y is an isomorphism.

We cannot use the shortcut used with other natural isomorphisms and write the hermitian structure as an equality $Z(\overline{Y}) = \overline{Z(Y)}$. For example the isomorphism associated to the empty set is the involution on R, which is not usually the identity. Also, on some simple domain categories a multiplicative functor has a natural hermitian structure. The one which comes with a hermitian TQFT is usually different from this, and so constitutes an additional piece of data for which we need a notation.

6.13. Examples

1. Let R be a commutative ring, and $u \in R$ a unit. Then the functor $Z^{\chi,u}$ of §3 is a TQFT.
2. Suppose B is a space with finite total homotopy. Then the functor Z^B defined in §4 is a TQFT with coefficient ring $\mathbf{Q}$. If "bar" is the trivial involution on $\mathbf{Q}$ then this is a hermitian TQFT with hermitian structure the identity $\theta = \mathrm{id} : Z(Y) \to Z(Y)$.
3. Suppose A is an abelian group (written multiplicatively) and $\alpha \in C^{d+1}(B, A)$ is a cocycle. Then the twisted theory Z^α defined in §5 is a

hermitian TQFT. The coefficient ring is the group ring $\mathbf{Q}[A]$ with involution $\overline{ra} = r(a^{-1})$. The hermitian structure is induced (via the inverse limit definition) by the hermitian structure in the precursor theory. The precursor theory has $\hat{Z}(Y, y, f) = \mathbf{Q}[A]$, and the hermitian structure is the involution on the coefficient ring.

Exercise 6.14.
1. Suppose $u \in R$ is a unit in a ring with involution and consider the the Euler theory defined using u. Show that the natural transformation $\theta =$ bar : $R \to R$ is a hermitian structure for this TQFT if and only if $\overline{u} = u$.
2. Show that the precursor TQFT $\hat{Z}$ of 6.13(3) is hermitian. Find an expression for the induced hermitian structure on Z^α in terms of the description given in (5.5).
3. Examples of complex-coefficient TQFT can be obtained from twisted finite homotopy theories with coefficient group $A = S^1$ via the natural ring map $\mathbf{Q}[S^1] \to \mathbf{C}$ (see the beginning of §5). Show that the involution on the group ring goes to the usual conjugation on $\mathbf{C}$.

LECTURE 7
Elementary Structure

Internal structure for a TQFT concerns algebraic consequences of the axioms, the particular domain category, and the particular TQFT. We describe simple examples, and apply this to calculations and classifications. External structure refers to manipulations of entire theories, for example change of coefficients, tensor products and sums. We will be primarily be concerned with internal structure in this section.

The first piece of structure is a canonical nondegenerate bilinear pairing on state modules. This is used to identify homomorphisms induced by spacetimes of the form $Y \times S^1$ as multiplication by the rank of $Z(Y)$. Then TQFT on the domain category of 1-manifolds, $S\text{Diff}^{0+1}$, are shown to correspond exactly to nondegenerate forms. Finally these forms are used to explore redundancy in the definition given in §6: the theory is shown to be determined by bordisms with all boundary outgoing. This is used to compare the definition here with the definition suggested by Atiyah [**A1**].

The second most common algebraic structures are ambialgebras. TQFT on the standard CW complex domain categories induce ambialgebra structures on every $Z(Y)$. Theories defined on $S\text{Diff}^{d+1}$ induce ambialgebra structures on $Z(S^d)$, or more generally $Z(S^k \times N)$ where N is a manifold of dimension $d-k$. In particular these ambialgebra structures characterize TQFT on graphs and surfaces: on the domain categories $F\text{Cx}^{0+1}$ and $S\text{Diff}^{1+1}$.

7.1. Pairings

Suppose Z is an R-coefficient TQFT on a domain category $\mathcal{M}$. For any boundary object Y there is a spacetime $Y \times I$, and we can obtain several bordisms from this by different assignments of incoming and outgoing boundary. Declaring all boundary to be incoming gives a pairing. More precisely, denote by λ_Y the homomorphism

$$Z(Y) \otimes Z(\overline{Y}) = Z(Y \sqcup \overline{Y}) \to Z_{(Y \times I; \overline{Y} \sqcup Y, \phi)} Z(\phi) = R.$$

If the TQFT is hermitian then composing with $\text{id} \otimes \theta \colon Z(Y) \otimes \overline{Z(Y)} \to Z(Y) \otimes Z(\overline{Y})$ gives a hermitian form on $Z(Y)$.

Pairings or forms for the examples described in §3–5 appear in the exercises: (3.4) for the Euler theory, (4.7) for the finite homotopy TQFT, and (5.19) for the twisted version.

Dually we can define a *copairing* Λ_Y by declaring all boundary to be outgoing and forming the composition

$$R = Z(\phi) \xrightarrow{Z_{(Y \times I; \phi, \overline{Y} \sqcup Y)}} Z(\overline{Y} \sqcup Y) = Z(\overline{Y}) \otimes Z(Y)$$

Note that since this is an R-homomorphism it is determined by the element $\Lambda_Y(1) \in Z(\overline{Y}) \otimes_R Z(Y)$.

Proposition 7.2.
1. The pairing and copairing λ_Y and Λ_Y are dual in the sense of (A2), so in particular $Z(Y)$ is a finitely generated projective R-module.
2. If $f: Y_1 \to Y_2$ is a morphism in $\partial\mathcal{M}$ then the homomorphisms associated to the mapping cylinders $Z_{\mathrm{Cyl}(f)}: Z(Y_1) \to Z(Y_2)$ and $Z_{\mathrm{Cyl}(\overline{f})}$ commute with the pairing and copairing.
3. The pairing for $\overline{Y}$ is $\lambda_{\overline{Y}} = \lambda_Y \psi$, where ψ switches factors. In particular if Z is a hermitian theory then $\lambda_Y(\mathrm{id} \otimes \overline{\theta_{\overline{Y}}})$ is a hermitian form.

Part of the significance of this is that we can think of Z as a functor to finitely generated projective R-modules, and for example use algebraic K-theory or traces to explore it. Or think of a hermitian theory as a form-valued functor, and use numerical invariants of forms.

Proof. The condition for "duality" of a pairing and copairing is given in A2.1. There are two conditions which are similar, and we consider only the one which can be expressed as $(\lambda \otimes \mathrm{id})(\mathrm{id} \otimes \Lambda) = \mathrm{id}$. Let $X_1: Y \to Y \sqcup \overline{Y} \sqcup Y$ and $X_2: Y \sqcup \overline{Y} \sqcup Y \to Y$ each be 2 copies of $Y \times I$ arranged as shown in (7.3).

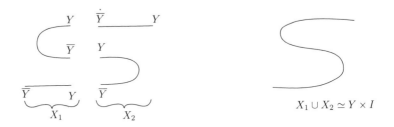

Figure 7.3 Duality of the pairing and copairing.

Then $Z_{X_1} = \mathrm{id} \otimes \Lambda$, $Z_{X_2} = \lambda \otimes \mathrm{id}$, and $Z_{X_2} Z_{X_2} = Z_{X_1 \cup X_2}$ is the composition $(\lambda \otimes \mathrm{id})(\mathrm{id} \otimes \Lambda)$. But $X_1 \cup X_2$ is equivalent as a bordism to $Y \times I$, by the collaring axiom. Since $Y \times I$ (with one boundary in and one out) is the identity morphism in the bordism category, $Z_{Y \times I} = \mathrm{id}$ as required.

A module which supports a form with dual coform is finitely generated projective, see (A2.3).

Next we prove statement 2., that $Z_{\text{cyl}(f)}$ commutes with pairings. Consider the diagram

$$\begin{array}{ccc} Z(Y_1) \otimes Z(\overline{Y}_1) & \xrightarrow{Z_{Y_1 \times I}} & R \\ \downarrow{\scriptstyle Z_{\text{cyl}(f)} \otimes Z_{\text{cyl}(\overline{f})}} & & \downarrow{\scriptstyle =} \\ Z(Y_2) \otimes Z(\overline{Y}_2) & \xrightarrow{Z_{Y_2 \times I}} & R \end{array}$$

The top is λ_{Y_1} and the bottom is λ_{Y_2}, so the goal is to show the diagram commutes. The composition $Z_{Y_2 \times I}(Z_{\text{cyl}(f)} \otimes Z_{\text{cyl}(\overline{f})})$ is induced by the bordism $(Y_1 \times I, (\text{id} \sqcup f_1)i) \sqcup_{Y_2} (Y_2 \times I, i) \sqcup_{Y_2} (\overline{Y}_1 \times I, (\text{id} \sqcup \overline{f}_1)i)$. But $\text{id} \sqcup (f_1 \times I) \sqcup \text{id}$ defines a morphism to this from $(Y_1 \times I, i) \sqcup_{Y_1} (Y_1 \times I, i) \sqcup_{Y_2} (\overline{Y}_1 \times I, i)$. Several applications of the collaring axiom shows this is equivalent to $(Y_1 \times I, i)$, which is the bordism which induces the other composition. Thus the diagram commutes.

We derive 3. Since a copairing determines the dual pairing, to show $\lambda_{\overline{Y}} = \lambda_Y \psi$ it is sufficient to show the right side of the expression is dual to $\Lambda_{\overline{Y}}$. The composition $(\lambda_Y T \otimes \text{id})(\text{id} \otimes \Lambda_Y)$ is described by the same picture 7.3, but replacing the top copies of $Y \times I$ in X_1 and X_2 by copies of $\overline{Y} \times I$. Here $T: Y \times I \simeq \overline{Y} \times I$ switches the ends of $Y \times I$. Straightening this out and using the collaring axiom shows this is obtained by glueing $Y \times I$ to $\overline{Y} \times I$, with opposite directions. We recognize this as a composition of images of the cylinder functor: $\text{cyl}(\text{id}_Y) \circ T\text{cyl}(\text{id}_{\overline{Y}})$. But as remarked in (6.10), these are inverses in the bordism category. Therefore the composition is the identity in the bordism category, and taken to the identity by Z.

Finally consider the hermetian case. We want to show

$$\lambda_Y(\text{id} \otimes \overline{\theta}_{\overline{Y}}) = \overline{\lambda_Y(\text{id} \otimes \overline{\theta}_{\overline{Y}})\psi}.$$

The previous identity for $\lambda_{\overline{Y}}$ shows the right side of this is $\overline{\lambda}_{\overline{Y}}(\theta_{\overline{Y}} \otimes \text{id})$. Therefore it is sufficient to show the diagram commutes:

$$\begin{array}{ccccc} Z(Y) \otimes \overline{Z(Y)} & \xrightarrow{\text{id} \otimes \overline{\theta}_{\overline{Y}}} & Z(Y) \otimes Z(\overline{Y}) & \xrightarrow{Z_{Y \times I}} & R \\ \downarrow{\scriptstyle =} & & \downarrow{\scriptstyle \theta_{\overline{Y}} \otimes \theta_Y} & & \downarrow{\scriptstyle \theta_\phi} \\ Z(Y) \otimes \overline{Z(Y)} & \xrightarrow{\theta_{\overline{Y}} \otimes \text{id}} & \overline{Z(\overline{Y})} \otimes \overline{Z(Y)} & \xrightarrow{\overline{Z}_{\overline{Y} \times I}} & R. \end{array}$$

The left square commutes by the involution property of θ, and the right commutes by the definition of hermitian. $\square$

This can be applied to a simple calculation. Define $Y \times S^1$ for any boundary object Y by glueing together the ends of $Y \times I$. This is a spacetime with empty boundary, so defines a bordism from the empty set to itself.

Proposition 7.4. $Z_{Y \times S^1}: R \to R$ is multiplication by the trace-rank of the module $Z(Y)$.

For example in the finite homotopy TQFT, $Z_{Y \times S^1}$ is multiplication by the homotopy order of $\text{Map}(Y \times S^1, B)$. This is a sum over homotopy classes $[f] \in [Y, B]$

of homotopy orders of $\mathrm{Map}\,(Y \times S^1, B)_{[f]}$. These, however are all 1, by an argument like the analysis for S^1 after 4.6. Thus we get multiplication by $\#[Y, B]$ which is the rank of the free module $Z(Y) = \mathbf{Q}[Y, B]$.

Proof. Regard $Y \times S^1$ as $Y \times I \cup_{Y \sqcup \overline{Y}} Y \times I$ using the collaring axiom. Assign the boundary of the first piece to be outgoing, and the boundary of the second to be incoming. Then this is a composition of bordisms, and in fact the ones used to define the pairing and copairing. Thus the composition property gives $Z_{Y \times S^1} = \lambda_Y \psi \Lambda_Y$. The image of 1 under this is the trace of the identity on $Z(Y)$, which is the definition of the trace-rank (A2.7).

The transposition ψ enters because we have to flip one of the copies of $Y \times I$ to get the Y end of one to match up with the $\overline{Y}$ end of the other. $\square$

Exercise 7.5. Suppose X is a bordism from Y to itself, ie. $X: Y \to Y$. Then the definition of bordism gives a morphism $\partial X \to \overline{Y} \sqcup Y$, and we can use this to form a glueing $\sqcup_Y X$. This is a spacetime with empty boundary. Show that $Z_{\sqcup_Y X}$ is multiplication by the trace of Z_X.

These pairings also give a classification for TQFT on the simplest nontrivial domain category.

Proposition 7.6. *Isomorphism classes of TQFT with coefficient ring R on the domain category* $S\mathrm{Diff}^{0+1}$ *correspond bijectively to isometry classes of nondegenerate pairings on R-modules. Hermitian TQFT correspond to hermitian forms.*

Proof. Suppose Z, Z' are modules with a nondegenerate pairing $\lambda \colon Z \otimes Z' \to R$. A boundary object Y in $S\mathrm{Diff}^{0+1}$ is a set of oriented points, say m with the natural orientation and n with the opposite orientation. Define $Z(Y) = (\otimes^m Z) \otimes (\otimes^n Z')$, then this trivially satisfies the tensor property for disjoint unions.

Next consider a bordism X. It is sufficient, by the tensor property, to specify the homomorphism induced by connected X. There are only four of these: S^1, and the unit interval with three different assignments of the boundary points as incoming or outgoing. If both ends are incoming, let Z_X be the form λ. If both are outgoing use the dual coform. If one is incoming and the other outgoing then $Z_X = \mathrm{id}$. Finally Z_{S^1} is the composition of the form and coform, so (as in 7.4) is multiplication by the trace-rank of Z. It is easy to see that this defines a TQFT on $S\mathrm{Diff}^{0+1}$, and that this construction gives an inverse for the function which assigns a module and form to a TQFT.

The fact that hermitian TQFT correspond to hermitian forms follows from 7.2(3). $\square$

The next application of pairings and copairings is to give formulas which convert boundary components from outgoing to incoming, and vice versa. Suppose X is a bordism from Y_1 to $Y_2 \sqcup Y_3$, and we want to relate this to the bordism defined by the same spacetime, from $Y_1 \sqcup \overline{Y}_2$ to Y_3. Figure 7.7 portrays a composition of bordisms, with $X_1 = X \sqcup Y_2 \times I$ and $X_2 = Y_3 \times I \sqcup Y_2 \times I$.

The composition $X_2 X_1$ is, by the collaring axiom, the same as the spacetime X with $Y_1 \sqcup \overline{Y}_2$ incoming. The disjoint union property for bordisms describes Z_{X_i} as tensor products of Z_X, the form for Y_2, and identity homomorphisms, so we see

LECTURE 7. ELEMENTARY STRUCTURE

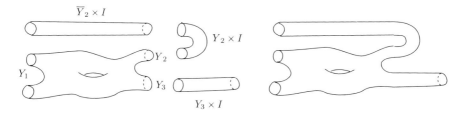

Figure 7.7 Changing outgoing to incoming.

that $Z_{(X;Y_1 \sqcup \overline{Y}_2, Y_3)}$ is the composition

$$Z(Y_1) \otimes Z(\overline{Y}_2) \xrightarrow{Z_{(X;Y_1, Y_2 \sqcup Y_3)} \otimes \mathrm{id}_{\overline{Y}_2}} Z(Y_3) \otimes Z(Y_2) \otimes Z(\overline{Y}_2)$$
$$\xrightarrow{\mathrm{id}_{Y_3} \otimes \lambda_{Y_2}} Z(Y_3) \otimes R = Z(Y_3).$$

Dually the coform can be used to convert a boundary component from incoming to outgoing.

These constructions show that the information in a TQFT is redundant, and is determined by bordisms with all boundary outgoing:

Corollary 7.8. *A TQFT is determined by the modules $Z(Y)$ and the elements $Z_{(X,\phi,\partial X)}(1) \in Z(\partial X)$.*

The form λ_Y can be reconstructed from the coform, which is the element corresponding to $Y \times I$. Then the form can be used to get bordisms with incoming boundary from ones with all boundary outgoing. □

In physical discussions these elements are sometimes called "vacuum vectors." $Z_{(X,\phi,\partial X)}(1)$ is the state obtained by propogating the "vacuum" $1 \in R = Z(\phi)$ across the spacetime X to ∂X.

We use this to compare the definition of TQFT given here with the original one advanced by Atiyah [1]. Atiyah defines a TQFT to be modules $Z(Y)$ for boundary objects Y, and elements $Z_X \in Z(\partial X)$ for spacetimes X. These are assumed to be functorial and multiplicative, which fits with our description. They are also assumed to be "involutory," which means there is an identity $Z(\overline{Y}) = Z(Y)^*$. These are not literally equal; rather this is a huristic shorthand for a natural isomorphism and is used to avoid having to spell out what "natural" means. This isomorphism is adjoint to a nondegenerate pairing $Z(Y) \otimes Z(\overline{Y}) \to R$, which is the pairing associated to Y. The argument of 7.2 shows this pairing is dual to the copairing $R \to Z(\overline{Y}) \otimes Z(Y)$. But this is the vacuum vector associated to $Y \times I$. Thus there is a redundancy: the "involutory" structure is already determined by the vacuum vectors, and all that is needed is the information that the vector associated to $Y \times I$ is nondegenerate as a copairing.

This redundancy sometimes leads to compatibility problems when the stuctures are defined independently. This occurred in some of the first attempts to construct examples, and deficient notation (the huristic use of equality in the involutory axiom) made it difficult to locate the problem.

Exercise 7.9.
1. Describe the "vacuum vectors" for the Euler and finite total homotopy TQFT in sections 3 and 4.
2. Carefully describe axioms for "vacuum vectors" which are equivalent to the TQFT axioms in §6. Begin with a functor $Z\colon \partial\mathcal{M} \to \mathrm{Mod}\,(R)$. Give axioms for hermitian TQFT beginning with Z and a natural isomorphism $\theta_Y\colon Z(\overline{Y}) \to \overline{Z(Y)}$ which is an involution in an appropriate sense.

7.10. Ambialgebras

We now describe some of the appearences and uses of ambialgebras in TQFT, beginning with 1-complexes and manifolds. These structures have appeared in exercises and examples, see (3.4), (4.7), (4.13), and (5.22). The definition and some algebraic structure are given in (A3).

Suppose Z is a TQFT on the domain category $F\mathrm{Cx}^{0+1}$, ie. defined for boundary objects finite sets of points, and spacetimes the finite graphs. Let T denote the "Triod": the cone on three points, with boundary the three points. Then letting two points be incoming and one outgoing defines a commutative product structure

$$m = Z_{(T;2\mathrm{pt},1\mathrm{pt})}\colon Z(\mathrm{pt}) \otimes Z(\mathrm{pt}) \to Z(\mathrm{pt}).$$

Similarly a coproduct is obtained by letting one point be incoming and two outgoing. The unit and counit are obtained from the 1-point spacetime $X = \partial X = \mathrm{pt}$, letting the boundary be outgoing or incoming respectively, as in (7.11).

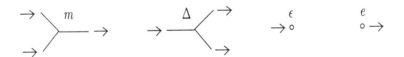

Figure 7.11 The structure bordisms.

The main point is the relation between the product and coproduct. This is the ambialgebra relation (A3.2), and is proved by the equivalence of the composite bordisms shown in (7.12).

Figure 7.12 The ambialgebra relation.

Recall that the equivalence relation on bordisms is generated by morphisms rel boundary in the domain category, and that morphisms in $F\mathrm{Cx}^{0+1}$ are homotopy equivalences obtained by a sequence of "deformations" (6.3). One of the deformations is changing the attaching map of a 1-cell by homotopy, which is how the graphs of (7.12) are related. Figure (7.12) is the same as (A3.5), which uses a graph notation to display the ambialgebra relation.

This construction associates a commutative ambialgebra to a TQFT defined on 1-complexes. Going further, if Z is a TQFT on finite $(d+1)$-complexes or on $F\mathrm{Hty}$, then any boundary object Y has an ambialgebra structure induced by $Y \times T$. So on these domain categories we could regard TQFT as ambialgebra-valued functors.

A similar construction works for manifolds. Suppose Z is a TQFT defined on the domain $S\mathrm{Diff}^{d+1}$ with boundaries d-manifolds and spacetimes $d+1$-manifolds. Let P denote the $(d+1)$-disk with two open balls removed (a "pair of Pants"). This has boundary three copies of S^d. Letting two be incoming gives a product on $Z(S^d)$, and letting one be incoming gives a coproduct. The unit and counit are obtained from the $(d+1)$-disk, as in (7.13).

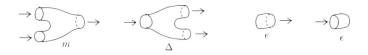

Figure 7.13 Ambialgebra operations on $Z(S^d)$.

Being a little less casual about detail, fix an embedding of two disks in the interior of the disk, $D^{d+1} \sqcup D^{d+1} \to D^{d+1}$ This is well-defined up to isotopy. Delete the interiors, then this gives a manifold P together with identifications of each boundary component with S^d. When specifying boundary components as incoming or outgoing, copies with opposite orientation $\overline{S}^d$ may appear. In these cases change the identification with S^d by a diffeomorphism of degree -1 to get the right orientations. This is how we arrange the product and coproduct to involve only $Z(S^d)$, and not $Z(\overline{S}^d)$.

Again the main point is the ambialgebra relation, which comes from the diffeomorphism shown in (7.14). These manifolds are both diffeomorphic to a sphere with four holes.

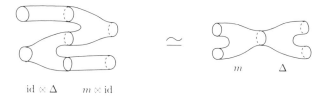

Figure 7.14 The ambialgebra relation.

As above this associates a commutative ambialgebra to a TQFT on $S\text{Diff}^{d+1}$. This can be extended by cartesian products: if N is an n-manifold with $n \leq d-1$ then we get an ambialgebra structure on $Z(N \times S^{d-n})$ using the product of N with the pants P of dimension $d-n+1$.

Exercise 7.15.
1. Suppose Z is a complex-coefficient TQFT on $F\text{Cx}^{d+1}$ or $F\text{Hty}$. Find a bordism $X\colon \text{pt} \to \text{pt}$ so that the ambialgebra structure on $Z(\text{pt})$ is semisimple as an algebra if and only if Z_X is an isomorphism. (See A5.2)
2. Suppose Z is a complex-coefficient TQFT on $S\text{Diff}^{d+1}$. Find a bordism $X\colon S^d \to S^d$ so that the ambialgebra structure on $Z(S^d)$ is semisimple as an algebra if and only if Z_X is an isomorphism.

Ambialgebra structures characterize TQFT on certain simple domains. In the hermitian case there are also ambialgebra involutions. If Z is a hermitian TQFT on graphs then there is a conjugate-linear involution on $Z(\text{pt})$ defined by

$$Z(\text{pt}) = Z(\overline{\text{pt}}) \xrightarrow{\theta_{\text{pt}}} \overline{Z(\text{pt})},$$

using the fact that the involution is trivial on the domain category. There is an analogous construction for manifolds: let $a\colon S^d \to S^d$ denote a diffeomorphism of degree -1, then the cylinder gives a bordism $\text{Cyl}(a)\colon S^d \to \overline{S}^d$. Composing this with the hermitian structure gives a conjugate-linear involution on $Z(S^d)$:

$$Z(S^d) \xrightarrow{\text{Cyl}(a)} Z(\overline{S}^d) \xrightarrow{\theta_{S^d}} \overline{Z(S^d)}$$

These involutions commute with the ambialgebra structure maps since they are derived from bordisms. This gives the definition of ambialgebra involutions (A3.15).

7.16. Classification on surfaces and graphs

We have described enough algebraic structure to classify TQFT in some simple cases.

Proposition. *The constructions above induce bijections from isomorphism classes of TQFT on $S\text{Diff}^{1+1}$, or on $F\text{Cx}^{0+1}$, to isomorphism classes of commutative ambialgebras over R. Hermitian TQFT correspond to hermitian ambialgebras.*

For example the semisimple commutative ambialgebras over $\mathbf{C}$ are simply products of copies of $\mathbf{C}$, with coproducts twisted by units in $\mathbf{C}$. So the TQFT can be described very explicitly in this case (see A5).

Sketch of proof. Begin with 1-complexes. Up to deformation we can assume these are trivalent graphs, with endpoints the boundary. A trivalent graph can be decomposed into pieces, each being a triod, arc, or single point. Therefore a bordism X in this category can be described as a composition of pieces obtained by disjoint unions of copies of these pieces. The data in an ambialgebra gives homomorphisms to use for the elementary pieces: use the product for a triod with two incoming endpoints, for example. Use tensor products to associate homomorphisms to disjoint unions of these, and compose to obtain a homomorphism Z_X.

If this is well-defined then it clearly gives an inverse for the construction which associates an ambialgebra to a TQFT, and so this association is a bijection. The involved part of the proof is to show that this is well-defined, ie. independent of the choice of trivalent graph in the equivalence class, and of the decomposition of it into pieces. We discuss this below.

Now consider surfaces. Boundary objects are unions of copies of circles, so associate to this a tensor product of copies of the underlying module of the ambialgebra. A bordism is a surface with boundary divided into incoming and outgoing components. Such a surface can be decomposed into cylinders, disks, and twice-punctured disks (pairs of pants). Again the ambialgebra structure associates a homomorphism to each elementary piece, so the whole can be reconstructed by tensor products and composition. Again this gives an inverse to the ambialgebra construction if it is well-defined, and the hard part is to show it is well-defined.

In both cases an ambialgebra involution defines a hermitian structure on the resulting TQFT. We explain why these involutions appear explicitly in this classification theorem, but not in the classification on $S\text{Diff}^{0+1}$ given in 7.6. In that case the boundary objects have canonical orientations, and there is no orientation-reversing morphism. Thus initially we get two modules $Z(\text{pt})$ and $Z(\overline{\text{pt}})$, dually paired by λ_{pt}. In the hermitian case the hermitian structure and the orientation (and end) reversing automorphism of an interval gives an isomorphism between these. Thus the hermitian structure gives an equivalence between otherwise distinct things. By contrast in the CW categories, and in $S\text{Diff}^{1+1}$ there are orientation-reversing morphisms: the identity on pt $\in \partial F\text{Cx}^{0+1}$ and the antipodal map on S^1. These morphisms can be used to get isomorphisms between the structures on $Z(Y)$ and $Z(\overline{Y})$ independently of the hermitian structure. Using these to identify the two leaves the hermitian structure as an independent piece of data.

Now consider the problem of showing these constructions are well-defined. In each case one shows that two decompositions into elementary pieces can be related by a sequence of "moves." One then verifies that the moves do not change the associated homomorphism. In the 1-complex case the moves come essentially from the definition of deformation: adding or deleting arcs attached by one end, and changing the attaching map of an arc by homotopy. Adding or deleting arcs corresponds to the unit and counit identities in the ambialgebra. Changing an attaching map by homotopy can be further decomposed into pieces in which the attaching map moves through a single trivalent vertex. This move appears in (7.12), and the fact that it does not change the end result is exactly the ambialgebra relation between product and coproduct. The conclusion is that the identities used to define an ambialgebra are exactly what is needed to show the homomorphisms Z_X are independent of moves in deformations of X, so Z is well-defined.

The proof for surfaces is the same in outline, but the details are conveniently described in terms of Morse functions. (See Milnor [**M**] for an elementary treatment.) Choose a Morse function $X \to [0,1]$ with the incoming boundary going to $\{0\}$ and the outgoing boundary going to $\{1\}$. There are a finite number of critical levels in $[0,1]$. If $[s,t]$ is an interval containing only a single critical level, in the interior, then the inverse image of $[s,t]$ is a disjoint union of annuli and a single pair of pants, or a single disk. Conversely any decomposition into pants, disks and annuli can be obtained from a Morse function after adding more annuli. Now we can relate two decompositions. Two Morse functions can be joined by a path of functions, all Morse except for a finite number of parameter values. At these points two critical levels may cross, or two critical points may cancel (a "death") or be generated (a "birth"). The decomposition changes only at these exceptional points, so elementary "moves" relating decompositions can be derived from what happens there. Death points correspond to using a disk to fill in one of the holes in a pair of pants. Algebraically this corresponds to the unit or counit identity for the product or coproduct. Birth points are the reverse of this. Crossing critical levels often does not change the decomposition. The only case in which it does is shown in (7.14), and corresponds to the ambialgebra relation between product and coproduct. So again the identities defining an ambialgebra are exactly what is needed to show Z_X is well-defined. □

7.17. Thickenings

It may seem curious that the answer is the same for the two cases in (7.16), and in particular there is a bijection between TQFT on graphs and those on surfaces. This is explained by a "thickening functor" from complexes to manifolds.

Given a boundary object Y in FCx^{0+1}, choose an embedding in $\mathbf{R}^2$ and take a (closed) regular neighborhood. Take the boundary of this to get a boundary object in $SDiff^{1+1}$. Y is a finite set of points, so the neighborhood is a finite collection of 2-disks, and the boundary is a union of circles. This extends to spacetimes: given an embedding of ∂X in $\mathbf{R}^2$, it can be extended to an embedding of X in the half space $\mathbf{R}^3_+$. Again take a regular neighborhood, and take the frontier of this. The frontier is the intersection of the closure of the neighborhood and the closure of the complement. This is a 2-manifold with boundary the 1-manifold associated to ∂X.

Compose this construction with a TQFT on $SDiff^{1+1}$ to get a TQFT on FCx^{0+1}. This induces a bijection between the sets of TQFT on the two domains.

The description is somewhat huristic because the construction is not well defined. Technically one constructs categories of all possible choices of embeddings and regular neighborhoods, and gets an induced TQFT by taking inverse limits over these categories.

The construction extends to a "functor" $FCx^{d+1} \to SDiff^{n+1}$ whenever $2d + 1 \geq n$. This is not well-defined as a functor, but does induce a function of TQFT by "composition." When $d = 0$ this takes the ambialgebra structure on $Z(D^n)$ described in (7.13) to the ambialgebra on $Z(\text{pt})$. Thickenings $FCx^{1+1} \to SDiff^{3+1}$ take the Andrews-Curtis phenomena to unstable differentiable properties of 4-manifolds.

LECTURE 8
Modular Examples

In this section we give a rough idea of what a modular TQFT is, and show that the examples described in sections 3–5 can be extended to modular TQFT. A formal definition is given in section 9.

8.1. The idea

In a modular theory the incoming and outgoing boundaries need not be disjoint, but are allowed to intersect in carefully controlled "corners." Further, the incoming boundary is allowed to be subdivided into pieces:

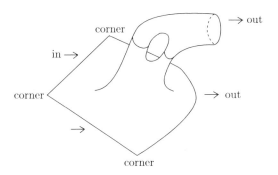

Figure 8.2 A modular bordism.

These decompositions take place in the context of "modular domain categories." We give examples without spelling out the properties they are supposed to exemplify. The rules are given in the next section. The first example is in some sense the smallest nontrivial modular domain, the modular version of the surface category $S\text{Diff}^{1+1}$. It has

Corners: Compact oriented 0-manifolds (finite sets);
Boundaries: Compact oriented 1-dimensional manifolds; and
Spacetimes: Compact oriented 2-dimensional manifolds.

We obtain subdivided boundaries by glueing together boundary objects along corners. Glueings which use up all the corners give 1-manifolds without boundary, and these may occur as boundaries of spacetimes in the usual way.

Now we go up one dimension, and describe the modular version of $S\text{Diff}^{2+1}$. Here we see the first instance of a common theme: the boundary objects are usually rigid. $S\text{Diff}^{2+1}$ has

Corners: Compact oriented *parameterized* 1-manifolds;

Boundaries: Compact oriented 2-manifolds with parameterized boundary; and

Spacetimes: Compact oriented 3-dimensional manifolds.

A parameterized 1-manifold is one with a choice of diffeomorphism of each component with S^1. This parameterization need not preserve orientation. When glueing boundaries together the glueing maps must preserve parameterizations. There is a unique parameterization-preserving map between connected parameterized 1-manifolds, so these are specified simply by a correspondence between components.

Similarly the corners in the $(3+1)$-dimensional analog are 2-manifolds together with a diffeomorphism to a fixed representative surface of the same genus. Alternatively one can use Riemann surfaces as corners: 2-manifolds with a fixed conformal equivalence class of complex structures. There is no useful generic extension of this pattern to dimension $4+1$. We don't mind because we are mainly looking for low-dimensional invariants. Also, theories usually come with internally specified ways to rigidify corners so there can be high-dimensional theories without a generic context.

8.3. Bordisms

A modular bordism $X: Y_1 \to Y_2$ is a spacetime X together with a morphism $f: \partial X \to (\sqcup_{W_1} \overline{Y}_1) \sqcup_{W_2} Y_2$.

This notation indicates that there are morphisms of corner objects $\partial Y_1 \to W_1 \sqcup \overline{W}_1 \sqcup W_2$ and $\partial Y_2 \to W_2$, the copies of W_1 are identified to get $\sqcup_{W_1} Y_1$, and then the conjugate of this is attached to Y_2 along W_2. In other words we allow the incoming and outgoing boundary to intersect in a corner, and allow the incoming boundary to be further subdivided; see 8.2. We do not allow subdivision of the outgoing boundary for reasons explained in the next section.

Bordisms can be composed by glueing along the common boundary, as in 6.8. Equivalence classes again form a category.

Definition 8.4. A *modular* TQFT with coefficient ring R takes boundary objects to R-modules and bordisms to homomorphisms. More explicitly, $X: Y_1 \to Y_2$ induces $Z_X: Z(Y_1) \to Z(Y_2)$. As before we assume this satisfies a composition property (ie. is a functor on the bordism category), and is multiplicative with respect to disjoint unions. Finally we require the *modularity condition* to be satisfied. This is: for a corner object W the natural ring structure on $Z(W \times I)$ has a unit; the unit in $Z(\partial Y \times I)$ acts as the identity via the natural module structure on $Z(Y)$; and the natural homomorphism

$$(8.5) \qquad Z(Y_1) \otimes_{Z(W \times I)} Z(Y_2) \to Z(Y_1 \sqcup_W Y_2)$$

is an isomorphism.

The ring and module structures and homomorphism (8.5) are obtained from the previous hypotheses, as we now explain. Let W be a corner object, and consider $W \times I^2$. This is a spacetime with boundary the union (along copies of W) of four copies of $W \times I$. Glue two of these together and reverse the orientations of two to get a bordism $W \times I^2 \colon W \times I \sqcup W \times I \to W \times I$, see 8.6. The induced morphism

$$Z_{W \times I^2} \colon Z(W \times I) \otimes_R Z(W \times I) \to Z(W \times I)$$

is a product which gives $Z(W \times I)$ a ring structure. It is often not commutative.

To see the module structure, suppose Y is a boundary object. Then $\partial(Y \times I) = (Y \times \{0\}) \cup (\partial Y \times I) \cup (Y \times \{1\})$. Regard this as a bordism $Y \sqcup (\partial Y \times I) \to Y$ and apply Z to get

$$Z(Y) \otimes Z(\partial Y \times I) \to Z(Y).$$

This is a *right* module structure on $Z(Y)$. The first part of the modularity hypothesis is that these rings have units, which act correctly on the modules.

Figure 8.6 The product and module structures.

Note that there is an isomorphism $W \times I \simeq \overline{W} \times I$ which interchanges the ends. This gives a ring isomorphism to the opposite ring $Z(\overline{W} \times I) \simeq Z(W \times I)^{\mathrm{op}}$. Recall (A1.2) that a right $Z(W \times I)^{\mathrm{op}}$-module structure corresponds to a left $Z(W \times I)$ structure. Thus if $W \subset \partial Y_1$ and $\overline{W} \subset \partial Y_2$, so $Y_1 \cup_W Y_2$ is defined, then $Z(Y_1)$ has a right $Z(W \times I)$ structure, $Z(Y_2)$ has a left structure, and the tensor product of (8.5) is defined.

We next describe the source of the homomorphism (8.5). Consider a union of boundaries $Y_1 \cup_W Y_2$. Multiply by I and use the collaring property to absorb the $\partial \times I$ term into one of the pieces. This gives a spacetime with boundary $\overline{(Y_1 \cup_W Y_2)} \cup (Y_1 \cup_W Y_2)$. Regard this as a bordism from the disjoint union $Y_1 \sqcup Y_2$ to the glueing $Y_1 \cup_W Y_2$ then it induces a homomorphism we denote by ρ; $\rho \colon Z(Y_1) \otimes_R Z(Y_2) \to Z(Y_1 \cup_W Y_2)$. This factors through the tensor over $Z(W \times I)$ to give (8.5). To see this it is sufficient to see that terms of the form $ar \otimes b$ and $a \otimes rb$ have the same image. This follows from the observation that we can get from $Y_1 \cup_W W \times I \cup_{\overline{W}} Y_2$ to $Y_1 \cup_W Y_2$ by absorbing the collar into either Y_1 or Y_2.

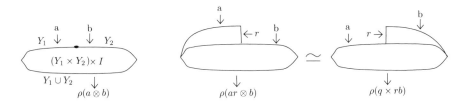

Figure 8.7 ρ and its factorization.

We also introduce some structure which will be used in the next section. There are pairings and coproducts

(8.8)
$$\lambda_Y \colon Z(Y) \otimes \overline{Z(Y)} \to Z(\partial Y \times I)$$
$$\Delta_W \colon Z(W \times I) \to Z(W \times I) \otimes Z(W \times I)$$

defined using $Y \times I$ with the copies of Y incoming and $\partial Y \times I$ outgoing, and by $\Delta_W(g) = \lambda_{W \times I}(g, 1)$. See 8.9. We see in the next section that this coproduct makes $Z(W \times I)$ an ambialgebra. On the modular domain $S\text{Diff}^{1+1}$ the entire TQFT is determined by this ambialgebra structure on $Z(\text{pt})$.

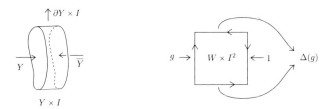

Figure 8.9 Pairing and coproduct.

In this picture of $W \times I^2$ the reversed orientation of the incoming boundary is obtained by reversing direction in the I coordinate, as indicated by the direction of the arrows on the boundaries.

8.10. The Euler theory

This is too simple to display much of the structure. The domain for this has corners finite CW complexes (without the usual rigidity), boundaries finite CW pairs, and spacetimes again finite CW pairs. Morphisms in all cases are homotopy equivalences.

As in §3 let R be a commutative ring and fix a unit $u \in R$. For every boundary object (Y, W) define $Z(Y, W) = R$. For a bordism $X \colon Y_1 \to Y_2$ define Z_X to be multiplication by u to the exponent $\chi(X, \cup_{W_1} Y_1)$. Here X is a modular bordism so has a morphism $\partial X \to (\cup_{W_1} Y_1) \cup_{W_2} Y_2$. The definition of Z_X uses the relative Euler characteristic with respect to the image of Y_1 in X, rather than Y_1 itself.

Exercise 8.11. Let Z be the Euler TQFT on finite CW complexes.
1. Verify that the ring structure induced on $Z(W \times I) = R$ is the usual one, and the module structure on $Z(Y, W)$ over $Z(W \times I)$ is just the usual R-module structure on R.
2. Use this to verify the modularity conditions of 8.4.
3. Evaluate the bilinear form $Z(Y) \otimes Z(Y) \to Z(\partial Y \times I)$ defined in 8.8.
4. Evaluate the coproduct on $Z(W \times I)$ defined in 8.8.

We verify the composition property. Suppose $X_1 \colon Y_1 \to Y_2$ and $X_2 \colon Y_2 \to Y_3$ are modular bordisms. The composed bordism is not quite the union of the two: identifications occur in X_1 corresponding to the identifications in Y_2 in ∂X_2.

Figure 8.12 A composition.

Specifically the composition spacetime is $(\cup_{W_2} X_1) \cup_{(\cup_{W_2} Y_2)} X_2$. But W_2 is part of the boundary of Y_2, so is identified to something in the incoming boundary Y_1 of X_1. Therefore the new identifications take place in the part ignored by the relative Euler characteristic, and this invariant is unchanged:

$$\chi(X_1, \cup_{W_1} Y_1) = \chi(\cup_{W_2} X_1, \cup_{(W_1 \sqcup W_2)} Y_1).$$

The composition formula $Z_{X_1 \cup X_2} = Z_{X_2} Z_{X_1}$ now follows from the additivity of the Euler characteristic (3.2).

8.13. The finite total homotopy TQFT

Now we describe a modular version of the theory defined in §4. Let B be a space with finite total homotopy. The domain for this TQFT has

Corners: Finite CW complexes W together with choices of representatives $g_* \colon W \to B$ for the homotopy classes $[W, B]$;

Boundaries: finite CW pairs (Y, W) together with choices $\{g_*\}$ for W; and

Spacetimes: finite CW pairs (X, Y).

The morphisms are homotopy equivalences, of pairs for boundaries and spacetimes, and which commute with the maps to B on corners.

The commutativity on corners gives the rigidity. For example it will often happen that the maps give an embedding of W into a product of copies of B, and in that case there is at most one morphism $W' \to W$ when W' is another corner object.

We can give such structures to objects in the generic manifold domains using universal choices and the parameterizations. For example $SDiff^{2+1}$ has corner objects parameterized closed 1-manifolds. Choose representatives $g_*\colon S^1 \to B$. A corner object is $W = W_1 \sqcup \cdots \sqcup W_n$ together with $h_i\colon S^1 \simeq W_i$. Then the compositions $\{g_* h_i^{-1}\}$ give representatives natural with respect to the morphisms in $SDiff^{2+1}$.

If $(Y, W, \{g_*\})$ is a boundary object, then define $[Y, g_*, B]$ to be the set of homotopy classes rel W of maps $Y \to B$ which agree with one of the g_* on W. The state modules for the TQFT are the rational vector spaces generated by these sets:

$$(8.14) \qquad Z(Y) = Z(Y, W, \{g_*\}) = \mathbf{Q}[Y, g_*, B].$$

Now suppose $X\colon Y_1 \to Y_2$ is a bordism, and $g\colon Y_1 \to B$ represents a generator of $Z(Y_1)$, ie. restricts to one of the given representatives on ∂Y. Then define the induced homomorphism by

$$(8.15) \qquad Z_X(h) = \sum_{G \in [X, h; B]} \#^\pi(\mathrm{Map}_h(X, B), G)[G|Y_2]$$

As in the nonmodular case (4.6) this is a sum over homotopy classes of extensions of h to all of X, of rational weighting factors times the restriction of the extension to Y_2. The weighting factors are homotopy orders as defined in (4.5).

Note that h is defined on the incoming boundary before the identifications $\cup_{W_1} Y_1$ are made. Therefore the first condition required for $Z_X(h)$ to be nonzero is that it define a map on this quotient, ie. the restrictions to the two copies of W_1 in ∂Y_1 be equal. These restrictions are required to be one of the choices of representatives g_* for homotopy classes $[W_1, B]$. Since there is exactly one representative in each homotopy class it follows that these restrictions are equal if and only if they are homotopic. This conclusion, that homotopy implies equality on corners, is needed several times in careful verifications of the TQFT axioms.

The composition property is obtained by modifying the proof in §4 in the same way as the proof for the Euler theory was modified just above. The composition bordism is a union $(\cup_{W_2} X_1) \cup_{(\cup_{W_2} Y_2)} X_2$, but the identifications take place in the incoming boundary of X_1 where g is specified. The argument after (4.6) applies to this union.

Exercise 8.16.
1. Suppose W is a corner object with choices $\{g_*\}$. The boundary object $W \times I$ has boundary two copies of W with choices $\{g_i \sqcup g_j\}$ for $i, j \in [W, B]$. Thus $Z(W \times I)$ is the vector space with basis the homotopies between the g_*. Show that the product in the natural ring structure is induced by composition of homotopies. Find an identity for this product.
2. Show the $Z(\partial Y \times I)$-module structure on $Z(Y)$ is given by the action of $[\partial Y \times I, \{g_* \sqcup g_*\}; B]$ on the set $[Y, \{g_*\}; B]$ defined by glueing homotopies on ∂Y.
3. Show the homomorphism of (8.5) is induced by the function of basis sets given by glueing maps $h_i\colon Y_i \to B$ together along W. The kernel is therefore

generated by $h_1 \otimes h_2 - h_1' \otimes h_2'$, where all restrict to g_i on W and there is a homotopy $h_1 \cup_W h_2 \sim h_1' \cup_W h_2'$. Show this can be written as $h_1' r \otimes h_2 - h_1' \otimes r h_2$, where $r \in [W \times I, g_i \cup g_i; B]$, and so conclude the modular axiom is satisfied.

Think of homotopies $W \times I \to B$ as paths in the space of maps from W to B, then $[W \times I, g_i \cup g_i; B] = \pi_1(\operatorname{Map}(W, B), g_i)$. The first part of the exercise identifies the ring as a sum of group rings

$$Z(W \times I) = \oplus_{i \in [W,B]} \mathbf{Q}[\pi_1(\operatorname{Map}(W, B), g_i)].$$

We describe the coproduct on $Z(W \times I)$ defined in (8.8). $\lambda_{W \times I}: Z(W \times I) \otimes Z(W \times I) \to Z(W \times I) \otimes Z(W \times I)$ is defined by letting two edges of $W \times I^2$ be incoming, and two be outgoing, see (8.17).

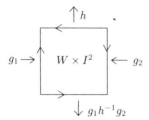

Figure 8.17 The bilinear form for $W \times I$ structures.

Let h_1, h_2 be basis elements, then $\lambda(h_1, h_2) = 0$ unless the restrictions to all four corners are the same. Suppose this is $g_i: W \to B$. Since $W \times I^2$ deformation retracts to the union of three edges (say the two sides and the top) the space of extensions of $h_1 \sqcup h_2$ to $W \times I^2$ is homotopy equivalent to the space of extensions of $g_i \sqcup g_i$ to the top edge $W \times I$. Denote the restriction to the top by f, then the bottom edge is $h_1 f^{-1} h_2$ Subsituting this into (8.15) gives

(8.18) $$\lambda(h_1, h_2) = \sum_{f \in \pi_1(\operatorname{Map}(W,B), g_i)} \mu(f)\, h_1 h^{-1} h_2 \otimes f$$

where $\mu(f) = \#^\pi(\operatorname{Map}_{(g_i \sqcup g_i)}(W \times I), f)$ as in (4.6). Note this number is 1 if $B = B_G$ for a finite group G. In general it only depends on i since the mapping space is a loop space. Denote it by μ_i, then the coproduct is

$$\Delta_W(h) = \lambda(h, 1) = \mu_i \sum_f h f^{-1} \otimes f = \mu_i \sum_j j \otimes j^{-1} g.$$

This coproduct is a twisting of the standard coproduct described in (A4.5) on a sum of group rings.

8.19. The twisted finite homotopy TQFT

This is a modular version of the theory described in §5. Fix a space B with finite total homotopy, an abelian coefficient group A, and a $(d+1)$-cocycle $\alpha \in C^{d+1}(B; A)$.

The appropriate domain category is obtained from the one described in 8.13 by adding orientations. The corner objects are rigid in this regard too: a choice of orientation cycle is required, not just a homology class.

Corners: Finite CW complexes W with a chain $w \in C_{d-1}(W)$, $H_{d+1}(W) = 0$, and representatives $g_* \colon W \to B$ for $[W, B]$;

Boundaries: finite CW pairs (Y, W) with $[Y] \in H_d(Y, W)$, a chain w representing $\partial[Y]$, $H_{d+1}(Y, W) = 0$, and choices $\{g_*\}$ for W; and

Spacetimes: finite CW pairs (X, Y) with $[X] \in H_{d+1}(X, Y)$, and $H_{d+1}(Y) = 0$.

The TQFT on these objects has coefficient ring the rational group ring of the coefficient group, $\mathbf{Q}[A]$. It is given by the same definitions as the nonmodular version in §5, but with some refinement of the ingredients.

The expressions (5.5) and (5.6) give "answers": explicit formulas but dependent on choices of representatives for equivalence classes. For the state module expression to make sense we must extend the definition of holonomy groups given in 5.4, modifying it to allow for corners. If $(Y, W, [Y], w, \{g_*\})$ is a boundary object and $f \colon Y \to B$ is a map whose restriction to W is equal to some g_i, then define the holonomy subgroup to be

$$\operatorname{Hmy}(f) = \{F^*[\alpha]([S^1 \times Y \cup D^2 \times W]),$$
$$\text{where } F \colon S^1 \times Y \cup D^2 \times W \to B, F|Y = f, F|W \times D^2 = g_i\}.$$

The only other modification required is in the index set for the summation. In (5.5) these are representatives for homotopy classes $[Y, B]$. When W is nonempty use representatives for $[Y, \{g_*\}; B]$, which is the notation for homotopy classes rel W of maps which agree with some g_i on W.

8.20. The construction

These choice-dependent formulas are derived from a more categorical construction, exactly as in §5. We define a precursor theory on a domain category with more data, then use inverse limits and integration to remove dependence on the extra data. The precursor theory in 5.24 requires choice of a cycle representing orientation classes of boundaries, and a map to B. The modular precursor is obtained by making similar modifications to the domain described in (8.19). The corners do not require changes.

Corners: Finite CW complexes W with a chain $w \in C_{d-1}(W)$, $H_{d+1}(W) = 0$, and representatives $g_* \colon W \to B$ for $[W, B]$;

Boundaries: finite CW pairs (Y, W) with a chain $y \in C_d(Y)$, $H_{d+1}(Y, W) = 0$, choices $\{g_*\}$ for W, and a map $f \colon Y \to B$ which agrees with one of the g_* on W; and

Spacetimes: finite CW pairs (X, Y) with $[X] \in H_{d+1}(X, Y)$, $y \in \partial[X]$, $H_{d+1}(Y) = 0$, and a rel-Y homotopy class of maps $F \colon X \to B$.

On this domain we define the precursor theory $\hat{Z}^\alpha$ exactly as in (5.25). The state modules are the constant modules $\mathbf{Q}[A]$, and the induced homomorphisms are multiplication by characteristic numbers obtained by pulling α back to X and evaluating on the orientation class. Enough data has been specified to make this operation well-defined.

As in 5.24 the state module $Z^\alpha(Y)$ is defined to be the inverse limit of the precursor functor $\hat{Z}^\alpha$ over a category $\mathcal{D}(Y)$. This category is constructed from the extra data required to make Y an object of the precursor domain category.

The induced homomorphisms are given as in (5.32) by the weighted sum of precursor homomorphisms (5.33). The weightings are the same homotopy orders used in the untwisted version (8.15). The proof this is well-defined, multiplicative, etc. is a straightforward combination of the proofs in the untwisted modular version and the twisted nonmodular version. The only interesting point in the untwisted modular version was the observation that some additional identifications in one of the bordisms do not make any difference because they occur in corners where maps, etc. are already specified. Here we observe that these identifications do not effect the characteristic numbers either, because they involve the orientation cycles of corners which are two dimensions below the dimension in which the evaluation takes place.

8.21. The ring $Z^\alpha(W \times I)$

We describe this ring as a central homomorphism algebra (A4.5), and determine the ambialgebra structure. A consequence is that these have units, which is the first modularity axiom.

Let $(W, w, \{g_*\})$ be a corner object, and suppose $f_j \colon W \times I \to B$ are representatives of the rel-boundary homotopy classes $[W \times I, \{g_* \sqcup g_*\}; B]$. Define a group

$$\hat{G}_i = (A/\mathrm{Hmy}\,(g_i \times I)) \times \pi_1(\mathrm{Map}\,(W, B), g_i)$$

with multiplication given as a central extension of the first factor by the second. Define the extension using the cocycle $\alpha \colon \pi_1 \times \pi_1 \to A/\mathrm{Hmy}$ defined by: (f_1, f_2) goes to $F^*\alpha(W \times I^2)f_3$ where: F is an extension of $f_1 \cup f_2$ to $W \times I^2$ (unique up to homotopy) f_3 is the restriction of F to the rest of $\partial(W \times I^2)$, and w is a $(d-1)$-chain representing the orientation of W. Recall α is the fixed $d+1$-cocycle.

Since $\mathrm{Hmy}\,(g_i \times I)$ and $\pi_1(\mathrm{Map}\,(W, B), g_i)$ are finite the homomorphism $A \to \hat{G}_i$ is central and has finite kernel and cokernel. In this case there is the "central homomorphism" ambialgebra structure on $\mathbf{Q}[\hat{G}_i]$ over the ring $\mathbf{Q}[A]$ (see A4.5).

Proposition. *As $\mathbf{Q}[A]$-algebras, $Z(W \times I) \simeq \oplus_i \mathbf{Q}[\hat{G}_i]$.*

The coproduct is determined below.

Proof. First, (5.5) describes $Z(W \times I)$ as a sum

$$Z(W \times I) \simeq \sum_{f_j} \mathbf{Q}[A/\mathrm{Hmy}\,(f_j)][f_j].$$

Each f_j restricts to some g_i on the ends, and is a homotopy between the ends. Since there is only one g_i in each homotopy class f_j must be a homotopy of g_i to itself. Or equivalently a loop in Map (W, B) based at g_i. This identifies the index set as a union of fundamental groups:

$$[W \times I, \{g_* \sqcup g_*\}; B] = \cup_i \pi_1(\text{Map}(W, B), g_i).$$

Further the holonomy groups of all loops at g_i are the same, so we can substitute Hmy $(g_i \times I)$ for Hmy (f_j). Using this we can rewrite the state module as

$$(8.22) \qquad Z(W \times I) \simeq \sum_{i \in [W,B]} \mathbf{Q}[(A/\text{Hmy}(g_i \times I)) \times \pi_1(\text{Map}(W, B), g_i)].$$

We write the basis elements as $[a]f$, where $[a]$ is the equivalence class of $a \in A$ in the quotient by the holonomy, and f is a loop.

Now consider the multiplication on this ring. The product $m([a_1]f_1, [a_2]f_2)$ is trivial unless both loops are based at the same g_i. Thus the decomposition in (8.22) is also a sum of rings. If they are based at the same g_i then the product is given by $Z_{W \times I^2}$ where two adjacent copies of $W \times I$ form the incoming boundary.

Figure 8.23 The multiplication.

The space of extensions of $f_1 \cup f_2$ to $W \times I^2$ is contractible. Therefore there is a single equivalence class $[F]$ in the sum in (5.6), and the homotopy order is 1. The average over the holonomy group also drops out because all the holonomy groups are the same. Therefore we get

$$(8.24) \qquad m([a_1]f_1, [a_2]f_2) = [a_1 a_2 F^* \alpha(w \times I^2)]f_3$$

where f_3 is the representative for the composition of homotopies $f_1 f_2$. Recall that w is a $(d-1)$-chain representing an orientation for W, so $w \times I^2$ is a $(d+1)$-chain in $W \times I^2$. This formula (8.24) is exactly the description of the product structure in a central extension given in (5.14), provided the function $F^* \alpha(w \times I^2)$ satisfies the cocycle relation (5.11). This is straightforward, so $Z(W \times I)$ is a sum of group algebras of central extensions. □

In particular since the sum in (8.22) is finite, and G and Hmy $(g_i \times I)$ are finite, this ring has a unit.

We can also identify the ambialgebra structure on this algebra. It must be obtained from the standard central homomorphism structure described in A4.5 by twisting by a central unit. The twisting can be determined from the counit in the ambialgebra structure, which is induced by $W \times I^2$ with the whole boundary considered as one copy of $W \times I$, all incoming. In the summand corresponding to a fixed $g \in [W, B]$ the standard counit from A4.5 is

$$\epsilon_0([a]f) = \begin{cases} 1/\#\mathrm{Hmy}\,(g \times I) \sum_{a \in \mathrm{Hmy}} ab & \text{if } f = g \times I \\ 0 & \text{otherwise.} \end{cases}$$

Comparison with (5.6) in this case shows the counit is the standard counit multiplied by

(8.25) $$\sum_{F \in \pi_2(\mathrm{Map}\,(W,B),g)} (\#^\pi(\Omega^2(\mathrm{Map}\,(W,B),F))F^*\alpha(w \times [S^2]).$$

Exercise 8.26.
1. Verify the cocycle relation for the function $F*\alpha(w \times I^2)$ appearing in (8.24).
2. Verify the tensor product modularity axiom for Z^α (see 8.16).
3. Suppose $B = B_G$, for G a finite group. Show the holonomy groups in 8.22 are trivial. Identify the $\pi_1(\mathrm{Map}\,(W, B), g)$ term as the centralizer of g_* (the set of elements which commute with the image of $g_* \colon \pi_1(W) \to G$, see (4.12)). Show that the twisting unit (8.25) for the ambialgebra structure is trivial.
4. Recall the situation in (5.7) where $B = B_G$ for finite G, $d = 1$ so α is a 2-cocycle, and consider the $(1+1)$-manifold modular TQFT defined by this (ie. on $S\mathrm{Diff}^{1+1}$). Show that the ambialgebra structure on $Z(\mathrm{pt} \times I)$ is the central extension ambialgebra structure on $\mathbf{Q}[\hat{G}]$ over $\mathbf{Q}[A]$, where $\hat{G}$ is the extension of G by A determined by α.
5. Let $B = B_G$, $d = 2$, and consider the resulting $(1+1)$-dimensional manifold modular TQFT (ie. on $S\mathrm{Diff}^{2+1}$). Show that the ambialgebra associated to the corner object S^1 is $\sum_{g \in G/\mathrm{conj}} \hat{G}_g$, where $\hat{G}_g$ is a central extension

$$1 \to A \to \hat{G}_g \to \mathrm{centralizer}(g) \to 1.$$

8.27. The endomorphism TQFT

As a final example we describe a modular theory on $(1+1)$-dimensional manifolds, ie. on $S\mathrm{Diff}^{1+1}$. This example illustrates the approach used to construct $(2+1)$-dimensional TQFT in section 10, but does not itself produce an interesting theory. The associated nonmodular theory is constant, and the modular theory is Morita equivalent to a constant in the sense of 9.24.

Let R be a commutative ring with involution, $u \in R$ an invariant unit, and A, B R-modules with a nondegenerate pairing $\lambda \colon A \otimes B \to R$.

The modular boundary objects of $S\mathrm{Diff}^{1+1}$ are oriented 1-manifolds with boundary, so are unions of copies of I and S^1. Define $Z(S^1) = R$ and $Z(I) = B \otimes A$, and extend to arbitrary manifolds by tensor products. We think of the B factor as associated with the 0 end of the interval and the A factor as associated with 1.

Now we define the homomorphism induced by a modular bordism $X\colon Y_1 \to Y_2$ using a product of copies of λ and identity maps. There are isomorphisms $\partial X \simeq (\cup_{W_1} \overline{Y}_1) \cup_{W_2} Y_2$, $\partial Y_2 = W_2$, and $\partial Y_1 = W_1 \sqcup \overline{W}_1 \sqcup W_2$ given as part of the modular bordism structure. The module $Z(Y_1)$ is a product of copies of R, which we ignore since they vanish in the tensor product, and copies of A, B. The ones corresponding to points in $W_1 \sqcup \overline{W}_1$ appear in pairs $A \otimes B$. Apply λ to these, to map to copies of R which again we neglect. What remains is copies of A, B corresponding to the points in W_2. But the product of these is the definition of $Z(Y_2)$, so these identifications give $Z(Y_1) \to Z(Y_2)$. Define this to be Z_X. Some examples are shown in figure 8.28.

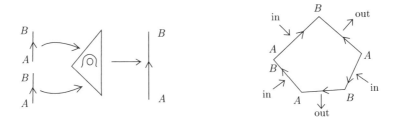

Figure 8.28 An example.

In the first example Z_X is

$$Z(Y_1) = (\overline{A}_1 \otimes A_1) \otimes (B_2 \otimes A_2) \xrightarrow{\mathrm{id} \otimes \lambda \otimes \mathrm{id}} B_1 \otimes A_2 = Z(Y_2).$$

The second example is given by

$$(B_1 \otimes A_1) \otimes (B_2 \otimes A_2) \otimes (B_3 \otimes A_3) \xrightarrow{\mathrm{id} \otimes \lambda \otimes 3\mathrm{id}} B_1 \otimes A_2 \otimes B_3 \otimes A_3 \xrightarrow{=} (B_1 \otimes A_3) \otimes (B_3 \otimes A_2).$$

Exercise 8.29.
1. Show that the ambialgebra structure on $Z(\mathrm{pt} \times I)$ is the endomorphism ring $\mathrm{End}(B)$ (see A4.1).
2. Verify the modular composition property for this TQFT.
3. Show that the corresponding nonmodular TQFT on $S\mathrm{Diff}^{1+1}$ is constant.
4. Suppose Z is a TQFT with $\mathbf{C}$ coefficients (or some field). Show that a hermitian structure $\theta\colon Z\mathrm{bar} \simeq \mathrm{bar}\, Z$ for the endomorphism TQFT corresponds to an isomorphism $B \simeq \overline{A}$ so that the resulting form $\lambda\colon A \otimes \overline{A} \to R$ is hermitian. (See A4.5 and note θ is determined by its action on $Z(I)$.)

LECTURE 9
Modular Axioms and Structure

This section is the modular analog of sections 6 and 7: we outline axioms for modular theories, then give some abstract structure.

In a modular theory state modules are defined for boundary objects with some sort of "corners." Homomorphisms are induced by spacetimes whose boundaries are divided into chambers which intersect in corners. The basic ideas are described in the beginning of section 8. The first objective here is to formalize the properties required of these boundaries and corners, in the definition of a "modular domain category."

Exploration of the algebraic structures generated by a modular TQFT begins in 9.7 with the special ambialgebra structure on $Z(W \times I)$ for each corner object W. The $1+1$ dimensional modular manifold category $S\text{Diff}^{1+1}$ has a single connected corner object (up to isomorphism) and TQFT on this category are seen to be characterized by the associated ambialgebra. Comparison with the nonmodular version in section 7 shows that most TQFT on this category are not modular.

Next we consider boundary objects, and show in 9.14 that $Z(Y)$ has a natural module structure and nondegenerate pairing over the ring $Z(\overline{\partial Y} \times I)$. The associated "rank" TQFT on the boundary category is defined in 9.20 by $Z^{\text{rk}}(W) = Z(W \times S^1)$ for a corner object W. This TQFT is shown to encode the rank of $Z(Y)$ as a $Z(W \times I)$ module.

In 9.23 Morita equivalence is used to change corner ambialgebras without changing the associated nonmodular TQFT. This in particular gives a way to associate a "conformal field theory" to a modular TQFT on 2+1-dimensional manifolds.

Finally we apply this to describe some of the structure of modular TQFT on 2+1 dimensional manifolds, ie. on $S\text{Diff}^{2+1}$. This includes a version of the "Verlinde theorem" concerning "fusion rule algebras" associated to such theories.

9.1. Modular domain categories

As with ordinary domain categories in section 6 it is useful to approach this axiomatically because the examples are so diverse. The axioms are essentially abstracted from the examples described in the previous section, see 8.1, 8.13, 8.19. We will not give all the details.

We have formalized "glueing" of objects in a domain category in 6.1. Now we want to describe objects whose *boundaries* are obtained by glueing. To do this we recycle the glueing technology, by supposing that the boundary objects are spacetimes in a second domain category. Specifically suppose that $(\mathcal{M}, \partial \mathcal{M})$ is a domain category, $(\partial^+ \mathcal{M}, \partial^2 \mathcal{M})$ is another domain category, and there is an inclusion $\partial \mathcal{M} \subset \partial^+ \mathcal{M}$. This inclusion is required to commute with the union operations and involutions, and $\partial^2 = \phi$.

In this situation an "$\mathcal{M}$ object with subdivided boundary" is an object $X \in \mathcal{M}$ whose boundary is given as a glueing in $\partial^+ \mathcal{M}$: there is $Y \in \mathcal{M}$, a morphism $\partial Y \to W \sqcup \overline{W}$ in $\partial^2 \mathcal{M}$, and a morphism $\partial X \to \cup_W Y$. Pieces of Y are called faces of X, and pieces of W are called corners.

To this basic framework we add some structure (with comments following):
1. The cylinder functor $\times I \colon \partial \mathcal{M} \to \mathcal{M}$ extends to a functor $\partial^+ \mathcal{M} \to \mathcal{M}$. When Y has boundary then the boundary of $Y \times I$ has "sides" $(\overline{\partial Y}) \times I$ as well as top and bottom $\overline{Y} \sqcup Y$. More generally there are "pinched" cylinders with sides corresponding to some components of ∂Y but not others, see figure 9.3. Specifically if $\partial Y = W_1 \sqcup W_2$ then $(Y, W_1) \times I$ has boundary $(\overline{Y} \cup_{W_1} Y) \cup_{\overline{W}_2 \sqcup W_2} (\overline{W}_2 \times I)$.
2. The glueing operation in $\mathcal{M}$ is extended to allow glueing along faces with boundary in an $\mathcal{M}$-object with subdivided boundary.
3. The collaring axioms of both domain categories are refined. In $\mathcal{M}$ we assume collaring for cylinders glued onto a face of an object with corners. In $\partial^+ \mathcal{M}$ we assume that the collaring morphisms are well-defined and natural up to concordance.

Figure 9.2 Pinched cylinders.

9.3. Comments

In 1. the topological model for a pinched cylinder is $(Y, W_1) \times I = Y \times I / \sim$, where $\sim$ is the equivalence relation $(y, s) \sim (y, t)$ if either $s = t$ or $y \in W_1$. The boundary of this is $Y \times \{0, 1\} \cup W_2 \times I$.

The data for glueing faces, in 2., looks much like the nonmodular data in 6.1(3): a morphism is given $\partial X \to \cup_W (Y \sqcup \overline{Y} \sqcup Y')$ and then the two copies of Y are identified. There is a compatibility condition required on the corners, however. To see the difficulty consider a glueing which around a corner object looks schematically like a vertex in a triangulation of a surface. See figure 9.4.

The identification of Y to $\overline{Y}$ induces a cycle of identifications of components of W, leading to an automorphism in the category $\partial^2 \mathcal{M}$. The compatibility condition is essentially that these automorphisms are identities.

Figure 9.4 A modular glueing.

We will not be more precise about this condition because it is at least generically unnecessary in many cases. Say that a modular domain category has "rigid corners" if for any two indecomposable objects W_1, W_2 of $\partial^2 \mathcal{M}$ there is at most one morphism $W_1 \to W_2$, and it is an isomorphism. When corners are rigid the cycles described above automatically give the identity.

The generic manifold categories described in 8.1 have rigid corners. The domains for the finite total homotopy TQFTs described in 8.13 and 8.19 are generically rigid. Part of the data is a collection of maps $W \to B$ representing homotopy classes, and morphisms must commute with these. Usually these maps give an embedding of W into a product of copies of B, and at most one map between such objects can commutes with an embedding. $(3+1)$-dimensional manifolds in which the corners are Riemann surfaces are also generically rigid.

Next consider the collaring morphisms in the boundary category, $Y \sqcup \partial Y \times I \to Y$. Two such are concordant if there is a $\mathcal{M}$ morphism of the cylinders $(Y \sqcup \partial Y \times I) \times I \to Y \times I$ which is the identity on the side $(\partial Y) \times I$ and restricts to the given morphisms on the ends. We assume there is a well-defined concordance class of collaring morphisms specified for each Y, and this is consistent with unions, glueing, etc. In the examples the collaring morphisms are *unique* up to concordance, so are automatically well-defined and natural.

Finally we consider pinched collars in the spacetime category. Suppose $\partial X = Y_1 \cup_W Y_2$ and W decomposes as $W_1 \sqcup W_2$. Then we can attach the pinched collar $(Y_1, W_1) \times I$ to X along the copy of Y_1. We suppose there is a morphism from the result back to X: $X \cup_{Y_1} (Y_1, W_1) \times I \to X$. This should be the identity from the other end of the collar to $Y_1 \subset \partial X$. On the rest of the boundary we find a morphism $Y_2 \cup_{W_2} (W_2 \times I) \to Y_2$. We assume this is a collaring morphism for $W_2 \subset Y_2$, ie. is in the standard concordance class. In the examples this is routine.

9.5. Modular bordisms and TQFT

The definitions of 8.3 and 8.4 now make sense in a modular domain category. A modular bordism $X: Y_1 \to Y_2$ is defined to be a spacetime X together with a morphism $\partial X \to (\cup_{W_1} \overline{Y}_1) \cup_{W_2} Y_2$, see 9.6. Note the incoming boundary is allowed to be subdivided, but the outgoing boundary is not. For an explanation of why not, see 9.11. For a picture of a modular bordism see Figure 8.2.

The extended glueing operation can be used to compose modular bordisms. If $X_2: Y_2 \to Y_3$ then $X_2 \circ X_1 = (X_1 \cup_{Y_2} X_2): Y_1 \to Y_3$. This gives a category, with identity morphisms provided by the pinched cylinders.

Definition 9.6. As in 8.4 we define a *modular* TQFT to be a functor on the modular bordism category. Thus a boundary object (possibly itself with boundary) Y is taken to an R-module $Z(Y)$, and a modular bordism is taken to an R-homomorphism Z_X. This functor is required to take unions to tensor products and satisfy the *modularity conditions*:
1. The algebra structure on $Z(W \times I)$ has a unit which acts by the identity on state modules $Z(Y)$, and
2. the natural homomorphism of (8.5) is an isomorphism:

$$Z(Y_1) \otimes_{Z(W \times I)} Z(Y_2) \to Z(Y_1 \cup_W Y_2).$$

These conditions are explained in 8.4, with more detail given in 9.14. We emphasize that the ring and module structures used in the modularity condition are constructed from functorial properties, and are not new assumptions. Verifying these conditions is a calculation, not a construction.

As in 6.11, a TQFT is *hermitian* if it has a hermitian structure: a multiplicative natural transformation of order two, $\theta \colon Z\text{bar} \to \text{bar}\, Z$.

9.7. The corner ambialgebras

We begin exploration of modular theories with corner objects.

Proposition. *Suppose Z is a modular TQFT on a domain category $\mathcal{M}$ and $W \in \partial^2 \mathcal{M}$ is a corner object. Then there is a natural special ambialgebra structure on $Z(W \times I)$. There is a natural isomorphism $J \colon Z(W \times I) \simeq Z(\overline{W} \times I)$ which reverses the order of the product and coproduct. If θ is a hermitian structure for Z then θJ is a hermitian involution on the ambialgebra.*

Proof. The first item is the "involution" $J \colon W \times I \to \overline{W} \times I$ which interchanges the ends. Begin with $W \times I \cup_{\overline{W}} \overline{W} \times I$, obtained by glueing the $\overline{W}$ end of $W \times I$ to the W end of $\overline{W} \times I$. We can consider this as a collar in two ways, so there are collaring morphisms from this to both pieces:

$$W \times I \leftarrow W \times I \cup_{\overline{W}} \overline{W} \times I \to \overline{W} \times I.$$

For TQFT purposes we can think of this sequence of morphisms as an isomorphism: the pinched cylinders give a bordism (rel boundary) from $W \times I$ to $\overline{W} \times I$. This is an "involution" in the bordism category in the sense that there is a $\mathcal{M}$ morphism from the composition of this with itself, to the cylinder of the identity map. To avoid a lot of notation we think of this as an orientation-reversing involution on $W \times I$. Denote the induced homomorphism by $J_W \colon Z(W \times I) \to Z(\overline{W} \times I)$, then $J_{\overline{W}} J_W = \text{id}$.

Now consider $W \times I^2$. As it comes to us the boundary of this has two copies of $W \times I$ and two of $\overline{W} \times I$. Using the involution above to change orientations, and by glueing together some of the pieces, we can get a number of different bordisms. For example the algebra structure used in 8.4 is obtained as follows: glue two segments together to give a "triangle," identify two sides as incoming copies of $\overline{W} \times I$ and the other as an outgoing copy of $W \times I$. This gives a bordism $W \times I \sqcup W \times I \to W \times I$, see figure 9.8. Applying Z gives a homomorphism $Z_{W \times I^2} \colon Z(W \times I) \otimes Z(W \times I) \to Z(W \times I)$. This is the product for the ring structure. The existence of a unit for this ring is one of the modularity axioms.

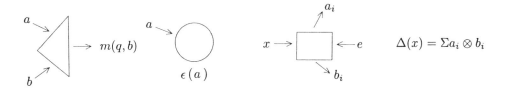

Figure 9.8 Ambialgebra structure maps for $Z(W \times I)$.

Glueing all the the boundary together to get a single copy of $\overline{W} \times I$ gives the counit $\epsilon \colon Z(W \times I) \to R$. The unit is used to define the coproduct. Note we cannot simply declare two faces of the triangle to be outgoing because the outgoing boundary is not allowed to be subdivided. To avoid this we separate the two by an incoming segment and feed in the unit. More explicitly consider $W \times I^2$ as a bordism $W \times I \sqcup W \times I \to W \times I \sqcup W \times I$ and define $\Delta(x) = Z_{W \times I^2}(x \otimes e)$.

The final piece of structure is an element which gives a unit for the trace ring. Define $W \times S^1$ by glueing together the ends of $W \times I$. The second modularity condition gives an expression for the state module: $Z(W \times I)$ is a left and right bimodule over itself (via glueing on the two ends) and $Z(W \times S^1)$ is obtained by identifying these two structures:

$$(9.9) \qquad Z(W \times S^1) = \otimes_{Z(W \times I)} Z(W \times I).$$

This identifies $Z(W \times S^1)$ with the trace group of $Z(W \times I)$, see A2.4. The special element is obtained as follows: Glue all the corners in the boundary of $W \times I^2$ to get a copy of $W \times S^1$, and declare this to be outgoing. Then this induces $Z_{W \times I^2} \colon R \to Z(W \times S^1)$. Define $c \in Z(W \times I)$ to be a preimage of $Z_{W \times I^2}(1)$.

Most of the verifications that these give a special ambialgebra structure are straightforward consequences of subdividing copies of $W \times I^2$ in various ways. Or more precisely, glueing together copies of $W \times I$ in various ways and identifying the results using the collaring axiom. For instance the ambialgebra relation for the product and coproduct come from subdivisions of a pentagon, as shown in figure 9.10. This figure establishes a convention about which edges in the pictures correspond to the variables, for example in the product $m(a, b)$ the "a" variable appears on the upper face, and "b" on the Lower. The figure also shows that the homomorphism induced by to the square with two in and two outgoing edges is given by $(m \otimes \mathrm{id})(\mathrm{id} \otimes \psi)(\Delta \otimes \mathrm{id})$.

Next consider glueing two opposite edges of a square to edges of a triangle, as in figure 9.11.

The figure on the right in 9.11 gives an action of $Z(W \times S^1)$ on $Z(W \times I)$. The construction, and the formula for the induced homomorphism of a square from 9.10

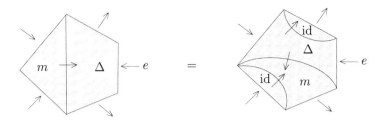

Figure 9.10 The ambialgebra relation.

Figure 9.11 The action of the trace algebra.

identify this with the action of the trace ambialgebra $TZ(W \times I) = Z(W \times S^1)$ on $Z(W \times I)$, see A4.7. Glueing in a disk with outgoing boundary leaves a pinched cylinder which induces the identity on $Z(W \times I)$. This shows that the element c constructed above does indeed give a good unit for the trace algebra, so we conclude that c makes $Z(W \times I)$ special in the sense of A4.7.

Finally we consider the involutions. Since the involution on $W \times I$ interchanges ends this is easily seen to reverse the product and coproduct. We can consider this an isomorphism to the opposite ambialgebra

$$J \colon Z(W \times I) \to Z(\overline{W} \times I)^{\mathrm{op}}.$$

If there is a hermitian structure $\theta \colon Z(\overline{Y}) \to \overline{Z(Y)}$ then since it commutes with induced homomorphisms it commutes with J. Composing the two gives a conjugate-linear anti-involution bar $= \theta J \colon Z(W \times I) \to Z(W \times I)$ which is a hermitian structure on the ambialgebra. □

9.12. Subdivision of outgoing boundary

We make one more use of figure 9.10, namely to explain why the outgoing boundary is not allowed to be subdivided in the axioms. The construction of Δ suggests a way to add this property: if a corner W separates two outgoing regions, then use the collaring axiom to insert a copy of $W \times I$, declare it to be incoming, and evaluate on some fixed element of $Z(W \times I)$. The problem comes in choosing the fixed element. When constructing compositions which leave the new incoming boundary beside an incoming piece, as in the right side of 9.11, the correct element to use is the identity element $e \in Z(W \times I)$. When doing compositions which leave the new piece in the interior, as in 9.10, one needs to use a preimage of a trace unit. Thus we see both are possible exactly when the ambialgebra is superspecial in the sense of A4.7: the image of the identity of $Z(W \times I)$ is the identity of the trace ring.

The conclusion is that Z extends to allow subdivision of the outgoing boundary, with the straightforward extensions of the composition property for bordisms, if and only if the corner ambialgebras are all superspecial. Most of the examples in §8 are not superspecial, so this axiom is excessively restrictive.

9.13. Classification on surfaces

The smooth oriented $1 + 1$ dimensional modular theories are classified by the special ambialgebras associated to a single point. (Note the corner category consists of finite sets of oriented points.) This is the modular analog of 7.16. Special ambialgebras over the complex numbers are easily understood, and are described in A5.1.

Theorem. *The construction above gives a bijection, from isomorphism classes of modular TQFT on $SDiff^{1+1}$ to isomorphism classes of special ambialgebras. Hermitian TQFT correspond to hermitian special ambialgebras.*

According to 7.16 nonmodular TQFT on this domain correspond to commutative ambialgebras. Forgetting the modular structure of a TQFT corresponds to passing to the trace ambialgebra as described in A4.7.

It follows from this that there are many nonmodular TQFT on this domain. Over $\mathbf{C}$ special ambialgebras are semisimple, and the trace ring is simply a product of copies of $\mathbf{C}$. But there are lots of nontrivial commutative ambialgebras over $\mathbf{C}$ and these must correspond to TQFT which cannot be extended to modular theories.

Sketch of the proof. We describe an inverse construction, which gives a TQFT from a special ambialgebra. The state modules are easy: to an oriented interval associate a copy of the algebra, and to a circle associate the trace algebra. Extend to general oriented 1-manifolds by taking disjoint unions to tensor products.

Now consider associating a homomorphism to a bordism. Suppose $X \colon Y_1 \to Y_2$ is a surface with boundary divided into incoming and outgoing pieces. Add a vertex in each complete incoming or outgoing circle. Then choose a triangulation of the surface which extends the given division of the boundary into arcs. Arbitrarily choose a direction across each interior edge, or equivalently choose directions for the dual graph. The dual graph can be thought of as providing instructions for a huge composition of the structure maps of the ambialgebra. Namely, a vertex with two edges in and one out corresponds to the product. Two edges out and one in corresponds to the coproduct. Three edges in correspond to two multiplications followed by the counit. Three edges out correspond to two applications of the coproduct to the unit. This huge composition defines a homomorphism $Z(Y_1) \to Z(Y_2)$.

This process must be modified slightly to compensate for the problems with subdividing outgoing boundaries (see 9.12). Somewhere in each cycle in the dual graph corresponding to a vertex on the surface, introduce a multiplication by a trace unit preimage c. Also multiply by c at each point added to break up an outgoing circle. See Figure 9.13.5.

Define Z_X to be the composition of this modified diagram of structure maps. To see it is well-defined we must see that it is independent of the choices: triangulation,

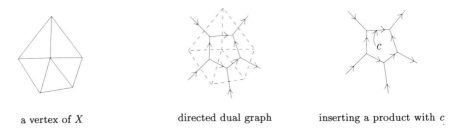

Figure 9.13.5 The dual graph, and products with c.

directions, and insertion points for the c products. Two choices are related by a sequence of simple changes: subdividing a single edge or simplex, changing a single direction, or moving the insertion point past a single vertex. These in turn follow from the ambialgebra identities and the fact that c is special ($m\psi\Delta(c) = e$). We omit the details.

The composition property for Z follows from the fact it is well-defined, because glueing together triangulations of X_1 and X_2 gives a triangulation of $X_1 \cup_Y X_2$. The modularity conditions are essentially obvious.

Thus this procedure defines a modular TQFT on surfaces. We claim it is an inverse for the functor which assigns an ambialgebra to a TQFT, so the functor induces a bijection of isomorphsim classes. Choosing simple triangulations of I^2 and implementing the procedure shows immediately that the ambialgebra associated to the constructed TQFT is the original one. Conversely there is an obvious transformation from the state modules of the constructed TQFT to the original one. The use of the ambilagebra structure from the original TQFT implies this transformation commutes with certain basic induced homomorphisms. It is then straightforward to use the composition property to show it commutes with all induced homomorphisms, so induces an isomorphism of TQFT.

Finally suppose there is a hermitian involution "bar" on the ambialgebra. Define $\theta\colon Z(\bar{I}) \to \overline{Z(I)}$ by bar J where J is induced by an orientation-reversing involution $I \to I$. This extends to θ for S^1 by passing to the trace algebra, and then to any 1-manifold Y by tensor products. We claim this gives a hermitian structure for the TQFT.

First think of the involution on the ambialgebra as giving an isomorphism to the conjugate opposite ambialgebra. Let $\hat{Z}$ denote the TQFT defined using the opposite ambialgebra, then "bar" gives an isomorphism $\hat{Z} \to \overline{Z}$. It remains to show that "orientation reversing involutions" give an isomorphism $Z\mathrm{bar} \to \hat{Z}$, ie. commute with homomorphisms induced by bordisms. To see this we decompose a bordism into a composition of simpler ones. If the boundary is subdivided we can recognize the bordism as a composition of one with fewer corners, and a triangle. The commutativity is easily checked for Z associated to triangles because these are the structure maps of the ambialgebra. Therefore by induction we can reduce to the case where there are no corners, ie. the nonmodular case. In this case the argument is given in 7.16. At this point we are in the trace ring which is commutative and therefore equal to it's opposite. And an orientation-reversing involution given on boundary components extends to an involution on the surface. □

9.14. Module structures on $Z(Y)$

Ambialgebras give a satisfactory description of the structure associate to corner objects, so we turn to a consideration of the module structure on $Z(Y)$ over the ambialgebra $Z(\partial Y \times I)$. Some care is required to get the order right, since the coefficient rings may be noncommutative.

First, $Z(Y)$ is a left $Z(\overline{\partial Y} \times I)$-module. The cylinder $\overline{Y} \times I$ has boundary $Y \cup (\partial Y \times I) \cup \overline{Y}$, so we can consider it as a bordism $(\overline{\partial Y} \times I) \cup Y \to Y$ (recall orientations of incoming boundary components are reversed), see (9.15). Applying Z gives the product
$$Z(\overline{\partial Y} \times I) \otimes Z(Y) \to Z(Y).$$

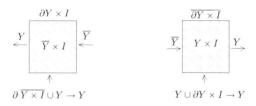

Figure 9.15 Module structures.

It might seem more natural to use the cylinder $Y \times I$. this has boundary $\overline{Y} \cup (\partial Y \times I) \cup Y$, so gives a bordism $Y \cup (\partial Y \times I) \to Y$. However careful attention to the conventions we are using shows this defines a *right* module structure
$$Z(Y) \otimes Z(\partial Y \times I) \to Z(Y).$$

Since we prefer left module structures we work with $Z(\overline{\partial Y} \times I)$. Note that a right module structure corresponds to a left structure over the opposite algebra, and in 9.7 the opposite algebra to $Z(W \times I)$ is identified with $Z(\overline{W} \times I)$.

Lemma 9.16. *Suppose $X: \cup_{W_1} Y_1 \to Y_2$ is a modular bordism. Then $Z_X: Z(Y_1) \to Z(Y_2)$ is a left $Z(\overline{\partial Y} \times I)$-module homomorphism.*

Proof. First we see the module structure on $Z(Y_1)$. Part of the data of a modular bordism is a morphism $\partial X \to \overline{\cup_{W_1} Y_1} \cup_{W_2} Y_2$, and the unions on the boundary involve morphisms $\partial Y_1 \to W_1 \sqcup \overline{W}_1 \sqcup W_2$ and $\partial Y_2 \to \overline{W}_2$. The first morphism gives $Z(Y_1)$ a module over $Z(\overline{\partial Y_1} \times I) \simeq Z(\overline{W}_1 \times I) \otimes Z(W_1 \times I) \otimes Z(W_2 \times I)$. The second morphism gives an isomorphism $Z(\overline{\partial Y_1} \times I) \simeq Z(W_2 \times I)$. Putting these together gives the required structure on $Z(Y_1)$.

The module structure on Y_i is defined using cylinders. Attach cylinders on X along both Y_2 and $\cup_{W_1} Y_1$, and compare via collaring morphisms; see figure 9.17.

The collaring morphisms show $r \cdot s \cdot Z_X(a) = r \cdot Z_X(s \cdot a) = Z_X(r \cdot s \cdot a)$, where $r, s \in Z(W_2 \times I)$, $a \in Z(\cup_{W_1} Y_1)$, and the dot indicates the module and ring multiplications. This is the desired module structure on Z_X. $\square$

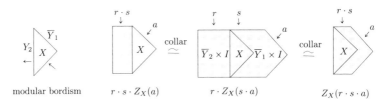

Figure 9.17 Module homomorphisms.

9.18. Nondegenerate pairings

Now we consider the modular analog of 7.1, which gives natural nondegenerate pairings on state modules.

$\partial(Y \times I) = \overline{Y} \cup Y \cup \overline{\partial Y} \times I$, so we may regard $Y \times I$ as a bordism $Y \sqcup \overline{Y} \to \overline{\partial Y} \times I$. Then we define

$$\lambda_Y = Z_{Y \times I}: Z(Y) \otimes Z(\overline{Y}) \to Z(\overline{\partial Y} \times I)$$

Conversely we can consider $Y \times I$ as a bordism $\partial Y \times I \to \overline{Y} \sqcup Y$. Define a copairing by applying this to a preimage of the trace ring unit:

$$\Lambda_Y = Z_{Y \times I}(c) \in Z(\overline{Y}) \otimes_{Z(\overline{\partial Y} \times I)} Z(Y).$$

Note we are regarding $Z(\overline{Y})$ here as a right $Z(\overline{\partial Y} \times I)$-module.

Lemma. *Suppose Y is a boundary object, ie. in $\partial^+ \mathcal{M}$. Then*
1. *λ_Y is a nondegenerate pairing over $Z(\overline{\partial Y} \times I)$, and Λ_Y is the dual copairing (see A2). In particular $Z(Y)$ is a finitely generated projective module over this ring.*
2. *Morphisms in $\partial^+ \mathcal{M}$ preserve pairings and copairings.*
3. *$\lambda_{\overline{Y}} = J \lambda_Y \psi$, and similarly for $\Lambda_{\overline{Y}}$. Thus if Z is hermitian with hermitian structure θ, then $\lambda(\mathrm{id} \otimes \theta)$ is a hermitian form on $Z(Y)$.*

Proof. To show the pairing and copairing are dual we want $m(\lambda \otimes \mathrm{id})(\mathrm{id} \otimes \Lambda) = \mathrm{id}$, where m indicates the action of the ring on the module (scalar multiplication). Geometrically this corresponds to glueing together three copies of $Y \times I$, as depicted in figure 9.19.

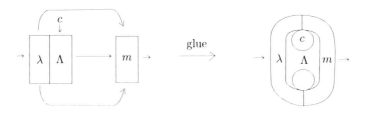

Figure 9.19 Duality of the pairing and copairing.

We end up with a pinched cylinder with a hole. But since c is a trace unit preimage, evaluating this on c corresponds to glueing a disk into the hole. The result is a pinched cylinder, which induces the identity map on Z. □

9.20. The rank TQFT

An immediate consequence of the lemma is a description of ranks of state modules. We extend this to a TQFT on the boundary category and relate it to the "fusion rule algebras" used in conformal field theory.

Corollary. *Consider $Y \times S^1$ as a bordism from the empty object to $\overline{\partial Y} \times S^1$. Then the image of the unit*

$$Z_{Y \times S^1}(1) \in Z(\overline{\partial Y} \times S^1) = TZ(\overline{\partial Y} \times I)$$

is the trace-rank of $Z(Y)$ as a projective $Z(\overline{\partial Y} \times I)$-module.

We identify the state module of $\overline{\partial Y} \times S^1$ with the trace group of the ring using 9.9. In this case the definition of the trace-rank (A2.7) is $\lambda \psi \Lambda$, which by the composition property is $Z_{Y \times S^1}$.

Now define a TQFT on the boundary category $\partial \mathcal{M}$ by

(9.21)
$$Z^{\mathrm{rk}}(W) = Z(W \times S^1), \text{ for } W \in \partial^2 \mathcal{M}$$
$$Z_Y^{\mathrm{rk}} = Z_{Y \times S^1} \text{ for } Y \text{ a } \partial^+ \mathcal{M} \text{ bordism}.$$

We call this the "rank" theory because the state modules are trace algebras $Z^{\mathrm{rk}}(W) = Z(W \times S^1) = TZ(W \times I)$, and the induced homomorphisms can be interpreted in terms of ranks of the projective modules $Z(Y)$. One consequence of this is that the coefficient ring can be considerably reduced, to the subring which contains ranks (often the integers). See 9.22 (2) for a special case.

Exercise 9.22.
1. Suppose $Y \colon W_1 \to W_2$ is a bordism in $\partial^+ \mathcal{M}$. Describe Z_Y^{rk} in terms of the projective left $Z(\overline{W}_2 \times I) \otimes Z(\overline{W}_1 \times I)^{\mathrm{op}}$-module $Z(Y)$. (For this assume the trace rings are generated by ranks of projective modules.)
2. Suppose B is a space of finite total homotopy, and $Z^B \colon F\mathrm{Cx} \to \mathrm{Mod}\,(\mathbf{Q})$ is the modular TQFT defined in 8.13. Describe the associated rank TQFT. (Hint: consider the nonmodular TQFT associated to the space $\Lambda B = \mathrm{Map}\,(S^1, B)$.)

Now suppose Z is a **C**-coefficient modular TQFT on $\mathcal{M}$. The corner ambialgebras are semisimple (A5.1) so (left) modules over them are sums of irreducibles. Let Φ_W denote the set of irreducibles over $Z(W \times I)$ and V_i the irreducible corresponding to $i \in \Phi$. Define a **Z**-coefficient "fusion" TQFT on $\partial^+ \mathcal{M}$ by: $Z^\Phi(W) = \mathbf{Z}[\Phi_{\overline{W}}]$. If $Y \colon W_1 \to W_2$ is a bordism in $\partial^+ \mathcal{M}$ we have a right $Z(\overline{W}_1 \times I)$-module structure and a left $Z(\overline{W}_2 \times I)$-module structure on $Z(Y)$, and these commute. Thus there is a decomposition into irreducibles $Z(Y) \simeq \oplus N_i^j V_i \otimes V_j^*$, where the sum is over $i \in \Phi_{\overline{W}_2}$, $j \in \Phi_{\overline{W}_1}$, and V_j^* denotes the right module dual to V_j. The N_i^j are nonnegative integers. Define $Z_Y^\Phi : \mathbf{Z}[\Phi_{\overline{W}_1}] \to \mathbf{Z}[\Phi_{\overline{W}_2}]$ by

$$Z_Y^\Phi([j]) = \sum_i N_i^j [i].$$

3. Show Z^Φ is a **Z**-coefficient TQFT on $\partial^+\mathcal{M}$. (In the (2+1)-manifold case Z^Φ is a (1+1)-manifold TQFT so is determined by the ambialgebra structure on $\mathbf{Z}[\Phi_{S^1}]$. This is called the "fusion rule algebra, since it describes how the irreducibles ("primary fields") fuse together.)

4. Show that the complexification of the fusion TQFT is the rank TQFT. Specifically define

$$Z^\Phi(W) = \mathbf{Z}[\Phi_{\overline{W}}] \to TZ(\overline{W} \times I) = Z^{\mathrm{rk}}(W)$$

by $[i] \mapsto \mathrm{rk}(V_i)$. Show this is a morphism of TQFT, and when tensored with **C** gives an isomorphism.

9.23. Morita equivalence and rational conformal field theories

A "rational conformal field theory" is, more-or-less, the state modules of a modular TQFT on 3-manifolds. We explain how a TQFT determines a conformal field theory and refer to [**W**], [**C**] for a discussion of the converse.

The usual description of a conformal field theory ([**S**], [**MS1**], [**MS2**], [**V**], [**K**]) goes as follows: There is a finite set Φ of "labels," or "colors," or "primary fields." Then if Y is a surface with boundary, or equivalently a closed surface with marked points, and I is a function assigning a label to each boundary component (or marked point) then there is defined a complex vector space $E_I(Y)$. In particular a sphere with 3 holes (pants) has associated spaces E_{ijk} for $i,j,k \in \Phi$. The modularity property describes how these are related when the surface is decomposed. Suppose $Y = Y_1 \cup_W Y_2$ with W a circle. Let I_1, I_2 be functions which associate labels to each boundary component of Y_1, Y_2 *except* W. Let $I \cup \{i\}$ denote the labeling which extends these by assigning i to W. Then

$$E_{I_1 \cup I_2}(Y_1 \cup Y_2) = \oplus_{i \in \Phi} E_{I_1 \cup \{i\}}(Y_1) \otimes_{\mathbf{C}} E_{I_2 \cup \{i\}}(Y_2).$$

To make this look more like the modularity condition here, define R to be $\mathbf{C}[\Phi]$ with algebra structure a sum of copies of $\mathbf{C}$. A module E over R corresponds exactly to a sequence of vector spaces E_i; modules over the component copies of $\mathbf{C}$. The sum $\oplus_i E_i$ is a module over R. More generally $E(Y) = \oplus_I E_I(Y)$, where the sum runs over all possible labelings, is a module over a tensor product of copies of R, one for each boundary component. The module associated to $S^1 \times I$ is isomorphic in a natural way to the ring R. Thus the modularity property becomes

$$E(Y_1 \cup_W Y_2) = E(Y_1) \otimes_{E(W \times S^1)} E(Y_2).$$

This leads us to expect that rational conformal field theories should be state modules of those modular 2+1-dimensional TQFT such that $Z(S^1 \times I)$ is a sum of copies of **C**. But in fact *any* modular TQFT gives a conformal field theory via Morita equivalence.

Suppose Z is a complex-valued modular TQFT on some modular domain category. The corner ambialgebras are special, so by the classification theorem A5.1 they are semisimple. Morita equivalence (A5.5) asserts that the category of modules over a semisimple **C**-algebra is equivalent to the category of modules over its center. We can therefore define the "Morita reduction" TQFT as follows: assign to each corner object W the center $CZ(W \times I)$. Then for each boundary object Y define $\hat{Z}(Y)$ to be a $CZ(\overline{\partial Y} \times I)$-module corresponding to $Z(Y)$ under Morita equivalence. For a modular bordism $X \colon Y_1 \to Y_2$ define $\hat{Z}_X \colon \hat{Z}(Y_1) \to \hat{Z}(Y_2)$ to be the $CZ(\overline{\partial Y_2} \times I)$-module homomorphism corresponding to Z_X. Then
  1. $\hat{Z}$ (when properly functorially defined, and under reasonable hypotheses on the corner category, see 9.24) gives a modular TQFT,
  2. if W is a corner object then $\hat{Z}(W \times I) = CZ(W \times I)$, and as an algebra this is a product of copies of **C**, and
  3. the associated nonmodular TQFT are naturally isomorphic.

So there is a "rational conformal field theory" associated to an arbitrary Z: the state module functor of the Morita reduction $\hat{Z}$.

Exercise 9.24. Say that a category with an operation $\sqcup$ has "natural maximal decompositions" if each object can be given as a union $W = W_1 \sqcup \cdots \sqcup W_n$ so that the components cannot be nontrivially further decomposed, and every morphism $W \to W'$ decomposes as a union of morphisms of components.
  1. State carefully what it means for two modular TQFT to correspond under Morita equivalence (this involves choices of irreducible $Z(W)$ modules for certain W, see A5.6).
  2. If the corner category has natural maximal decompositions then make precise the definition of $\hat{Z}$ given above, and show it is a modular TQFT with commutative corner ambialgebras.
  3. Suppose V is a module over a commutative ring R, and has a symmetric nondegenerate bilinear form. Then endomorphism TQFTs are defined on $1+1$-dimensional manifolds in 8.27, and on $2+1$ dimensional manifolds and $1+1$ dimensional complexes in §10 which associate the endomorphism algebra $\text{End}(V)$ to connected corner objects. Show that the Morita reductions of these TQFT are constant.

9.25. The Verlinde basis and link invariants

The state modules $Z(W \times S^1)$, for W a corner object, have canonical bases. These were described in the context of conformal field theories in [**V**]. We use them here to form invariants of "colored links," and later to describe certain structure maps as matrices.

We continue to assume that Z is a complex-coefficient modular TQFT on some domain category.

The corner ambialgebra $Z(W \times I)$ is semisimple so is described by dimension and twisting data (Φ_W, n_*, u_*). Here Φ_W is a finite set, n_* is a positive integer-valued function on Φ_W, and u_* is a nonzero complex function on Φ, see A5.1. The state module $Z(W \times S^1)$ is identified with the trace ambialgebra of this, so as an algebra is a sum of copies of $\mathbf{C}$. The coproduct on the ith factor is twisted by u_i^2. The algebra isomorphism to a sum of copies of $\mathbf{C}$ is well-defined up to permutation of factors, so gives canonical coordinates for $Z(W \times S^1)$ as a vector space. The standard basis vectors are the primitive idempotents in the ring. Denote these by e_i, then we modify this slightly by using

$$x_i = u_i e_i.$$

The benefit of the modification is that this basis is orthonormal with respect to the ambialgebra form ϵm. Also the ranks of irreducible $Z(W \times I)$-modules are positive integer multiples of these elements (see 9.22).

In the constructions of TQFT from representations of Hopf algebras (§10) the set Φ_{S^1} can be identified with the set of indecomposable representations of nontrivial rank. In the physics literature these are sometimes referred to as "primary fields."

Now suppose X is a spacetime object. A *framed link* in X is a decomposition $X \simeq X_0 \cup_{W \times S^1} (W \times D^2)$. In the $(2+1)$-dimensional manifold case the corner objects are unions of (parameterized) circles, and this definition is the classical idea of a framed link in a 3-manifold.

A "coloring" for a link is a choice of an element in Φ_W. Note Φ is multiplicative under unions: $\Phi_{W_1 \sqcup W_2} = \Phi_{W_1} \times \Phi_{W_2}$, so this can also be thought of as a choice of element in Φ_{W_j} for each component W_j. In the classical case this amounts to a choice in Φ_{S^1} for each component of the link.

The boundary of X_0 is $\partial X \sqcup \overline{W \times S^1}$. Declare the $W \times S^1$ part to be incoming, the rest outgoing. Then the induced homomorphism is

(9.26) $$Z_{X_0} \colon Z(W \times S^1) \to Z(\partial X).$$

Suppose the link is "colored" with the element $i \in \Phi_W$, then the invariant associated to the colored link is the image of the associated basis element $Z_{X_0}(x_i) \in Z(\partial X)$.

Note since the elements x_i form a basis for the state module of the incoming boundary, the homomorphism Z_{X_0} is determined by these invariants. In the classical $2+1$-dimensional case this means the theory on manifolds with torus boundary components is completely determined by these "link invariants." Some early treatments of TQFT (eg. [**W1**] and [**RT**]) used this to avoid discussion of state modules of surfaces of higher genus. Closed 3-manifolds can be described in terms of manifolds with torus boundary via the "Kirby calculus."

9.27. The structure of $Z(T^2)$

Suppose that Z is a modular $(2+1)$-dimensional complex-coefficient TQFT. We describe some structure of the state module of the 2-torus $T^2 = S^1 \times S^1$. This has been the object of a great deal of activity under the heading "fusion rule algebras" (to mention only a very few references, see [**V**], [**MS1**], [**MS2**], and [**CPR**]). The point here is that much of the structure seen in the examples is a consequence of the modularity axioms, not of the specific constructions.

We caution that the interesting examples are actually defined on a central extension of the $2+1$ dimensional manifold category, rather than on $S\mathrm{Diff}^{2+1}$ itself. This does not effect the structure considered here.

We further suppose that Z is a *hermitian* TQFT. This means there is a natural isomorphism of order 2, $\theta\colon Z(\overline{Y}) \to \overline{Z(Y)}$. Without this assumption we would restrict to orientation-preserving morphisms. With it we can allow orientation-reversing ones and regard them as inducing conjugate-linear homomorphisms of state modules. The morphisms considered are:

9.28. Maps

1. $J\colon S^1 \times I \to S^1 \times I$ which interchanges the ends of the I coordinate (and changes the orientation);
2. $T\colon S^1 \times I \to S^1 \times I$ which twists once in the S^1 coordinate, compared to the fixed parameterization on the ends; and
3. $S\colon S^1 \times I \to I \times S^1$ which interchanges the coordinates (and changes orientation).

If we think of $S^1 \times I$ as obtained from $\mathbf{R} \times \mathbf{R}$ by dividing by $\mathbf{Z}$ in the first factor and restricting to I in the second, then J is given by $\begin{bmatrix} 1 & 0 \\ 0 & -1 \end{bmatrix}$. Similarly T is given by $\begin{bmatrix} 1 & 0 \\ 1 & 1 \end{bmatrix}$ and S is given by $\begin{bmatrix} 0 & 1 \\ 1 & 0 \end{bmatrix}$. Relations among these are $J^2 = \mathrm{id}$, $S^2 = \mathrm{id}$, $JTJ = T^{-1}$, and $STS = TJST$.

9.29. Homomorphisms

Applying Z to the mapping cylinders of these morphisms, we get
1. $\theta J\colon Z(S^1 \times I) \to Z(S^1 \times I)$ a hermitian involution on the ambialgebra (conjugate-linear, and reverses product and coproduct);
2. $T\colon Z(S^1 \times I) \to Z(S^1 \times I)$ a $Z(S^1 \times I)$-bimodule isomorphism; and
3. $\theta S\colon Z(T^2) \to Z(T^2)$ a conjugate linear map which takes the trace ambialgebra structure on $TZ(S^1 \times I)$ to the rank ambialgebra structure on $Z(S^1 \times S^1)$ (see 9.20), and is a conjugate-isometry with respect to the natural hermitian form.

In 3. the natural form is the one described in 7.2(3), and is related to the ambialgebra form by $\lambda = \epsilon m(\mathrm{id} \otimes J)$. Conjugate-isometry means $\lambda(\theta S(a), \theta S(b)) = \overline{\lambda(a,b)}$; the conjugation being necessary because the homomorphism is conjugate-linear. The other morphisms in the list also preserve the form, but this follows from the other properties.

9.30. Matrices

We go further and describe these in terms of matrices with respect to the Verlinde basis described in 9.25. Recall that Φ denotes the set of $\mathbf{C}$ summands of the ring $Z(T^2)$, and if $i \in \Phi$ then the corresponding basis element is $x_i = u_i e_i$. Here e_i is the idempotent corresponding to the summand and u_i is the twisting coefficient. This basis is orthonormal with respect to the ambialgebra form ϵm, and the ranks of irreducible $Z(S^1 \times I)$-modules are positive integer multiples of these elements (see 9.22).

There are
1. an involution bar : $\Phi \to \Phi$,
2. nonzero complex numbers t_i for $i \in \Phi$, and
3. a matrix $[s_{ij}]$ indexed by $i, j \in \Phi \times \Phi$.

These determine the homomorphisms above by

(9.31)
$$\theta J(\sum c_i x_i) = \sum \bar{c}_i x_{\bar{i}}$$
$$T(\sum c_i x_i) = \sum t_i c_i x_i$$
$$\theta S(\sum c_i x_i) = \sum_{i,j} \bar{c}_i s_{ij} x_j$$

The form of θJ comes from the description of anti-involutions in A5.4. T is a bimodule homomorphism so preserves the $\mathbf{C}$ factors of $S(T^2)$. Relations among these numbers are
1. $n_{\bar{i}} = n_i$ and $u_{\bar{i}} = \overline{u_i}$, from $(\theta J)^2 = \mathrm{id}$ and A5.4,
2. $\bar{t}_{\bar{i}} = t_i^{-1}$, from $JTJ = T^{-1}$,
3. $[s_{ij}]^{-1} = [\overline{s_{\bar{j}i}}] = [\overline{s_{ij}}]$, from the fact that S is an isometry and $S^2 = \mathrm{id}$, and
4. $t_i t_j \bar{s}_{ij} = \sum_k \bar{t}_k \bar{s}_{ik} s_{kj}$, from $STS = TJST$.

There is also a Diophantine condition: S interchanges the two ambialgebra structures. The second-coordinate structure is "diagonal" in the sense that the basis vectors are multiples of the primitive idempotents. The first-coordinate structure is the rank TQFT of 9.20, so the structure maps can be expressed in terms of ranks of modules over $Z(S^1 \times I)$. These are integral combinations of the Verlinde basis elements. This implies that certain combinations of the s_{ij} and the twisting units u_i are integral.

The coefficients in the matrix for multiplication in the first-coordinate ambialgebra structure are the "fusion rules", see 9.22. The fact that S takes this to the standard diagonal product is stated in the conformal field theory literature as "the S matrix diagonalizes the fusion rules", and is referred to as the "Verlinde theorem." Various other constraints are believed to hold, for example the twisting units are cyclotomic integers [**DeBG**].

The possible data satisfying these conditions have been classified in very special cases, see eg. [**CPR**].

Exercise 9.32.
1. Describe the first-coordinate ambialgebra structure on $Z(T^2)$ explicitly in terms of the data given above. In particular obtain a formula for the "fusion rules" (see 9.22). Beware that the homomorphism induced by θS is conjugate-linear.
2. Describe structure on $Z(T^2)$ when Z is not hermitian, induced by the orientation-preserving automorphisms. Use T and $\hat{S} = JS$, with the relation $T\hat{S}T\hat{S}T = \hat{S}$.

LECTURE 10
Categorical Constructions

In this section we give an introduction to the use of categories to construct modular TQFT on some low-dimensional domain categories. This is done by working out an example, the theory on $1+1$-complexes associated to representations of $SL(2)$ mod 5. The section ends with a brief description of the use of similar techniques to construct $2+1$-dimensional manifold theories.

A particular feature of the TQFT being constructed is that they are insensitive to 1-complexes. This means that disjointly adding points to a graph does not change the state module, and attaching a 1-complex to a 2-complex does not change the induced homomorphism. This is not an enormous restriction: many finite complex TQFT can be "twisted" by tensoring with an Euler TQFT to be insensitive to 1-complexes. To see this note the restriction of the TQFT to $0+1$-complexes is characterized by an ambialgebra structure on $Z(\text{pt})$ (see 7.16). If $\phi \to \text{pt}$ induces an isomorphism $Z(\phi) \to Z(\text{pt})$ then this ambialgebra is obtained by twisting the standard structure on the coefficient ring by some unit. We can undo this twisting, and obtain an insensitive theory, by tensoring with the Euler TQFT based on the inverse unit (see 3.3).

The abstract construction being illustrated associates a TQFT to a "finite semisimple tensor category." We have already seen some of these TQFT: if R is a field of characteristic prime to the order of a finite group G then the category of R-representations of G give such a category. The resulting TQFT is the finite total homotopy TQFT associated to the space B_G (sections 4, 8), tensored with an Euler TQFT to make it insensitive to 1-complexes. There are two points to the categorical view. First there are some categories which are not representation categories, so give new examples of TQFT. Second this gives a different view of calculations in the representation TQFT. Roughly the earlier approach used elements of the group as a basis for the group ring, where here we decompose the ring into indecomposable representations.

There is an extended exercise in which a TQFT associated to another category is constructed in parallel with the main example. There are two tracks in this exercise: the category of representations of the cyclic group $\mathbf{Z}/2$ for those with only pencil and paper, and representations of $\mathbf{Z}/3$ if a computer algebra program (eg. Mathematica or Maple) is available. Most of the larger example is still most easily

10.1 Tensor categories

A tensor category is an additive category with a product operation which is associative and commutative up to canonical isomorphism, and has a unit object. The examples to have in mind are categories of representations of a finite group, with product operation given by ordinary tensor product.

To be more precise about tensor categories we need some notation. An *arrangement* of a collection of objects $A_1, \ldots A_n$ in the category is an ordering and a grouping (using parentheses) so that each group contains two elements: either two smaller groups, one group and a single object, or two objects. For example

$$A_3((A_2, A_1), A_4)$$

is an arrangement. The point is that an arrangement describes a particular way to repeatedly apply a binary operation to a collection of objects to get a single object. For instance if the product of objects A, B is denoted by $A \diamond B$ then the arrangement above specifies the object

$$A_3 \diamond ((A_2 \diamond A_1) \diamond A_4).$$

If α is an arrangement of objects A_* then we denote by $\diamond_\alpha A_*$ the iterated product specified by α.

The information that comes with a tensor category is an assignment to any two arrangements α and β of objects A_* an isomorphism

$$\Phi_{\alpha,\beta} \colon \diamond_\alpha A_* \to \diamond_\beta A_*.$$

These isomorphisms are required to compose properly: if α, β, γ are three arrangements, then

(10.2) $$\Phi_{\beta,\gamma} \Phi_{\alpha,\beta} = \Phi_{\alpha,\gamma}$$

We could formalize this by saying Φ is a functor on the category whose objects are the arrangements, and which has exactly one morphism between any two objects. For this reason we refer to (10.2) as "functoriality" of Φ.

Note that we can get between any two arrangements by a sequence of "elementary" steps, each of which either interchanges two adjacent elements, or associates three adjacent elements. The functoriality (10.2) then implies we can obtain an arbitrary rearrangement isomorphism by composing isomorphisms associated to elementary steps. In many treatments of tensor categories only the elementary isomorphisms are assumed to be given (eg. MacLane [**McL 1**] where they are called "symmetric monoidal categories", and Deligne-Milne [**DM**]). The functoriality property then is replaced by rather elaborate identities involving only elementary isomorphisms (the pentagon and hexagon identities).

A tensor category also has a unit object, which we will denote by "1", with given isomorphisms $1 \diamond A \simeq A \simeq A \diamond 1$. These isomorphisms are compatible in an appropriate sense with the $\Phi_{\alpha,\beta}$. We will not need to be explicit about this, however.

Exercise 10.3. Let $R[G]$ denote the group ring of a finite group. If A and B are modules over this ring (ie. representations of G) then the tensor product $A \otimes_R B$ is given the module structure in which $g \in G$ acts by $g(a \otimes b) = (ga) \otimes (gb)$. Define rearrangement isomorphisms on this category by permutation: Suppose α and β are arrangements of A_i, $i = 1, \ldots n$. We suppose that in α they are ordered as they are enumerated, and let σ be the permutation of $1, \ldots n$ which takes them to the order used in β. Then a basis element $a_1 \otimes \cdots \otimes a_n$ in $\otimes_\alpha A_*$ is taken by $\Phi_{\alpha,\beta}$ to $a_{\sigma(1)} \otimes \cdots a_{\sigma(n)}$ in $\otimes_\beta A_*$.
1. Show that $\Phi_{\alpha,\beta}$ is a morphism of $R[G]$-modules.
2. Show that the functoriality property (10.2) is satisfied.

10.4 Semisimple tensor categories

We now restrict to finite semisimple categories. This essentially means there are finitely many irreducible objects, and any object is a sum of copies of the irreducibles. To be more precise let 1 denote the unit object and let R denote the ring of endomorphisms Morph $(1,1)$. The product in the ring is defined by composition of morphisms, and is also given by $f, g \mapsto \tau(f \diamond g)$, where $\tau \colon 1 \diamond 1 \to 1$ is the canonical isomorphism given as part of the structure of 1 as a unit object. (See [**DM**]). This second description shows that R is a commutative ring. This ring acts on all the morphism sets: if $g \colon 1 \to 1$ and $h \colon A \to B$ then gh is defined by the composition
$$A \xrightarrow{\simeq} 1 \diamond A \xrightarrow{g \diamond h} 1 \diamond B \xrightarrow{\simeq} B.$$

The conditions required on the category are:
1. there are objects (which we will denote by numbers $1, \ldots n$ for convenience, and with 1 the identity object) such that
2. (irreducibility) $R = \text{Morph}\,(1,1)$ is a field, and Morph (i, j) is 0 if $i \neq j$ and if $i = j$ is free of rank one over R, generated by the identity map;
3. every object is a sum of copies of these objects,
4. (duality) there is an involution $i \mapsto \bar{i}$ so that $i \diamond j \to 1$ has a summand isomorphic to 1 if and only if $j = \bar{i}$, and in that case it is unique, and the projection $\lambda \colon i \diamond \bar{i} \to 1$ is a nondegenerate form (in the sense of A2); and
5. (symmetry) every morphism $f \colon i \diamond i \to 1$ satisfies $f\Phi = f$, where Φ is the rearrangement isomorphism which interchanges the two copies of i.

The result being illustrated in this section is:

10.5 Theorem. *A semisimple tensor category satisfying these conditions induces a modular TQFT on finite $1+1$-complexes, with coefficient ring the endomorphism ring* Morph $(1,1)$.

The symmetry condition 5. reflects the fact that the bar involution on complexes is the identity (complexes are not "oriented"). The semisimplicity assumption may seem very strong, but we will see the structure of the category is reflected directly in the structure of the corner ambialgebra (9.7), and when the ground ring is a field this ambialgebra must be semisimple (A5). In particular I have been unable to construct an example using a non-semisimple category: the corner ambialgebras fail to be special, cf. the examples in A4.12.

We remark very briefly on the proof of Theorem 10.5. The rest of this section basically describes how to define the invariant. We think of a 2-complex as a 1-parameter family of graphs. As the parameter varies certain changes take place in the 1-complexes, and the definition follows from describing what the TQFT should do with each of these changes. But by far the largest part of the proof involves showing that the result is well-defined. To do this the 2-complex is represented as two different 1-parameter families. Then there is a 2-parameter family giving a deformation of one family to the other. Studying the singularities in the 2-parameter family gives a list of primitive changes in 1-parameter families. If the definition is invariant under these primitive changes then it is well-defined. Each primitive change gives an identity that must be verified. In a sense we are lucky that the construction using an ordinary tensor category does satisfy all these identities. It would have been entirely reasonable for one of these identities to have imposed nontrivial additional conditions, but this does not happen.

10.6 Examples of finite semisimple tensor categories

The object is to describe the rearrangement isomorphisms explicitly by giving matrices, and for this we need to use a choice of bases for spaces of summands. Suppose for each triple of irreducibles a, b, c we have a basis for the space $\text{Morph}(c, a \diamond b)$. The conditions on the category imply that each nonzero morphism maps onto a direct summand, so this parameterizes the c summands of $a \diamond b$. We assume these bases satisfy some normalizations:

1. The basis element for the 1-dimensional space $\text{Morph}(a, a \diamond 1)$ is the inverse of the standard isomorphism $a \diamond 1 \to a$, and
2. if $a \neq b$ then the commutation isomorphism $a \diamond b \to b \diamond a$ takes the basis for $\text{Morph}(c, a \diamond b)$ to the basis for $\text{Morph}(c, b \diamond a)$.

A base for $(a \diamond b) \diamond c$ is obtained by iteration: suppose $e \to a \diamond b$ is the ith basis element in the e summands of $a \diamond b$, and $d \to e \diamond c$ is the jth summand in the basis. Then the composition

$$d \to e \diamond c \to (a \diamond b) \diamond c$$

gives a basis element for the d summands. These elements are ordered lexicographically as strings (e, i, j) (recall we are denoting the irreducibles by numbers, and in particular have chosen an ordering). A basis for $a \diamond (b \diamond c)$ is constructed similarly.

Since all rearrangement isomorphisms are compositions of associations and commutations, it suffices to describe the matrices taking the base for $d \to a \diamond b$ to the base for $d \to b \diamond a$, and the base for $d \to (a \diamond b) \diamond c$ to the basis for $d \to a \diamond (b \diamond c)$. The normalization conditions somewhat reduce the data. The first condition implies that any association or commutation involving the unit object, eg. $(a \diamond 1) \diamond b \to a \diamond (1 \diamond b)$ is given by the identity matrix. We paraphrase this by saying that the unit associates and commutes freely. The second normalization implies that the commuters $a \diamond b \to b \diamond a$ are given by identity matrices unless $a = b$.

Entries in the association matrices are called "$6j$ symbols" in the theoretical physics literature.

10.7 Example: $SL(2)$ mod 5

We will give the data without deriving it from anything. The connection to $SL(2)$ will be described in section 10.51. This is a category over the ring $\mathbf{Z}/5$ with two irreducibles, denoted "1" and "2". The data not determined by the normalizations is:

1. $2 \diamond 2 \simeq 1 \oplus 2$;
2. the commutation of $2 \diamond 2$ is given by

$$\begin{array}{ccc} 2\diamond 2 & \xrightarrow{\Phi} & 2\diamond 2 \\ \downarrow\simeq & & \downarrow\simeq \\ 1\oplus 2 & \xrightarrow{[1]\oplus[4]} & 1\oplus 2 \end{array}$$

3. the associator for $2\diamond 2\diamond 2$ is given by

$$\begin{array}{ccc} (2\diamond 2)\diamond 2 & \longrightarrow & 2\diamond(2\diamond 2) \\ \downarrow\simeq & & \downarrow\simeq \\ (1\oplus 2)\diamond 2 & \longrightarrow & 2\diamond(1\oplus 2) \\ \downarrow\simeq & & \downarrow\simeq \\ 2\oplus 1\oplus 2 & \longrightarrow & 2\oplus 1\oplus 2 \end{array}$$

where the bottom map is [1] on the 1 summand and $\begin{bmatrix} 2 & 4 \\ 3 & 3 \end{bmatrix}$ on the 2 summands. Coefficients in the 2 summands are written as column vectors, and the matrix acts by multiplication on the left.

Products of objects can be conveniently summarized in a "multiplication table":

Exercise 10.8.
1. Verify that the standard projection $\lambda: 2\diamond 2 \to 1$ is a nondegenerate bilinear form, and find the dual coform Λ. (It is the multiple of the standard inclusion $1 \to 2\diamond 2$ which makes the composition the identity.
2. Show that the category 10.7 is not a category of representations. (If so each irreducible has an integer dimension. Derive a system of equations for these dimensions from the multiplication table for objects, and solve it.)
3. Verify that the data given in 10.7 does give a tensor category: look up the relations required in [**DM**] or [**McL2**] and verify them. (Be careful about the ordering in bases of decompositions of 4-fold products).
4. Show that this is the only tensor category with two irreducibles and $2\diamond 2 \simeq 1 \oplus 2$. (Consider the entries in the matrices described in 10.7 as unknowns and the relations used in 2. as equations in these unknowns. Use eg. Mathematica or Maple to find solutions. Set the modulus to be variable to

see that this only works mod 5. A 1-parameter set of solutions is expected, obtained from the given example by changing bases of the injections $1 \to 2 \diamond 2$ and $2 \to 2 \diamond 2$.)

5. Find the structure matrices for the category of complex representations of $\mathbf{Z}/2$ if you are working by hand, and $\mathbf{Z}/3$ if you have computer assistence. Use:

   a) $\mathbf{C}[\mathbf{Z}/2]$ has two irreducibles. The unit is $\mathbf{C}$ on which it acts trivially, and the other is $\mathbf{C}$ with $\mathbf{C}[\mathbf{Z}/2]$ acting by multiplication by -1. The only product not determined by unit properties is $2 \otimes 2 \simeq 1$: take the map $\mathbf{C} \to \mathbf{C} \otimes \mathbf{C}$ generated by $1 \mapsto 1 \otimes 1$ as the generator.

   b) $\mathbf{C}[\mathbf{Z}/3]$ has three irreducibles. The unit is $\mathbf{C}$ with the group acting trivially. Let ρ be a third root of unity, then the other two irreducibles are $\mathbf{C}$ with the generator of $\mathbf{Z}/3$ acting by multiplication by ρ and ρ^2. Denote these by "2" and "3" respectively. Show that the isomorphism $\mathbf{C} \to \mathbf{C} \otimes \mathbf{C}$ which takes the unit to the product with itself gives isomorphisms $1 \to 2 \otimes 3$, $2 \to 3 \otimes 3$, and $3 \to 2 \otimes 2$. Determine the associativity and commutivity matrices with respect to these bases.

10.9 State modules

A modular $1+1$-complex TQFT associates a state module to a graph with a distinguished finite subset (the corners). We begin with contractible graphs. Later the general case is reduced to this by cutting closed loops.

We draw contractible graphs as rooted planar trivalent trees, with corners the endpoints of the branches:

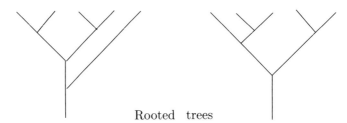

Figure 10.10 Rooted trees.

Trivalent means three edges at each (non-endpoint) vertex. Planar means a subset of the plane. Rooted means there is a distinguished endpoint which is drawn at the bottom, and all the others (the branches) are drawn at the top. We define state modules for=20this class of graphs. Any contractible graph is equivalent to one of these up to deformation, and the general definition is obtained by an inverse limit over these equivalent rooted trees. Here we are not concerned with the general definition, and think of these trees as convenient for calculation.

The basic idea is to exploit the connection between iterated products in a category and planar trivalent rooted trees.

Suppose $\mathcal{C}$ is a finite semisimple tensor category as in 10.5. Let A denote the sum of the irreducibles, in order: $A = 1 \oplus 2 \cdots \oplus n$. To a rooted planar trivalent tree we associate an iterated product of A with itself, one copy for each branch endpoint. The planar trivalent branching structure gives an arrangement of the copies of A and therefore give a well-defined way to do this by iterating the binary operation $\diamond$. For instance the left-hand tree in 10.10 specifies the product $((A \diamond A) \diamond (A \diamond A)) \diamond A$. If the tree specifies the arrangement α then the relative state module associated to the tree is the maximal trivial summand of $\diamond_\alpha A$. In the example $Z(Y, W) = \text{Morph}(1, ((A \diamond A) \diamond (A \diamond A)) \diamond A)$.

10.11 Bases for state modules

Basis for these modules are given by labelings of the trees: label each edge with an irreducible, and each vertex with an integer. The root edge is labeled "1". These must satisfy a condition: suppose at a vertex the ascending edges are labeled with a, b and the lower edge is labeled with c. Then the integer on the vertex must lie between 1 and $\dim(\text{Morph}(c, a \diamond b))$. This integer specifies one of the chosen basis elements of the space $\text{Morph}(c, a \diamond b)$. Note that a, b, c cannot occur as labels around a vertex if $\text{Morph}(c, a \diamond b) = 0$.

In the examples we are considering the spaces of morphisms all have dimension 0 or 1. This allows a considerable simplification: we agree that all vertex labels are 1, so can be omitted.

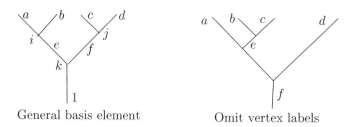

General basis element   Omit vertex labels

Figure 10.12

For example we find the bases for the right-hand tree in figure 10.12, with category $SL(2)$ mod 5. Using the multiplication table for objects (10.7) we see $c = d$. If $c = 1$ then $a = b$, and if $c = 2$ then one of a, b must be 2. Listing the possibilities as (a, b, c, d) in lexicographic order we get:

$$(1,1,1,1), (1,2,2,2), (2,1,2,2), (2,2,1,1), (2,2,2,2)$$

10.13 Glueing trees

The procedure above defines relative state modules $Z(\text{tree}, \text{endpoints})$. To get state modules of non-contractible graphs we algebraically "glue" together pairs of endpoints using the modularity property. For this we need the algebra structure on $Z(I, \partial I)$ and the module structures on $Z(\text{tree}, \text{endpoints})$ over this, one for each endpoint. This structure will be a special case of a description of how to glue two trees to get a larger tree.

Suppose we have two trees, one with a right-hand branch emerging directly from the root, and an unknown subtree to the left (labeled "X" in figure 10.14), and one with a specified branch emerging from an unknown subtree (labeled "Y" in 10.14). We want to glue the specified branches together to obtain a tree with base Y and X attached to it.

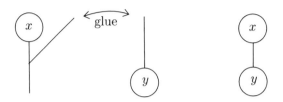

Figure 10.14

This is supposed to product a homomorphism

(10.15) $$Z(X \cup \text{branch}) \otimes Z(Y \cup \text{branch}) \to Z(X \cup Y).$$

We describe this in terms of basis elements. Choose labelings of the edges of the trees corresponding to basis elements of the state modules, with a and b on the edges to be glued. Join the two trees at the roots, and move the root of the special tree onto the branch labeled b. Recall that the state spaces correspond to morphisms from 1 to iterated products. This move corresponds to taking the product of two morphisms and then rearranging. The fact we can move the location where the root is joined comes from the fact that the unit object can be commuted and associated freely. Next change the association to move the a branch over onto the b branch (see 10.16). The result may not be a single basis element, but a linear combination of basis elements with various values of c on the stem of the a-b fork. Recall that the fork corresponds to a morphism $c \to a \diamond b \subset A \diamond A$. We then apply a bilinear form $\lambda \colon A \diamond A \to 1$.

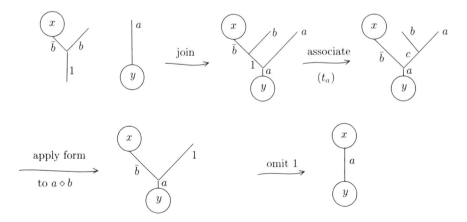

Figure 10.16

This form is defined on the $a \diamond b$ summand of $A \diamond A$ by: it is zero unless $b = \bar{a}$, and in that case it is the right inverse for the chosen basis element $1 \to a \diamond b$. Thus the composition with $c \to a \diamond b$ is zero unless $b = \bar{a}$ and $c = 1$, in which case it is the identity. The final step, of omitting the branch with label 1, corresponds to the canonical isomorphism $B \simeq B \diamond 1$ for appropriate B.

In this process a basis element goes to either 0 or a multiple of another basis element. The nonzero case is when $b = \bar{a}$, and the coefficient comes from the association step. Define t_a to be the coefficient in the association matrix shown in 10.17. In the example $SL(2)$ mod 5, from 10.7(3) we see $t_1 = 1$ and $t_2 = 2$.

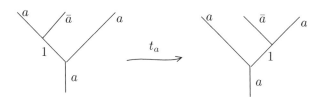

Figure 10.17

10.18 The corner ambialgebra

The state space $Z(I, \partial I)$ has basis given by labelings of the tree with two branches. Suppose the label on the left branch is a, then the duality hypothesis 10.43 implies that the label on the other branch is $\bar{a}$ (recall the root is labeled by 1). The correspondence $(a, \bar{a}) \to a$ gives an isomorphism from the ambialgebra to the free module generated by the set of irreducibles in $\mathcal{C}$. The product in the ring is obtained by glueing two such trees together as above. The coproduct is given by reversing this: use the coform to introduce a new fork, then associating it to two forks and splitting:

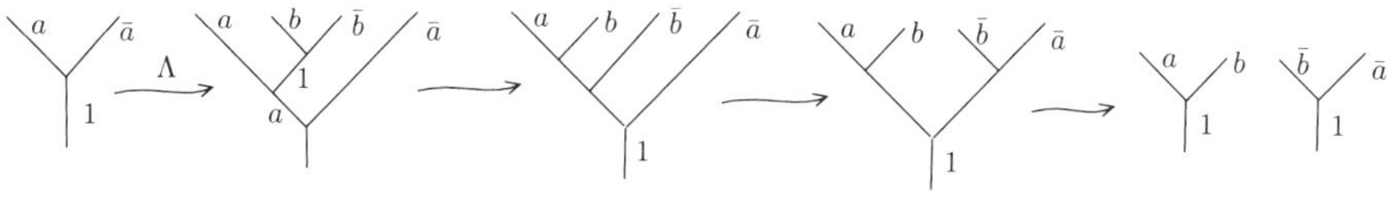

Figure 10.19

The result is $\Delta(a) = a \otimes a$. The reason is that one of the associations introduces a coefficient t_a, but the coform is t_a^{-1} times the standard $1 \to a \diamond \bar{a}$ (see 10.81.), and the two coefficients cancel. Summarizing and determining the unit and counit and involution we get:

$$m(a \otimes b) = \begin{cases} 0 & \text{if } b \neq a \\ t_a a & \text{if } b = a \end{cases}$$

(10.20)
$$\begin{aligned} e &= \sum t_a^{-1} a \\ \Delta(a) &= a \otimes a \\ \epsilon(a) &= 1 \\ \overline{a} &= \overline{a} \end{aligned}$$

Exercise 10.21.
1. Find the unit and counit for the ambialgebra structure on $Z(I, \partial I) = R[\text{irreducibles}]$
2. Find the trace ambialgebra (A4.7).
3. The trace ambialgebra determines the TQFT on $1+1$-manifolds. In particular Z_{S^2} is given by $\epsilon_{tr} e_{tr}$. Give a formula for this in general, and evaluate it for $SL(2)$ mod 5.
4. Find the corner and trace ambialgebras for modules over $\mathbf{C}[\mathbf{Z}/2]$ or $\mathbf{C}[\mathbf{Z}/3]$, and evaluate Z_{S^2}.

10.22 State modules of graphs

Suppose Y is a tree with endpoints $W \cup \{y_0, y_1\}$. Glueing together y_0 and y_1 gives a graph with a closed circuit. We assert that a basis for the state module of this graph is given by the subset of the basis of the unglued one in which the labels on y_0 and y_1 are dual.

To see this, recall that the modularity condition asserts that this state module is given by

$$Z(Y/y_0 = y_1, W) = Z(Y, W \cup \{y_0, y_1\})/\simeq$$

where $\simeq$ is the equivalence relation obtained by identifying two module structures over $Z(I, \partial I)$. The module structures are obtained as follows: glueing intervals onto these endpoints give two left module structures on $Z(Y, W \cup \{y_0, y_1\})$ over the corner ambialgebra. Denote these by m_0 and m_1. Convert one of these (say m_1) to a right module structure using the bar involution and denote the result by $\overline{m}_1$. Then the equivalence relation is generated by $m_0(a, X) = \overline{m}_1(X, a)$ for X some basis label on the tree and a a basis element in the algebra.

Now suppose a is the label on y_0. Every basis element in the ambialgebra except a kills this, and multiplication by a changes it by a scalar. Applying the bar involution on the ambialgebra takes a to $\overline{a}$. This acts on the y_1 endpoint to kill all basis elements except those labeled with $\overline{a}$, and these are changed by scalars. Therefore all labelings with non-dual endpoints are equivalent to 0, and there are no other relations.

We illustrate this by finding bases for some graphs we will be working with in the computations. The first represents a graph with two closed loops, and the second a graph with three loops (a is glued to $\overline{a}$, etc.).

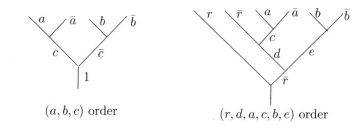

(a, b, c) order $\qquad$ (r, d, a, c, b, e) order

Figure 10.23

Using the multiplication table for objects in $SL(2)$ mod 5 the basis for the left graph (listed as (a, b, c) is:

(10.24) $\qquad (1,1,1), (1,2,1), (2,1,1), (2,2,1), (2,2,2)$

For the graph on the right we list the labels in the order (r, d, c, a, e, b), for reasons which will become clear later. Further, in the table we leave blank (r, d) entries or (c, a) entries which are the same as the line above. This displays a block structure we will use later.

r	d	c	a	e	b
1	1	1	1	1	1
				1	2
		1	2	1	1
				1	2
1	2	2	2	2	2
2	1	2	2	2	2
2	2	1	1	1	1
				1	2
				2	2
		1	2	1	1
				1	2
				2	2
		2	2	1	1
				1	2
				2	2

(10.25)

Exercise 10.26.
1. Find bases for the state modules of graphs shown in 10.23 for the category $\mathbf{C}[\mathbf{Z}/2]$ or $\mathbf{C}[\mathbf{Z}/3]$.

2. Find the "fusion" TQFT corresponding to these TQFT (See 9.22). The fusion TQFT is a 0+1-complex theory with **Z** coefficients, defined so that if Y is a graph regarded as a 0+1-complex bordism from the empty set to itself then the homomorphism $Z_Y^\Phi \colon \mathbf{Z} \to \mathbf{Z}$ is multiplication by the dimension of the state module $Z(Y)$. More generally $Z^\Phi(\text{pt})$ is free abelian generated by the irreducibles of $\mathcal{C}$. The bordism $S^1 \colon \text{pt} \to \text{pt}$ induces an endomorphism of $Z^\Phi(\text{pt})$, and the a, b entry in the matrix representing this bordism is the number of admissible labelings of this graph split open into trees:

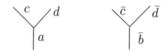

Figure 10.27

3. Find the dimension of the state module $Z(\vee^{10} S^1)$ for $SL(2)$ mod 5, and for $\mathbf{C}[\mathbf{Z}/2]$ or $\mathbf{C}[\mathbf{Z}/3]$.
4. Find the eigenvalues and eigenvectors of the matrix described in 2., and use this to give a formula for the dimension of $Z(\vee^n S^1)$ as a function of n.

10.28 Slices through 2-complexes

We now have sufficient information about the state modules to begin consideration of the homomorphisms induced by 2-complexes. For our purposes a 2-complex is obtained by attaching 2-disks to a 1-point union of circles. We may subsitute for the 1-point union the product of the 1-point union with an interval; this only changes the 2-complex by a deformation which does not affect the TQFT. For example consider the complex with two 1-cells, denoted a and b, and one 2-cell attached by the word $aba^{-1}b$. We use the notation for presentations of groups,

$$\langle a, b \mid aba^{-1}b \rangle$$

and refer to the 2-cell as a "relation". When drawing and analysing this we attach the relations to (1-cells)$\times I$ rather than just the 1-cells. The attaching map can then be spread out and drawn more clearly:

Now consider vertical slices through this picture. Going from left to right we see: first the two generator circles appear from nowhere. When we reach the beginning of the attaching map we see a tiny new circle (the slice through the relation cell) grow out of the central axis. Moving on, we see one leg of this new arc move around the a circle, and back to the axis. Then it moves around b, backwards around a, around b again. As we pass beyond the end of the attaching map the relation circle shrinks to a point. Finally the generator circles suddenly vanish.

The calculation of the induced homomorphism decomposes correspondingly into pieces:
1. begin or end the complex by introducing or deleting the generator circles;

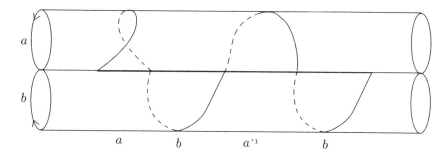

Figure 10.29

2. begin or end a relation by introducing or squeezing out the relation arc;
3. passing one end of the relation arc around one of the generator circles.

These induce homomorphisms among the state modules of the empty set, and wedges of two and three circles. If these homomorphisms are determined, then we can calculate invariants of the example, and more generally of any complex with only two 1-cells.

10.30 Beginning and ending a presentation

For purposes of computing examples we will concentrate on presentations with two generators (2-complexes with one 0-cell and two 1-cells). Nothing qualitatively different happens in presentations with more generators, and the state modules are much larger.

To begin we go from the empty set to a point (which has the same state module) to $\vee^2 S^1$ represented as a glued tree on the left in figure 10.23. $Z(\text{pt})$ is 1-dimensional with basis the tree with a single branch and label (1). This maps to the basis element represented by the labels $(1,1,1)$ in $Z(\vee^2 S^1)$. This basis element is first in the standard ordering (10.24) and the space has dimension 5. Thus the homomorphism is represented by the 1×5 matrix $(1,0,0,0,0)$.

The endpresentation homomorphism is dual to this: it takes a basis element (a,b,c) in $Z(\vee^2 S^1)$ to 0 unless $a = b = c = 1$, and this goes to (1). This homomorphism is represented by the 5×1 matrix with 1 in the first row and the rest zeros.

Suppose X is a 2-complex with two 1-cells. If we regard this as a bordism from the 1-skeleton $\vee^2 S^1$ to itself the induced homomorphism is represented by a 5×5 matrix. Regarding X as a bordism $\phi \to \phi$ gives a 1×1 matrix, obtained from the larger one by conjugating by the homomorphisms determined above. The description above shows the smaller matrix is the $(1,1)$ entry in the larger.

10.31 Beginning and ending a relation

These are homomorphisms

$$Z(\vee^2 S^1) \to Z(\vee^3 S^1) \to Z(\vee^2 S^1).$$

We represent these graphs as glued trees as in figure 10.23, and order the bases as in 10.24 and 10.25.

The central part of beginning a relation is the bordism $D^2 : \phi \to S^1$, encountered as the slices first touch the relation disk. This is part of the restriction of the TQFT to $1+1$-manifolds. According to 9.13 this restriction is characterized by the trace amibialgebra of the corner ambialgebra. The homomorphism induced by D^2 is the unit in this trace ambialgebra. The corner ambialgebra is described in 10.20, and the definition of traces is given in A4.7. Combining these we see that the trace unit is $(1) + 4(2) \in Z(S^1)$.

The other ingredients of beginning a relation are associations required to get the trees in the standard forms shown in 10.23. These are shown in figure 10.32:

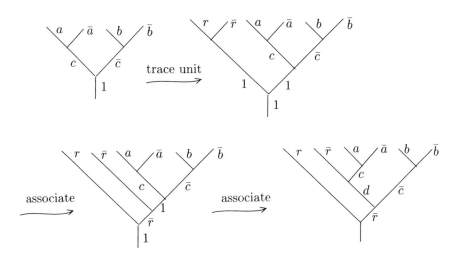

Figure 10.32

The output is a 5×15 matrix giving the homomorphism $Z(\vee^2 S^1) \to Z(\vee^3 S^1)$ in the bases 10.24, 10.25. For instance the last basis element in $Z(\vee^2 S^1)$ is $(2,2,2)$. The trace unit takes this to $(1,2,2,2) + 4(2,2,2,2)$ in the basis (r,a,b,c) for the second tree in 10.32. The first association is the identity since one of the objects being associated is the unit. The second association takes $(1,2,2,2)$ to $(1,2,2,2,2,2)$, and $(2,2,2,2)$ to $2(2,1,2,2,2,2) + 3(2,2,2,2,2,2)$, in the basis (r,d,c,a,e,b) for labels on the final tree. The associations come from 10.7, and in particular the image of $(2,2,2,2)$ is determined by the first column in the associator for the 2-summands given in 10.7(3). Putting these together shows $(2,2,2)$ goes to $(1,2,2,2,2,2) + 3(2,1,2,2,2,2) + 2(2,2,2,2,2,2)$. Locating these basis elements in the ordering given in 10.25 gives the last column of the beginrelation matrix. The other columns are found similarly, and the result is

$$
(10.33) \qquad \text{Beginrel} = \begin{bmatrix} 1 & 0 & 0 & 0 & 0 \\ 0 & 1 & 0 & 0 & 0 \\ 0 & 0 & 1 & 0 & 0 \\ 0 & 0 & 0 & 1 & 0 \\ 0 & 0 & 0 & 0 & 1 \\ 0 & 0 & 0 & 0 & 3 \\ 4 & 0 & 0 & 0 & 0 \\ 0 & 4 & 0 & 0 & 0 \\ 0 & 0 & 0 & 0 & 0 \\ 0 & 0 & 4 & 0 & 0 \\ 0 & 0 & 0 & 4 & 0 \\ 0 & 0 & 0 & 0 & 0 \\ 0 & 0 & 0 & 0 & 0 \\ 0 & 0 & 0 & 0 & 0 \\ 0 & 0 & 0 & 0 & 2 \end{bmatrix}
$$

The endrelation matrix is obtained by reversing these steps: associate from the fourth back to the second tree in 10.32. We then cut off the r branch and apply the counit in the trace ambialgebra. Note that that this reversed association may produce labels different from 1 on the branches adjacent to the root in the second tree. These are taken to 0 in the cutting process. So for example any basis element (r, d, c, a, e, b) with $c \neq \bar{e}$ must go to zero. The counit in the trace ambialgebra takes every basis element to $1 \in \mathbf{Z}/5$. The final result is:

$$
(10.34) \qquad \text{Endrel} = \begin{bmatrix} 1 & 0 & 0 & 0 & 0 & 0 & 1 & 0 & 0 & 0 & 0 & 0 & 0 & 0 & 0 \\ 0 & 1 & 0 & 0 & 0 & 0 & 0 & 1 & 0 & 0 & 0 & 0 & 0 & 0 & 0 \\ 0 & 0 & 1 & 0 & 0 & 0 & 0 & 0 & 0 & 1 & 0 & 0 & 0 & 0 & 0 \\ 0 & 0 & 0 & 1 & 0 & 0 & 0 & 0 & 0 & 0 & 1 & 0 & 0 & 0 & 0 \\ 0 & 0 & 0 & 0 & 1 & 2 & 0 & 0 & 0 & 0 & 0 & 0 & 0 & 0 & 4 \end{bmatrix}
$$

Exercise 10.35.
1. Find the begin- and endrelation matrices for the category $\mathbf{C}[\mathbf{Z}/2]$ or $\mathbf{C}[\mathbf{Z}/3]$.
2. The matrix product (Endrel)(Beginrel) represents the homomorphism induced by the bordism

$$(\vee^2 S^1) \vee S^2 \colon \vee^2 S^1 \to \vee^2 S^1$$

Recall from the introduction that the TQFT is insensitive to 1-complexes, which means the 1-point union induces the same homomorphism as the disjoint union. By multiplicativity this should be Z_{S^2} times the identity of $Z(\vee^2 S^1)$. Verify this (see 10.21).

10.36 The circulator

The most important ingredient in the calculation is the "circulator", the homomorphism $Z(\vee^3 S^1) \to Z(\vee^3 S^1)$ induced by moving one end of the relation loop around the "a" loop. We begin with some strategy for organizing the calculation. In the standard labeling shown in 10.32, the circulator involves only the part above the edge labelled "d". The move requires r and d as data, but they do not change. c and a do change, and e, b are neither needed as data nor changed by the move. With the basis labels written as (r, d, c, a, e, b) and ordered lexicographically as in 10.34, the matrix breaks into blocks. Each choice of r, d determines a block, and the whole matrix is obtained by putting these on the diagonal. Each block decomposes further: it is a matrix describing how the c, a labels transform, then tensored (on the right) by the identity matrix to reflect the independence of the e, b labels.

Reference to the list of basis elements in 10.34 gives the sizes of the blocks: there are four blocks (possible values of r, d). In the first block the c, a variables are 2-dimensional, as are the e, b variables. Thus in this block we calculate a 2×2 matrix and tensor with the 2×2 identity. The second and third blocks are 1-dimensional. The last block, with $r, d = 2, 2$, has three c, a dimensions, and three e, b dimensions. So we must find a 3×3 matrix, and tensor it with the 3×3 identity.

We now begin consideration of the move itself. This proceeds in three steps: an "improvement" to move the relation arc into a slightly more convenient place, then the "core", and finally the inverse of the improvement to return to the initial position. In terms of calculation, both the improvement and core give matrices. The final matrix is obtained by conjugating the core by the improvement. This suggests a final bit of strategy: calculate the core first, since if it turns out to be the identity the conjugated matrix is also the identity, and it is not necessary to compute the improvement matrix.

The improvement and core moves are shown in figure 10.37 and 10.38 respectively. Only the part which changes, the branches above the "d" edge in 10.32, are shown. We explain the core move. To move the r branch to the other side of the a loop we cut the loop on the edge labeled x, then glue the a ends. In order to do the glueing the a, r, x tree must be moved into the standard form used in 10.13. The result of the glueing needs an association step to return to the initial configuration. Recall from 10.13 that glueing branches with labels $a, \bar{a}$ picks up a coefficient t_a, and conversely cutting picks up a coefficient t_a^{-1}. For the current category $t_1 = 1$ and $t_2 = 2$ (see 10.16).

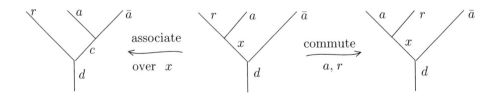

Figure 10.37

LECTURE 10. CATEGORICAL CONSTRUCTIONS

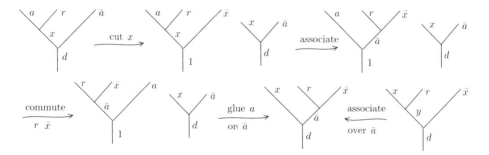

Figure 10.38

We go through the core calculation for the largest block, where $r = d = 2$. The variables are the labels (a, x), and the possible values are $(1, 2)$, $(2, 1)$, and $(2, 2)$. Cutting gives the diagonal matrix with t_x^{-1} on the diagonal. This diagonal is $(3, 1, 3)$. The association and commuting gives the identity matrix, and the glueing of the a branches is gives the diagonal matrix with diagonal $(1, 2, 2)$. We now reorder the basis from lexicographic as (a, x) to lexicographic as (x, a). This is represented by the matrix which interchanges the first two basis elements. The final move is the association which moves the r branch back to the left side. This takes the basis element $(x, a) = (1, 2)$ to $(x, y) = (1, 2)$ by the identity. The other two basis elements are mapped by the inverse of the 2-summand association in 10.7(3). This matrix is its own inverse mod 5, so the association matrix is $[1] \oplus \begin{bmatrix} 2 & 4 \\ 3 & 3 \end{bmatrix}$. Putting these pieces together gives

(10.39)
$$\begin{bmatrix} 0 & 2 & 0 \\ 1 & 0 & 4 \\ 4 & 0 & 3 \end{bmatrix}$$

Going through the same steps for the other blocks shows that they are all given by identity matrices.

Next we derive the conjugating matrix coming from the "improvement". Note this is only necessary for the last block since the core matrices in the other blocks are the identity. In this block $r = d = 2$. The first move is the inverse of an association. Going from (c, a) coordinates to (a, x) coordinates this is given by $[1] \oplus \begin{bmatrix} 2 & 4 \\ 3 & 3 \end{bmatrix}$. The second move commutes the r and a branches. This is represented by the diagonal matrix with diagonal $(1, 1, 4)$. Putting these together gives conjugating matrix

(10.40)
$$\begin{bmatrix} 1 & 0 & 0 \\ 0 & 2 & 4 \\ 0 & 2 & 2 \end{bmatrix}$$

The circulator matrix for this block is 10.39 conjugated by 10.40, which is

$$(10.41) \qquad \operatorname{Circ}_4 = \begin{bmatrix} 0 & 4 & 3 \\ 1 & 2 & 2 \\ 1 & 1 & 1 \end{bmatrix}$$

The full circulator matrix is then $I_6 \oplus (\operatorname{Circ}_4 \otimes I_3)$, where I_n denotes the $n \times n$ identity matrix. Note that tensoring on the right with the identity amounts to replacing each matrix entry $a_{i,j}$ by the 3×3 matrix $a_{i,j} I_3$.

Exercise 10.42.
1. Find the circulator matrix for the category $\mathbf{C}[\mathbf{Z}/2]$ or $\mathbf{C}[\mathbf{Z}/3]$.
2. Evaluate Z_{X_n} where $X_n = \langle a \mid a^n \rangle$. Note that by insensitivity to 1-complexes this is the same as the homomorphism induced by $\langle a, b \mid a^n \rangle$, which is the $1,1$ entry in

$$(\text{Endrel})(\operatorname{Circ})^n(\text{Beginrel})$$

3. One of the most remarkable experimental results is that the circulator matrices tend to have special orders (often the prime p). Verify that the circulator for $SL(2)$ mod 5 has order 5 (ie. the fifth power is the identity. Remember we are working mod 5, and note that this calculation reduces to checking the order of 10.39 or 10.41).
4. The fact that the circulator has order 5 gives $Z(\vee^3 S^1)$ the structure of a module over the group ring $\mathbf{Z}/5[\mathbf{Z}/5]$. The structure of such modules is very simple: they are sums of irreducibles, and there are five irreducibles, of the form R/I^n, where R is the group ring, I is the kernel of the augmentation ideal, and $0 \leq n \leq 4$. Determine the decomposition of $Z(\vee^3 S^1)$ into irreducibles over this ring. (Do this by reducing the computation to some manipulation of the circulator block matrix 10.41).
5. Find the order of the circulator for the category $\mathbf{C}[\mathbf{Z}/2]$ or $\mathbf{C}[\mathbf{Z}/3]$, and if appropriate determine its decomposition into irreducibles as in 4.

10.43 The interchange

The final move we need is to slide an end of the relation loop around the b generator. We achieve this by first interchanging the a and b loops, applying the circulator move determined in 10.36, and then interchanging back. The new move required is therefore the "interchange" which interchanges the two generator loops. This move is obtained by associating the e edge over to the c edge, commuting c and e, and then associating the c edge back onto the r edge, as shown in figure 10.44.

We divide this into blocks in the same way as with the circulator. The labels r, c, e do not change, but their values determine the association and commutation matrices. The label d does change, and a, b are independent of the move. Thus if we write the labels as (r, c, e, d, a, b) and order them lexicographically we get a block diagonal matrix. The number of blocks is the number of possible values of (r, c, e), and each block is a matrix describing how d changes, tensored (on the right) by the identity matrix in the (a, b) variables.

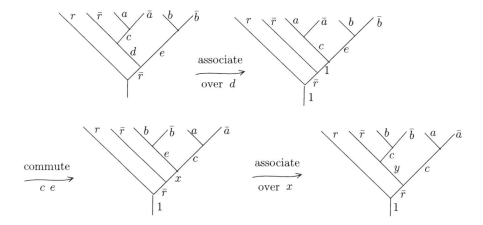

Figure 10.44

We now determine the block matrices. Both the associator and commuter are identity matrices unless $r = c = e = 2$, so all matrices are the identity except in the last block. In this block the commuter is the diagonal matrix with $(1, 4)$ on the diagonal, corresponding to the commuters for $x = 1$ and $x = 2$. This is conjugated by the associating matrix in 10.7(3). The result is $\begin{bmatrix} 2 & 1 \\ 2 & 3 \end{bmatrix}$. Finally in this block $a = b = 2$ so this matrix is tensored with the 1×1 identity. The entire matrix is therefore given by

$$(10.45) \qquad I_{13} \oplus \begin{bmatrix} 2 & 1 \\ 2 & 3 \end{bmatrix}$$

This however is not yet the interchange matrix. It must be multiplied by permutation matrices to account for the fact that the ordering in the basis has been changed. For the circulator the order was lexicographic when written as (r, d, c, a, e, b), whereas the interchange begins with the lexicographic order from (r, c, e, d, a, b). The permutation taking circulator order to interchange order is

$$(10.46) \qquad (1, 2, 3, 4, 5, 14, 6, 7, 10, 8, 9, 11, 12, 13, 15).$$

This is conveniently found mechanically as follows: generate the list of labels written in interchange form, and sort lexicographically. Append the number i to the ith label. Next go through this list and rewrite each label sequence in circulator form, leaving the new index last. Sort this new list. The last term in the ith sequence is the position in the interchange basis of the ith element in the circulator basis, so gives the desired permutation.

Similarly a permutation is required to go from the output of the interchange back to circulator order. The output is lexicographic when written as (r, e, c, d, b, a) (recall the two loops have been interchanged), and we want to go back to the circulator ordering.

The final result, when (10.45) is multiplied on the right by the permutation (10.46) and on the left by the permutation immediately above, is

(10.47) $$\text{Inter} = \begin{bmatrix} 1 & 0 & 0 & 0 & 0 & 0 & 0 & 0 & 0 & 0 & 0 & 0 & 0 & 0 & 0 \\ 0 & 0 & 1 & 0 & 0 & 0 & 0 & 0 & 0 & 0 & 0 & 0 & 0 & 0 & 0 \\ 0 & 1 & 0 & 0 & 0 & 0 & 0 & 0 & 0 & 0 & 0 & 0 & 0 & 0 & 0 \\ 0 & 0 & 0 & 1 & 0 & 0 & 0 & 0 & 0 & 0 & 0 & 0 & 0 & 0 & 0 \\ 0 & 0 & 0 & 0 & 1 & 0 & 0 & 0 & 0 & 0 & 0 & 0 & 0 & 0 & 0 \\ 0 & 0 & 0 & 0 & 0 & 2 & 0 & 0 & 0 & 0 & 0 & 0 & 0 & 0 & 1 \\ 0 & 0 & 0 & 0 & 0 & 0 & 1 & 0 & 0 & 0 & 0 & 0 & 0 & 0 & 0 \\ 0 & 0 & 0 & 0 & 0 & 0 & 0 & 0 & 0 & 1 & 0 & 0 & 0 & 0 & 0 \\ 0 & 0 & 0 & 0 & 0 & 0 & 0 & 0 & 0 & 0 & 0 & 1 & 0 & 0 & 0 \\ 0 & 0 & 0 & 0 & 0 & 0 & 0 & 1 & 0 & 0 & 0 & 0 & 0 & 0 & 0 \\ 0 & 0 & 0 & 0 & 0 & 0 & 0 & 0 & 0 & 0 & 1 & 0 & 0 & 0 & 0 \\ 0 & 0 & 0 & 0 & 0 & 0 & 0 & 0 & 0 & 0 & 0 & 0 & 1 & 0 & 0 \\ 0 & 0 & 0 & 0 & 0 & 0 & 0 & 1 & 0 & 0 & 0 & 0 & 0 & 0 & 0 \\ 0 & 0 & 0 & 0 & 0 & 0 & 0 & 0 & 0 & 1 & 0 & 0 & 0 & 0 & 0 \\ 0 & 0 & 0 & 0 & 0 & 2 & 0 & 0 & 0 & 0 & 0 & 0 & 0 & 0 & 3 \end{bmatrix}$$

Exercise 10.48.
1. Verify that the square of this matrix is the identity (interchanging loops twice should have no effect).
2. Find the interchange matrix for the TQFT associated with $\mathbf{C}[\mathbf{Z}/2]$ or $\mathbf{C}[\mathbf{Z}/2]$.

10.49 Doing the calculation

To find the invariant associated to a presentation with two generators, for instance

$$\langle a, b \mid aba^3b, a^{-1}ba \rangle$$

we scan through the the relations left to right, and build up a product of matrices right to left. Begin a relation with 10.33. If at some point the next term in the relation is a^n, then multiply (on the left) by the nth power of the circulator matrix from 10.41. Remember that this matrix has order 5 (see 10.42(3)), so n may be replaced by $0 \leq j < 5$ congruent to n mod 5. In particular one may avoid having to invert the matrix when n is negative. If the next term in the relation is b^n then multiply by the interchange 10.47, apply the nth power (mod 5) of the circulator, and apply the interchange again. When the end of a relation is reached, apply the endrelation matrix 10.34. If there is another relation, repeat the whole process. The end result of all this is a 5×5 matrix. Finally, according to 10.30, the homomorphism $Z_X : Z(\phi) \to Z(\phi)$ is given by multiplication by the $(1,1)$ entry in this matrix.

Exercise 10.50.
1. Implement this on a computer, and calculate the invariants for the presentations
$$\langle a, b \mid abab^{-1}a^{-1}b^{-1}, a^n b^m \rangle$$

as a function of m and n. Compare the results with the homology of the 2-complexes. If there are k relations then the homology is obtained as follows: define a $2 \times k$ integer matrix with $(1,j)$ entry the sum of the exponents on a terms in the jth relation, and $(2,j)$ entry the sum of the exponents on b terms in the jth relation. Then $H_2(X)$ is the kernel of this matrix in $\mathbf{Z}^k$, and $H_1(X)$ is the cokernel.
2. Repeat for the category $\mathbf{C}[\mathbf{Z}/2]$ or $\mathbf{C}[\mathbf{Z}/2]$.
3. Find some other presentations with two generators, and calculate the invariants. For example try $\mathbf{Z}/m \oplus \mathbf{Z}/n$ with presentation

$$\langle a, b \mid a^n, b^m, aba^{-1}b^{-1}\rangle.$$

10.51 Extensions

In this section we describe how the example fits into the general situation, first of TQFT on $1+1$-complexes, then on $2+1$-manifolds.

According to Theorem 10.5 any finite semisimple tensor category with a "duality" involution and summetric nondegenerate forms defines a modular TQFT on $1+1$-complexes. Most of these categories are representations of finite groups: $R[G]$-modules where G is finite and R is a field of characteristic prime to the order of G. Indeed a theorem of Deligne asserts that all such categories over fields of characteristic 0 are representation categories.

When the construction in this section is done using a representation category then the resulting TQFT is the same as the "finite total homotopy" TQFT described in Section 4 and 8.13 using the classifying space B_G, and tensored with an Euler TQFT to make it insensitive to 1-complexes. There are several conclusions. First it extends to a much larger domain category (all finite complexes). Second, the invariants are essentially invariants of the fundamental group, recording information about homomorphisms $\pi_1(X) \to G$. The view seen from this chapter is quite different from the earlier discription: roughly it is a change of coordinates from the basis $G \subset R[G]$ to a basis corresponding to a decomposition of $R[G]$ into irreducibles. The outcomes can be interesting, describing homomorphisms in terms of irreducibles. But still the information is classical in nature.

There is a source of such categories which are not representation categories. These are constructed from representations of algebraic groups by Gelfand and Kazhdan [**GK**]. If G is an algebraic group and p is a sufficiently large prime, then the category is a subquotient of the category of mod p representations of G. The example presented in this section is the simplest of these: the category for the rank one group $SL(2)$ and prime 5. At the time of writing calculations have been done for $SL(2)$ up to prime 19, and for the rank 2 groups $SL(3)$, $SP(4)$, and $G(2)$ for primes up to 11 or 13 [**Q2**]. These TQFT are qualitatively similar to the example, but the calculations are much larger (some of the state spaces $Z(\vee^3 S^1)$ have dimension larger than 10^6). In particular all of them have $Z_{S^2} = 0$, which as explained in Section 2 is necessary for a TQFT which might detect counterexamples to the Andrews-Curtis conjecture. So far, however, no counterexamples have been found.

TQFT on $(2+1)$-manifolds have received much more attention. The algebra used here is similar to the that used by Reshetikhin and Turaev [**RT**], though they approach the invariants through link descriptions of 3- and 4-manifolds. A reference which uses topology closer to that used here, is Walker [**Wk**]. At this time this paper is still an incomplete draft, but still seems to be the most useful reference for the construction. There is a large literature on these theories: [**A2**], [**CPR**], [**C**], [**D1**], [**D2**], [**KM**], [**Ko**], [**RT**], [**Y1**], [**Y2**], [**Wk**] is an incomplete list, quite out of date.

The first step in the $1+1$-complex theory is the definition of the relative state module of a finite set of points in a tree, using an iterated product of objects in a category. Since the tree can be changed up to deformation we can think of it as the cone on the finite set. This has automorphism group the symmetric group, and the heart of the construction is the construction of an action of this symmetry group on the iterated product using associativity and commutativity isomorphisms. The $2+1$-manifold theory begins similarly, with a finite set of points in a 2-sphere (with fixed framings of the associated tangent spaces). Again the state module is an iterated product of objects, one for each point. Again the key is to construct an action of the symmetry group on the state module. Now the symmetry group is a variant on the braid group. Again we need associating and commuting isomorphisms, but not quite as natural as in a tensor category: the iteration of commuting isomorphisms $A \diamond B \to B \diamond A \to A \diamond B$ is no longer required to be the identity. This reflects the fact that in a surface two right-hand twists interchanging two points gives a full twist which is not isotopic to the identity (holding the points fixed). A category with this structure is called a "quasitensor" or "braided monoidal" category.

The construction proceeds similarly to the $1+1$-complex theory. The glueing construction which gives closed loops for graphs gives 1-handles (torus summands) for surfaces. There are several surface moves which do not have graph analogs, and these require a bit more structure in the category.

The input for this construction is a finite semisimple quasitensor category (with a little more data). The main source of such categories is representations of Hopf algebras. The coproduct $\Delta \colon H \to H \otimes H$ of a Hopf algebra induces a product in the category of representations. If the coproduct is coassociative then the category product is associative. If the coproduct is "triangular" in the terminology of Drinfeld [**D**] then the category product is commutative up to natural isomorphism and forms a tensor category. A slightly weaker condition, "quasitriangular", gives a quasitensor category.

One class of quasitriangular Hopf algebras are the Drinfeld doubles of other Hopf algebras. If R is a field of characteristic prime to the order of a finite group G, then $R[G]$ has a Hopf algebra structure which induces the usual tensor product on representations. The Drinfeld double of this algebra has a finite semisimple quasitensor category of representations. Carrying through the categorical construction of a TQFT using this category gives the finite total homotopy TQFT based on the space B_G, restricted to $2+1$-manifolds, and twisted by an Euler TQFT [**Y2**].

The other important class of quasitriangular Hopf algebras are deformations of Lie algebras ("quantum groups"). The category of representations is a quasitensor category, but is not finite semisimple. In very special cases (when the deformation parameter is a root of unity) these categories have finite semisimple subquotients. These are very closely related to the Gelfand-Kazhdan categories used for $1 + 1$-complexes. Indeed a connection is conjectured in [**GK**], which has been verified in many cases by xxxinkelberg (1993 Harvard thesis) using recent work of Kazhdan and Lustig. On e consequence is that the category associated to the quantum deformation of G at a pth root of unity, and the the one associated to mod p representations of G have the same K_0 rings (irreducibles correspond, and products decompose as sums of irreducibles in the same ways). Presumably the TQFT constructed from these categories are closely connected, but this is not yet understood.

So far there are no useful categorical constructions of TQFT on $3+1$-manifolds, though as explained in Section 2 this is one of the most promising eventual applications of these techniques.

APPENDIX A
Algebra

Here we discuss algebraic structures which figure in the structure of TQFT: tensor products, nondegenerate pairings, traces, and ambialgebras. Tensor products appear in the most basic characteristic property of the theories; products over commutative rings in the basic case, and over noncommutative rings in modular theories. Consequently tensor products appear everywhere. This material is reviewed in §A1.

Nondegenerate pairings are almost as basic as tensor products since state modules have natural pairings. Pairings in turn lead to traces, which are used to evaluate some induced homomorphisms. Pairings and traces are described in §A2.

Ambialgebras are algebras with a certain type of coproduct. The prefix "ambi," as in ambidextrous, means "both," and refers to having both a product and coproduct. Bialgebras and Hopf algebras also have both products and coproducts, but which satisfy different conditions. Fortunately ambialgebras have a simpler structure than Hopf algebras. The definition is given in §A3, and examples (endomorphism-, group-, and trace ambialgebras) are described in §A4. Then a characterization and classification of semisimple ambialgebras over the complex numbers is given in §A5.

For more information and detail see [**R**] for a discussion of tensor products, and [**B**] for traces in the generality used here. For other applications of ambialgebras to TQFT see [**Q**].

A1. Tensor products

If A and B are abelian groups then the (barebones) tensor product $A \otimes B$ is defined by: take the free abelian group on the symbols $a \otimes b$, and impose the relations

(A1.1)
$$(a + a') \otimes b = a \otimes b + a' \otimes b$$
$$a \otimes (b + b') = a \otimes b + a \otimes b'.$$

By "impose relations" we mean divide by the subgroup generated by the difference of the two sides of each relation. The relation then holds in the quotient. The free abelian group has elements the finite formal sums $\Sigma_i n_i(a_i \otimes b_i)$, and these are added by adding coefficients. Or we can think of this as the integer-valued functions on the cartesian product $A \times B$. In the relations the addition on the left takes place in the group A or B, while the addition on the right is the formal addition in the free group.

Now suppose A and B are left modules over a commutative ring R (we discuss the noncommutative case later). Then $A \otimes_R B$ is defined to be $A \otimes B$ with the further identifications $ra \otimes b = a \otimes rb$ for $r \in R$. There is an R-module structure on this defined by $r(a \otimes b) = (ra) \otimes b$.

We point out two easy cases. First there is an isomorphism $R \otimes_R R \simeq R$, by the function $r \otimes s \mapsto rs$. More generally if S_1 and S_2 are finite sets, then the free R-module generated by the product is a tensor product:

$$R[S_1] \otimes_R R[S_1] \simeq R[S_1 \times S_2]$$

the isomorphism being given by $s_1 \otimes s_2 \mapsto (s_1, s_2)$.

A1.2. Noncommutative rings

Now consider modules over general rings. We can no longer simply identitify two left structures because this also divides out the difference between left and right multiplication in the ring: if $ra \otimes b = a \otimes rb$ then

$$rsa \otimes b = sa \otimes rb = a \otimes srb = sra \otimes b.$$

This forces $rsa = sra$ which is undesirable. Instead one usually assumes A is a *right* R-module, B is a left module, and define $A \otimes_R B$ by imposing the relation $ar \otimes b = a \otimes rb$.

We will find it convenient to work mostly with left module structures. For this we define the *opposite* ring R^{op} to have the same underlying set as R, but a different product. Denote the product in R by m, then define $m^{\text{op}}(r, s) = m(s, r)$. A right R-module structure is easily seen to be equivalent to a left R^{op}-module structure. With this convention if A is a (left) R^{op} module and B is a R-module then $A \otimes_R B$ is defined by imposing identities which look the same as in the commutative case (A1.1).

Another view of this is that $A \otimes B$ is a $R^{\text{op}} \otimes R$ module, and the R-tensor product is obtained by identifying the two structures. We will encounter a slight generalization suggested by this: if A is a (left) $R^{\text{op}} \otimes R$-module, then $\otimes_R A$ is defined by identifying the two operations: $(r \otimes 1)a = (1 \otimes r)a$. This is the algebraic analog of glueing together two boundary components of a connected spacetime.

If the ring is not commutative then the tensor product destroys structure: $A \otimes_R B$ is not naturally an R-module. More generally if A is an $R^{\text{op}} \otimes R$ module then $\otimes_R A$ does not have a natural R or R^{op} module structure. What we commonly encounter is an $R_1^{\text{op}} \otimes R_1 \otimes R_2$ module A, and then there is a natural R_2-module structure on $\otimes_{R_1} A$.

A1.3. Antiinvolutions

If R is a ring then an antiinvolution is a homomorphism $r \mapsto \bar{r}$ which has order two and reverses the multiplication: $\bar{\bar{a}} = a$ and $\overline{ab} = \bar{b}\bar{a}$. We also use the notation bar $(a) = \bar{a}$, and bar: $R \to R$. In this notation the conditions become bar^2 = id and bar $m = m\psi$(bar $\otimes$ bar), where ψ is the transposition $\psi(a \otimes b) = b \otimes a$.

A ring together with an antiinvolution is sometimes called a "hermitian ring." The example to have in mind is a group ring $R[G]$ with antiinvolution induced by $\bar{g} = g^{-1}$. The complex numbers is the standard example of a commutative ring with involution. The examples are all commutative (in which case an antiinvolution is the same as a ring involution) until the modular material of §8.

In the terminology of A1.2 an antiinvolution is a ring isomorphism of R with the opposite ring R^{op}. Thus if R has an antiinvolution and A is an $R \otimes R$ module we can define $\otimes_R A$ by using the involution to convert one of the structures to an R^{op} structure. This amounts to imposing the relation $(r \otimes 1)a = (1 \otimes \bar{r})a$.

Suppose A is a module over R, and denote the scalar multiplication explicitly by $m: R \otimes A \to A$. If R has an antiinvolution we denote by $\overline{A}$ the "conjugate" module structure on the same abelian group A. This is defined by A-multiplication by the conjugate. When R is not commutative this is a *right* module structure: $\overline{m}(a, r) = m(\bar{r}, a)$.

We will encounter one further refinement. Suppose S is a commutative ring with involution, and R is an algebra over S. Then R is a *hermitian* algebra if it has an antiinvolution which is conjugate linear with respect to the involution on S. This antiinvolution defines an algebra isomorphism of R with the conjugate opposite algebra: bar: $R \to \overline{R}^{\text{op}}$. Or it can be considered an isomorphism from the conjugate algebra to the opposite algebra: $\overline{R} \simeq R^{\text{op}}$. This means, for example, a tensor product $\otimes_R A$ can be constructed from a $\overline{R} \otimes R$ module structure on A.

A2. Pairings and traces

In the general noncommutative situation pairings are defined as follows. R is an S-algebra, A is a left R-module, and B is a right R-module. A *pairing* is then an R-bimodule homomorphism $A \otimes_S B \to R$. For example there is the standard evaluation pairing of A with its dual. Define the dual to be the left R-homomorphisms hom$_R(A, R)$, with right R-module structure given by $(f \cdot r)(a) = f(a)r$. (The multiplication is on the right so $r \cdot f$ is still a left homomorphism). The evaluation pairing is then defined by $a \otimes r \mapsto f(a)$.

If R is hermitian and A is a left module then a *hermitian form* is a form $\lambda: A \otimes_S \overline{A} \to R$ such that $\lambda(a, b) = \overline{\lambda(b, a)}$.

A *dual copairing* for a pairing λ is an element $\Lambda \in B \otimes_R A$ with the following property: if $\Lambda = \sum y_i \otimes x_i$ then for any $a \in A$ $b \in B$,

(A2.1) $$a = \sum \lambda(a, y_i) x_i, \quad b = \sum \overline{\lambda(x_i, b)} y_i.$$

We will say here that a pairing is *nondegenerate* if it has a dual copairing. This implies the usual (algebraic) definition of nondegeneracy, that the adjoint $A \to B^*$ is an isomorphism. If A is projective then the two definitions are equivalent.

This is a convenient time to introduce a graphic notation for some algebraic manipulations. These are close in spirit to a TQFT on graphs, and some of them can be interpreted literally that way. Reading left to right, denote the form as a graph with two incoming ends labeled with "A" and "B," and one outgoing end

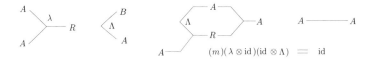

Figure A2.2 Pairing, copairings, and the relation.

labeled with the ring "R." We display this, and the graphic version of the dual copairing identity (A2.1), in figure (A2.2).

Exercise A2.3.
1. Let A be the (finitely generated) free module R^n with basis e_i. Show that the standard pairing $\lambda(re_i, se_j) = r\bar{s}\delta_{i,j}$ is a nondegenerate hermitian form on P (here $\delta_{i,j}$ is the "delta function" which is 1 if $i = j$ and 0 otherwise). Find the dual coform.
2. Show that a module has a nondegenerate pairing with some other module if and only if it is finitely generated projective. (hint: factor the identity map through a finitely generated free module).
3. Show that a nondegenerate pairing has a unique copairing, and conversely the copairing determines the pairing.

A2.4. Traces and ranks

If R is a ring, define the *trace group* TR to be the quotient of R by the (additive) subgroup generated by commutators $\{ab - ba\}$. This is $\otimes_R R$ in the tensor notation above. Denote the quotent map by $\mathrm{tr}: R \to TR$. This is the universal central function: the universal homomorphism with the property $\mathrm{tr}(ab) = \mathrm{tr}(ba)$.

If R is commutative then $R = TR$, and in particular TR is a ring. We caution that in general TR does not have a natural ring structure. When R is an ambialgebra we will define a ring structure on TR, but then the trace is usually a counit, not a ring morphism.

This definition extends to a notion of traces for functions: suppose A has a nondegenerate pairing λ with B, and with dual copairing Λ. Suppose $f: A \to A$ is an R-homomorphism. Define

(A2.5) $$\mathrm{Tr}(f) = \mathrm{tr}\,\lambda(f \otimes \mathrm{id})(\psi\Lambda).$$

Written out, if $\Lambda = \sum x_i \otimes y_i$ then this becomes $\mathrm{Tr}(f) = \sum \mathrm{tr}\,\lambda(f(y_i), x_i)$. Beware that the trace function on the right side of (A2.5) is necessary to make this well-defined on Λ as an element of $B \otimes_R A$.

In fact the trace of a function is independent of the form, so is a well-defined invariant of endomorphisms of finitely generated projective R-modules.

Exercise A2.6.
1. Show that if λ is the standard form on R^n, and f is represented by a matrix with entries $f_{i,j}$ then this definition gives the usual sum of diagonal entries: $\mathrm{Tr}(f) = \sum \mathrm{tr}(f_{i,i})$.

2. Show that Tr (f) is independent of the nondegenerate pairing on A (compare with the evaluation pairing).

The trace on functions defines a central function on the endomorphism ring of A, namely Tr : End $(A) \to TR$. Since End (A) is a ring there is the universal central function tr : End $(A) \to T(\text{End}\,(A))$. Since this is universal and the function trace is central there is a factorization of Tr through a homomorphism $T(\text{End}\,(A)) \to TR$. When the base ring is a field, or more generally if A has a free summand, then this is an isomorphism and the two traces can be identified.

Finally we define the *trace-rank* of a finitely generated projective module to be the trace of the identity map:

(A2.7.) $$\text{rk}_R(A) = \text{Tr}\,(\text{id}\,) = \text{tr}\,\lambda(\psi\Lambda).$$

Exercise A2.8.
1. Show that the trace-rank of a free module of dimension n is the image of n in TR, namely tr (ne).
2. Show that $\text{rk}_R(A_1 \oplus A_2) = \text{rk}_R(A_1) + \text{rk}_R(A_2)$.

A3. Ambialgebras

Suppose S is a commutative ring. Recall that an *algebra* over S is an S-module R with an associative product which commutes with the module structure, and a unit. We interpret the product as a homomorphism $m\colon R \otimes_S R \to R$, and the unit as $e\colon S \to R$.

Dually a coalgebra is a module with a comultiplication $\Delta\colon R \to R \otimes_S R$ and a counit $\epsilon\colon R \to S$. The coassociativity and counit relations are dual to the usual relations for products and units. The counit, for example, satisfies $(\epsilon \otimes \text{id}\,)\Delta = \text{id} = (\text{id} \otimes \epsilon)\Delta$.

Definition A3.1. An *ambialgebra* is defined to be $(R, m, e, \Delta, \epsilon)$, both an algebra and coalgebra structure, which satisfy the ambialgebra relation

(A3.2) $$\Delta m = (m \otimes \text{id}\,)(\text{id} \otimes \Delta) = (\text{id} \otimes m)(\Delta \otimes \text{id}\,)$$

and the symmetry relation

(A3.3) $$\begin{cases} \psi\Delta e = \Delta e \\ \epsilon m = \epsilon m\psi \end{cases}$$

Here ψ is the transposition $\psi(a \otimes b) = b \otimes a$. The ambialgebra relation means that $\Delta\colon R \to R \otimes_S R$ is a left R-homomorphism with respect to left multiplication in R and in the left copy of $R \otimes R$, and is a right homomorphism with respect to right multiplication. This is to be contrasted with the bialgebra and Hopf algebra relation which requires that Δ be a ring homomorphism.

We use a graph notation like that of 2.2 to manipulate these relations. Denote the multiplication by a graph with two incoming branches and one outgoing (reading from left to right) as shown in figure A3.4, then the ambialgebra and symmetry

Figure A3.4 Ambialgebra operations.

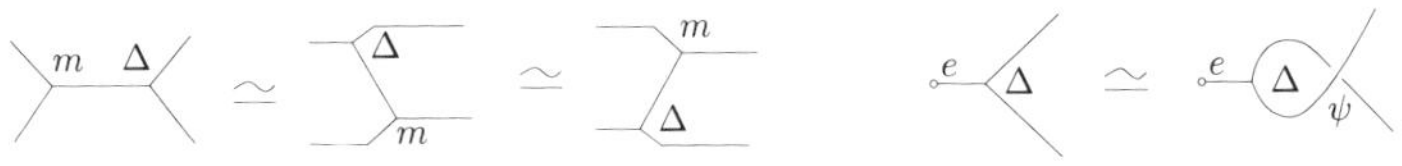

Figure A3.5 Ambialgebra relations.

Figure A3.6 Coassociativity.

relations are shown in figure A3.5 and the coassociative and counit conditions are in figure A3.6.

Vertical lines intersect these diagrams in points, each corresponding to a copy of R. When written in the symbolic notation, eg. $R \otimes R \otimes R$, left to right in the symbols corresponds to bottom to top on the vertical line. In effect this notation corresponds to thinking of the operations as coming from a sort of TQFT with graphs as spacetimes, so manipulations of graphs corresponds to manipulation of the corresponding operations. This is literally correct for commutative ambialgebras, and could be made precise in more generality using planar graphs with labels, etc.

Exercise A3.7.
1. Show that $\epsilon m \colon R \otimes R \to S$ is a nondegenerate symmetric bilinear form, with coform Δe. In particular conclude R must be a finitely generated projective S-module.
2. Use this to show the trace-rank of R, as an S-module, is $\epsilon m \Delta e$.
3. Complete the graph proof in figure A3.8 that $((m\psi) \otimes \text{id})(\text{id} \otimes \Delta) = (\text{id} \otimes m)(\psi \otimes \text{id})(\text{id} \otimes \Delta)$

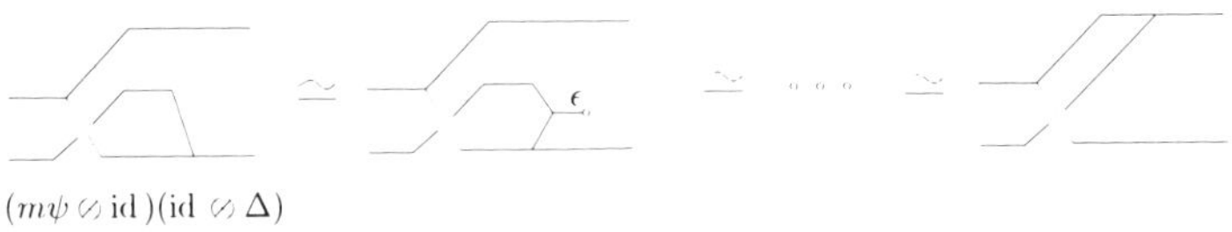

$(m\psi \otimes \text{id})(\text{id} \otimes \Delta)$

Figure A3.8

A3.9. Basic facts

We include these here with indications of proofs because there seems to be no published reference on ambialgebras.

First, the data given for an ambialgebra is redundant. For example it is sufficient to have an associative product m, a homomorphism $\epsilon\colon R \to S$, and the information that ϵm is a nondegenerate symmetric bilinear form. To reconstruct the rest of the data note that the dual coform to ϵm is Δe, and Δ and e can be recovered from this, m, and ϵ. This identifies the algebras which support an ambialgebra structure as "symmetric algebras," see Curtis and Reiner [**CR**].

In practice, when trying to locate ambialgebra structures, a different set of data is more convenient. The structure can also be reconstructed from an algebra structure on R, an element $\Delta e \in R \otimes R$ such that $r\Delta e = (\Delta e)r$ and $\psi \Delta e = \Delta e$, and a homomorphism $\epsilon\colon R \to S$ so that $(\epsilon \otimes \mathrm{id})(\Delta e) = e$.

Lemma A3.10. *Suppose $(R, m, e, \Delta, \epsilon)$ is an ambialgebra over S. Then*
1. *m is associative and Δ is coassociative;*
2. *m is commutative if and only if Δ is cocommutative; and*
3. *any other ambialgebra structure on the algebra (R, m, e) is obtained by twisting Δ, ϵ by a central unit.*

If u is a central unit in R then the coproduct and counit on R by obtained by "twisting" are

(A3.11)
$$(\Delta u)(x) = \Delta(m(u, x))$$
$$(\epsilon u^{-1})(x) = \epsilon(m(u^{-1}, x))$$

Proof. Suppose Δ', ϵ' gives another ambialgebra structure on the algebra. Then the two structures commute in the sense that

(A3.12)
$$(\mathrm{id} \otimes \Delta)\Delta' = (\Delta' \otimes \mathrm{id})\Delta.$$

Figure (A3.13) gives a graphical proof of this.

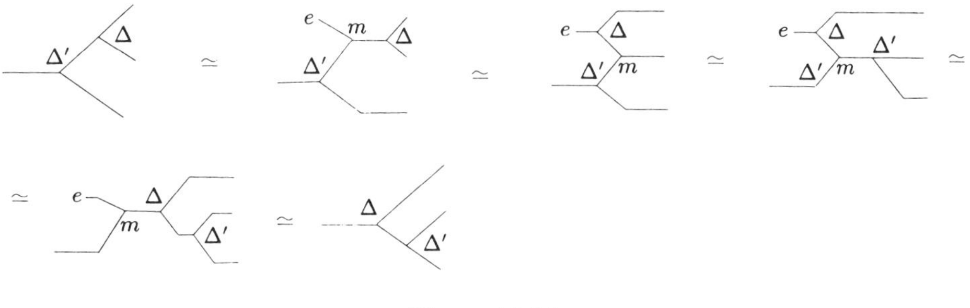

Figure A3.13

Setting $\Delta' = \Delta$ in (A3.12) we see that Δ is coassociative. The dual of this proof shows that m is associative.

For statement (2) of the lemma, if m is commutative then $(\psi\Delta, \epsilon)$ is also an ambialgebra structure on R. But a structure is determined by the counit, so since the counit is unchanged we see $\psi\Delta = \Delta$. The converse, that cocommutativity implies commutativity, follows by duality.

For (3) compose the identity (A3.12) on the left with $(\text{id} \otimes \epsilon \otimes \text{id})$ to get

(A3.14) $$\Delta' = \big(((\text{id} \otimes \epsilon)\Delta') \otimes \text{id}\,\big)\Delta.$$

Compose this with $\text{id} \otimes \epsilon'$ to get

$$\text{id} = ((\text{id} \otimes \epsilon)\Delta')((\text{id} \otimes \epsilon')\Delta).$$

This implies that $(\text{id} \otimes \epsilon)\Delta'$ is an isomorphism $R \to R$. Since it is also a left R-homomorphism it is right multiplication by the unit $u = (\text{id} \otimes \epsilon)\Delta'(e)$. Subsitute this into (A3.14) and apply to e to get $\Delta'(e) = \Delta(u)$. Thus $\Delta'(x) = x \cdot \Delta(u) = \Delta(x \cdot u)$, where we have abbreviated $m(x,y)$ to $x \cdot y$.

We can see u is central by taking x out on the other side in the last expression: $\Delta'(x) = \Delta'(e) \cdot x = \Delta(u \cdot x)$, then apply $\epsilon \otimes \text{id}$ to get $x \cdot u = u \cdot x$.

Comparing with the definition (A3.11) shows that Δ' is obtained by twisting by u, as required. □

To illustrate how the lemma is used, suppose we can show that the algebra structure is an endomorphism algebra. Then the coalgebra structure must be a twisting of the one described in A4.1. It is usually not difficult to determine twisting units. For example the composition $m\Delta: R \to R$ is multiplication by a central element of R, and twisting multiplies this by the twisting unit. In some cases (eg. endomorphism algebras) there are few enough central elements that twistings can be determined by comparing the two counits.

A3.15. Hermitian ambialgebras

We add an involution to the above data. Suppose S has a ring involution "bar" then a *hermitian* ambialgebra over S has the additional data of an involution bar : $R \to R$ satisfying:
1. bar is conjugate-linear as a homomorphism of S-modules (ie. induces an isomorphism $R \to \overline{R}$);
2. it reverses the the order of the product and coproduct: $\text{bar}\, m = m\psi(\text{bar} \otimes \text{bar})$ and $\Delta\text{bar} = (\text{bar} \otimes \text{bar})\psi\Delta$, and
3. the unit and counit commute with the involutions.

Suppose (R, bar) is a hermitian ring. Since ambilagebra structures are determined by $\Delta(e)$ or ϵ, it follows that an ambialgeba structure is hermitian if and only if $\overline{\Delta(e)} = \Delta(e)$, or $\epsilon\text{bar} = \text{bar}\,\epsilon$. Using this we can refine conclusion (3) of lemma A3.10: all hermitian ambialgebra structures on a ring can be obtained from a given one by twisting by bar-invariant central units.

A4. Examples of ambialgebras

We describe ambialgebra structures on endomorphism rings, group rings, and trace groups. Suppose S is a commutative ring, with involution as above.

The very first example is a structure on S as an ambialgebra over itself. The product and unit are the given ones, the product being understood as the isomorphism $m\colon S \otimes_S S \to S$ and the unit is the identity $S = S$. The coproduct is the inverse of the product, m^{-1}, and the counit is again the identity. Other examples can be obtained from this by twisting (A3.11).

A4.1. Endomorphism rings

Let λ be a nondegenerate pairing $\lambda\colon A \otimes B \to S$ and let $\Lambda\colon S \to B \otimes A$ denote the dual copairing. Define an ambialgebra structure on $R = B \otimes A$ by:

$$m = \mathrm{id} \otimes \lambda \otimes \mathrm{id} : B \otimes A \otimes B \otimes A \to B \otimes A$$
$$e = \Lambda\colon S \to B \otimes A$$
$$\Delta = \mathrm{id} \otimes \psi\Lambda \otimes \mathrm{id} : B \otimes A \to B \otimes A \otimes B \otimes A$$
$$\epsilon = \lambda\psi\colon B \otimes A \to S.$$

This ring is isomorphic to the endomorphism ring of B. An element $b \otimes a \in B \otimes A$ defines a homomorphism $B \to B$ by $c \mapsto \lambda(a,c)b$. This defines an isomorphism $B \otimes A \to \mathrm{End}(B)$ which preserves multiplication. Another view of this is that the adjoint of λ gives an isomorphism with the dual $A \simeq B^*$, and then

$$B \otimes A \simeq B \otimes B^* \simeq \mathrm{End}(B).$$

This nondegenerate pairing formulation is more symmetric with respect to domain and range than endomorphisms, and is often more convenient.

Exercise A4.2.
1. Write out the structure maps in terms of the standard basis $\{e_{i,j}\}$ for $\mathrm{End}(S^n)$ when λ is the standard form of (A2.3).
2. Show that the ring isomorphism $B \otimes A \simeq \mathrm{End}(B)$ takes the trace $\mathrm{tr}\colon \mathrm{End}(B) \to S$ defined in (A2.5) to $\lambda\psi\colon B \otimes A \to S$.
3. Show that if $B = \overline{A}$, so λ is a sesquilinear form $A \otimes \overline{A} \to S$, then the ambialgebra structure in A4.1 is hermitian with respect to the involution

$$\mathrm{bar} = \psi\colon \overline{A} \otimes A \to A \otimes \overline{A} = \overline{\overline{A} \otimes A}.$$

The graph notation introduced above gives a convenient way to manipulate this structure. For example the product and coproduct are shown in figure A4.3.

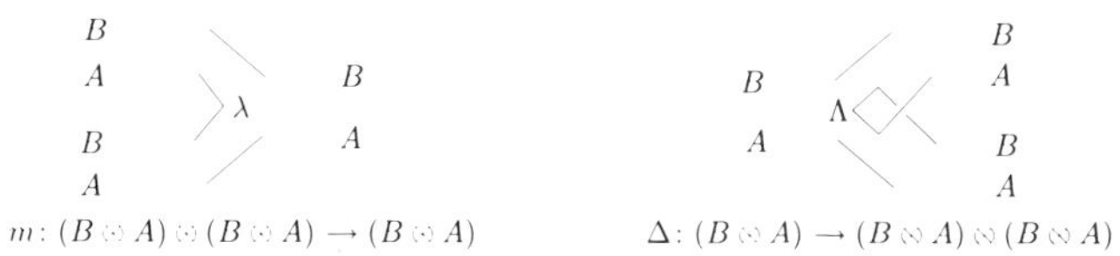

Figure A4.3

Exercise A4.4.
1. (Highly recommended) Use the graph notation to verify that the product and coproduct on $B \otimes A$ satisfy the ambialgebra relation (A3.2).
2. Show that the composition $m\Delta: B \otimes A \to B \otimes A$ is multiplication by the trace-rank of A.

A4.5. Group rings

Suppose G is a finite group, then the group ring $S[G]$ has elements of the form $\sum_g s_g g$, where the coefficients s_g are in S. An ambialgebra structure on this is given (in terms of basis elements $g, h \in G$) by:

$$m(g, h) = gh$$
$$e = e \quad \text{(the unit of } G\text{)}$$
$$\Delta(g) = \sum_h h \otimes h^{-1}g$$
$$\epsilon(g) = \delta_{g,e} = \begin{cases} 1, & \text{if } g = e \\ 0, & \text{otherwise} \end{cases}$$
$$\text{bar}(sg) = \bar{s}g^{-1}.$$

Exercise A4.6.
1. Verify the ambialgebra relation A3.2.
2. Evaluate the composition $m\Delta: S[G] \to S[G]$.
3. Show that the trace group $TS[G]$ is the free module $S[G/\text{conj}]$ generated by the conjugacy classes of G. Or alternately, is the set of central functions $G \to S$.
4. Suppose R is an ambialgebra over S, and G is finite. Modify the definition above to give an S-ambialgebra structure on $R[G]$.

We encounter an extension of this in §8. Suppose $\beta: A \to \hat{G}$ is a homomorphism into the center of $\hat{G}$, A is abelian, the kernel and cokernel of β are finite, and the order of the kernel is invertible in S. Then the group ring $S[\hat{G}]$ is an $S[A]$-algebra via the ring homomorphism $S[A] \to S[\hat{G}]$ into the center of $S[\hat{G}]$ induced by β. The "central homomorphism" ambialgebra structure on this algebra is given by

$$\Delta(g) = \sum_{h \in G} \hat{h} \otimes \hat{h}^{-1}g$$

$$\epsilon(g) = \begin{cases} \frac{1}{\#\ker\beta} \sum_{a \in \ker\beta} ab, & \text{if } \beta(b) = g \\ 0, & \text{if } g \notin \text{im}\beta. \end{cases}$$

Here $G = \hat{G}/\text{im}\beta$ is the cokernel of β, and $\hat{h}$ indicates an element in $\hat{G}$ with $\beta(\hat{h}) = h$. Since the tensor product is over $S[A]$ the element $\hat{h} \otimes \hat{h}^{-1}g$ is independent of the choice of $\hat{h}$.

Note that since an ambialgebra is necessarily finitely generated, $S[G]$ can only be an ambialgebra over S if G is finite. If G is a compact Lie group and S is the real or complex numbers everything but the unit can be extended. Consider

$\mathbf{R}[G]$ as the continuous or L_2 real-valued functions on G, then multiplication is the convolution product ([**BtD, II.1**]), and the definition above for the coproduct extends if interpreted as an integral over G. The counit is evaluation at the identity. The unit is however only a distribution: the delta function concentrated at the identity. If we extend to distributions to get the unit then other things go wrong, as they must since $R[G]$ is not finite dimensional. This example sheds some light on why we have trouble with units in ambialgebras.

A4.7. Trace rings

An ambialgebra structure on a ring induces an ambialgebra structure on the trace group, except for a unit whose existence must be assumed. These trace rings are important in modular TQFT.

Define an ambialgebra to be *special* if there is an element $c \in R$ such that $m\psi\Delta(c) = e$. Below we define all the structure of an ambialgebra on the trace group, except for a unit for the product structure. "Special" is exactly the hypothesis that the trace ring has a unit, which acts by the identity on R. In §9 we see that modular $(1+1)$-dimensional TQFT correspond exactly to special ambialgebras. There are non-special examples given in A4.12, and in §A5 we see that "most" ambialgebras are not special.

We say an ambialgebra is *superspecial* if in fact $c = e$ works in the above, to give a unit for the trace ring. Equivalently R is superspecial if $m\Delta \colon R \to R$ is the identity.

Lemma. *If R is a special ambialgebra over S then TR is a commutative ambialgebra with structure*

$$m_t(\mathrm{tr}\,(r), \mathrm{tr}\,(s)) = \mathrm{tr}\,(m(m(-,r) \otimes \mathrm{id}\,)(\Delta(s))$$
$$e_t = \mathrm{tr}\,(c), \text{ where } m\psi\Delta(c) = e$$
$$\Delta_t(\mathrm{tr}\,(r)) = (\mathrm{tr} \otimes \mathrm{tr}\,)\Delta(r)$$
$$\epsilon_t(\mathrm{tr}\,(r)) = \epsilon(r)$$
(A4.8) $$\mathrm{bar}\,(\mathrm{tr}\,(r)) = \mathrm{tr}\,(\bar{r}).$$

Further, $m\psi\Delta\colon TR \to R$ is a ring map into the center of R which makes R an ambialgebra over TR. Finally if R is superspecial then $\mathrm{tr} \colon R \to TR$ is a ring map which an isomorphism on the center of R.

The expressions in (A4.8) are to be understood as, eg. "if $a \in TR$ then $\epsilon_t(a) = \epsilon(r)$ for any $r \in R$ with $\mathrm{tr}\,(r) = a$." To see these are well-defined functions on TR we check that the right-hand sides are central functions of r, s. For example the centrality of m_t in the r variable is shown in figure A4.9.

The proof of the lemma is a reasonably straightforward manipulation in the graph notation. For example the product and coproduct are shown in figure A4.10.

There is a dual analog of the "special" condition, which is always satisfied. The function $(\mathrm{tr} \otimes \mathrm{id}\,)\Delta \colon R \to TR \otimes R$ defines a *coaction* of TR on R. The counit described in (A4.8) automatically coacts trivially on R.

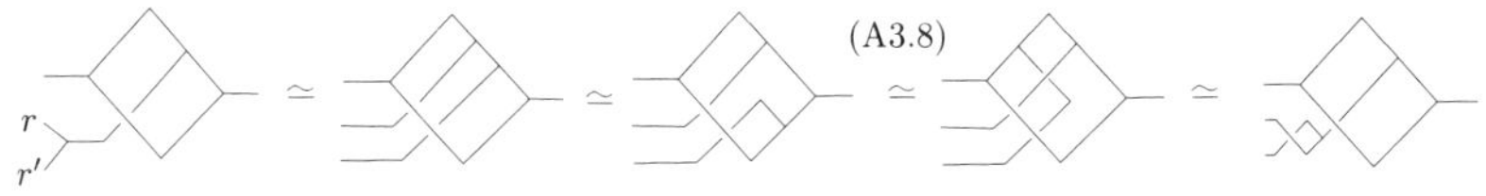

$$\text{(A3.8)}$$

Figure A4.9

Figure A4.10

Exercise A4.11.
1. Show that m_t is commutative.
2. Verify the ambialgebra relation (A3.2).
3. Describe the product on $TS[G]$, G a finite group.
4. Suppose S is a commutative ring, $\lambda\colon A \otimes B \to S$ is a nondegenerate form, and consider the identification of ambialgebras $B \otimes A \simeq \operatorname{End}(A)^{\operatorname{op}}$ described in 4.1. Show that the functional trace of A2.5 tr $\colon \operatorname{End}(A)^{\operatorname{op}} \to S$ is identified with $\lambda\psi\colon B \otimes A \to S$, and that the induced homomorphism $T(\operatorname{End}(A)) \to S$ is an ambialgebra morphism.
5. Suppose R is a *commutative* ambialgebra, so the quotient $R \to TR$ is a bijection. Show the multiplication in TR is given by $m_t(r,s) = (m\Delta(e))rs$ (where multiplication in R is indicated by juxtaposition). Deduce that in the commutative case the following are equivalent: (a) R is special; (b) $m(\Delta(e))$ is a unit (c) there is a twisting of the ambialgebra structure which is superspecial.

A4.12. Non-special ambialgebras

In A5.1 it is shown that a special ambialgebra over a field is semisimple. However there are lots of non-semisimple ambialgebras, so in a sense "most" ambialgebras are not special. We give a reference for a systematic construction, then describe two examples. The examples come from "mutant" modular TQFT which are not everywhere defined. Since they are not special they cannot occur in a complete TQFT.

An algebra has associated to it a "quiver" which is a finite directed graph. The vertices of the graph are the simple R-modules, and edges correspond to nontrivial extensions. Green [**G**] has shown that when the base ring is a field, any finite directed graph occurs as the quiver of some symmetric algebra. A symmetric algebra is one which supports an ambialgebra structure (A3.9). On the other hand a semisimple algebra has no nontrivial extensions, so the quiver must consist of a finite set of points. Thus all the examples with nontrivial quivers are non-special ambialgebras.

For an explicit example let R be $S[y]/y^2$, so R is 2-dimensional over S with generators e and y, and $y^2 = 0$. Then define

$$\Delta(e) = e \otimes y + y \otimes e \quad \Delta(y) = y \otimes y$$
$$\epsilon(e) = 0 \quad \epsilon(y) = e.$$

The trace algebra has generators a, b, such that $a^2 = 2b$ and $ab = b^2 = 0$. (This follows from the exercise just above, A4.11(5).) In particular it does not have a unit so R is not special.

The second example is noncommutative. R has dimension 8, with generators indexed by elements of $(\mathbf{Z}/2)^3$. Represent $\mathbf{Z}/2$ as $\{0,1\}$, then the generators are denoted x_α where $\alpha = (\alpha_1, \alpha_2, \alpha_3)$ and $\alpha_i \in \{0,1\}$. Define

$$m(x_\alpha, x_\beta) = \begin{cases} 0 & \text{if } \alpha_3 = \beta_1 \text{ or } \alpha_2 = \beta_2 = 0 \\ x_{(\alpha_1, 0, \beta_3)} & \text{if } \alpha_3 \neq \beta_1 \text{ and } \alpha_2 \neq \beta_2 \\ x_{(\alpha_1, 1, \beta_3)} & \text{if } \alpha_3 \neq \beta_1 \text{ and } \alpha_2 = \beta_2 = 1 \end{cases}$$

$$e = x_{(1,1,0)} + x_{(0,1,1)}$$

$$\Delta(e) = \sum_\alpha x_\alpha \otimes x_{\alpha + (1,1,1)}$$

$$\epsilon(x_\alpha) = \begin{cases} 1 & \text{if } \alpha = (1,0,0) \text{ or } (0,0,1) \\ 0 & \text{otherwise} \end{cases}$$

$$\text{bar}(x_{(\alpha_1, \alpha_2, \alpha_3)}) = x_{(\alpha_3, \alpha_2, \alpha_1)}.$$

The trace algebra of this example is isomorphic to that of the first example, so again is not special. There is a little more structure in this example which which qualitatively explains the non-specialness. It is $\mathbf{Z}$ graded with the degree of x_α given by $\alpha(1) + \alpha(2) + \alpha(3) - 2$. The product and unit have degree 0 with respect to this grading, while the coproduct has degree -1 and the counit degree $+1$. The induced grading on the trace ambialgebra has the generator a of degree 0, and b of degree -1. The product in the trace ring (A4.7) uses the coproduct, so has degree -1. An identity element for the product would have to have degree $+1$, and there are no nonzero elements in this degree.

A5. Complex ambialgebras

We give a classification and characterization of special ambialgebras over the complex numbers.

Proposition A5.1. *An ambialgebra over $\mathbf{C}$ is special if and only if it is semisimple as an algebra and if and only if $m\Delta: R \to R$ is an isomorphism. These are isomorphic to sums of twisted matrix (endomorphism) ambialgebras and are determined by:*
1. *a finite set Φ,*
2. *a positive function $n: \Phi \to \mathbf{Z}$, and*
3. *a nonzero function $u: \Phi \to \mathbf{C}$.*

The ambialgebra associated to the data (Φ, n_*, u_*) is obtained as follows: For each $i \in \Phi$ let B_i be a vector space of dimension n_i, and set

(A5.2) $\qquad (R, m, e, \Delta, \epsilon) = \oplus_{i \in \Phi}(\text{End}\,(B_i), m_i, e_i, u_i\Delta_i, u_i^{-1}\epsilon_i),$

where m_i etc. is the structure described in (A4.1), with coproduct and counit twisted by the central unit $u_i\text{id}_i$, as in (A3.10). One choice of element $c \in R$ which descends to a unit in the trace ambialgebra is $c = \Sigma \frac{1}{u_i n_i} e_i$.

Exercise A5.3. Let R be the ambialgebra over $\mathbf{C}$ corresponding to the data (Φ, n_*, u_*).

1. Show that $\epsilon(e) = \sum_i n_i u_i^{-1}$ and that $\epsilon m \Delta e = \sum n_i^2$. (see A3.7(2)).
2. Show that the trace ambialgebra is described by the data $(\Phi, 1, u_*^2)$. (Caution: the identification $T(\text{End}\,(V_i)) \simeq \mathbf{C}$ using the function trace of A4.9 is not a ring map. Use the first expression in 2. to evaluate the twisting coefficients.)

Now let $R = \mathbf{C}[G]$, the group ring of a finite group with the ambialgebra structure described in A4.6.

3. Verify that $m\Delta$ is an isomorphism, so the theorem applies (see A4.7(2)).
4. Use A4.7(3) to identify the index set Φ with the set of conjugacy classes in G. (A great deal of the classical representation and character theory of finite groups can be read off from the ambialgebra structure: the trace ambialgebra $T\mathbf{C}[G]$ for example is the character ring.)

We briefly prove (A5.1). $m\Delta$ is multiplication by a central element of R, and is an isomorphism if and only if this element is a unit. In this case the inverse gives an element c so that $m\Delta(c) = e$, so R is special. In a semisimple algebra described as in (A5.2) $m\Delta$ is multiplication by $n_i u_i$ in the ith component, so is an isomorphism because $n_i u_i$ is a unit in $\mathbf{C}$.

To complete the proof it is necessary to show that a special ambialgebra is semisimple. This is very like Maskie's theorem [**CR, p. 420**] or more precisely Ikeda's theorem [**CR, p. 425**]. We work on the element level to simplify the notation. Suppose R is an ambialgebra over S, and let $\Delta(e) = \sum a_i \otimes b_i \in R \otimes R$. Suppose $f \colon A \to B$ is an S-homomorphism of R-modules, and define $\hat{f}$ by $\hat{f}(x) = \sum a_i f(b_i x)$. Then $\hat{f}$ is an R-homomorphism, and this operation satisfies: if g is an R-homomorphism then $\widehat{fg} = \hat{f}g$, or $\widehat{gf} = g\hat{f}$, whichever is defined.

This gives a retraction of the S-homomorphisms to the R-homomorphisms. This implies that an R-module is projective if and only if it is projective as an S-module. To see this suppose A is S-projective, $g \colon A \to B$ and $h \colon C \to B$ are R-homomorphisms with h onto. Then there is a factorization as S-homomorphisms, $g = hf$. But $g = \hat{g} = \widehat{hf} = h\hat{f}$, so $\hat{f}$ provides a factorization as R-homomorphisms. Thus A is projective.

In the case at hand $S = \mathbf{C}$ so all modules are projective. Thus all R-modules are projective, from which it follows every module is a sum of irreducibles. This implies that R is an endomorphism ring, and semisimple. $\square$

A5.4. Involutions

A hermitian ambialgebra has an antiinvolution which reverses the order of both product and coproduct, and is conjugate-linear over the base ring. In the situation of the classification of A5.1 these can be explicitly described. Let R be a semisimple ambialgebra with data (Φ, n_*, u_*) and suppose $\lambda_i \colon A_i \otimes B_i \to \mathbf{C}$ are pairings used to describe $B_i \otimes A_i$ as $\operatorname{End}(B_i)$ as in A4.1. Then a hermitian involution bar $\colon R \to R$ corresponds to

1. an involution bar $\colon \Phi \to \Phi$ so that $n_{\bar{i}} = n_i$, and $u_{\bar{i}} = \overline{u_i}$, and
2. isomorphisms $\tau_i \colon \overline{B_i} \to A_{\bar{i}}$ satisfying: if $b \in B_i$, $c \in B_{\bar{i}}$ then $\lambda_i(\tau_{\bar{i}}(c), b) = \overline{\lambda_{\bar{i}}(\tau_i(b), c)}$.

The involution itself is given by

$$\operatorname{bar} = \sum_i \psi(\tau_i \otimes \tau_{\bar{i}}^{-1}).$$

We are abusing notation here. A conjugate-linear homomorphism $B \to A$ corresponds to a homomorphism $f \colon \overline{B} \to A$. The conjugate-linear version should be denoted $f \circ \operatorname{bar}$, where $\operatorname{bar} \colon B \to \overline{B}$ is the identity map. With this notation the antiinvolution is given by $\sum \psi(\tau_i \operatorname{bar} \otimes \operatorname{bar} \tau_{\bar{i}}^{-1})$. Omitting the identity map "bar" causes trouble in calculations because the dual of this is not the identity. Instead, $\operatorname{bar}^* \colon \overline{B}^* \to B^*$ is induced by composition with the conjugation map on the coefficient ring; $\operatorname{bar} \colon \mathbf{C} \to \mathbf{C}$.

Exercise A5.5.
1. Describe involutions on endomorphism rings $\hom_{\mathbf{C}}(B_i, B_i)$. Show they correspond to nondegenerate forms $\beta_i \colon B_i \otimes B_{\bar{i}} \to \mathbf{C}$ which are skew-symmetric in the sense $\overline{\beta_i} = \beta_{\bar{i}} \psi$. The corresponding involution $f \mapsto \overline{f}$ is determined by $\beta_{\bar{i}}(f(x), y) = \beta_i(x, \overline{f}(y))$ for $x \in B_i$ and $y \in B_{\bar{i}}$.
2. Describe involutions on commutative special ambialgebras (specified by data of the form $(\Phi, 1, u_*)$)

A5.6 Morita equivalence

This is an equivalence between modules over an endomorphism ring and over the base ring. It can be used to simplify modular TQFT, and gives a connection with the "modular functors" used by Segal [S] to describe conformal field theories. We continue to work over the complex numbers.

Suppose B is a finite dimensional complex vector space. Then the endomorphism ring $\operatorname{End}(B)$ acts on the left on B, and on $B \otimes E$ by the action on the first factor. This gives a functor

$$B \otimes \colon \operatorname{Mod}(\mathbf{C}) \to \operatorname{Mod}^\ell(\operatorname{End}(B)).$$

The Morita equivalence theorem is that this is an equivalence of categories. (We are using here the fact that all modules over $\mathbf{C}$ are projective.) Thus every left $\operatorname{End}(B)$-module is of the form $B \otimes E$, and a $\operatorname{End}(B)$-homomorphism $B \otimes E_1 \to B \otimes E_2$ is the identity of B tensored with a $\mathbf{C}$-homomorphism $E_1 \to E_2$.

This extends to semisimple algebras by sums. If $R = \oplus_i \operatorname{End}(B_i)$ then a left R-module is of the form $\oplus_i B_i \otimes E_i$. There is a more compact way to write this. Let

CR denote the center of R. Then R is a CR-algebra in an obvious way, $B = \oplus_i B_i$ is a CR-module, and $R = \operatorname{End}_{CR}(B)$. The Morita theorem can then be formulated as:
$$B \otimes_{CR} \colon \operatorname{Mod}(CR) \to \operatorname{Mod}^\ell(R)$$
is an equivalence of categories.

We describe an inverse functor, for use in modifying TQFT. For each R-module A we want a functorial CR-module $\hat{A}$ with a natural isomorphism $B \otimes \hat{A} \to A$. Consider the category $\mathcal{C}_A$ of all (C, f) where C is a CR-module, $f \colon B \otimes_{CR} C \to A$ is an R-isomorphism. Morphisms are CR morphisms $C \to C'$ whose tensors with the identity on B commute with the maps to A. Define a functor $\mathcal{C}_A \to \operatorname{Mod}(CR)$ by $(C, f) \mapsto C$. Then let $\hat{A}$ be the inverse limit of this functor. According to the Morita theorem there is a unique morphism between any two objects of $\mathcal{C}_A$ so the inverse limit also is a Morita module for A. The assignment $A \mapsto \hat{A}$ is the required functor.

A5.7. Tensor products and Morita equivalence

$\operatorname{End}(B)$ acts on the dual B^* on the *right*, by composition. The same analysis as above gives, for semisimple R, an equivalence
$$\otimes_{CR} B^* \colon \operatorname{Mod}(CR) \to \operatorname{Mod}^r(R).$$

Now consider a tensor product $A_1 \otimes_R A_2$. For this to be defined we need a right R-module structure on A_1 and a left structure on A_2. From the above these are given by $A_1 \simeq E_1 \otimes_T V^*$ and $A_2 \simeq V \otimes_T E_2$. The tensor product is then just $E_1 \otimes_T E_2$:

(A5.8) $\qquad \left(E_1 \otimes_{CR} B^*\right) \otimes_{\operatorname{End}_{CR}(B)} \left(B \otimes_{CR} E_2\right) \simeq E_1 \otimes_{CR} E_2.$

This can be used to systematically replace semisimple algebras by commutative ones, often of much smaller dimension.

BIBLIOGRAPHY

[Ab] E. Abe, *Hopf algebras*, Cambridge tracts in Math. Vol. 74, Cambridge Univ. Press, Cambridge and New York, 1980.

[A1] M. F. Atiyah, *Topological quantum field theories*, Publ. Math. Inst. Hautes Etudes Sci. (Paris) **68** (1989), 175–186.

[A2] _____, *New topological invariants of 3- and 4-dimensional manifolds*, Proc. Symp. Pure Math., vol. 48, Amer. Math Soc., 1988, pp. 285–299.

[A3] _____, *On framings of 3-manifolds*, Topology **29** (1990), 1–7.

[AB] M. F. Atiyah and R. Bott, *The Yang-Mills equations over Riemann surfaces*, Phil. Trans. R. Soc. London **A308** (1982), 532–615.

[ADPE] S. Axelrod, S. Della Pietra, and E. Witten, *Geometric quantization of Chern-Simons gauge theory*, J. Diff. Geometry **33** (1991), 787–902.

[B] H. Bass, *Euler characteristics and characters of discrete groups*, Invent. Math. **35** (1976), 155–196.

[Br] J. Birman, *Braids, links and mapping class groups*, Annals of Math Studies, vol. 82, Princeton University Press, 1974.

[CPR] M. Caselle, G. Ponzano and F. Ravanini, *Towards a classification of fusion rule algebras in rational conformal field theories*, Preprint November 1991.

[C] L. Crane, *2-d Physics and 3-d Topology*, Comm. Math. Phys. **135** (1991), 615–640.

[CR] C. Curtis and I. Reiner, *Representation theory of finite groups and associative algebras*, Interscience, 1962.

[DebG] J. DeBoer and J. Goeree, *Markov traces and II_1 factors in conformal field theory*, Communications in Math. Phys. **139** (1991), 267–304.

[DW] R. Dijkgraaf and E. Witten, *Topological gauge theories and group cohomology*, Comm. Math. Physics **129** (1990), 393–429.

[D1] V. G. Drinfeld, *Quantum groups*, Proc. Int. Cong. Math. 1986, vol. 1, Amer. Math Society, 1987, pp. 798–820.

[D2] _____, *Quasi-Hopf algebras and Knizhnik-Zamolodchikov equations*, Problems of modern quantum field theory (A. Belavin et al., ed.), Springer, 1989, pp. 1–13.

[F] D. Freed, *Classical Chern-Simons theory*, Adv. Math. (to appear).

[FG] D. Freed and R. Gompf, *Computer calculations of Witten's 3-manifold invariant*, Comm. Math. Physics **141** (1991), 79–117.

[FQ] D. Freed and F. Quinn, *Chern-Simons theory with finite gauge group*, Comm. Math. Physics **156** (1993), 435–472.

[G] E. Green, *Frobenius algebras and their quivers*, Can. J. Math., **30** (1978), 1029–1044.

[HT] A. Hatcher and W. Thurston, *a presentation for the mapping class group of a closed orientable surface*, Topology **19** (1980), 221–237.

[H] N. Hitchin, *Flat connections and geometric quantization*, Comm. Math. Physics **131** (1990), 347–380.

[JW] L. C. Jeffrey and J. Weitsman Half density quantization of the moduli space of flat connections and Witten's semiclassical manifold invariants, Topology **32** (1993), 509–530.

[KM] R. Kirby and Melvin, *The 3-manifold invariants of Witten and Reshetikhin-Turaev for sl(2,C)*, Invent. Math. **105** (1991), 473–545.

[Ko] T. Kohno, *Topological invariants for 3-manfiolds using representations of mapping class groups I*, Topology **31** (1992), 203–230.

[K] M. Kontsevich, *Rational conformal field theory and invariants of 3-dimensional manifolds*, preprint 1988.

[McL1] S. MacLane, *Homology*, Springer-Verlag, 1963.

[McL2] _____, *Categories for the working mathematician*, Graduate Texts in Mathematics, vol. 5, Springer-Verlag, 1971.

[M] J. Milnor, *Lectures on the h-cobrodism theorem*, Princeton Mathematical Notes, Princeton University Press, 1965.

[MS1] G. Moore and N. Seiberg, *Lectures on RCFT*, to appear in the Proc. 1989 Banff Summer School.

[MS2] G. Moore and N. Seiberg, *Classical and quantum conformal field theory*, Commun. Math. Phys. **123** (1989), 177–254.

[R] J. J. Rotman, *An introduction to homological algebra*, Pure and Applied Math., vol. 85, Academic Press, 1979.

[Q] F. Quinn, *Topological foundations of topological quantum field theory*, preprint 1991.

[RT] N. Yu. Reshetikhin and V. G. Turaev, *Invariants of 3-manifolds via link polynomials and quantum groups*, Invent. Math. **103** (1991), 547–598.

[S] G. Segal, *The definition of conformal field theory*, (preprint 1989-91).

[Sw] M. E. Sweedler, *Hopf algebras*, Benjamin, New York, 1969.

[V] E. Verlinde, *Fusion rules and modular transformations in 2d conformal field theory*, Nucl. Phys. **B300** (1988), 360–376.

[Y1] D. N. Yetter, *Quantum groups and representations of monoidal categories*, Math. Proc. Camb. Phil. Soc. **108** (1990), 261–290.

[Y2] _____, *Topological quantum field theories associated to finite groups and crossed G-sets*, J. Knot Theory and its Ram. **1** (1992), 1–20.

[Wk] K. Walker, *On Witten's 3-manifold invariants*, Preprint draft 1991.

[Wa] C. T. C. Wall, *Formal deformations*, Proc. London Math. Soc. **(3) 16** ((1966), 342–352.

[Wi1] E. H. Witten, *Quantum field theory and the Jones polynomial*, Commun. Math. Phys **121** (1989), 351–399.

[Wi2] _____, *Topological quantum field theory*.

[Wi3] _____, *Geometric quantization of Chern-Simons gauge theory with complex gauge group*, Commun. Math. Phys (1991).

[Wi4] _____, *The central charge in three dimensions*, preprint 1989.

INDEX OF NOTATIONS

Notations are listed in order of appearance. The number given is that of the largest numbered item preceeding the appearance.

$\chi(X)$, Euler characteristic, 3.1.
Map, Space of maps, 4.4.
$\#^\pi$, homotopy order, 4.5.
Hmy, Holonomy group, 5.4.
$[h, g]_G$, commutator in G, 5.17.
$\varprojlim$, inverse limit, 5.28.
$\varinjlim$, direct limit, 5.31.
$\pi_0 \mathcal{D}$, components of a category, 5.28.
"bar", involution, 6.1.
$S\text{Diff}^{d+1}$, Domain category of smooth manifolds, 6.2.
$F\text{Hty}$, $S\text{Hty}^{d+1}$ homotopy domain categories, 6.3.
$F\text{Cx}^{d+1}$, finite CW domain category, 6.3.
$\diamond$, Product in a tensor category, 10.1.
$SL(2)$ mod 5, 10.7.
$\otimes$, tensor product, A1.
tr, Tr, trace, A2.5.
rk_{tr}, Trace-rank, A2.7.
End (B), endomorphism ring, A4.1.

INDEX OF TERMINOLOGY

Terms are listed alphabetically. The number given is that of the largest numbered item preceeding the appearance.

Additive (category) 9.3.
Ambialgebra, A3.1.
Andrews-curtis conjecture, 2.0.
Antiinvolutions, A1.3.
Arrangement, 10.1.

Bar construction (B_G), 5.7.
Beginrelation matrix, 10.31.
Bordism
   category, 6.8.
   modular, 8.3.
Boundary, 1.0, 6.1.
Braided monoidal category, 10.51.

Cellular chains, 5.9.
Central homomorphism (ambialgebra) A4.6.
Characteristic numbers, 5.2.
Chern-Simons TQFT, 5.36.
Circulator matrix, 10.36.
Cocycle relation, 5.11.
Collars, 6.1.
Colorings, of links, 9.25.
Commutator, 5.17.
Comultiplication, A3.
Conjugate (module) A1.3.
Copairing, A2.
Corners, 8.1.
   ambialgebras, 9.7.
Cylinders, in a domain category, 6.1.

Direct limit, 5.31.
Domain category, 6.1.

modular, 9.1.

Endomorphism ring, A4.1.
Endrelation matrix, 10.31.
Euler (TQFT, characteristic) 3.1.

Four-manifolds, 2.0.
Finite total homotopy, 4.2.
Framed link, 9.25.

Gelfand-Kazhdan category, 10.51.
Glueing
    in a domain category, 6.1.
    endpoints in a grapn, 10.13.
Group ring, A4.5.

Hermitian
    ambialgebra, A3.15.
    form, A2.
    structure, 6.11.
    TQFT, 6.11.
Holonomy, 5.4, 5.28.
Homotopy order, 4.5.
Hopf algebra, 10.51, A3.3.

Incoming (boundary) 1.0, 6.8.
Interchange matrix, 10.43.
Inverse limit, 5.27.
Isomorphism (of TQFT), 3.6.

Mapping cylinders, 6.10.
Modular
    bordism, 8.2, 9.5.
    domain category, 9.1.
    TQFT, 9.5.
Monoidal (category) 10.2.
Morita equivalence
    of TQFT, 9.23.
    of categories, A5.6.

Natural maximal decompositions, (in a category) 9.24.
Nondegenerate (pairings), A2.1.

Opposite ring, A1.2.
Oriented complexes, 5.1.
Outgoing (boundary) 1.0, 6.8.

Pairing, A2.

Quantum group, 10.51.
Quasitensor category, 10.51.

Rank